T0320427

Biofuels

Offering a comprehensive overview of biofuels and bioenergy systems, *Biofuels: Technologies, Policies, and Opportunities* describes advances in technologies and global policies around biofuels as a renewable energy source. It discusses the basics of biofuel and bioenergy systems and current status and potential challenges in developed and developing countries. The book also highlights the questions that should be asked, available options, and processes (conventional and advanced) to enhance biofuel production.

- Details how technological interventions can influence the operation of an effective bioenergy system.
- Presents information regarding renewable energy directives and global policies related to energy chains, energy models, market status, and appropriateness of technology selection for different generations of biofuel generation.
- Covers socioeconomic aspects and technoeconomic feasibility as well as a detailed life cycle simulations (LCS) approach, revealing the real constraints being faced in the biofuels sector.
- Helps bioenergy professionals to prepare a roadmap for day-to-day operations.
- Describes recent advances such as biohythane, advanced oxidation process, and nanocatalyzed pretreatment for biofuel generation from wastewater.
- Addresses the most commonly discussed issues in the biofuel sector and the rationales underpinning them.

Written for professionals, academic researchers, decision makers, and policymakers in the biofuel sector, this book provides readers with a wide-ranging review of current research and developments in the respective field.

Biofuels

Technologies, Policies, and Opportunities

Edited by
Rena and Sunil Kumar

CRC Press is an imprint of the
Taylor & Francis Group, an **informa** business

First edition published 2023
by CRC Press
6000 Broken Sound Parkway NW, Suite 300, Boca Raton, FL 33487-2742

and by CRC Press
4 Park Square, Milton Park, Abingdon, Oxon, OX14 4RN

CRC Press is an imprint of Taylor & Francis Group, LLC

ISBN: 978-1-032-05482-7 (hbk)
ISBN: 978-1-032-05484-1 (pbk)
ISBN: 978-1-003-19773-7 (ebk)

DOI: 10.1201/9781003197737

Typeset in Times
by codeMantra

Contents

PART V Anaerobic Biotechnology

PART VI Advance Green Fuel

Editors

Dr. Rena is an avid researcher with a demonstrated history of working in the research and development industry. She has a Doctorate in Environmental Engineering from Royal Melbourne Institute of Technology University (RMIT University, Melbourne-Australia)-AcSIR (India) Joint (Cotutelle Ph.D.). She has worked in several national and international projects. She has also worked as a senior research fellow at CSIR—National Environmental Engineering Research Institute (CSIR—NEERI, India) and Korean Energy Technology Evaluation Planning (KETEP, Korea) project which dealt with Bio-HCNG from agricultural by-product for transport fuel market in developing countries. Her areas of specialization are hydrogen energy, renewable energy, waste management, climate change, circular economy, and sustainable development. She has published numerous reviews and research papers in high-impact journals based on her areas of expertise.

Dr. Sunil Kumar is a Senior Principal Scientist and Head of Waste Reprocessing Division at CSIR—National Environmental Engineering Research Institute (CSIR—NEERI, India). He has more than 22 years of experience in the field of environmental engineering with emphasis on waste management and associated energy aspects. Dr. Kumar is also a senior Humboldt Senior Fellow. He has handled many national and international projects. Dr. Kumar has received numerous awards and honors based on his expertise. He has published more than 300 research and review papers in the field of waste management. He has more than 11,252 citations, and his i10 index is more than 201.

Contributors

Shashi Arya
Imperial College
London, UK

Ali Asger
Department of Chemical Engineering
Institute of Chemical Technology
Jalna, India

S. Baranidharan
NIT
Tadepalligudem, India

Young-Cheol Chang
Chemical and Biological Engineering
Muroran Institute of Technology
Hokkaido, Japan

Gowardhan Kumar Chouhan
Institute of Environment and Sustainable Development
Banaras Hindu University
Varanasi, India

Pranav Prashant Dagwar
Bharathidasan University
Tiruchirappalli, Tamil Nadu

Shreyansh P. Deshmukh
Department of Chemical Engineering
Institute of Chemical Technology
Jalna, India

Sukhendu Dey
Department of Environmental Science
The University of Burdwan
Raiganj, India

Brajesh Kumar Dubey
Department of Civil Engineering
IIT
Kharagpur, India

Apurba Ratan Ghosh
Department of Environmental Science
The University of Burdwan
Burdwan, India

Divya Gupta
Environmental Science and Engineering Department
Indian Institute of Technology Bombay
Mumbai, India

Samuel Jacob
Department of Biotechnology
SRM Institute of Science and Technology
Kattankulathur, India

Rohit Jambhulkar
CSIR—National Environmental Engineering Research Institute (CSIR—NEERI)
Nagpur, India

Ramalingam Kayalvizhi
Department of Biotechnology
SRM Institute of Science and Technology
Kattankulathur, India

M. Shahbaz Khan
CSIR—National Environmental Engineering Research Institute (CSIR—NEERI)
Nagpur, India

Parth M. Khandagale
Department of Chemical Engineering
Institute of Chemical Technology
Jalna, India

Dhamodharan Kondusamy
School of Energy and Environment
Thapar Institute of Engineering and Technology
Patiala, India

Akhilesh Kumar
Institute of Environment and Sustainable Development
Banaras Hindu University
Varanasi, India

Mukesh Kumar
Department of Mechanical Engineering
Indian Institute of Technology
Ropar, India

Ranjeet Kumar
Civil Engineering Department
National Institute of Technology
Patna, India

Sunil Kumar
Waste Reprocessing Division
CSIR—National Environmental Engineering Research Institute (CSIR—NEERI)
Nagpur, India

Archana Kumari
CSIR—National Environmental Engineering Research institute (CSIR—NEERI)
Nagpur, India

Debajyoti Kundu
Waste Reprocessing Division
CSIR—National Environmental Engineering Research Institute (CSIR—NEERI)
Nagpur, India

C. K. Madhubalaji
Ozone Research Applications India Private Limited
Nagpur, India

Sukanta Mahavidyalaya
Department of Environmental Science
University of North Bengal
West Bengal, India

Sameena N. Malik
Department of Chemical Engineering
Institute of Chemical Technology
Jalna, India

Ameya S. Mantri
Department of Chemical Engineering
Institute of Chemical Technology
Jalna, India

Aneesh Mathew
Department of Civil Engineering
National Institute of Technology
Tiruchirappalli, India

Nityanand Singh Maurya
Civil Engineering Department
National Institute of Technology
Patna, India

Amit Kumar Mishra
Department of Environmental Sciences
Dr. Ram Manohar Lohia Avadh University
Ayodhya, India

S. Venkata Mohan
Department of Energy and Environmental Engineering
CSIR—Indian Institute of Chemical Technology (CSIR—IICT)
Hyderabad, India

Abhishek Mote
Department of Chemical Engineering
Institute of Chemical Technology
Jalna, India

Mritunjay
Civil Engineering Department
National Institute of Technology
Patna, India

Krishna R. Nair
Environmental Science and Engineering Department
Indian Institute of Technology Bombay
Mumbai, India

Anudeep Nema
Civil Engineering Department, School of Engineering
Eklavya University
Damoh, India

Vijay Nimkande
CSIR—National Environmental Engineering Research Institute (CSIR—NEERI)
Nagpur, India

Krutarth H. Pandit
Department of Chemical Engineering
Institute of Chemical Technology
Jalna, India

Aman N. Patni
Department of Chemical Engineering
Institute of Chemical Technology
Jalna, India

Arpan Patra
Environmental Science and Engineering Department
Indian Institute of Technology
Mumbai, India

Rajnikant Prasad
G. H. Raisoni Institute of Business Management
Jalgaon, India

Loganath Radhakrishnan
Institute for Globally Distributed Open Research and Education
Indian Green Service
Delhi, India

Palas Samanta
Department of Environmental Science
University of North Bengal
West Bengal, India

J. Santosh
Department of Energy and Environmental Engineering
CSIR—Indian Institute of Chemical Technology (CSIR—IICT)
Hyderabad, India

Omprakash Sarkar
Department of Civil, Environmental, and Natural Resources Engineering
Luleå University of Technology
Luleå, Sweden

Sudipta Sarkar
Department of Civil Engineering
IIT Kharagpur,
India

Adwait T. Sawant
Department of Chemical Engineering
Institute of Chemical Technology Mumbai,
Jalna, India

Dayanand Sharma
Sharda University
Uttar Pradesh, India

Nidhi Sharma
Institute of Environment and Sustainable Development (IESD)
Banaras Hindu University
Varanasi, India

Shruti Sharma
Environmental Science and Engineering Department
Indian Institute of Technology Bombay
Mumbai, India

Deval Singh
Environmental Science and Engineering Department
Indian Institute of Technology Bombay
Mumbai, India

Saurabh Singh
Institute of Environment and Sustainable Development
Banaras Hindu University
Varanasi, India

Dipesh Kumar R. Sonaviya
Department of Civil Engineering
C. S. Patel Institute of Technology
Charusat, India

Gujjala Lohit Kumar Srinivas
Waste Reprocessing Division
CSIR—National Environmental Engineering Research Institute (CSIR—NEERI)
Nagpur, India

Ramachandran Devasena Umai
Department of Biotechnology
SRM Institute of Science and Technology
Kattankulathur, India

Jay Prakash Verma
Institute of Environment and Sustainable Development
Banaras Hindu University
Varanasi, India

Part I

Global Outlook of Biofuels

1 Transition of Biofuels from the First to the Fourth Generation

The Journey So Far

Gujjala Lohit Kumar Srinivas
CSIR—National Environmental Engineering Research Institute (CSIR—NEERI)

Deval Singh
Indian Institute of Technology

Sunil Kumar
CSIR—National Environmental Engineering Research Institute (CSIR—NEERI)

CONTENTS

1.1 INTRODUCTION

Fossil reserves still contribute a majority fraction of energy toward the production of fuels, chemicals and materials, and considering the fact that these sources are nonrenewable in nature, it is not a sustainable option. In addition, the build-up of greenhouse gases in the atmosphere emanating mainly during the combustion of fossil fuels is aggravating the issue of climate change (Foster et al. 2007; Cherubini and Strømman 2011). Thus, there is an urgent need to adopt renewable sources, and thereby, it will give access to energy security and ensure food security. Bioenergy represents a significant portion of the renewable energy having the capability to

DOI: 10.1201/9781003197737-2

supplement the conventional sources in terms of fulfilling the energy needs. Further, its use would provide a string case for sustainable development (Cotula et al 2008). Biofuel can be broadly defined as the fuel produced from organic matter and can be calssified into four generations based on the type of feedtsocks used for the synthesis and technological progress involved in that specific category.

1.2 DIFFERENT GENERATIONS OF BIOFUEL

1.2.1 First Generation

The first generation refers to the utilization of edible crops for synthesis of biofuels. Feedstocks most commonly used for synthesis of biofuels are sugarcane, corn, whey, barley, potato residues, sugarbeets, etc. (Lee and Lavoie 2013). Brazil is one of the leaders in the utilization of sugarcane as a crop for bioethanol production. The processing steps involve crushing the sugarcane in water to release sucrose into the solution that is subsequently fermented and purified (distilled and membrane sieved) to obtain fuel-grade bioethanol. One of the associated problems of bioethanol production from sugarcane is the fact that there is a subsequent increase in the price of sugar, which affects the economics of the process. According to Lee and Lavoie (2013), cost of bioethanol from raw sugar was estimated as 0.68 US$/L. Corn is also one of the favorable sources for production of ethanol, and the process involves hydrolysis of starch (catalyzed by amylase) to release glucose molecules, which are then fermented to yield ethanol. According to Lee and Lavoie (2013), one ton of corn yields 400–450 L of ethanol along with by-products such as residual biomass left over after the distillation step, which can be valorized as fodder for grazing animals to obtain favorable economics.

Biodiesel is another such biofuel that is being produced at industrial scale and uses oil-containing crops. The process flow involves extraction of oil followed by transesterification with methanol to obtain fatty acid methyl esters, and this reaction is catalyzed by acid/alkali/enzymes (lipase). One of the major contributors in the cost of biodiesel is from the feedstock. According to Lee and Lavoie (2013), market price of soybean oil is 1,230 US$/ton, palm oil is 931 US$/ton, and Canola oil is 1,180 US$/ton. One ton of oil results in 1,000–1,200 L of biodiesel, and the market price of biodiesel has been estimated to be 0.85 US$/L. Considering the major contribution of feedstock in the economics of biodiesel, cheaper alternatives such as used cooking oil (331 US$/ton) and nonedible crop-based oil (350–500 US$/ton) are being increasingly used. Further, the market price of vegetable oils is influenced by the international markets, and hence there is always volatility in their prices. In comparison, the market price of used cooking oil and nonedible tree-based oils is not influenced by the international markets, and hence there is an obvious interest in the shift from first- to second-generation feedstocks. In addition, first-generation biofuels involve the controversy of food versus fuel owing to the use of edible crops. A pictorial representation of the feedstocks required for biofuel production is shown in Figure 1.1.

1.2.2 Second Generation

Second-generation biofuels are fuels produced from a wide range of feedstocks, *viz.*, wood chips, agricultural residues, municipal solid waste, nonedible tree-based

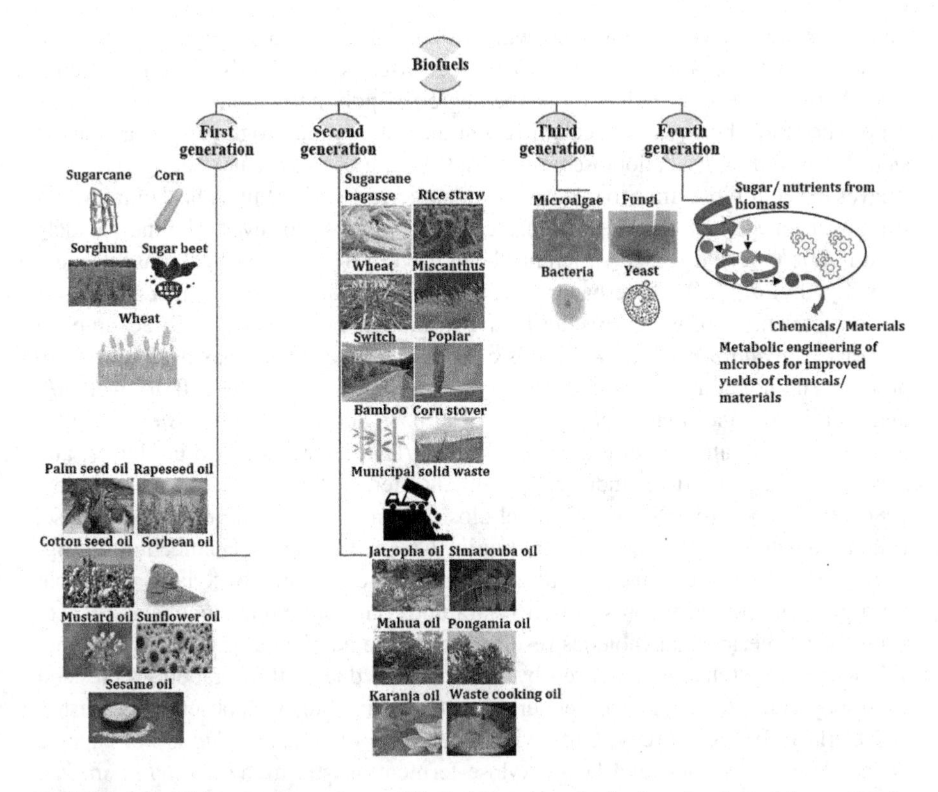

FIGURE 1.1 Different generations of biofuel based on the feedstock adopted for synthesis. (Images have been sourced from https://publicdomainvectors.org/; https://pixabay.com/.)

oil, used cooking oil, etc. One of the major advantages is a significant reduction in the price of feedstock; however, the challenging aspect is the conversion of these resources to obtain the biofuel. Further, there is always a scope of producing multiple products from the second-generation feedstocks in the form of a biorefinery, considering their diverse composition. Bioethanol production from second-generation feedstock generally follows the steps, *viz.*, delignification, saccharification, fermentation, distillation, and membrane sieving. Delignification of biomass is considered as the pretreatment step that allows easy access of the sugar reserves (holocelluloses) to the hydrolytic enzymes and thus increasing the efficiency of sugar release from the biomass. Some of the associated changes are the breakage of lignin–hemicellulose and pectin complex, and an increase in the porosity of biomass (Mosier et al. 2005). Techniques adopted for delignification can be categorized into physical (milling, irradiation, etc.), chemical (dilute acid, alkali, wet oxidation, solvolysis, ozonolysis, steam, liquid hot water, ammonia fiber explosion, supercritical fluid, etc.), and biological (laccase, lignin peroxidase, manganese peroxidase, etc.). Saccharification of holocelluloses is conducted through enzymatic (cellulases and xylanases) and acid-mediated hydrolysis, while the fermentation of released sugars is carried out by yeast. Enzymes employed for breaking the glycosididc bonds between the sugar molecules are termed as cellulases, which is a combination of exoglucanases, endoglucanases, and beta-glucosidases. Endoglucanases act randomly within the cellulose chain, resulting in reduction of the degree of polymerization, and they mainly act on the amorphous parts of cellulose. On the other hand, exoglucanases act on the side chains and cleave cellobiose units from the cellulose molecule. Beta-glucosidase cleaves the cellobiose units to release the monomeric glucose unit. A host of microbes (fungus, bacteria, actinomycetes, etc.) secrete cellulases, however. The most widely used is *Trichoderma reesei*, which has been used in the industrial cellulase production (Lynd et al. 2002; Esterbauer et al. 1991). Fermentation of sugar molecules is the subsequent step, which results in the formation of bioethanol. To accomplish this task, *Saccharomyces cerevisiae* is conventionally used and has been reorted to produce ethanol at a yield of 0.45 g/g and a specific rate of 1.3 g/g cell dry weight/h under optimal conditions (Verduyn et al. 1990). In addition, *S. cerevisiae* has been reported to have an ethanol tolerance of >100 g/L and is resistant to inhibitors, and hence have been used for industrial bioethanol fermentation (Casey and Ingledew 1986). In addition to cellulose, lignocellulosic biomass also contains hemicellulose, which is a heterogeneous polymer containing C6 and C5 sugars. Hemicellulose originating from hardwoods and agricultural residues is dominated by xylans, while that from grasses and softwoords is rich in arabinoxylan and galactomannans, respectively. Monomeric sugar moieties resulting from the hemicellulosic fraction are not able to be simultaneously utilized by *S. cerevisiae* due to the catabolic repression caused by glucose, and, hence, sequential utilization of sugars is observed (Johnston and Carlson 1992). Moreover, the wild-type strain of *S. cerevisiae* is not capable of metabolizing xylose, and hence xylose-fermenting strains, *viz.*, *Pitchia stipitis*, *Candida shehatae*, etc., have been used to increase the yield of bioethanol (Toivola et al. 1984). Apart from yeasts, bacteria, *viz.*, *Zymomonas mobilis*, etc., have been used for fermentation of sugars (Swings and De Ley 1977; Ingram et al. 1987).

Conventional techniques use sequential hydrolysis and fermentation; however, in the economics point of view (20% reduction in the overall capital expenditure of the process has been reported by, simultaneous saccharification and fermentation (SSF) has gained popularity (Bothast and Schlicher 2005). One of the major reasons behind the development of the simultaneous process is the feedback inhibition of cellulases by glucose and cellobiose released during the hydrolysis, thereby leading to reduction in the efficiency of the hydrolysis step (Gauss et al. 1976). However, there are also disadvantages associated with the association. One of the major disadvantages is the difference in the optimum temperature of hydrolysis and fermentation. Further, the enzymes and yeast cannot be reused, owing to difficulty in separation from lignin post fermentation. Other reasons for reduction in the efficiency of SSF process are deactivation of enzymes, adsorption of cellulase on unproductive sites, limited access of the cellulose chain ends for the exoglucanases, etc. (Ooshima et al. 1990).

Second-generation biodiesel is termed as the process of using nonedible tree-based oils (*such as.*, Jatropha seed oil, Karanja seed oil, Mahua seed oil, Pongamia seed oil, Waste cooking oil, etc.) for the production of biodiesel. It has the associated benefits of utilizing waste residues, in addition to reduction of the overall greenhouse gas emissions, when compared to fossil fuels (IEA 2010). One of the drawbacks of second-generation biodiesel is the lack of availability of feedstock for industrial level

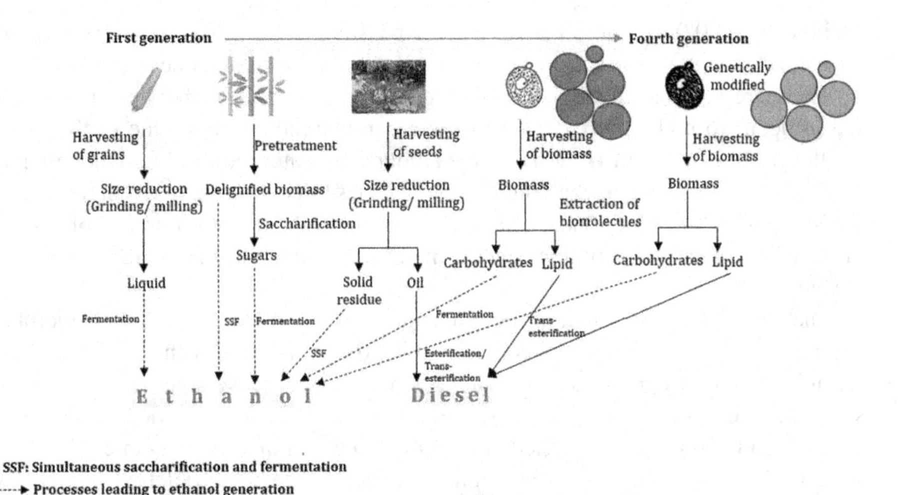

FIGURE 1.2 Routes for biofuel production from the first to the fourth generation.

of production. In addition, additional level of purification (*viz.*, degumming) is often needed to utilize crude feedstocks, *viz.*, used cooking oil. A pictorial representation of the routes adopted for bioethanol and biodiesel production (first to fourth generation) is shown in Figure 1.2.

1.2.3 Third Generation

Third generation of biofuels utilize components of microbial (oleaginous microalgae, yeast, fungus, bacteria, etc.) biomass, *viz.*, lipids for biodiesel production and carbohydrates for bioethanol production. Oleaginous microbes can be defined as the organisms that have the capability to store lipids less than 20% of their dry cell weight (Meng et al. 2009). Potential microalgal candidates adopted for biodiesel production are *Chlorella vulgaris*, *Chlorella minutissima*, *Scenedesmus quadricauda*, *Tetraselmis chuii*, *Dunaliella salina*, *Dunaliella tertiolecta*, and *Pyrocystis lunula* (Li et al. 2010). Some of the oleaginous yeasts reported in the literature are *Rhodotorula* sp., *Cryptococcus* sp., *Rhodosporodium* sp., *Sporidiobolus* sp., *Trichosporon* sp., *Yarrowia lipolytica*, *Candida* sp., etc. (Ageitos et al. 2011; Polburee et al. 2015). Kumar and Banerjee (2019) have reported biodiesel production from lipid extracted from *Trichosporon* sp., which was able to utilize crude glycerol and agricultural residue, and have reported the maximum of 43% w/w of lipids. Among the oleaginous filamentous fungi, some of the oleaginous candidates reported in the literature are *Mortierella alpine*, *Mucor circinelloises*, *Umbelopsis isabelline*, *Aspergillus oryzae*, etc. (Li et al. 2011; Miguel et al. 1997; Meeuwse et al. 2011; Adachi et al. 2011). Bacteria employed for biodiesel production majorly belong to the actinomyces group, *viz.*, *Mycobacetrium* sp. (Barksdale and Kim 1977), *Nocardia* sp. (Alvarez et al. 1997), *Rhodococcus* sp. (Alvarez et al. 1996), *Dietzia* sp. (Alvarez and

Steinbuchel 2002), *Gordonia* sp. (Alvarez and Steinbuchel 2002), *Micromonospora* sp. (Hoskisson et al. 2001), and *Streptomyces* sp. (Olukoshi and Packter 1994).

There are multiple challenges related to biodiesel production that need to be studied in detail to make the overall process a commercial success. One of the major challenges lies in the utilization of large amounts of water required for cultivation of microalgae, which can be challenging for countries having sub-zero temperatures for a major part of the year. One of the operational issues is the development of an efficient lipid extraction technique, since a major component of this operation includes dewatering.

The defatted biomass remaining after lipid extraction from oleaginous microbes can be used for bioethanol production to make the economics of the overall process feasible. The unit operations employed for the desired conversion are (1) hydrolysis of defatted biomass to release fermentale sugars, (2) fermentation of sugars to ethanol, and (3) purification of ethanol to meet the fuel quality. Fetyan et al. (2022) have studied the potential of bioethanol production from the defatted biomass of *Nannochloropsis oculata* grown using the combination of aqueous extract of sugarcane bagasse and carbon dioxide as substrates (mixotrophic conditions). Acid hydrolysis followed by enzymatic treatment was adopted for releasing the sugars where a maximum yield of 232 mg/g defatted biomass was obtained. The sugars obtained were fermented using *S. cerevisiae*, and a maximum ethanol yield of 6.17 g/L was obtained. Details about the bioethanol and biodiesel production reported in the literature utilizing first- to fourth-generation feedstocks are shown in Tables 1.1 and 1.2, respectively.

1.2.4 Fourth Generation

The fourth generation of biofuel production can be defined as process systems utilizing microbes/biomass that are genetically engineered to enhance the overall yield. Some of the strategies adopted are improving the photosynthetic efficiency of microalgae, enhancing the light penetration, reduction in the photoinhibition, increasing the lipid content of oleaginous microbes, decreasing the lignin content in lignocellulosic biomass, etc. (Tandon and Jin 2017). Techniques such as the use of truncated chlorophyll antenna of the chlorophyll have been reported to improve the light penetration, whereas expanding the absorption spectrum range of microalgae has been reported to enhance the photosynthesis efficiency (Wolf et al. 2018). Some of the commonly used genetic engineering tools for enhancing biofuel yield are zinc finger nucleases, transcription activator like effector nucleases, and CRISPR/Cas9 (clustered regularly interspaced palindromic sequences) (Abdullah et al. 2019). Niu et al. (2013) have reported overexpression of *DGAT*1 gene in *Phaeodactylum tricornutum*, which resulted in 35% increase in the neutral lipid content. Hung et al. (2013) have reported overexpression of *DGAT*2 gene in *Chlamydomonas reinhardtii*, which resulted in ninefold increase in the triacylglycerol content. Hsieh et al. (2012) have reported overexpression of *G3P*DH, *GPAT*, *LPAA*T, *PAP*, and *DGA*T genes in *C. minutissima*, which led to twofold increase in the lipid content compared to the wild strain. Tabulation of the literary articles related to fourth-generation biofuel production is done in Tables 1.2 and 1.3.

TABLE 1.1
Bioethanol from Different Generations of Production

	Substrate	Ethanol Productivity (g/L.h)	Ethanol Yield (g/g)	Ethanol Titer (g/L)	Reference
First generation	Sugarcane molasses	9.2	0.47	54.3	Kumar et al. (2017)
	Sugar beet molasses	–	0.50	83.2	Razmovski and Vucurovic (2012)
	Sugar beet thick juice	–	0.4	132.4	
	Sweet sorghum stalk juice	–	–	33	Liu and Shen (2008)
	Corn	–	0.4	130.5	Zabed et al. (2016)
Second generation	Corn stover (10% loading) + corn (20% loading)	–	–	99.3	Yu et al. (2019)
	Rice straw	–	–	75.3	Takano and Hoshino (2018)
	Sugarcane bagasse	0.63	–	75.5	Gao et al. (2018)
	Cotton stalk	0.61	0.44	–	Keshav et al. (2016)
Third generation	*Arthrospira platensis* (Spirulina)	–	0.16	–	Markou et al. (2013)
	Chlamydomonas reinhardtii	–	0.29	–	Nguyen et al. (2009)
Fourth generation	CO_2 (*Synechocystis* sp. PCC 6803 – engineered microalgae)	–	0.69	–	Dexter and Fu (2009)
	Rice straw (Engineered *Saccharomyces cerevisiae*)	–	0.41	8.2	Sakamoto et al. (2012)

1.3 TECHNO-ECONOMICS OF DIFFERENT GENERATIONS OF BIOFUEL

Bioethanol produced from first-generation feedstocks has been reported to have a selling price of 1.39–1.66 US$/L with substrates like sugarcane, corn, and wheat (Table 1.3). The major contributor of the selling price is the cost of the raw materials (NREL 2000). Thus, there has been an obvious shift to the second-generation technology where agricultural residues have been used, which resulted in reduction of the selling price of bioethanol up to 0.97 US$/L, as reported by Macrelli et al. (2012) with sugarcane baggase and leaves as substrates. Details of the reported works on techno-economics of bioethanol production are tabulated in Table 1.3.

TABLE 1.2
Transesterification Efficiency and Enhancement of Lipid Content for First- to Fourth-Generation Biodiesel Production Processes

	Substrate	Efficiency of Transesterification (%)	Enhancement of Lipid Content	Reference
First generation	Unrefined rice bran oil	90.0	–	Kattimani et al. (2014)
	Waste soybean oil	95.0	–	Lin et al. (2021)
	White mustard oil	96.5	–	Yesilyurt et al. (2019)
	Rapeseed oil	99.5	–	Rezki et al. (2020)
	Sesame oil	87.8	–	Dawodu et al. (2014)
Second generation	Jatropha oil	90.1	–	Jain and Sharma (2010)
	Karanja oil	97.0	–	Patel and Sankhavara (2017)
	Pongamia oil	81.0	–	Ortiz-Martínez et al. (2016)
	Waste cooking oil	96.0	–	Degfie et al. (2019)
Third generation	*C. pyrenoidosa*	98.4	–	D'oca et al. (2011)
	Isochrysis zhangjiangensis Chaetoceros sp., *Chlorella Vulgaris*, and *Nannochloropsis* sp.	90.1	–	Liu et al. (2015)
Fourth generation	*Thalassiosira pseudonana* (strains 1A6 and 1B1)	–	2.4 (1A6) and 3.3 (1B1) folds higher than wild type in exponential phase 4.1 (1A6) and 3.2 (1B1) folds higher under silicon starvation	Trentacoste et al. (2013)
	Phaeodactylum tricornutum	–	2.5-fold enhancement in neutral lipid (overexpression of PtME)	Xue et al. (2015)

TABLE 1.3
List of Studies on Technoeconomic Analysis of Bioethanol Production (First to Fourth Generation)

	Substrate	Bioethanol Selling Price (US$/L)	Reference
First generation	Sugarcane	1.66	Bain (2007)
	Corn	1.29	
	Wheat	1.39	
Second generation	Sugarcane bagasse and leaves	0.97	Macrelli et al. (2012)
	Rice straw	0.627	Ranganathan (2020)
	Agricultural residues	1.17	Bain (2007)
Third generation	*Chlorella* sp.	2.61 US$/kg	Wu et al. (2018)
Fourth generation	Recombinant strain of *Zymomonas mobilis* used for ethanol fermentation	2.98–4.06 US$/gal	Li et al. (2018)

Studies on technoeconomic analysis of biodiesel suggest that the selling price of first-generation technologies is between 0.3 and 0.85 US$/L, while that from second-generation feedstocks are between 0.69 and 1.56 US$/L (Table 1.4). Details of the reported works on techno-economics of biodiesel production are tabulated in Table 1.4.

1.4 CONCLUSION

First-generation biofuel technologies have been well implemented all around the world and are sustainable alternatives to the fossil fuels. However, their adoption brings forward issues like food versus fuel controversy, utilization of agricultural lands, fluctuations in the food prices, etc. Similarly for biodiesel, utilization of vegetable oils leads to fluctuations in their prices, and thus the efforts are being shifted toward second-generation feedstocks (comparatively less expensive and no involvement with the food chain). Technologies to utilize these recalcitrant substrates have been developed thoroughly but still have some drawbacks that need to be tackled through the biorefinery route by utilizing all the components of the biomass for biofuels/biocommodities production. Third-generation biofuels are technologies related mostly with the utilization of microalgae, which presents the most sustainable option owing to their carbon accumulation ability. Although the biomass production rate is faster in this case compared to the terrestrial counterparts, it faces technical challenges likedewatering, lipid extraction, etc., which need to be overcome before its commercial exploitation. Fourth-generation biofuels utilize genetically modified microbes/biomass to improve the overall yield of biofuel conversion. Thus, the future of biofuels may not rest with any one particular category but a combination of these categories so as to meet the ever-increasing global demand of energy.

TABLE 1.4
List of Studies on Technoeconomic Analysis of Biodiesel Production (First to Fourth Generation)

	Substrate	Biodiesel Selling Price (US $/L)	Reference
First generation	Canola oil	0.85	Lee and Lavoie (2013)
	Soybean oil	0.30	Weber (1993)
	Sunflower oil	0.63	Weber (1993)
Second generation	Rapeseed oil	0.69	Noordam and Withers (1996)
	Jatropha oil	0.99	Ofori-Boateng and Lee (2011)
	Pongamia oil	1.56	Prasad and Singh (2020)
Third generation	Microalgal lipids (cultivation – open pond, 400 ha); biogas was the byproduct produced from the anaerobic digestion lagoon	0.5–0.82	Benemann and Oswald (1996)
	Microalgal lipids (cultivation – photobioreactor)	2.80	Delrue et al. (2012)
	Microalgal lipids (cultivation – raceway pond + photobioreactor)	2.69	Delrue et al. (2012)
Fourth generation	*Scenedesmous obliquus*, *Chlorella* sp.	2.17 US $/L (wild type) ® 1.74 US $/L (engineered microalgae)	Zewdie and Ali (2022)

REFERENCES

Abdullah, B., Muhammad, S.A.F.A.S., Shokravi, Z., Ismail, S., Kassim, K.A., Mahmood, A.N. and Aziz, M.M.A. "Fourth generation biofuel: A review on risks and mitigation strategies." *Renewable and Sustainable Energy Reviews* 107 (2019): 37–50.

Adachi, D., Hama, S., Numata, T., Nakashima, K., Ogino, C., Fukuda, H., and Kondo, A. "Development of an Aspergillus oryzae whole-cell biocatalyst coexpressing triglyceride and partial glyceride lipases for biodiesel production." *Bioresource Technology* 102, no. 12 (2011): 6723–6729.

Ageitos, J.M., Vallejo, J.A., Veiga-Crespo, P. and Villa, T.G. "Oily yeasts as oleaginous cell factories." *Applied Microbiology and Biotechnology* 90, no. 4 (2011): 1219–1227.

Alvarez, H. and Steinbüchel, A. "Triacylglycerols in prokaryotic microorganisms." *Applied Microbiology and Biotechnology* 60, no. 4 (2002): 367–376.

Alvarez, H.M., Kalscheuer, R., and Steinbüchel, A. "Accumulation of storage lipids in species of Rhodococcus and Nocardia and effect of inhibitors and polyethylene glycol." *Lipid/Fett* 99, no. 7 (1997): 239–246.

Alvarez, H.M., Mayer, F., Fabritius, D. and Steinbüchel, A. "Formation of intracytoplasmic lipid inclusions by Rhodococcus opacus strain PD630." *Archives of Microbiology* 165, no. 6 (1996): 377–386.

Bain, R.L. *World Biofuels Assessment; Worldwide Biomass Potential: Technology Characterizations (milestone report).* No. NREL/MP-510-42467. National Renewable Energy Lab.(NREL), Golden, CO, 2007.

Barksdale, L., and Kim, K.-S. "Mycobacterium." *Bacteriological Reviews* 41, no. 1 (1977): 217.

Benemann, J.R., and Oswald, W.J. "Systems and economic analysis of microalgae ponds for conversion of CO_2 to biomass." *Nasa Sti/recon Technical Report N* 95 (1994): 19554.

Bothast, R.J., and Schlicher, M.A. "Biotechnological processes for conversion of corn into ethanol." *Applied Microbiology and Biotechnology* 67, no. 1 (2005): 19–25.

Casey, G.P., and Ingledew, W.M. "Ethanol tolerance in yeasts." *CRC Critical Reviews in Microbiology* 13, no. 3 (1986): 219–280.

Cherubini, F., and Strømman, A.H. "Principles of biorefining." In *Biofuels*, pp. 3–24. Academic Press, 2011. Edited by: Ashok Pandey, Steven C. Riche, Edgard Gnansounou. Elsevier Inc. https://doi.org/10.1016/C2010-0-65927-X

Cotula, L., Dyer, N., and Vermeulen, S. Fuelling Exclusion? The Biofuels Boom and Poor People's Access To Land, IIED, 2008, London. ISBN: 978-1-84369-702-2

D'oca, M.G.M., Viêgas, C.V., Lemoes, J.S., Miyasaki, E.K., Morón-Villarreyes, J.A., Primel, E.G., and Abreu, P.C. "Production of FAMEs from several microalgal lipidic extracts and direct transesterification of the Chlorella pyrenoidosa." *Biomass and Bioenergy* 35, no. 4 (2011): 1533–1538.

Dawodu, F.A., Ayodele, O.O., and Bolanle-Ojo, T. "Biodiesel production from Sesamum indicum L. seed oil: An optimization study." *Egyptian Journal of Petroleum* 23, no. 2 (2014): 191–199.

Degfie, T.A., Mamo, T.T., and Mekonnen, Y.S. "Optimized biodiesel production from waste cooking oil (WCO) using calcium oxide (CaO) nano-catalyst." *Scientific Reports* 9, no. 1 (2019): 1–8.

Delrue, F., Setier, P.-A., Sahut, C., Cournac, L., Roubaud, A., Peltier, G., and Froment, A.-K. "An economic, sustainability, and energetic model of biodiesel production from micro-algae." *Bioresource Technology* 111 (2012): 191–200.

Dexter, J., and Fu, P. "Metabolic engineering of cyanobacteria for ethanol production." *Energy & Environmental Science* 2, no. 8 (2009): 857–864.

Esterbauer, H., Steiner, W., Labudova, I., Hermann, A., and Hayn, M. "Production of Trichoderma cellulase in laboratory and pilot scale." *Bioresource Technology* 36, no. 1 (1991): 51–65.

Fetyan, N.A.H., El-Sayed, A.E.-K.B., Ibrahim, F.M., Attia, Y.A., and Sadik, M.W. "Bioethanol production from defatted biomass of Nannochloropsis oculata microalgae grown under mixotrophic conditions." *Environmental Science and Pollution Research* 29, no. 2 (2022): 2588–2597.

Forster, P., Ramaswamy, V., Artaxo, P., Berntsen, T., Betts, R., Fahey, D.W., Haywood, J. Lean, J., Lowe, D.C., Myhre, G., Nganga, J., Prinn, R., Raga, G., Schulz, M., and Van Dorland, R. Changes in Atmospheric Constituents and in Radiative Forcing. In: Climate Change 2007: The Physical Science Basis. Contribution of Working Group I to the Fourth Assessment Report of the Intergovernmental Panel on Climate Change [Edited by: Solomon, S., D. Qin, M. Manning, Z. Chen, M. Marquis, K.B. Averyt, M.Tignor and H.L. Miller]. Cambridge University Press, Cambridge, United Kingdom and New York, NY, USA, 2007.

Gao, Y., Xu, J., Yuan, Z., Jiang, J., Zhang, Z., and Li, C. "Ethanol production from sugar-cane bagasse by fed-batch simultaneous saccharification and fermentation at high solids loading." *Energy Science & Engineering* 6, no. 6 (2018): 810–818.

Gauss, W.F., Suzuki, S., and Takagi, M. "*Manufacture of Alcohol from Cellulosic Materials Using Plural Ferments.*" U.S. Patent 3,990,944, issued November 9, 1976, United States

Hoskisson, P.A., Hobbs, G., and Sharples, G.P. "Antibiotic production, accumulation of intracellular carbon reserves, and sporulation in Micromonospora echinospora (ATCC 15837)." *Canadian Journal of Microbiology* 47, no. 2 (2001): 148–152.

Hsieh, H.-J., Su, C.-H., and Chien, L.-J. "Accumulation of lipid production in *Chlorella minutissima* by triacylglycerol biosynthesis-related genes cloned from Saccharomyces cerevisiae and Yarrowia lipolytica." *Journal of Microbiology* 50, no. 3 (2012): 526–534.

Hung, C.-H., Ho, M.-Y., Kanehara, K., and Nakamura, Y. "Functional study of diacylglycerol acyltransferase type 2 family in *Chlamydomonas reinhardtii*." *FEBS Letters* 587, no. 15 (2013): 2364–2370.

IEA. *Sustainable Production of Second-Generation Biofuels – Potential and Perspectives in Major Economies and Developing Countries*. International Energy Agency, Paris, 2010. www.iea.org/papers/2010/second_generation_biofuels.pdf.

Ingram, L.O., Conway, T., Clark, D.P., Sewell, G.W., and Preston, J.F. "Genetic engineering of ethanol production in *Escherichia coli*." *Applied and Environmental Microbiology* 53, no. 10 (1987): 2420–2425.

Jain, S., and Sharma, M.P. "Biodiesel production from Jatropha curcas oil." *Renewable and Sustainable Energy Reviews* 14, no. 9 (2010): 3140–3147.

Johnston, M. "Regulation of carbon and phosphate utilization." In: *The Molecular and Cellular Biology of the Yeast Saccharomyces: Gene Expression*. Edited by: Jones, E.W., Pringel, J.R., Broach, J. Cold Spring Harbor Laboratory Press, Cold Spring Harbor, NY, 1992, 193–281.

Kattimani, V., Venkatesha, B., and Ananda, S. "Biodiesel production from unrefined rice bran oil through three-stage transesterification." *Advances in Chemical Engineering and Science* 4 (2014), 361–366. https://doi.org/10.4236/aces.2014.43039.

Keshav, P.K., Naseeruddin, S., and Rao, L.V. "Improved enzymatic saccharification of steam exploded cotton stalk using alkaline extraction and fermentation of cellulosic sugars into ethanol." *Bioresource Technology* 214 (2016): 363–370.

Kumar, R., Ghosh, A.K., and Pal, P. "Fermentative energy conversion: Renewable carbon source to biofuels (ethanol) using *Saccharomyces cerevisiae* and downstream purification through solar driven membrane distillation and nanofiltration." *Energy Conversion and Management* 150 (2017): 545–557.

Kumar, S.P.J., and Banerjee, R. "Enhanced lipid extraction from oleaginous yeast biomass using ultrasound assisted extraction: A greener and scalable process." *Ultrasonics Sonochemistry* 52 (2019): 25–32.

Lee, R.A., and Lavoie, J.-M. "From first-to third-generation biofuels: Challenges of producing a commodity from a biomass of increasing complexity." *Animal Frontiers* 3, no. 2 (2013): 6–11.

Li, P., Miao, X., Li, R., and Zhong, J. "In situ biodiesel production from fast-growing and high oil content *Chlorella pyrenoidosa* in rice straw hydrolysate." *Journal of Biomedicine and Biotechnology* 2011, Article ID 141207, 8 pages. https://doi.org/10.1155/2011/141207.

Li, W., Ghosh, A., Bbosa, D., Brown, R., and Wright, M.M. "Comparative techno-economic, uncertainty and life cycle analysis of lignocellulosic biomass solvent liquefaction and sugar fermentation to ethanol." *ACS Sustainable Chemistry & Engineering* 6, no. 12 (2018): 16515–16524.

Li, Y., Han, D., Hu, G., Dauvillee, D., Sommerfeld, M., Ball, S., and Hu, Q. "Chlamydomonas starchless mutant defective in ADP-glucose pyrophosphorylase hyper-accumulates triacylglycerol." *Metabolic Engineering* 12, no. 4 (2010): 387–391.

Lin, C.-H., Chang, Y.-T., Lai, M.-C., Chiou, T.-Y., and Liao, C.-S. "Continuous biodiesel production from waste soybean oil using a nano-Fe_3O_4 microwave catalysis." *Processes* 9, no. 5 (2021): 756.

Liu, J., Chu, Y., Cao, X., Zhao, Y., Xie, H., and Xue, S. "Rapid transesterification of micro-amount of lipids from microalgae via a micro-mixer reactor." *Biotechnology for Biofuels* 8, no. 1 (2015): 1–8.

Liu, R., and Shen, F. "Impacts of main factors on bioethanol fermentation from stalk juice of sweet sorghum by immobilized *Saccharomyces cerevisiae* (CICC 1308)." *Bioresource Technology* 99, no. 4 (2008): 847–854.

Lynd, L.R., Weimer, P.J., Van Zyl, W.H., and Pretorius, I.S. "Microbial cellulose utilization: Fundamentals and biotechnology." *Microbiology and Molecular Biology Reviews* 66, no. 3 (2002): 506–577.

Macrelli, S., Mogensen, J., and Zacchi, G. "Techno-economic evaluation of 2nd generation bioethanol production from sugar cane bagasse and leaves integrated with the sugar-based ethanol process." *Biotechnology for Biofuels* 5, no. 1 (2012): 1–18.

Markou, G., Angelidaki, I., Nerantzis, E., and Georgakakis, D. "Bioethanol production by carbohydrate-enriched biomass of Arthrospira (Spirulina) platensis." *Energies* 6, no. 8 (2013): 3937–3950.

McAloon, A., Taylor, F., Yee, W., Ibsen, K., and Wooley, R. *Determining the Cost of Producing Ethanol from Corn Starch and Lignocellulosic Feedstocks.* No. NREL/TP-580-28893. National Renewable Energy Lab.(NREL), Golden, CO (United States), 2000.

Meeuwse, P., Tramper, J., and Rinzema, A. "Modeling lipid accumulation in oleaginous fungi in chemostat cultures: I. Development and validation of a chemostat model for *Umbelopsis isabellina.*" *Bioprocess and Biosystems Engineering* 34, no. 8 (2011): 939–949.

Meng, X., Yang, J., Xu, X., Zhang, L., Nie, Q., and Xian, M. "Biodiesel production from oleaginous microorganisms." *Renewable Energy* 34, no. 1 (2009): 1–5.

de Miguel, T., Calo, P., Díaz, A., and T.G. Villa. "The genus rhodosporidium: A potential source of beta-carotene." *Microbiologia (Madrid, Spain)* 13, no. 1 (1997): 67–70.

Mosier, N., Wyman, C., Dale, B., Elander, R., Y.Y. Lee, M. Holtzapple, and M. Ladisch. "Features of promising technologies for pretreatment of lignocellulosic biomass." *Bioresource Technology* 96, no. 6 (2005): 673–686.

Nguyen, M.T., Choi, S.-P., Lee, J.-W. Lee, J.-H., and Sim, S.-J. "Hydrothermal acid pretreatment of *Chlamydomonas reinhardtii* biomass for ethanol production." *Journal of Microbiology and Biotechnology* 19, no. 2 (2009): 161–166.

Niu, Y.F., Zhang, M.H., Li, D.W., Yang, W.D., Liu, J.S., Bai, W.B., and Li, H.Y. "Improvement of neutral lipid and polyunsaturated fatty acid biosynthesis by overexpressing a type 2 diacylglycerol acyltransferase in marine diatom *Phaeodactylum tricornutum.*" *Marine Drugs* 11, no. 11 (2013): 4558–4569.

Noordam, M., and Withers, R. "Producing biodiesel from canola in the inland northwest: An economic feasibility study." *Idaho Agricultural Experiment Station Bulletin* 785 (1996): 12.

NREL (2000) https://www.nrel.gov/news/press/2000/41bioenergy.html

Ofori-Boateng, C, and Lee, K.T. "Feasibility of Jatropha oil for biodiesel: Economic analysis." In *World Renewable Energy Congress-Sweden*, no. 057, Linköping University Electronic Press, Linköping, Sweden, pp. 463–470, 8–13 May 2011. ISBN: 978-91-7393-070-3, http://dx.doi.org/10.3384/ecp11057463.

Olukoshi, E.R., and Packter, N.M. "Importance of stored triacylglycerols in Streptomyces: Possible carbon source for antibiotics." *Microbiology* 140, no. 4 (1994): 931–943.

Ooshima, H., Burns, D.S., and Converse, A.O. "Adsorption of cellulase from *Trichoderma reesei* on cellulose and lignacious residue in wood pretreated by dilute sulfuric acid with explosive decompression." *Biotechnology and Bioengineering* 36, no. 5 (1990): 446–452.

Ortiz-Martínez, V.M., Salar-García, M.J., Palacios-Nereo, F.J., Olivares-Carrillo, P., Quesada-Medina, J., Pdl De Los Ríos, A., and Hernández-Fernández, F.J. "In-depth study of the transesterification reaction of *Pongamia pinnata* oil for biodiesel production using catalyst-free supercritical methanol process." *The Journal of Supercritical Fluids* 113 (2016): 23–30.

Patel, R.L., and Sankhavara, C.D. "Biodiesel production from Karanja oil and its use in diesel engine: A review." *Renewable and Sustainable Energy Reviews* 71 (2017): 464–474.

Polburee, P., Yongmanitchai, W., Lertwattanasakul, N., Ohashi, T., Fujiyama, K., and Limtong, S. "Characterization of oleaginous yeasts accumulating high levels of lipid when cultivated in glycerol and their potential for lipid production from biodiesel-derived crude glycerol." *Fungal Biology* 119, no. 12 (2015): 1194–1204.

Prasad, S.S., and Singh, A. "Economic feasibility of biodiesel production from pongamia oil on the island of Vanua Levu." *SN Applied Sciences* 2, no. 6 (2020): 1–9.

Ranganathan, P. "Preliminary techno-economic evaluation of 2G ethanol production with co-products from rice straw." *Biomass Conversion and Biorefinery* (2020): 1–14. https://doi.org/10.1007/s13399-020-01144-8.

Razmovski, R., and Vučurović, V. "Bioethanol production from sugar beet molasses and thick juice using *Saccharomyces cerevisiae* immobilized on maize stem ground tissue." *Fuel* 92, no. 1 (2012): 1–8.

Rezki, B., Essamlali, Y., Aadil, M., Semlal, N., and Zahouily, M. "Biodiesel production from rapeseed oil and low free fatty acid waste cooking oil using a cesium modified natural phosphate catalyst." *RSC Advances* 10, no. 67 (2020): 41065–41077.

Sakamoto, T., Hasunuma, T., Hori, Y., Yamada, R., and Kondo, A. "Direct ethanol production from hemicellulosic materials of rice straw by use of an engineered yeast strain codisplaying three types of hemicellulolytic enzymes on the surface of xylose-utilizing *Saccharomyces cerevisiae* cells." *Journal of Biotechnology* 158, no. 4 (2012): 203–210.

Swings, J., and De Ley, J. "The biology of zymomonas." *Bacteriological Reviews* 41, no. 1 (1977): 1–46.

Takano, M., and Hoshino, K. "Bioethanol production from rice straw by simultaneous saccharification and fermentation with statistical optimized cellulase cocktail and fermenting fungus." *Bioresources and Bioprocessing* 5, no. 1 (2018): 1–12.

Tandon, P., and Jin, Q. "Microalgae culture enhancement through key microbial approaches." *Renewable and Sustainable Energy Reviews* 80 (2017): 1089–1099.

Toivola, A., Yarrow, D., Van Den Bosch, E., Van Dijken, J.P. and Alexander Scheffers, W. "Alcoholic fermentation of D-xylose by yeasts." *Applied and Environmental Microbiology* 47, no. 6 (1984): 1221–1223.

Trentacoste, E.M., Shrestha, R.P., Smith, S.R., Glé, C., Hartmann, A.C., Hildebrand, M., and Gerwick, W.H. "Metabolic engineering of lipid catabolism increases microalgal lipid accumulation without compromising growth." *Proceedings of the National Academy of Sciences* 110, no. 49 (2013): 19748–19753.

Verduyn, C., Postma, E., Scheffers, W.A., and van Dijken, J.P. "Physiology of *Saccharomyces cerevisiae* in anaerobic glucose-limited chemostat culturesx." *Microbiology* 136, no. 3 (1990): 395–403.

Weber, J. A. *The Economic Feasibility of Community Based Biodiesel Plants.* PhD diss., University of Missouri-Columbia, Columbia, 1993.

Wolf, B.M., Niedzwiedzki, D.M., Magdaong, N.C.M., Roth, R., Goodenough, U., and Blankenship, R.E. "Characterization of a newly isolated freshwater Eustigmatophyte alga capable of utilizing far-red light as its sole light source." *Photosynthesis Research* 135, no. 1 (2018): 177–189.

Wu, W., Lin, K.-H., and Chang, J.-S. "Economic and life-cycle greenhouse gas optimization of microalgae-to-biofuels chains." *Bioresource Technology* 267 (2018): 550–559.

Xue, J., Niu, Y.-F. Huang, T., Yang, W.-D. Liu, J.-S. and Li, H.-Y. "Genetic improvement of the microalga *Phaeodactylum tricornutum* for boosting neutral lipid accumulation." *Metabolic Engineering* 27 (2015): 1–9.

Yesilyurt, M.K., Arslan, M., and Eryilmaz, T. "Application of response surface methodology for the optimization of biodiesel production from yellow mustard (Sinapis alba L.) seed oil." *International Journal of Green Energy* 16, no. 1 (2019): 60–71.

Yu, J., Xu, Z., Liu, L., Chen, S., Wang, S., and Jin, M. "Process integration for ethanol production from corn and corn stover as mixed substrates." *Bioresource Technology* 279 (2019): 10–16.

Zabed, H., Faruq, G., Sahu, J.N., Boyce, A.N., and Ganesan, P. "A comparative study on normal and high sugary corn genotypes for evaluating enzyme consumption during dry-grind ethanol production." *Chemical Engineering Journal* 287 (2016): 691–703.

Zewdie, D.T., and Yimam Ali, A. "Techno-economic analysis of microalgal biofuel production coupled with sugarcane processing factories." *South African Journal of Chemical Engineering* 40 (2022): 70–79.

2 Intervention in Biofuel Policy through Indian Perspective

An Effort toward Sustainable Development Goals

Rohit Jambhulkar
CSIR—National Environmental Engineering Research Institute (CSIR—NEERI)

Nidhi Sharma
Banaras Hindu University

CONTENTS

ABBREVIATIONS

BSFC Brake-Specific Fuel Consumption
CAGR Compound Annual Growth Rate
CO Carbon Monoxide
CPCB Central Pollution Control Board

DOI: 10.1201/9781003197737-3

DTE	Down to Earth
EBP	Ethanol Blending Program
FDI	Foreign Direct Investment
GDP	Gross Domestic Productivity
GHG	Greenhouse Gas
GIS	Geographic Information System
GoI	Government of India
IISR	Indian Institute of Spices Research
INDC	Intended Nationally Determined Contribution
ISMA	Indian Sugar Mills Association
ISRO	Indian Space Research Organization
MAFW	Ministry of Agriculture and Farmers' Welfare
MCAFPD	Ministry of Consumer Affairs, Food, and Public Distribution
MEF	Ministry of Economy and Finance
MNRE	Ministry of New and Renewable Energy
MoA	Ministry of Agriculture
MoEFCC	Ministry of Environment, Forest, and Climate Change
MoF	Ministry of Finance
MoPNG	Ministry of Petroleum and Natural gas
MoPR	Ministry of Panchayati Raj
MoRD	Ministry of Rural Development
MoS&T	Ministry of Science and Technology
MoTA	Ministry of Tribal Affairs
MRTH	Ministry of Road Transport and Highways
MSW	Municipal Solid Waste
Mts	Million Tonne
NBM	National Mission on Biodiesel
NITI Aayog	National Institution for Transforming India
NOX	Nitrogen Oxides
NPB	National Policy on Biofuels
OMCs	Oil Marketing Companies
OPEC	Organization of Petroleum Countries
R&D	Research and Development
SBI	State Bank of India
SDGs	Sustainable Development Goals
SIAM	Society of Indian Automobile Manufacturers
SOX	Sulfur Oxides
TEC	Total Energy Consumption
TERI	The Energy and Research Institute
USA	United States of America
VOCs	Volatile Organic Compounds
w.r.t.	With Respect To

2.1 INTRODUCTION

Energy is a crucial parameter that plays a pivotal role in the growth of emerging economies globally, including India. With a growing population in India, energy consumption is increasing, and energy security has become a critical challenge. India stands at the third-largest energy importer globally after the USA and China, importing crude oil to meet 80% of its energy needs (Das, 2019). Besides, excessive use of crude oil results in greenhouse gas (GHG) emissions, which affect climate change. This has resulted in India's shift toward clean energy options, such as energy through renewable resources, and focused attention has been given to the biofuels by Indian policymakers. The policies related to biofuels are connected with the intervention programs, implementation strategies, and goals for the creation and consumption of ethanol and biodiesel (Kumar et al., 2013).

India's intended nationally determined contribution (INDC) commitment under Paris Climate Agreement helped to build efficient and effective energy management area in the country (Gautam et al., 2019). India committed to achieving the reduction in GHG emissions by limiting the world average temperature by 2°C. To meet the commitment of achieving sustainable development goals (SDGs), India has actively started working toward its clean energy agenda. Alternatives to fossil fuels that are both realistic and strategic include renewable energy sources like biofuels and biodiesel.

India's biofuels journey started in 2001 with the launching of 5% ethanol blending program (EBP). Just after this, the Government of India adopted a biofuel mission in 2003 by launching the National Mission on Biodiesel. By the year 2011–2012, the program aimed to blend 20% biodiesel into a diesel fuel. Later in 2009, the Government of India launched the National Policy on Biofuels (NPB). The NPB 2009 envisaged 20% optional blending target for both petrol and diesel by 2017, which also revealed a vision for biofuel development with the involvement of various institutions and technologies that in long run helped in the expansion of feedstock for biofuel making (Das, 2018). The NPB was then reframed in 2018 as a National Policy on Biofuels (NPB) 2018. The objective is to reduce reliance on imported crude oil, enhance farmers' income, empower youth through employment generation, optimally use dry land, and contribute to sustainability. The NPB 2018 predicted denotative blending target of 20% and 5% of bioethanol and biodiesel in petrol and diesel, respectively, by 2030. Recently, the Government of India cabinet approved the amendments to the NPB 2018 on 18 May 2022. Due to advancements in the field of biofuels, various decisions were taken to increase biofuels production. The goals are to increase farmers' income, empower young people by creating jobs, reduce reliance on imported crude oil, make the best use of dry land, and promote sustainability.

2.2 ENERGY DEMAND OF INDIA

Conventional and nonrenewable energy resources such as crude oil, coal, natural gas are the main contributors to meet India's energy demand. Being an emerging economy, India is currently facing a momentous challenge to meet its energy demand in sustainable and accountable manner. In India, two leading sectors with high energy demand are transportation and industrial sectors, among all other sectors.

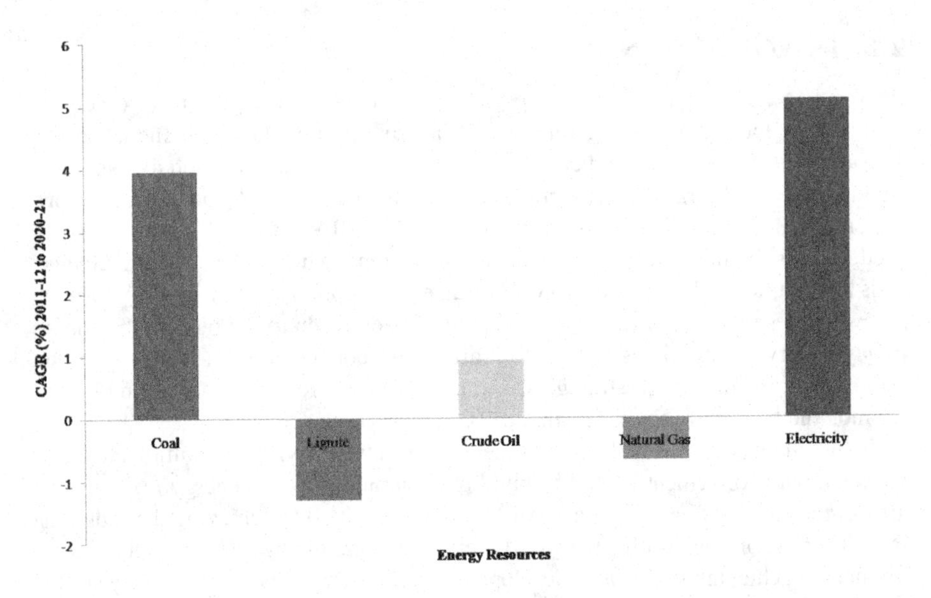

FIGURE 2.1 Growth rate of consumption of energy resources in India.

2.2.1 Energy through Fossil Fuel

Based on several research studies, it has been found that the contributions of fossil fuels to the current energy production are approximately 85%, which can consider being as a major energy resources (Pugazhendhi et al., 2019, Abraham et al., 2020). It is the known fact that overutilization of nonrenewable energy resources leads to energy crisis in upcoming decades (Sindhu et al., 2016, Kumar et al., 2019).

India's growing energy demand indicates that the requirement of equivalent fossil fuel is about 1,516 million tonnes by 2030 (Joshi et al., 2017). The International Energy Association said that from an economic perspective, total energy consumption (TEC) is a useful indication of efficient or inefficient end-use in economic activities and may propose sustainable course-correction actions. India's energy usage has increased in a healthy way. As per the Energy Statistics Report 2022, a significant or the minimal increment in consumption of energy resources has been seen in the last decade from 2011–2012 to 2020–2021. During this decade, the estimated percentage compound annual growth rates (CAGR) in the consumption of energy resources such as coal, lignite, crude oil, natural gas, and electricity were 3.96, −1.30, 0.93, −0.67, and 5.09%, respectively, as shown in Figure 2.1. Due to COVID-19 pandemic, the energy consumption rate in India has been slightly reduced as compared to that in 2019–2020.

2.2.2 Negative Impact on SDGs

In the last few decades, numerous negative effects of fossil matrix have been felt by the environment with increased development activities. The petroleum supply chain consumes a considerable amount of inputs and feedstock, which ultimately results in waste, discharge, and air pollutants. The usage of petroleum derivatives causes the emission of several polluting gases, including dust, volatile organic compounds (VOCs), carbon monoxide (CO), sulfur oxides (SO_x), and nitrogen oxides (NO_x) (Grist, 2016). As a result, this industry influences the quality of surface waters and groundwater, the frequency of acid rain, and the prevalence of soil pollution, and greenhouse gas (GHG) emissions (Bruckberger et al., 2019; Grist, 2016; Şahin et al., 2019). These results have a negative impact on the climate and are either directly or indirectly related to the SDGs.

2.3 INTERVENTION IN BIOFUEL POLICIES IN INDIA

The Government of India has implemented a number of programs and regulatory changes to promote the use of biofuels in the nation during the past 20 years (Table 2.1). Among them, the Ethanol Blending Program, the National Biodiesel Mission, and the National Biofuel Policy, 2008, were the most important ones (National policy on Biofuel, 2008). After reviewing these programs in view of the foregoing experiences, the Government of India has issued the National Policy on Biofuels, 2018, in order to create new targets consistent with the developments in other renewable energy sectors (National Policy on Biofuel, 2018).

2.3.1 Ethanol Blending Program

In 1940, the practice of blending ethanol with petrol was first introduced after several years since 2001, the Government of India was urged to develop the domestic biofuel production industry by worries about energy crisis and environmental problems. The Indian Government launched ethanol blending program (EBP) in 2002 with the goal of achieving ethanol blending. In its early stages, the government imposed 5% ethanol blending in nine significant sugarcane-growing states and four union territories, with the program being optional on a pilot basis. However, the necessary requirement was made optional in 2004, since ethanol was not readily available in the desired quantity. With the exception of few north eastern states and Jammu & Kashmir, the scope of the second EBP was expanded all over the country in 2006, and the idea of set fixed procurement price was introduced. At a time, 1.05 billion L of ethanol was needed to reach this goal, compared to a supply of only 440 million L. In 2013, the Government of India was chosen to go through open bidding. Ethanol was free from the 12.5% excise charge in 2015–2016. Additionally, OMCs were given instructions of 10% blending target throughout the majority of India as soon as practicable (Roy and Chandra, 2019; Das, 2020). Ethanol is produced in distilleries using a variety of feedstocks, including sugarcane juice, bagasse, sugar beet, rotting potatoes, and molasses (a dark brown syrupy liquid that is left over after sugar is removed from sugarcane juice). At that time, molasses was mostly used in India for the production of ethanol. Because of this, the production of sugarcane was crucial to the success of this initiative. For every ton of crushed sugarcane, 107 kg of sugar and 46 kg of molasses are produced, which together may make 11 L of ethanol.

As a result, only 2% ethanol blending was possible in 2007 after the policy's implementation. This number dropped to 0.67% in 2012 before rising to 3.5% in 2015

TABLE 2.1
Policy Intervention in Biofuels from 1948 to 2018

Act/Policy/Program	Year	Ministry Involved	Vision	Why Failed?	Reference
Power Alcohol Act,	1948		• This act mandated 20% (E20) blending of alcohol with petroleum to achieve energy security of India and to free from the oversupply of molasses from sugar cultivation.	The act was unable to implement due to: Lesser supply of ethanol and due to relatively stable global oil supplies and prices.	Indian Power Alcohol Act (2000), Ray (2012), Saravanan (2020)
Ethanol Blending Program (EBP)	2002	MoPNG	• Mandates 5% ethanol blending with petrol from January 2003.	Unable to implement due to lack of feedstock.	Saravanan (2018)
	2004		• Government made 5% E5 optional due to increasing cost of ethanol production and due to acute shortage of blended biofuel.	Failure to adopt ethanol pricing formula; GoI deferred to implement amendments of 2004, 2007, and 2008 due to the shortage in sugar molasses, and due to delay in the procurement.	Ram (2006)
	2007		• Government revised the ethanol blending norms and made E5 mandatory across the Country except J&K & N–E States.		Saravanan (2020)
	2008		• Government pushed the country toward 10% blending mandates.		

(Continued)

TABLE 2.1 (*Continued*)
Policy Intervention in Biofuels from 1948 to 2018

Act/ Policy/ Program	Year	Ministry Involved	Vision	Why Failed?	Reference
National mission on biodiesel (NBM)	2003	Planning Commission of India recommended its launch in 2003	This mission aimed at the plantation of Jatropha on wastelands from 2003 to 2007 in order to achieve 5% (B5) blending of Jatropa oil with diesel till 2007 and gradually increase to 20% (B20) till 2011–2012.	Fails due to lack of confidence building among agricultural and industrial sectors. The act was deficient in jatropha seed production, collection and extraction; deficient in jatropha plantation.	Saravanan (2018), Ram (2006), Saravanan (2020)
	2006	MoRD (Nodal Agency); MoPNG	MoPNG in Oct. 2005 announced Biodiesel Purchase Policy to provide biodiesel for oil marketing companies @ Rs. 26.50 /liter from Jan 2006.	Unrealistic Purchase price as cost of biodiesel production was 20%–50% higher than price set. Unenthusiastic response from OMCs.	
National biofuel policy	*2009*	MNRE MoPNG MoA MoRD MoPR MoTA MoS&T MoEF MoF	Set a target of 20% blending in both diesel and petrol till 2017. Policy strived to use recognized 400 species of trees bearing nonedible oilseeds for biofuels production in the country. Government declared biodiesel and bioethanol as 'goods' in order to ensure their unrestricted movement across the country. Policy allowed 100% FDI to attract foreign investment in the biofuel sectors.	Policy failed to provide incentives for blenders and retailers	Saravanan (2018) Ram (2006) Bandyopadhyay (2014) Kumar (2013) Saravanan (2020)
	2018		Aims to reduce India's oil import dependency; to provide better income opportunities to farmers by helping them dispose of their surplus stock in economic manner.		

and 2.07% in 2016. The sugarcane crop's output of sugar and molasses in 2017–2018 was 300 million tonnes (Mt), while the blending rate for that same period was about 4%. However, there are signs that it might increase to 7%–8% in the near future, largely due to the OMCs' higher prices (GoI, 2018). Due to higher sugarcane output than in the prior years and higher ethanol procurement prices, there has been a little increase over the past few years (Das, 2020).

Ministries, businesses, and research institutions all played a significant role in the effective implementation of EBP. The role of government in EBP included several ministries, including MNRE, MRTH, MAFW, MEF, MCAFPD, NITI Ayog, and state governments, and it was specified that each ministry's role ranged from setting targets, pricing, and marketing, to supporting research in the field of biofuels, establishing emission norms, R&D in sugarcane and other biofuels, monitoring the environmental benefits of EBP, setting standards for quality control of biofuels, recommending policy, etc. Industries like SIAM, ISMA, and OMC must optimize engines in accordance with pollution standards, maintain reliable supply to OMCs, and meet prescribed blending rates. The development of high-yielding varieties, R&D technologies to increase sugarcane productivity, and next-generation technologies for ethanol production are meant to be the focus of research institutions like SBI, IISR, TERI, and other state institutes and universities (Roy and Chandra, 2019).

2.3.2 Biodiesel Blending Program

Biodiesel as an alternative fuel is broadly used for blending with diesel to power automobiles. The raw materials used to make biodiesel include different types of vegetable oil, seeds from nonedible oils, animal fats, used cooking oil, etc. After the oil crisis of 1970, biodiesel has gained attraction and its production has boomed. Predominantly today, the production of biodiesel mostly involves the transesterification of vegetable and animal fats (Gautam et al., 2019). The NPB 2018 seeks to boost the biodiesel blend rate in diesel to 5% by 2030, even though it is now less than 0.5% (Das, 2020).

Several studies have been conducted to examine the operation characteristics of engines using biodiesel blends, and it has been shown that the braking power produced by the engine reduces as the amount of biodiesel increases (Madiwale et al., 2018). Additionally, it has been determined that as the proportion of biodiesel grows, the brake-specific fuel consumption (BSFC) rises due to the reduced calorific value of biodiesel fuels compared to petro-based diesel (Paul et al., 2014). Due to the blend's increased oxygen content, which promotes efficient combustion, the BTE also rises as the quantity of biodiesel in the blend does (Efe et al., 2018). The generation of more NO_X with an increase in the proportion of biodiesel in the mix is a significant issue related to the usage of biodiesel (Nalgundwar et al., 2016; Paul et al., 2014). In order to solve this issue, methanol or ethanol additives are typically added to blends in modest amounts to both decreased NO_x production and CO emissions. The problem of greater viscosity of biodiesel fuel, which influences its spray properties, is also solved by the use of alcohol additive. According to the research done by (Gautam and Kumar, 2015), using more alcohol with jatropha ethyl ester reduces viscosity while also lowering NO_x and CO emissions. The experimental analysis of jatropha blended with palm oil was carried out in 2014 (Paul et al., 2014), and a similar analysis of jatropha with turpentine oil was carried out in 2016 (Nalgundwar et al., 2016); the results of such dual biodiesel blends are similar to those with petro-based diesel. The research has also been carried out through mixing two biodiesel obtained from different sources with petro-based diesel and studying the corresponding performance characteristics. Biodiesel produced from several sources has been compared for performance in reference (Efe et al., 2018). It compares the performance of biodiesel made from sources such as sunflower, canola, maize, soybean, and hazelnut with various mix percentages of 20%, 50%, and 100%. According to comparison results, blends with 20% biodiesel content produce the best results of all biodiesel fuels.

2.3.3 National Biodiesel Mission

The Planning Commission of India in July 2002 set up a committee for the development of biofuels. The responsibility given by the committee was to stimulate the emergence of a coherent biofuel industry in India by devising the plan. The committee priorities the biodiesel and recommended the launching of the National Commission on Biodiesel in April 2003 (Planning Commission, 2003). For this mission, Ministry of Rural Development was designated as the nodal ministry (https://policy.asiapacificenergy.org/node/2777; Hirvonen, 2009). This initiated the journey of National Biodiesel Mission (NBM) in country. However, it was not an official biodiesel policy for country, but taking cue from this new biodiesel scheme, a number of state governments began to devise their own biodiesel policies.

The focus of National Mission on Biodiesel was not to use edible crops for this mission; hence they identified a nonedible feedstock—*Jatropa curcas* for biodiesel development. The NBM took place in two phases—from 2003 to 2007 and from 2007 to 2012 (Planning Commission, 2003).

The period of first phase (2003–2007) aimed to convert 4 lakh hectares of wasteland to Jatropa farming. In this phase Government wanted to demonstrate the benefits of a comprehensive biodiesel industry; hence they concentrated on proper management of biodiesel mission from Jatropa plantation to Jatropa oil blending with diesel and focused on marketing of Jatropa-blended diesel across the country. The government entirely funded the setting up of Jatropa plantation, whereas seed collection centers and oil extraction units were supposed to be financed with "margin money," i.e., money from private sector investments, government subsidies, and bank loans. Processing of Jatropa oil and its blending with diesel, i.e., biofuel manufacturing units, were seen as a completely commercial venture. The government also raised awareness among public for biodiesel acceptance as an automotive fuel through public events (Malode et al., 2021).

The second phase of NBM was from 2007 to 2012. The goal of second phase was to make biodiesel industry as a self-sufficing industry backed by capitals from private sectors and international donors. The shared impact of both the phases of NBM was to produce appropriate quantities of Jatropa oil to attain 20% blending of biodiesel by 2020 (Hirvonen, 2009). Although the demand for diesel in India is five times higher than for the petrol, the biodiesel business is still in its infancy. Still the Indian Government is unable to satisfy its ambitious goal of achieving 20% mandatory diesel blending by 2020 because of lack of Jatropa seeds in the industry (Zafar, 2022). In 2014, 2015, and 2016, the production of biodiesel from various feedstocks was 130, 135 and 140 million liters, respectively, and was predicted to reach up to 150 million liters in 2017. This was due to the effective implementation started by NBM. Through 2018, it would add an additional 10 million liters (DTE).

2.3.4 National Policy on Biofuels

NPB was enacted in 2009 with an unrestricted blending objective to achieve a 20% blend of ethanol and biodiesel by 2017 (Das, 2018). According to Arjune (2017), NPB 2009 was mainly focused on the cultivation of oilseed-bearing plants on wasteland, but it did not look at biofuels from agricultural waste. Research carried out by Sorda et al. (2010) revealed that the cost of producing biodiesel was greater than the cost of purchasing it, making the blending requirements challenging to comply with. Another analysis on NPB 2009 revealed that the biofuels policy was biased toward the supply side and recommended promoting flex fuels to increase demand for biofuels; the paper, however, did not address the limitations on the interstate transportation of ethanol and molasses that are now in place (Basavaraj et al., 2012).

The Government of India reformed the biofuels policy by putting into effect NPB 2018 with the aim of promoting the use of biofuels in the mobility and energy sector. Currently, biodiesel makes up less than 0.1% of diesel fuel, while ethanol is mixed into gasoline to the tune of 2.0%. By 2030, NPB 2018 wants to boost the percentages of biodiesel and ethanol in gasoline and diesel to 5% and 20%, respectively, via the EBP and BBP programs. One of the highlights of NPB 2018 is how it divides biofuels into three categories, including 1G, 2G, and 3G biofuels. First-generation (1G) biofuels are the most basic kind of biofuels, and they comprise ethanol from molasses and biodiesel from nonedible oilseed crops. Second-generation (2G) biofuels, on the other hand, are sophisticated biofuels made from municipal solid waste (MSW), agricultural waste, etc. However, Bio-CNG is classified as third-generation (3G) biofuels. With the consent of the National Biofuel Coordination Committee (NBCC), it also permitted the use of extra food grains for the manufacturing of ethanol. It suggested a higher purchase price for 2G biofuels compared to 1G biofuels, as well as a VGF program, for the construction of 2G ethanol biorefineries worth Rs. 5,000 crores over a 6-year period. The policy allows 100% Foreign Direct Investment (FDI) through the automated approval system in the biofuel industry as long as biofuel is exclusively utilized for internal demand. The policy prohibited the import and export of biofuels, but imports of feedstock would be allowed. Through tax credits, early depreciation on plant expenses, and increased support for the construction of 2G bio refineries, it aims to offer financial aid. The OMCs would be in charge of marketing, distributing, and storing biofuels.

NPB 2018 intends to provide indigenous feedstock for biofuel production and is a paradigm-shifting biofuel policy for India's next ten years. The strategy prioritizes biofuels strategically, as shown in Figure 2.2, and aims to achieve SDGs.

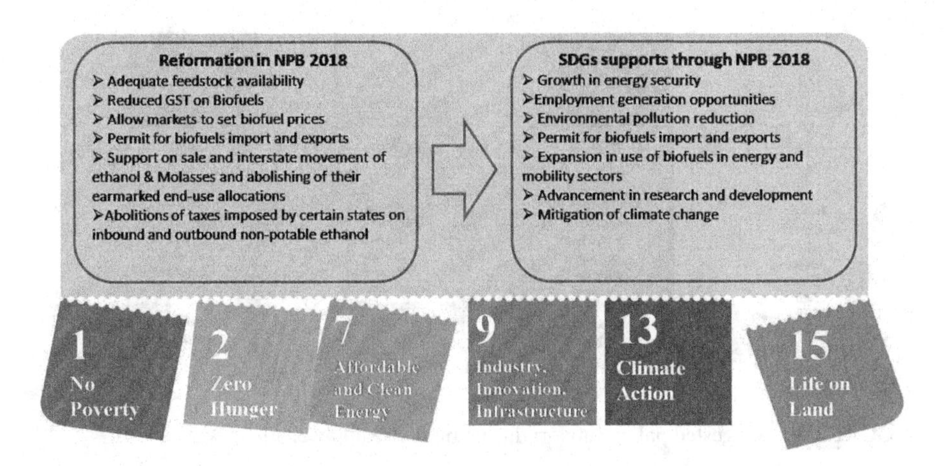

FIGURE 2.2 Reformation in NPB 2018 to achieve SDGs.

2.4 CHALLENGES AND OPPORTUNITIES

Policies relating to biofuels have a significant impact on how the energy sector develops. Biofuel regulations influencing various sectors, such as agriculture, research, industries, and commerce, have a substantial effect on the profitability of biofuel production. Because there are so many different types of policy instruments (taxes, subsidies, price support, etc.) and uses of biofuels, it is challenging to identify pertinent policies and quantify their effects on particular situations (ONF LOSS, 2019).

For instance, different phases of the biofuel industry might be impacted by subsidies (Kutas et al., 2007). Assigning policies to one category or the other may be somewhat artificial because of the interconnected nature of the different locations in the biofuel supply chain where direct and indirect policy actions may boost the industry (ONF LOSS, 2019). The background for the discussion of the biofuel policy in India, its implications, and its distortions at various stages of the biofuel supply chain in production, commercialization, and sustenance in the promotion of the biofuel sector is provided in Figure 2.3, which was modified from the Global Subsidies Initiative of Steenblik.

Following the 2018 National Biofuel Policy, excess food grains may be converted to ethanol with the clearance of the previous National Biofuel Coordination Committee. However, presently there is an absence of such committee in place; it is unclear who has authority to approve the conversion of extra food grain stocks into biofuel. Additionally, the strategy has not adequately presented the problems with technical viability.

The authors state once again that the aggressive biofuel targets outlined in the 2018 strategy appear to be "unrealistic," given the country's stumbling gasoline prices and the Government of India's attempt to meet those goals despite inadequate infrastructure. In 2016, to establish second-generation (2G) biorefineries, the state-owned OMCs, Hindustan Petroleum Corporation Limited and Bharat Petroleum Corporation Limited signed memoranda of understanding with Punjab and Haryana

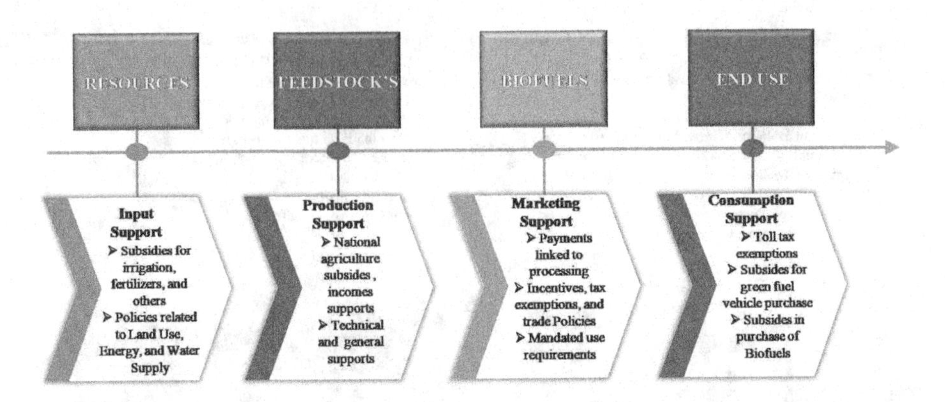

FIGURE 2.3 Suggested policy support during biofuels supply chain.

and eight other states. Wheat and rice straw were intended to be used as the biofuel feedstock for the proposed 2G biorefineries; however, they were ultimately unsuccessful. It amply demonstrates the supply–demand imbalance for biofuel feedstock; thus, India could fall short of its blending goals again.

Numerous obstacles mentioned above have not yet been removed due to the sluggish pace of biofuel development. The Indian Planning Commission had recommended necessary changes to the regulations at the time. According to Section 19 of the Standards of the Weights and Measures Act, 1976, the term "biofuels" needs to be adequately defined to avoid ambiguity regarding a specific specification, such as measure, performance, materials, and physical properties of biofuels. In order to maintain the limits for environmental pollutants emission and discharge, modifications must be made to the Air (Protection and Control of Pollution) Act, 1981, and the Environment (Protection) Act, 1986.

The volatility of oil prices also requires a careful assessment of opportunities for the production and use of biofuels. Governments have started showing interest in production and marketing of biofuels, even using subsidies to make them financially viable, due to the volatility of crude oil prices and overdependence on imports. In order to promote biofuels in the near future, it is crucial to address today's most significant challenges, including energy security, trade imbalances, greenhouse gas (GHG) emissions, rural livelihoods, and domestic agricultural commodities (Dufey, 2006). In addition, biofuels are viewed as a possible factor in the socioeconomic advancement of rural regions and have the ability to significantly reduce poverty by generating new job opportunities, therefore having an indirect or direct influence on the several United Nations sustainable development goals (Kammen, 2006). India is primarily known as an agriculturally rich nation; the biofuels sector might significantly increase employment there. There may be new employment possibilities in the biofuels industry, which includes second-generation ethanol, biodiesel, and related products from biorefineries. The major agro-industrial sector in the nation, which supports 50 million farmers and others who depend on them, is the cultivation and processing of sugarcane for the manufacturing of sugar and ethanol (approximately 7.5% of the rural population). Additionally, the employment in the sugar business contributed to significant foreign exchange savings, allowing India to enhance its Gross Domestic Productivity (GDP) and develop its infrastructure. Additionally, it will lessen reliance on members of the Organization of Petroleum Exporting Countries (OPEC) and pave the way for self-sufficiency in petroleum needs.

2.5 ACHIEVING SDGs THROUGH POLICIES

Holistic actions must be required to achieve SDGs through policy, which can be achieved through critical analysis and deconstruction of biofuel policies. There is a need for modification in industrial activities in favor of the SDGs. Biofuels production as a renewable energy resource is an exceptionally dynamic and exhilarating field, and the contribution of this industry to the SDGs is noteworthy. The main drivers behind biofuel are SDG 1 (to reduce poverty), SDG 2 and 3 (by using MSW and agriculture waste to promote the circular economy), SDG 7 (to provide clean energy by reducing dependency on fossil fuels), SDG 8 (employment generation in rural India through crop production and biofuel production), and SDG 13 (encourage use of land and marginal areas for growing energy crops and help to restore the land). All the biofuel-related SDGs are discussed in the following paragraphs.

Reduce Poverty (SDG 1): Countries like India can reduce their energy demand by adopting renewable energy approaches, as the petroleum market is quite unstable due to the scarcity of reserves. This approach can help the nation to eradicate poverty (Gielen et al., 2019). Developing countries like India are capable of generating renewable energy through biofuel production. But there is a need to support industrial development through biofuel policies, which should be done by the Government of India, which ultimately helps to eradicate poverty.

Zero Hunger (SDG 2): The vision of this goal is to eradicate hunger and all types of malnutrition by 2030. To achieve the vision, there is a need of sustainable agriculture practices, equal land access, innovative technologies and markets, as well as global collaboration on infrastructure and technological investments to increase agricultural output (https://www.theexplorer.no/goals/zero-hunger/; https://www.jordantimes.com). Shrestha et al. (2019) concluded with the economic models that food cost raised due to biofuel productions from crops like sugarcane, corns, oil grains, etc. Although food availability is unreliable to feed India's growing population, policy reformation and proper implementation can lead to ending hunger.

Clean Energy (SDG 7): India's paradigm shift in displacing conventional fossil fuels with renewable energy resources is remarkable as biofuel production has increased in the recent years. Clean energy production by utilizing biomass feedstock is seemed to be an authentic solution and positively promotes SDG. Biofuel policies can significantly contribute to meeting the demand for biofuels; the most sustainable approaches should be adopted to green energy production through biofuels without harming the environment – GHGs emissions like CH_4, CO_2, NO_x, etc. (Srirangan et al., 2012). Some questions may arise on GHG benefits, as biofuels derived from certain first-generation (1G) feedstock have GHG discharges comparable to those from fossil fuels (i.e., NO_x and CO_2) (Farrell et al., 2006). Nevertheless, biofuels produced from 2G feedstock are projected to decrease GHG emissions by 85%, and agriculture biomass may reduce by up to 50% (Cherubini and Ulgiati, 2010). While talking about biodiesel fuels, the contribution of biodiesel to NO_x emission is the same as that of petro-diesel. Still, some studies believe that its blending with petro-diesel has cleaner emission profiles. In a study reviewed by Sagar and Kartha (2007), the blending of biodiesel in vehicles may reduce the emissions of pollutants like NO_x by 10%, CO and other particulates by 45%, C–H compounds by 65%, and SO_2 by 100%.

Industrial Innovation and Infrastructure (SDG 9): Innovation in the biofuel production field has started with the shift of food to investigate new biomass feedstock. Biomass was considered to be the most direct resource of biofuels like bioethanol and biodiesel and possibly additional compounds (Duque et al., 2018; Pandey and Prakash, 2018). With increasing demand, bioethanol production increased with the innovations, resulting in the shift toward 2G bioethanol, also known as cellulosic ethanol. 2G bioethanol may be produced from agricultural wastes, including maize stover, sugarcane straw, wheat straw, and other stale grains, as well as the organic portion of MSW (Schwartz, 2010). This technology opens up substantial prospects for the whole production chain, including farmers, the ethanol business, new biotechnology firms, project developers, and investors, because it is based on by-products of other crops (i.e., food crops on arable land).

Climate Action (SDG 13): Climate is changing, and it is one of the greatest threat to sustainable developments, which has wide-ranging, unprecedented effects that will disproportionately affect the most disadvantaged. The achievement of the SDGs depends on swift action to address climate change and lessen its adverse effects (https://www.unoosa.org).

With the highest growth in fuel consumption, transportation is the biggest sector in terms of GHG emissions. In 2010, it was responsible for roughly 23% of worldwide CO_2 emissions and 27% of end-use energy emissions, with urban transportation accounting for almost 40% of end-use fuel consumption. CO_2 lingers in the environment for more than a century and causes long-term warming (IPCC 2014). With over 3% of the global emissions, India ranks sixth in terms of GHG emissions. By 2030, it is anticipated that India's per-person CO_2 emissions will rise to 1.6 tonnes. To address climate change and to reach the prescribed level, there is a necessity for GHG reduction (Prasad et al., 2019). The Government of India is committing to reduce emissions by 30%–35% by 2030 with respect to 2005 levels. The clean transportation sector will probably help GHG reduction objectives as we move toward green mobility and a low-carbon economy (Aradhey, 2018). India can achieve this goal by switching to clean energy sources like ethanol and biodiesel, and biofuel rules are also helping to reduce emissions by improving fuel efficiency requirements.

Life on Land (SDG 15): Conservation, restoration, and sustainable use of terrestrial and other ecosystems and preservation of a variety of living forms found on land requires consistent focus on conservation and restoration. SDG 15 is particularly concerned with maintaining forests sustainably, reclaiming degraded lands, successfully halting desertification, minimizing damaged natural habitats, and halting biodiversity loss (https://www.theexplorer.no/goals/life-on-land/). Somerville et al. (2010) examined the utilization of marginal soil for energy crop production; the method necessitates quick plant adaptation to water-limited lands and the choice of cultivars that do well in semi-arid/arid environments (Carroll and Somerville, 2009). Accurate projections of the energy density per hectare and the potential yield of biofuel are essential for successful implantation, but they must also consider another less visible factor; plants will move 20% of the CO_2 fixed to the soil, which will partially stabilize and serve as a carbon sink. This exudate carbon will help restore soils and provide nutrients for the growth of biodiversity (Timmis and Ramos, 2021). The growth of roots is a useful approach of carbon sequestration, and the exudates they produce, which include unconsumed carbon, may react with the minerals in the soil to speed up their sluggish metabolism. The cultivation of energy crops on marginal soils will increase soil availability and, in the medium run, may improve soils suitable for growing food crops. Specifically for the generation of biofuel, energy crop plants are produced; however, the crops differ by geographic location (Kocar and Civas, 2013; Hood et al., 2013).

2.6 CONCLUSION

The present chapter focuses on the roadmap for the achievements of SDGs through the policy intervention in biofuels sectors. The NPB 2018 with an amendment in 2022, helping sustainable development through six identified SDGs. This chapter aims to help policymakers, academics, and biofuel industry participants understand the impact of biofuel regulations on the industry's long-term growth. New policies and strategies may be developed, or current strategies may be adjusted, to address the particular sector's difficulties by removing or lowering the impact of substantial impediments. To grow or improve the sector, there is a strong need to promote biofuel consumption on a big scale by enacting obligatory rules requiring the inclusion of biofuels in conventional fuels. More incentives should be provided to the biofuel industry. In addition, tax exemption programs and policies must be developed in order to attract both domestic and foreign investors in the biofuels sector. There is a significant demand for investment in R&D-related activities to generate second- and third-generation biofuels and improve competitiveness in biofuel markets. Private players must be encouraged to generate income and contribute expertise to the growth of the case business on a large scale. Furthermore, through IEC, there is a need to promote the usage of biofuels among automotive users. The Government of India must promote the possibilities of biomass energy, i.e., bioenergy that is environmentally friendly, economical, fully functioning, and genuine. It should be highlighted that the sustainability of bioenergy is critical, and necessary precautions should be taken to achieve SDGs by lowering GHG emissions in bioenergy systems.

REFERENCES

Abraham, A., Mathew, A. K., Park, H., Choi, O., Sindhu, R., Parameswaran, B., Pandey, A., Park, J. H. and Sang, B. I. (2020). Pretreatment strategies for enhanced biogas production from lignocellulosic bio- mass. *Bioresource Technology*, 301, 122725.

Aradhey, A. (2018) India biofuels annual, In: *GAIN-Report no IN8085, Global Agricultural information Network (GAIN)*, US Department of Agriculture-Foreign Agricultural Services (FAS).

Bandyopadhyay, K. K., Pradhan, S., Sahoo, R. N., Singh, R., Gupta, V. K., Joshi, D. K., & Sutradhar, A. K. (2014). Characterization of water stress and prediction of yield of wheat using spectral indices under varied water and nitrogen management practices. *Agricultural Water Management*, 146, 115–123.

Basavaraj, G., Rao, P. P., Ravinder Reddy, C., Kumar, A. A., Srinivasa Rao, P., & Reddy, B. V. S. (2012). A review of national biofuel policy in India: A critique-need for promotion of alternative feedstocks. *Journal of Biofuels*, 3(2), 65–78.

Bruckberger, M. C., Morgan, M. J., Walsh, T., Bastow, T. P., Prommer, H., Mukhopadhyay, A., & Puzon, G. J. (2019). Biodegradability of legacy crude oil contamination in Gulf War damaged groundwater wells in Northern Kuwait. *Biodegradation*, 30(1), 71–85.

Carroll, A., & Somerville, C. (2009). Cellulosic biofuels. *Annual Review of Plant Biology*, 60(1), 165–182.

Cherubini, F., & Ulgiati, S. (2010). Crop residues as raw materials for biorefinery systems–A LCA case study. *Applied Energy*, 87(1), 47–57.

Das, S. (2018). Achievements and misses of the Indian national policy on biofuels 2009. *Economics and Policy of Energy and the Environment*, LX(2), 5–30.

Das, S. (2019). The early bird catches the worm - first mover advantage through IoT adoption for Indian public sector retail oil outlets. *Journal of Global Information Technology Management*, 22(4), 280–308.

Das, S. (2020). The national policy of biofuels of India–A perspective. *Energy Policy*, 143, 111595.

Dufey, A. (2006) *Biofuels Production, Trade and Sustainable Development: Emerging Issues*. International Institute for Environment and Development, London.

Duque, E., Daddaoua, A., Cordero, B. F., Udaondo, Z., Molina-Santiago, C., Roca, A., ... & Ramos, J. L. (2018). Ruminal metagenomic libraries as a source of relevant hemicellulolytic enzymes for biofuel production. *Microbial Biotechnology*, 11(4), 781–787.

Efe, Ş., Ceviz, M. A., & Temur, H. (2018). Comparative engine characteristics of biodiesels from hazelnut, corn, soybean, canola and sunflower oils on DI diesel engine. *Renewable Energy*, 119, 142–151.

Farrell, A. E., Plevin, R. J., Turner, B. T., Jones, A. D., O'hare, M., & Kammen, D. M. (2006). Ethanol can contribute to energy and environmental goals. *Science*, 311(5760), 506–508.

Gautam, R., Ansari, N. A., Thakur, P., Sharma, A., & Singh, Y (2019a). Status of bio fuel in india with production and performance characteristics: A review. *International Journal of Ambient Energy*, Doi: 10.1080/01430750.2019.1630298.

Gautam, R., & Kumar, N. (2015). Comparative study of performance and emission characteristics of Jatropha alkyl ester/butanol/diesel blends in a small capacity CI engine. *Biofuels*, 6(3–4), 179–190.

Gautam, R., Shyam, S., Reddy, B. R., Govindaraju, K., & Vinu, R. (2019b). Microwave-assisted pyrolysis and analytical fast pyrolysis of macroalgae: Product analysis and effect of heating mechanism. *Sustainable Energy & Fuels*, 3(11), 3009–3020.

Gielen, D., Boshell, F., Saygin, D., Bazilian, M. D., Wagner, N., & Gorini, R. (2019). The role of renewable energy in the global energy transformation. *Energy Strategy Reviews*, 24, 38–50.

Global Status Report "Renewables 2010", REN21 Secretariat, Paris. Available online at http://www.ren21.net/Portals/97/documents/GSR/REN21_GSR_2010_full_revised%20Sept2010.pdf.

National Policy on Biofuel (2018). Government of India. (2018). http://petroleum.nic.in/sites/default/fles/biofuelpolicy2018.1.pdf (Accessed on 2 January 2019).

National Policy on Biofuels (2008). Government of India. *Ministry of New & Renewable Energy*, 1–18.

Grist, M. (2016). Environmental management and technology in oil refineries. In *Environmental Technology in the Oil Industry* Stefan Orszulik (Ed.), Springer, Cham, Switzerland, 375–392.

Hirvonen, M. (2009). *National Mission on Biodiesel: A Study of Science, Development and Policy Processes in India.* Mphil Thesis, The University of Edinburgh, Scotland.

Hood, D. C., Raza, A. S., de Moraes, C. G. V., Liebmann, J. M., & Ritch, R. (2013). Glaucomatous damage of the macula. *Progress in Retinal and Eye Research*, 32, 1–21.

https://gain.fas.usda.gov/Recent%20GAIN%20Publications/Biofuels%20Annual_New%20Delhi_India_9-28-2018.pdf.

https://www.jordantimes.com.

https://policy.asiapacificenergy.org/node/2777.

https://www.theexplorer.no/goals/life-on-land/.

https://www.theexplorer.no/goals/zero-hunger/.

http://www.unoosa.org.

IPCC (2014). Summary for policymakers In: Field C. B. et al. (ed) *Climate Change 2014: Impacts, Adaptation, and Vulnerability. Part A: Global and Sectoral Aspects. Contribution of Working Group II to the Fifth Assessment Report of the Intergovernmental Panel on Climate Change* (pp. 1–32), Cambridge University Press, United Kingdom and New York, NY, USA.

Joshi, G., Pandey, J. K., Rana, S., & Rawat, D. S. (2017). Challenges and opportunities for the application of biofuel. *Renewable and Sustainable Energy Reviews*, 79, 850–866.

Kammen, D. M. (2006) Bioenergy in developing countries: Experiences and prospects. In: Hazell P., & Pachauri R. K. (eds) *Bioenergy and Agriculture: Promises and Challenges.* IFPRI, Washington, DC.

Koçar, G., & Civaş, N. (2013). An overview of biofuels from energy crops: Current status and future prospects. *Renewable and Sustainable Energy Reviews*, 28, 900–916.

Kumar, P., Kumar, V., Kumar, S., Singh, J., & Kumar, P. (2019) Bioethanol production from sesame (Sesamum indicum L.) plant residue by combined physical, microbial and chemical pretreatments. *Bioresource Technology*, 2019, 122484.

Kumar, S., Shrestha, P., & Salam, P.A., 2013. A review of biofuel policies in the major biofuel producing countries of ASEAN: Production, targets, policy drivers and impacts. *Renewable and Sustainable Energy Reviews*, 26, 822–836.

Kutas, G., Lindberg, C., & Steenblik, R. (2007). Biofuels–at What Cost?: Government Support for Ethanol and Biodiesel in the European Union (pp. 14–25). Prepared For the Global Subsidies Initiative (GSI) of the International Institute for Sustainable Development (IISD) Geneva, Switzerland.

Madiwale, S., Karthikeyan, A., & Bhojwani, V. (2018). Properties investigation and performance analysis of a diesel engine fuelled with Jatropha, Soybean, Palm and Cottonseed biodiesel using ethanol as an additive. *Materials Today: Proceedings*, 5(1), 657–664.

Malode, S. J., Prabhu, K. K., Mascarenhas, R. J., Shetti, N. P., & Aminabhavi, T. M. (2021). Recent advances and viability in biofuel production. *Energy Conversion and Management*: X, 10, 100070.

Nalgundwar, A., Paul, B., & Sharma, S. K. (2016). Comparison of performance and emissions characteristics of DI CI engine fueled with dual biodiesel blends of palm and jatropha. *Fuel*, 173, 172–179.

ONF LOSS. (2019). *The State of Food and Agriculture 2019.* Food and Agriculture Organization, Italy.

Pandey, A. K., & Prakash, R. (2018). Energy conservation opportunities in pulp & paper industry. *Open Journal of Energy Efficiency*, 7(04), 89.

Paul, G., Datta, A., & Mandal, B. K. (2014). An experimental and numerical investigation of the performance, combustion and emission characteristics of a diesel engine fueled with jatropha biodiesel. *Energy Procedia*, 54, 455–467.

Planning Commission. (2003). Report of the Committee on Development of Bio-fuel. http://planningcommission.nic.in/reports/genrep/cmtt_bio.pdf.

Prasad, S., Venkatramanan, V., Kumar, S., & Sheetal, K. R. (2019). Biofuels: A clean technology for environment management. In *Sustainable Green Technologies for Environmental Management*. Editors: Shachi Shah, V. Venkatramanan, Ram Prasad, Springer, Singapore, 219–240.

Pugazhendhi, A., Mathimani, T., Varjani, S., Rene, E. R., Kumar, G., Kim, S.-H., Ponnusamy, V.K., & Yoon, J.J. 2019. Biobutanol as a promising liquid fuel for the future-recent updates and perspectives. *Fuel*, 253:637–646.

Ram Mohan, M. P., Phillippe, T. G. T., & Mazhuvanchery, S. (2006). Biofuel laws in Asia: Instruments for energy access, security, environmental protection and rural empowerment. *Environmental Protection and Rural Empowerment. Asian Biotechnology and Development Review*, 8(2), 51–75.

Ray, S., Mun, E. Y., Buckman, J. F., Udo, T., & Bates, M. E. (2012). Memory for emotional picture cues during acute alcohol intoxication. *Journal of Studies on Alcohol and Drugs*, 73(5), 718–725.

Roy, M. M., & Chandra, A. (2019). Promoting biofuels: The case of ethanol blending initiative in India. *Clean Technologies and Environmental Policy*, 21(5), 953–965.

Sagar, A. D., & Kartha, S. (2007). Bioenergy and sustainable development? *Annual Review of Environment and Resources*, 32(1), 131–167.

Şahin, U. (2019). Forecasting of Turkey's greenhouse gas emissions using linear and nonlinear rolling metabolic grey model based on optimization. *Journal of Cleaner Production*, 239, 118079.

Saravanan, A. P., Mathimani, T., Deviram, G., Rajendran, K., & Pugazhendhi, A. (2018). Biofuel policy in India: A review of policy barriers in sustainable marketing of biofuel. *Journal of Cleaner Production*, 193, 734–747.

Saravanan, A. P., Pugazhendhi, A., & Mathimani, T. (2020). A comprehensive assessment of biofuel policies in the BRICS nations: Implementation, blending target and gaps. *Fuel*, 272, 117635.

Schwartz, J. R., Mark, S., & Wolfson, A. (2010). A first-order simulator to control dioxin emissions: NMCRC-ATMOS. *Waste Management & Research*, 28(5), 461–471.

Shrestha, D. S., Staab, B. D., & Duffield, J. A. (2019). Biofuel impact on food prices index and land use change. *Biomass and Bioenergy*, 124, 43–53.

Sindhu, R., Binod, P., Pandey, A. (2016). Biological pretreatment of lignocellulosic biomass–an overview. *Bioresource Technology*, 199, 76–82.

Somerville, C., Youngs, H., Taylor, C., Davis, S. C., & Long, S. P. (2010). Feedstocks for lignocellulosic biofuels. *Science*, 329(5993), 790–792.

Sorda, G., Banse, M., & Kemfert, C. (2010). An overview of biofuel policies across the world. *Energy Policy*, 38(11), 6977–6988.

Srirangan, K., Akawi, L., Moo-Young, M., & Chou, C. P. (2012). Towards sustainable production of clean energy carriers from biomass resources. *Applied Energy*, 100, 172–186.

Timmis, K., & Ramos, J. L. (2021). The soil crisis: The need to treat as a global health problem and the pivotal role of microbes in prophylaxis and therapy. *Microbial Biotechnology*, 14(3), 769–797.

Zafar, S., 2022. Biodiesel Program in India – an Analysis; National Biodiesel Mission https://www.bioenergyconsult.com/tag/national-biodiesel-mission/.

Part II

Technological Advances, Challenges & Opportunities of Biofuels

3 Conventional and the Recent Advances in Technologies for the Production of Different Generations of Biofuels

Anudeep Nema
Eklavya University

Rajnikant Prasad
G. H. Raisoni Institute of Business Management

Ranjeet Kumar
National Institute of Technology

Dayanand Sharma
Sharda University

Dipesh Kumar R. Sonaviya
C. S. Patel Institute of Technology

Nityanand Singh Maurya
National Institute of Technology

Archana Kumari
CSIR—National Environmental Engineering Research Institute (CSIR—NEERI)

DOI: 10.1201/9781003197737-5

CONTENTS

3.1 INTRODUCTION

Inflation in international oil prices, energy supply security, global warming, and the emergence of new agricultural prospects are driving the seek for clean, sustainable, and cost-competitive energy sources alternate to fossil fuels. These are the key driving reasons behind biofuels, which have emerged as promising energy sources to ensure a long-term energy supply. Biofuels are developed as a petroleum alternative because of their nontoxic, sulfur-free, biodegradable nature, and derivation from renewable sources. Renewable biomass sources are used to make biofuels. The increasing demand for fossil fuels has resulted in severe environmental challenges. With the current trend of increasing energy demand, renewable energy source dependency has increased and expected a 30% increase in the global energy demand by 2040 (Joshi et al. 2017). By 2050, road traffic is expected to contribute 87 gigatons of CO_2 equivalent to global greenhouse gas emissions. It can have a detrimental influence on natural resources and the environment, such as pollution and global climate change (Ambaye et al. 2021). One of the most important processes for the survival of a planet is the conversion of sunlight into chemical energy. The consumption of CO_2 and the generation of O_2 and plant resources are all parts of transforming solar energy into chemical energy for plant reproduction. The resources utilized to generate bioenergy are referred to as "biomass." Forests, crops, and leftovers from the agroforestry and animal sectors are primary biomass sources. Out of total global food production, about one-third was wasted or lost as per Food and Agriculture Organization in 2011 (Rena et al. 2020).

From the ancient times, biomass is a popular kind of source of energy used by humankind. Although exact data are unavailable, it is appraised that almost 33% of the world population relies on the conventional biomass as their primary energy source (agricultural, forestry residues, wood, livestock, etc.). As a result, conventional biomass accounts for over 90% of the global biomass use. Households use traditional biomass in various parts of Asia, Latin America, and Africa to satisfy their most energy needs, mostly for cooking. Biomass is inefficient in these situations, resulting in the degradation of natural resources and harm to the users. The energy supplied by these applications is frequently of low quality and collecting and transporting them involves much labor. Moreover, the production of agriculture and fruits also increases the waste generation during process like handling, transportation, processing, and preparation (Prashanth Kumar et al. 2019; Rena et al. 2020; Srikanth et al. 2009). This biomass represents a significant fraction of biomass that can be used as raw material for biofuels.

Moreover, producing fuels from conventional biomass sources can worsen the situation of deforestation, putting extra strain on the local ecology and increasing net greenhouse gas (GHG) emissions. Despite these disadvantages, billions of people use conventional biomass sources to meet their energy demands, since they are easily accessible and less costly. Dry biomass is easier to acquire and store, and it has a long history of use in many cultures. Furthermore, many nations would have to raise their energy imports without this feature, and many low-income people would have to spend more money on alternative energy sources (IEA 2011).

Knowing this, significant work has to be done in biofuels to increase economic viability. Biofuels may become a more competitive fuel, as technology progresses in their production and conversion. Chemical and biological advances and new crops for energy generation, novel enzymes, and artificial simulations of biological processes (anaerobic digestion, fermentation, and so on) can help to minimize their output. As a result, millions of dollars are being invested in developing these technologies. This review aims to look at some of the most current advancements in biofuel manufacturing.

3.2 METHODS OF EXTRACTION AND BASIC KNOWLEDGE ON BIOFUELS

Biofuels are used in the transportation sector, electricity, and biogas production. Biofuels contribute a significant fraction of the transportation sector worldwide. The advantages of using biofuels include easy extraction, sustainability, combustion based on the carbon dioxide cycle, and is environment friendly (Gaurav et al. 2017). India aims at national average target of 20% ethanol blended with gasoline by 2025 and 5% blending of biodiesel with conventional biodiesel by 2030 (Behera et al. 2014). The greenhouse gas can be tackled by using biofuels (biodiesel, biogas, and ethanol) as alternative fossil-based fuels. Biofuels such as biodiesel, ethanol, methane, and bio-oil are produced from biomass in the gaseous or liquid form. Biofuels are carbon neutral and environment friendly, since they neutralize the CO_2 generation through biofuels during their growth by absorbing CO_2 from the atmosphere.

The current prediction of 10%–50% energy consumption will be from biomass by 2050 (Kumar et al. 2015a). The biofuels production from biomass by adopting emerging technologies. This also reduces the burden of biomass handling and the issue of disposal. Biomass is the fourth largest available energy resource globally, which is natural, inexpensive, and available in abundance in a short duration of time. The annual global biomass production is about 220 billion tonnes on a dry weight basis, equivalent to 4,500 EJ of solar energy captured each year, corresponding to 270 EJ of the annual bioenergy market (Kumari & Singh 2018).

Similar to fossil fuels, biofuels exist in three phases of solid, liquid, and gas. Solid fuels include fuels like wood chips, wood pallets, and animal waste. Liquid biofuels include ethanol, biodiesel, green diesel, methanol, and green gasoline (Chandra 2021). Gaseous biofuels include biomethane and biohydrogen. The primary biomass includes food crops rich in sugar and starch, such as sugarcane and corn; non-food biomass includes crop residue and lignocellulosic material obtained from the municipal and industrial solid waste, and algal biomass (Ho et al. 2014). The energy from the biofuels is stabilized during biological carbon fixation, where carbon dioxide is converted into sugar that is present in the plant biomass and living organism (Alaswad et al. 2015). The biomass feedstocks used for biofuel production are categorized as dedicated crops and wastes and residues (Nikolić et al. 2016).

3.3 EXTRACTION METHOD

The biomass used to extract biofuels contains appreciable amounts of sugar content directly or in the other forms that can be converted into sugar forms like cellulose and starch. Lignocellulosic materials obtained from agricultural waste such as sugarcane bagasse, wheat straw, crops such as switch grass, and forest materials like sawmill and paper mill discard are considered viable raw materials for producing biofuels to its abundant availability that are renewable and producc low greenhouse gases than traditional fossil fuels.

The lignin content of the biomass complicates the production process. The digestibility of cellulose present in the lignocellulosic material is hindered by structural complications, compositional factors, and physicochemical composition. So, some treatment is needed to provide easy digestibility of cellulose called pretreatment. Pretreatment is the crucial step involved in the conversion of biomass to biofuel. However, biomass cannot be directly used for biofuel due to lignin in lignocellulose, acting as a hindrance to the fungi and bacteria toward conversion into biofuels. Lignocellulose contains carbohydrate polymers in cellulose, and hemicellulose and lignin as aromatic polymers. Carbohydrate polymer contains five to six carbon sugar tightly bonded to the lignin.

Lignocellulosic materials are categorized as virgin biomass, waste biomass, and energy crops. Virgin biomass includes terrestrial plants like trees and grasses, and waste biomass is waste products from industrial processes like straw and bagasse and energy crops like elephant grass, which contains high lignocellulose content. The extraction process is an intermediate step in the physicochemical process of biofuel production (Li et al. 2019b; Peng et al. 2019). It is categorized as physical, chemical, biological, and combined (Kadar et al. 2003; Gaurav et al. 2017). In the combined pretreatment method, temperature, pressure, or biological steps are combined with chemical treatment and, accordingly, called physicochemical or biochemical pretreatment methods (Agbor et al. 2011). Combined methods are generally effective, since they combine the benefits of both the methods by improving biomass digestibility and are employed in emerging technologies. The combined method includes microwave-assisted acid pretreatment and supercritical fluid extraction. Physical, chemical, biological, and combined pretreatment methods are explained hereafter.

3.3.1 Physical Method

The physical method includes milling, microwave irradiation, freezing, pyrolysis, and extrusion. This method mainly increases the surface area by shredding the material. However, this method is always used in addition to another method to be effective.

3.3.1.1 Milling/Comminution

Milling/comminution methods were mostly used for first-generation biofuels like oil extraction from seeds or parts of plants like groundnuts, cotton, etc. (Kumar et al. 2015a). After selecting material, milling is the first method applied in the pretreatment process. In this method, the size of the material is reduced through shredding, chipping, grinding, and milling to increase the digestibility of the material. The size reduction increases the bulk density, which helps in the densification process. It includes chopping, ball milling, crushing, colloid milling, hammer milling, grinding, and disk milling methods for particle size reduction. The type of method to be used depends on the size of the particle required. For example, coarse milling reduces the size from meters to a few centimeters; intermediate milling reduces the size from centimeter to millimeter; fine milling reduces the size to between 50 and 500 μm; and ultrafine milling reduces size less than 20 μm. Use of disk and milling pretreatment to improve the enzymatic hydrolysis of rice straw was carried out by Hideno et al. (2009). The result showed an increase in xylose from 78% to 89% and glucose yield from 41% to 54%. However, Da Silva et al. (2010) used ball milling and wet disk milling to pretreat sugarcane bagasse and straw.

The size reduction does not always yield improved performance. Biochemical methane potential assay of sunflower oil cake showed a low biogas production rate in the smallest particle size ranges from 0.355–0.55, 0.710–1.0 than 1.4–2.0 mm particle size (De la Rubia et al. 2011). The mechanical method's benefits include being chemical free, easy to extract, and economical. However, the mere application of the mechanical method does not yield a good result in the quantity of oil production. The drawback of this method includes high energy consumption and the high capital cost of equipment. For example, Gu et al. (2018) improved energy efficiency by using planetary ball milling on premilled wood fiber. The energy consumption ranged between 0.50 and 2.15 kWh/kg for milling for 7–30 minutes at a speed of 270 rpm. Moreover, low yields and high residue oil contents limit commercial usages.

3.3.1.2 Microwave Irradiation

It is a conventional technique where the material is heated to change the structure of cellulose and remove cellulose and lignin. The pretreatment process uses a microwave, gamma rays, ultrasound, and electron beam. This process helps reduce the sugar's enzymatic sensitivity by disrupting silicified waxy surfaces. It improves biomass digestibility through high energy (Bak et al. 2009). Gamma rays aid in the depolarization and irradiation of cellulose and lignin content of the material. Xiang et al. (2016) showed that gamma irradiating hybrid poplar sawdust with cobalt 60 improved enzymatic scarification. The result showed an increased reduction in sugar yield of 519 mg/g with irradiation of 300 kGy with 45°C enzymatic hydrolysis temperature, 84 hours of enzymatic hydrolysis time, and a 90 FPU/g enzyme loading. Irradiation pretreatment effectively decomposes lignocellulosic biomass to low molecular carbohydrates from cellulose (Liu et al. 2015). It includes benefits in terms of short time, moderate temperature requirement, and minimal formation of undesirable inhibitory substances.

Ultrasound helps disrupt the cell wall and increases the surface area. Yue et al. (2021) used ultrasound and microwave heating pretreatment for lipid and food waste to enhance methane production. Under the energy input of 50,000 kJ/kg total volatile solids, a higher soluble Chemical Oxygen Demand (COD) of 10,130 mg/L was observed for ultrasound compared with 1,910 mg/L with microwave heating. An increase in the methane yield of lipids waste pretreated with ultrasound was observed compared to the microwave. The overall energy conversion to methane for lipid waste pretreated with ultrasound was 69.89% and 58.98% for microwave pretreatment. He et al. (2017) investigated the effect of ultrasound pretreatment on the structural changes in wood. The 300 W of ultrasound power with a frequency of 28 kHz in the aqueous solution improved the physicochemical structure of wood.

Microwaves are magnetic waves with a frequency of 0.3–300 GHz and are applied to disrupt the plant cell (Luque-García & Luque De Castro 2004). The magnetic waves cause an increase in the temperature, which increases the kinetic energy, essential for the disintegration of the cell. Microwave heating is done with direct contact with plant material and is controlled by two phenomena—conduction through ion movement and dipole rotation acting simultaneously (Brachet et al. 2002). Gabhane et al. (2011) used microwave irradiation to disintegrate cellulosic substrates. Microwave pretreatment is generally assisted with other pretreatment methods. For example, Zhao et al. (2010) used microwave and alkali pretreatment for agricultural residue with *Trichoderma* sp. 3.2942 by solid-state fermentation. A significant increase in cellulosic sugar and reducing sugar was observed.

Chen et al. (2012) used microwave irradiation pretreatment to increase biodiesel yield efficiency from waste cooking oil. Under the microwave power of 750 Watt, with the reaction time of 3 minutes, It was observed that the biodiesel yield was increased 97.9% as compared to the conventional heating. The advantage of this method is that it includes a short process, is less energy intensive, and has high uniformity compared to the traditional heating method. However, for lignocellulosic material, microwave pretreatment assisted with acid and alkaline pretreatment is used to reduce the formation of inhibitors like phenolic compounds.

3.3.1.3 Freezing

Freezing is one of the recent pretreatment methods, where a significant increase in the enzyme digestibility of lignocellulosic biomass is observed. In this method, milled biomass is mixed with water and frozen at a temperature below −18°C for some time, and after that it is removed and thawed at room temperature (22°C). Rooni et al. (2017) reported 17.42% hydrolysis efficiency, with biomass frozen and thawed four times. There is a considerable promise for use as a low cost, somewhat effective pretreatment technology in the Nordic nations, but they are prohibitively expensive where cold temperatures require more energy. Chang et al. (2011) observed increased digestibility of rice straw from 48% to 84% using freezing with 417.27 g/kg sugar yield from 150 U cellulose and 138.77 g/kg reducing sugar yields for 100 U xylanase under 48-hour operation.

3.3.1.4 Pyrolysis

In this pretreatment process, thermal decomposition of biomass occurs in the absence of oxygen, resulting in char, bio-oil, and gaseous products, and operates between 400°C and 650°C. However, the water content of the biomass has to be removed before pyrolysis (Huber & Dumesic 2006). The char formed is further treated with acid, followed by leaching with water. The leachate water contains glucose, further used as a carbon source in biofuel production. Pyrolysis is categorized as slow pyrolysis and fast pyrolysis. In slow pyrolysis, the production of solid biochar takes a few hours for the process completion. In fast pyrolysis, enhanced bio-oil production generally operates at a high heating rate.

The maximum size of particles that can be used is 2 mm. After that, the gasification process occurs, where biomass is given high temperatures. Gasification is when biomass is converted at a temperature greater than 700°C under controlled oxygen. The final product form is syngas—a mixture of CO, CO_2, and H. These gases can be further used for electricity production via fuel cells or gas engines. The advantage of pyrolysis includes complete conversion of carbon present in the biomass to valuable products like bio-oil, hydrogen, and reduced CO_2 generation (Yang et al. 2019).

3.3.1.5 Extrusion

Extrusion is a thermo-mechanical process where screws (either single or double) spin into a tight barrel equipped with a temperature controller. This mechanism produces high shearing force with raw material, screw, and barrel. This pretreatment results in an increased cellulose exposure ratio, increased surface area, and porosity of the material (Duque et al. 2017). Yoo et al. (2011) improved cellulose to glucose conversion to 95% using 350 rpm screw speed, 80°C maximum barrel temperature, and 40% wet barrel moisture content.

Kuster Moro et al. (2017) investigated the use of a twin-screw extruder for pretreatment of sugarcane biomass, which included the addition of chemicals such as Tween, ethylene glycol, glycerol, and water. The result showed that the ratio of biomass to glycerol 1:0.75 and 1:0.53 for bagasse and straw increases the hydrolysis yield by inserting a reverse element in the screw configuration, increasing the straw hydrolysis yield 68.2%. The advantage of this method includes disruption of biomass structure, resulting in defibrillation and shortening of fiber with lower energy requirement compared to comminution.

3.3.2 Chemical Pretreatment

Chemical pretreatment offers promising ways to improve the biodegradability of material through removing hemicellulose and lignin and reducing polymerization degree in lignocellulose (Agbor et al. 2011). Chemicals used for pretreatment of lignocellulosic waste include oxidizing agents, acid, alkali, and ionic liquid.

3.3.2.1 Acid Pretreatment

In this method, milled biomass is soaked in water and kept in an acidic solution where acid solubilizes polysaccharides (especially hemicellulose) into monomers by enzymatic hydrolysis under the conditions of either high acid concentration and low temperature or low acid concentration and high temperature. The acid used includes organic and inorganic ones like sulfuric acid, hydrochloric acid, phosphoric acid, and nitric acid. HNO_3 and H_2SO_4 are widely used due to the economy in the pretreatment process. For lignocellulose, acid pretreatment offers a suitable structure required for enzymatic hydrolysis. Acid pretreatment depends on the acid type, acid concentration, temperature, and solid-to-liquid ratio. Generally, acid pretreatment requires high pressure and temperature for pretreatment. Zhu et al. (2009) obtained more than 90% cellulose conversion under the temperature of 180°C operated for 30 minutes and 1.8%–3.7% concentration of H_2SO_4 followed by disk milling. However, acid pretreatment forms inhibitory compounds with a corrosive nature, and requires high operation and maintenance costs. Hernández-Salas et al. (2009) used 1.2% (v/v) HCl for pretreatment of sugarcane and agave bagasse under 121°C, and temperature of 4 hours resulted in 37.21% reducing sugar for sugar cane depicting bagasse and 35.37% for sugarcane pith bagasse. However, during the pretreatment process, the formation of formic acid and levulinic acid under high temperature and pressure has a negative effect on the downstream process, which further increases the overall cost process (Behera et al. 2014). To decrease the formation of inhibitors, acid pretreatment parameters such as acid concentration, reaction temperature, and retention duration should be monitored.

3.3.2.2 Alkali Pretreatment

In alkali pretreatment, the biomass is mixed with a base like NaOH and aqueous ammonia under predefined pressure and temperature. The alkaline treatment removes hemicelluloses from polysaccharides, resulting in organic acids that decrease the pH. It reduces cellulose accessibility (for saccharification by enzymes) inhibitors like acetyl groups, lignin, and various uronic acid substitutions (Kumari & Singh 2018). This pretreatment method's main benefit includes removing the acetyl group under uronic acid substitutions in hemicellulose that improves the accessibility of cellulose and hemicellulose in hydrolytic enzymes (Zheng et al. 2009). It breaks the ester bond present between the lignin, cellulose, and hemicellulose, and prevents the fragmentation of hemicellulose polymers (Gáspár et al. 2007). The most widely used base for biofuel production is NaOH, with advantages like cost effectiveness and being widely used for pretreating lignocellulosic biomass for biofuel production. McIntosh and Vancov (2010) used 2% NaOH, the temperature of 60°C, and operated for 90 minutes, which resulted in a 4.3 times increase in total sugar from *Sorghum bicolour* straw. Gupta and Lee (2009) used aqueous ammonia pretreatment on hybrid poplar, and the result showed that more than 60% of the lignin was removed at a temperature of 60°C–120°C and 4 hours of reaction time. Aqueous ammonia improves the accessibility of hydrolytic enzymes by removing lignin without extensive oxidation (Sipponen & Österberg, 2019). Saha and Cotta (2006) used alkaline H_2O_2 pretreatment for ethanol production from wheat straw. The result showed formation of monomeric sugar of 8.6% (w/v) under 2.15% H_2O_2 (v/v), pH 11.5, and 35°C temperature for 24 hours. The drawback of this method includes not being suitable for woody biomass, and there is a loss of hemicelluloses and the formation of inhibitors and salt formation after neutralization of pretreatment that poses a challenge to its disposal.

3.3.2.3 Liquid Method

There are some methods available that are recently studied and are least explored. Ionic liquids are organic salts with a melting point of 100°C and are made up of different cations and anions, typically big organic cations and tiny inorganic anions (Jackowiak et al. 2011). Cations are composed of imidazolium, phosphonium, pyrrolidinium, and choline. Ionic liquid methods are usually divided into protic and aprotic. This includes using liquefied liquefied dimethyl ether as an organic solvent to extract biofuel from biomass (spent coffee grounds, soybean, and rapeseed cakes) (Sakuragi et al. 2016). It utilizes a solvent having a low melting point, high thermal stability, high polarities, and minimal vapor pressure (Agbor et al. 2011).

Alayoubi et al. (2020) pretreated cellulose-containing materials such as cotton, spruce, and oak sawdust with 1-ethyl-3-methylimidazolium acetate, followed by enzymatic hydrolysis of the untreated and pretreated materials. After processing, glucose yields were 70% for cotton and 60% for oak sawdust, but only 50% for spruce sawdust. The ethanol yields were consistently around 50%. After a gentle pretreatment with 1-ethyl-3-methylimidazolium acetate, the ethanolic yields from lignocellulosic substrates were 2.6–3.9 times higher. The ionic liquid pretreatment factors are the type of ionic liquid used, i.e., anions or cations, time, and temperature. Another concern is their vulnerability to moisture and the high cost of recovery. Ionic liquid pretreatment offers an advantage over volatile cellulose solvents in thermal stability, low toxicity, viscosity, and low flammable properties. The major drawback includes high cost and toxicity.

3.3.2.4 Oxidative Pretreatment

Oxidative pretreatment uses an oxidant for pretreatment where biomass undergoes oxidative polymerization. It includes using peracetic acid, H_2O_2, or wet oxidation as an oxidizing agent. In the oxidative pretreatment, reactions like oxidative cleavage of aromatic nuclei, electrophilic substitution, and displacement of side chain occurs (Gonçalves et al. 2014). H_2O_2 uses oxidative delignification, which detaches and solubilizes the lignin, and improves the digestibility of enzymes.

Peracetic acid is a strong oxidant used for biomass pretreatment. It oxidizes the hydroxyl group to the carbonyl group present in the lignin side chain. Moreover, peracetic acid provides stability and ease of preparation and recovery. Pretreatment with peracetic acid at 80°C improves cellulose digestibility. Zhao et al. (2007) used peracetic acid pretreatment for sugarcane bagasse and factors affecting the pretreatment. The result showed that under the peracetic acid charge of 50%, liquid to the solid ratio of 6:1, 80°C temperature, and 2-hour time, 80% of the cellulose is converted to glucose by enhancing the digestibility. The peracetic acid treatment is carried out at a low temperature and low atmospheric pressure, thus reducing inhibitor formation like furfural (Mussatto & Roberto, 2004). Yin et al. (2011) used peracetic acid for increasing the pretreatment efficiency. The result showed 45% improvement in the lignin removal. Duncan et al. (2010) used this prehydrolysis to make peracetic acid for the pretreatment of aspen wood. Pretreatment with peracetic acid releases 98% of glucose from cellulose, compared to 22% without pretreatment.

In the case of H_2O_2 pretreatment, due to its oxidative nature, the delignification process takes place, thus improving enzymes digestibility (Sheikh et al. 2015). The H_2O_2 pretreatment is generally used to improve the microbial conversion of biomass to ethanol. For example, Song et al. (2016) used H_2O_2 pretreatment to enhance the enzymatic hydrolysis and lignin removal from Jerusalem artichoke. The process yielded 84% of ethanol. However, the disadvantage of this method includes its explosive nature when used in concentrated form, and the process is cost intensive (Yin et al. 2011). The precautions related to the explosive nature, storage cost, and transportation cost limit its use in large-scale applications.

3.3.3 Biological Pretreatment

Biological pretreatment uses biological enzymes or microorganisms for the pretreatment process. It includes fungi, microbial consortium, and enzymes for pretreatment. The microorganism used includes white, brown, and soft-rot fungi and bacteria to degrade the lignin and hemicellulose, and minimize cellulose's polymerization present in the biomass. The enzymes secreted by different fungi and bacteria have a different effect on the pretreatment process. For example, white and soft-rot fungi release lignin peroxidases, laccases, polyphenol oxidases, manganese-dependent peroxidases, and xylanase as the primary components that aid in the delignification and help in the enzymatic saccharification rate (Sindhu et al. 2016).

The synergistic activity of a microbial consortium, which includes different bacteria and fungi, is responsible for the biodegradation of lignocellulosic biomass. Potumarthi et al. (2013) showed simultaneous pretreatment and saccharification of rice husk by a white-rot fungus (*Phanerochete chrysosporium*). Growing the fungus on rice husk resulted in effective delignification, and the pretreated biomass was then subjected to enzymatic hydrolysis. On the 18th day of fungal treatment, the fungal pretreated rice husk generated the most reducing sugars (895.9 mg/mL/2g of rice husk). This technique lowers the expenses of traditional pretreatment by eliminating the need for washing and inhibitor removal. Taha et al. (2015) reported increased straw saccharification by co-culturing lignocellulose degrading bacteria. Two times higher fungal isolation for enzyme activities were observed as compared to the bacteria. Co-culturing resulted in a seven times increase in saccharification rate. Co-culturing enhances saccharification, resulting in greater commercial possibilities for the utilization of microbial consortiums.

Sen et al. (2016) showed the conditions of pretreatment effect (acid concentration, enzyme additions, rice straw concentration, hydrolysis time, $FeCl_3$, and particle size). The result showed that under 0.8–1.0 M HCl condition, the yield of sugar, glucose, xylose reducing sugar from rice straw with 0.15 mm particle size at 20 minutes hydrolysis time was 52.9%, 2.8, 14.5, and 38.6 g/L, respectively. The yield of sugar was not affected by the $FeCl_3$. The enzymatic hydrolysis activity tends to increase the sugar yield reduction up to 49.8 g/L. Zhang et al. (2011) used species of microbes that coexisted in consortium mixed with distillery wastewater for pretreatment of cassava residue. The result showed an enhanced methane yield of 259.46 mL/g-VS of cassava residue under 12 hours of pretreatment, which was 96.63% higher than untreated.

The consortium has several advantages: increased productivity, improved enzymatic saccharification efficiency, pH control during sugar use, and increased substrate utilization (Kalyani et al. 2013). Its advantages include chemical-free treatment, low energy requirement, with minimal effect on the environment. The disadvantage includes low efficiency.

The drawback of enzyme extraction includes low filtration efficiency and large energy consumption, which is unsuitable for industrial applications (Chen et al. 2020). The time required for biological treatment varies from 10 to 14 days with the controlled condition and needs large space to limit its wide-scale usage. Moreover, in case of fungal treatment, much care is needed to prevent the contamination of one fungi to the other, since it reduces the performance of the treatment process. Biological treatment, when combined with other treatment processes, can improve performance. For example, Yan et al. (2017) evaluated bacteria-enhanced dilute acid pretreatment. The enzymatic hydrolysis increases the digestibility of rice straw by 70% in bacteria-enhanced dilute acid pretreatment compared to only dilute acid pretreatment. However, further research is needed to optimize the parameters needed for the biological pretreatment process. This method is still in the developing phase, as it requires costly enzymes with high incubation time and need of de-emulsifier in the downstream process (Ahmad et al. 2019).

3.3.4 Combined Pretreatment

The methods specified above has its advantage and disadvantages. The material selection determined the pretreatment methods to be used. Each pretreatment has its uniqueness and limitations, and cannot be used universally for all types of biomasses. Improved efficiency can be achieved by combining two or more pretreatment processes. For example, the comminution pretreatment is energy intensive; the biological pretreatment is time consuming with lower efficiency; oxidative pretreatment due to its explosive nature its use is limited; and alkaline pretreatment is limited to the agricultural residue and is expensive. Hence, the combined pretreatment method like physical–chemical, chemical–biological, and chemical–chemical pretreatment methods are efficient and improve the overall process. The combined treatment is used for industrial applications where high biofuels are yielded.

3.3.4.1 Microwave-Assisted Acid Pretreatment

To increase lignocellulose degradation, microwave radiation is combined with dilute acid pretreatment. The cellulose is decomposed through microwave, and hemicellulose is degraded in the acidic condition, followed by lignin decomposition at increased temperature and pressure (Hendriks & Zeeman 2009). The pretreatment duration and pressure application depend on the type of biomass used (Ma et al. 2009). Mikulski and Kłosowski (2020) and Hermiati et al. (2020) for wheat and rye stillages and sugarcane trash, respectively. In this, the acid pretreatment helps remove lignin and hemicellulose, and microwave pretreatment further helps to reduce the lignin content in the biomass. When combined with other pretreatments like acid, microwave increases the enzymatic hydrolysis. For example, Binod et al. (2012) used alkali, acid combined with microwave pretreatment for sugarcane bagasse yielded 83% of sugar in the enzymatic hydrolysis followed by 65% of sugar for alkaline and microwave, and 10% for acid and microwave pretreatment. Kanitkar et al. (2011) studied the optimization of oil extraction parameters using microwave-assisted pretreatment on soybeans and rice bran. The solvent to feedstock ratio of 3:1 resulted in 17.3% and 17.2% oil yield when operated for 20 minutes under 120°C, as compared to 11.3% and 12.4% for control. After pretreatment, the highest glucose concentration of 156 mg/g of dry weight and 75% of the cellular hydrolysis was yielded after 24 hours of the process. Combined pretreatment usually has higher pretreatment efficiency but is associated with increased capital and operation and maintenance costs. Before selecting a feasible pretreatment method, factors like operation time, economy, or any increase in performance should be analyzed.

3.3.4.2 Supercritical Fluid Extraction

A supercritical fluid is a substance that exhibits liquid and gas properties like density and compressibility, respectively. It is a process in which supercritical fluid extracts one component from the matrix. CO_2 is the most common supercritical fluid used. Because of its nonflammability, low toxicity, low cost, and mild critical temperature and pressure, supercritical CO_2 is the most commonly utilized SFE solvent (Falsafi et al. 2020). It is nontoxic, environmental friendly, recyclable, and requires low critical temperature and pressure.

This method is essentially used to extract essential oils from different plant materials. The most common supercritical fluids are CO_2 and water (Daza Serna et al. 2016). Its properties are nontoxic, nonflammable, cheaply available, and require ambient temperature and pressure conditions. Ambient temperature prevents the oil from thermal denaturation. This method extracts nonpolar chemicals like lipids from the biomass (Bjornsson et al. 2012), where a supercritical CO_2 extractor was used to isolate algal lipids to liquid hydrocarbon fuels. However, apart from CO_2, other cosolvents like ethanol or methanol can be used, depending upon the nature of the extract. According to Fadel et al. (1999) the extract analytes with supercritical fluids shows higher activity of antioxidation, lower viscosity, and gives higher yield as compared to the recovered extract with other technologies.

Maran and Priya (2015) used supercritical fluid extraction for oil extraction from muskmelon. Under the pressure of 44 MPa, at the temperature of 49°C, along with

CO_2 flow rate of 0.64 g/min, and extraction time of 81 minutes resulted in 48.11% oil yield, which was higher than Soxhlet extraction of (46.83%).

However, operational costs are too high for its actual application on a large scale with safety issues like operating pressure 7.39 MPa and temperature 304.25 K compared to conventional techniques. The factors that influence the fractionation process are the static time, flow rate, extraction pressure, extraction cosolvent, temperature, and dynamic time. The most important physical parameter is a pressure that affects the supercritical fluid extraction (Ni et al. 2015). Falsafi et al. (2020) evaluated the fractionation and extraction of petroleum biomarkers from tarballs and crude oil. The effect of pressure at 240, 300, and 360 atm for saturates isolation and 315, 330, and 345 atm for aromatic isolation showed the best efficiency for saturates and aromatic in 300 and 330 atm, respectively. For temperature, the maximum extraction was at 50°C and 90°C for saturates and aromatic, respectively. Dynamic and static times were checked in the range of 5–15 minutes and 20–40 minutes, respectively, with the 10 and 30 minutes were selected as the optimized dynamic and static extraction time to extract both aromatic and saturate biomarkers.

3.4 THE DEVELOPMENT OF NEW METHODS FOR PRODUCING BIOFUELS

3.4.1 New Feedstock for First-Generation Biofuels

First-generation biofuels are primarily derived from wheat, barley, maize, oilseeds, sugarcane potatoes, and other sources (Malode et al. 2021). There are some crops that are not grown commercially, but they are of immense importance in biofuel production. Sorghum, camelina, cassava, and jatropha are the crops that will be used to produce biodiesel and bioethanol, and they serve as first-generation fuels. The reason behind that is the low intensity of GHG as compared to other commercial crops. These crops can be grown on marginal land but deliver a higher yield. Crops like jatropha, cassava, and sorghum can be a good alternative for the production of biofuels.

Sorghum is an essential crop in dry regions because of its ability to withstand both drought and heat. Compared to cassava, its average output is only 1.4 t/ha (Babu et al. 2013). The sorghum yield may appear poor, since it is not optimized for commercial operations in the nations where it is grown in the highest amounts, such as India and Nigeria. Improved management and fertilization could lead to increased yields. Sorghum genetic change has also piqued researchers' curiosity. The genetic mapping of sorghum is now underway, which should give the scientific community with the means to increase sorghum production yields. Sorghum cultivars that generate high volumes of lignocellulosic biomass for second-generation biofuel production are also of interest to certain researchers in the United States and the grain utilized for first-generation bioethanol production (Mussatto et al. 2010).

Cassava is a tropical plant that needs a warm and humid environment and significant rainfall for its growth. The countries participating primarily in the production of cassava are Nigeria, Brazil, Thailand, and Indonesia, because the climatic conditions of these areas favor the growth of this plant. The roots of the cassava plant are rich in starch. The average yield for cassava is 12.2 t/ha, but it can be as high as 31 t/ha in some countries like India. China, Nigeria, Brazil, and Thailand are all interested in expanding cassava production for ethanol fuel (Food and Agriculture Organization (FAO), 2013).

Jatropha curcas is a poisonous crop whose seeds can be crushed to produce highly poisonous vegetable oil. It is a semiarid plant that grows best in warm, humid climates (Berni & Manduca 2014). This crop was historically grown for use in soap and candles, as well as medicinal. Due to the country's national policy prohibiting food oils as fuel, India is particularly interested in growing jatropha for fuel production and has ambitious plans to convert nearly 11 million hectares of wasteland into jatropha fields. Jatropha could also be farmed in Southern Africa, Southeast Asia, and Latin America (Bailis et al. 2014).

Some additional crops, such as *Camelina*, are not currently produced substantially but are being explored as potential new feedstocks, according to the USDA (*Camelina sativa*). Camelina is a plant that is native to Northern Europe and Central Asia, and it currently produces oil in quantities comparable to those of rapeseed. However, it has not been commercially produced for at least 50 years, and, as a result, there has been little attention paid to increasing yields, which is an area that needs to be researched more. Camelina has low tillage and weed management requirements, making it an excellent crop for organic farmers. However, because it has not been commercially grown too far, it has not yet proven to be a viable feedstock for biodiesel production, as has been the case with other plant-based feedstock.

Other raw materials include tallow, lard, and waste cooking oil, which can be turned to biodiesel if available at a reasonable price. Since it makes little sense to encourage cattle grazing to increase tallow production, these raw materials are available in limited quantities (Nogueira 2011). Triacylglycerol is obtained from soy, canola, and other oilseeds, as well as palm fruit mesocarp (Youngs & Somerville 2012). The average output of oil crops varies from hundreds to thousands of liters per hectare, depending on species, climate, and farming practices. Brazil often considers biodiesel crops, including soybean, castor, and palm oil. Their biodiesel yields might be 0.7, 0.5, and 5.0 tons/ha, respectively (Nogueira 2011). Economic incentives favor ethanol production, since temperate oilseeds provide less biofuel than corn.

3.5 COMPOSTABLE BIOMASS FOR THE PRODUCTION OF SECOND-GENERATION BIOFUELS

It is unwise to replace unsustainable petroleum production with unsustainable agricultural fuel production. Until far, lignocellulosic wastes have primarily been used to make second-generation biofuels. 10–25 million tonnes of dry cellulosic biomass will be required to produce 1 billion gallons of liquid fuel. To partially replace fossil fuels with renewable fuels, enormous feedstock, both cultivated and harvested, will be required. In the fall, agriculture leftovers are harvested; in the winter, woody residues or crops are harvested; in the spring, cover crops such as rye are harvested; and in the summer, energy crops such as sorghum or switchgrass are harvested.

Maize is the most important crop in terms of worldwide grain or seed output, generating around 880 million tonnes of grain and a similar number of stems and stripped cobs, generally known as stover, which can be used to create biofuel and other products. If half of the maize stover in the United States were converted to cellulosic ethanol, approximately 13.5 billion gallons of ethanol would be generated. Bagasse is an easily available biomass resource that contributes to one-third of sugarcane plants' biomass. According to the Brazilian government, if half of this biomass is used for ethanol production, the plant's ethanol yield may grow from 6,000 to 10,000 L/ha.

This is because the amount of land required to produce sufficient biofuel to have a significant impact on demand is entirely reliant on the productivity of a given feedstock on a certain parcel of land. This is regulated by various physiological aspects, including genetic diversity, agricultural practices, and environmental factors such as soil quality and water availability, as well as climate. Plant biologists face numerous challenges in their field of study, including identifying the most productive plant species that can be grown on various types of marginal or abandoned land, optimizing genetics and production practices, and assessing any environmental risks or benefits associated with encouraging widespread use of such species for energy production. To maximize the efficiency of land conversion to energy production, it is required to maximize "yield" – the amount of biomass created per unit of land area – in order to minimize land diverted from other purposes. This can be accomplished by optimizing management strategies or genetically modifying biomasses (Youngs & Somerville 2012).

Because water is a fundamental constraint on plant output, increasing water usage efficiency and drought tolerance will be a key goal of producing bioenergy crops. The semiarid area could be used to cultivate drought-resistant plants like *Agave* spp. Prairie cordgrass and Eucalyptus spp. can thrive in salinized soils. Tolerance to other environmental stresses, such as floods and colds, is also useful. Identifying noninvasive organisms or developing methods like conditional sterility that can limit invasive spread is crucial in this regard (Youngs & Somerville 2012).

Perennial species such as sugarcane, energy cane, elephant grass, switchgrass, and *Miscanthus* spp. have the potential to be exploited as sources of lignocellulosic biomass for bioethanol production due to their intrinsic high light, water, and nitrogen use efficiency. Additionally, woody biomass may be harvested sustainably for lumber and paper manufacture, and it may even be used as a biofuel feedstock in some regions. On a global scale, a significant percentage of agricultural land has been turned to forest, and the continuing trend toward electronic media consumption and paper recycling may decrease demand for pulp wood. This opens up the possibility of repurposing woody biomass for energy production (Youngs & Somerville 2012).

Remaining residues from the agricultural business, such as sugarcane bagasse, cassava bagasse, corn stover, wheat, and rice straw, are also important sources of lignocellulosic biomass. It has even proven possible to use municipal solid waste and paper industry wastes to create second-generation biofuels (Babu et al. 2013). However, other biomass crops have secondary metabolites that are poisonous to microorganisms, making them unsuitable for use as biofuel. Identifying and removing such chemicals through genetic approaches is a top objective for plant feedstock research if it can be accomplished without causing pest and pathogen problems in the process (Youngs & Somerville 2012).

3.6 RECENT ADVANCEMENTS IN THE DEVELOPMENT OF BIOFUELS THROUGH THE USE OF GENETIC ENGINEERING

Significant improvement has happened in the engineering of microbes to manufacture fuels to endeavor biofuels. The US Department of Energy's National Renewable Energy Laboratory (NREL) has prioritized lignin bioengineering to lower lignin content in biomass feedstocks. Lignin lowers sugar yields during saccharification and hydrolysis, and inhibits ethanol production. These polysaccharides and their abundance in plant cell walls are important, because they yield more ethanol and other biofuels when processed (Ambaye et al. 2021). To minimize net GHG emissions, microbially produced gasoline, diesel, and jet fuel alternatives can be adopted. However, fermentation rather than saccharification is the key technological hurdle to expanding total production of these fuels (Klein-Marcuschamer et al. 2012). The biological processes of biofuel producers must be optimized to attain high performance, which includes increased product output, concentration, productivity, and tolerance, among other things. Rather than the other way around, a system-wide optimization of midstream and downstream processes should be used to establish production strains (Rastogi et al. 2021). It has recently been discovered that traditional strain improvement approaches (e.g., mutagenesis and breeding) can result in increased biofuel productivities from algal strains, and that these methods are being used more frequently. However, there have been no significant advancements in the molecular engineering of eukaryotic algae for the production of liquid biofuels. As a result, despite certain breakthroughs, there are still obstacles to overcome to bring advanced biofuels to the market at a competitive price compared to fossil fuels for commercialization. Furthermore, the weak link in this technology is carbon capture and sequestration, which continues to escape the coal sector despite decades of research and development (Rastegari et al. 2019; Kumar et al. 2015b).

To increase the overall yield of bioethanol production from lignocellulosic biomass, genetic engineering of ethanol fermentation microorganisms to increase their substrate selectivity toward hemicellulosic sugars and macroalgal polysaccharides is probably the most plausible way. Recently, random mutagenesis and metabolic engineering have become conventional procedures for creating microbial strains that can be used in biofuel production, having been used for several decades. Because of the difficulty connected with identifying changed genes resulting from random mutation, the random mutation and selection technique is ineffective for further enhancing cellular performance. On the other hand, metabolic engineering is concerned with increasing cellular performance by considering the entire metabolic route and modifying particular genes. According to current expectations, this combined strategy will develop microorganisms capable of producing a variety of biofuels on an industrial scale at a competitive price (Li et al. 2019a).

3.7 MICROBIAL ELECTROLYSIS FOR HYDROGEN RECOVERY

Hydrogen is an ecofriendly biogas, cost effective and pollution free, and energy saving. Hydrogen is regarded as one of the ideal biofuel fuel for future transportation, because it can be converted to electric energy in fuel cells or burnt and converted to mechanical energy without the obvious production of CO_2 (Malhotra 2007). Hydrogen can be produced through chemical, physical, and biological mechanisms (Mohan et al. 2010, 2011). Hydrogen production through the biological route, especially through acidogenic fermentation, has attracted recent science due to the lower economics of the process. Moreover, considering wastewater as a substrate for hydrogen production is significant due to its dual benefits, *viz.*, hydrogen production and waste remediation (Mohan et al. 2008, 2010). Hydrogen can be produced biologically by algal and cyanobacterial biophotolysis of water or by photofermentation of organic substrates from photosynthetic bacteria. High hydrogen yields can be achieved by the use of thermophilic microorganisms such as *Caldicellulosiruptor saccharolyticus* or *Thermotoga elfii* (de Vrije et al. 2002; de Vrije and Claassen 2003; Claassen et al. 2004). Applying additional voltage to the in-situ potential generated by the bacterial cell allows bioenergy generation like hydrogen and methane or various products like hydrogen peroxide at the cathode (Liu et al. 2005) These fermentations can be operated in the liquid phase with immobilized cells or by enabling the formation of self-flocculated granular cells or sludge to prevent washout of the hydrogen-producing cells. These fermentations can be operated in the liquid phase with immobilized cells or by enabling the formation of self-flocculated granular cells or sludge to prevent washout of the hydrogen-producing cells.

Moreover, hydrogen is a common product in anaerobic bacterial fermentations and may be an interesting by-product in future large-scale industrial fermentation. As an example of this, around 40×106 M^3 of H and 60×106 M^3 of CO_2 were produced annually from the 1960s to the 1980s in a Russian biobutanol plant as a by-product but were not used at the time.

Methanogenic bioelectrochemical systems: Methanogens are the only group of microorganisms on earth producing significant amounts of methane. Anaerobic digestion of complex organic substrates proceeds through a series of parallel and sequential steps, with several groups of microorganisms involved. Biogas plants produce methane gas sustainably along with carbon dioxide from plant biomass, which may come from organic household or industrial waste or from specially grown energy plants. Cooperation of the population of microorganisms enables the synthesis of certain products, which are then used by another group of bacteria. The bacteria belonging to domain archaea, which are involved in the production of methane, exhibit synergistic relationships with other populations of microorganisms.

Methanogenic microorganisms can be divided into three functional groups: hydrolyzing and fermenting bacteria, obligate hydrogen-producing acetogenic bacteria, and methanogenic archaea. Hydrolytic acidogenic bacteria (HABs) hydrolyze complex organic polymers (carbohydrate, proteins, and lipids) into simple compounds (sugar, amino acids, and fatty acids) during the first step of the degradation of fermenting bacteria (FB) such as Bacteriocides, Clostridia, and Bifidobacteria. During the acidogenesis process, volatile fatty acids (VFA), alcohols, H_2, and CO_2 are produced. Similarly, acetic acid, H, and CO_2 are produced in the acetogenesis step by mixed microbial culture (Marshall et al. 2012), the obligate H-producing acetogens, Syntrophobacter (PUAs: propionateutilizing acetogens), and Syntrophomonas (BUAs: butyrate-utilizing acetogens) represent the major part of acetogens. A key factor in the degradation is that low-temperature anaerobic oxidation of butyrate and propionate occurs only in syntrophic association with H-utilizing methanogens (HUMs), consuming H and CO for methane (CH) production, preventing the accumulation of increasing H pressure in the digester.

Traditional farm biogas plants are run as a single- or two-stage process at around 37°C with an uncontrolled secondary fermentation in large storage tanks. Due to different optimal conditions specific for the hydrolytic and methanogenic bacteria, two-stage processes are increasingly applied, particularly in large industrial biogas plants. Most biogas fermentation tanks are run as liquid fermenters. The biogas fermentation tank may contain more than 12% (w/v) dry mass (so-called dry fermentation) or less (liquid fermentation). Further development of biogas technology is expected to increase production efficiency. Presently, only up to the maximum of about 70% of the organic matter in biomass is converted to CH and CO. In order for this to increase, the hydrolysis stage must be enhanced.

3.8 ADVANCEMENT IN FOURTH-GENERATION BIOFUELS

The concept of "fourth-generation algal biofuels" or "photosynthetic biofuels" has only lately been developed, and it refers to biofuels produced from algae. It is believed that the breakthrough in algal biofuels will be achieved through the metabolic engineering of photosynthetic microorganisms to create and secrete ethanol rather than through the use of genetic engineering. Combined with genetically optimized feedstocks meant to trap huge amounts of carbon, fourth-generation technology produces fuels that are as efficient as they are environmentally friendly. The capture and sequestration of CO_2 are critical to this process, which allows fourth-generation biofuels to be classified as a negative carbon source of energy (Ambaye et al. 2021).

3.9 ECONOMIC ASPECTS IN BIOFUEL

The studied countries used a mix of market-pull and technological-push policies to increase biofuel production, and use at various stages of technological and market development (Ebadian et al. 2020). It is feasible to avoid potential problems with food production by using low-value biomass for clean energy. Low-value biomass includes agricultural, forestry, and wild flora waste (Segneanu et al. 2013). Diesel's market share will expand in Germany and the US, underscoring the necessity of second-generation biodiesel. In contrast, Brazil and the US rely substantially on gasoline for transportation, making second-generation bioethanol vital. Figure 3.1 compares the cost of producing second-generation biofuels to diesel and gasoline. Fossil fuel taxes are expected to rise, increasing the economic feasibility of biofuels. Optimizing technology for second- and third-generation biofuels will also reduce manufacturing costs.

One of the most important difficulties confronting the world today, as depicted in Figure 3.1, is the development of economically viable biofuels. Despite this, there is a limited information on the economics of these revolutionary processing technologies, because the studies are particularly site specific, and might vary substantially based on the current and future economic and market conditions in the individual places. Biomass planting, growth maintenance, harvesting, transportation, and processing

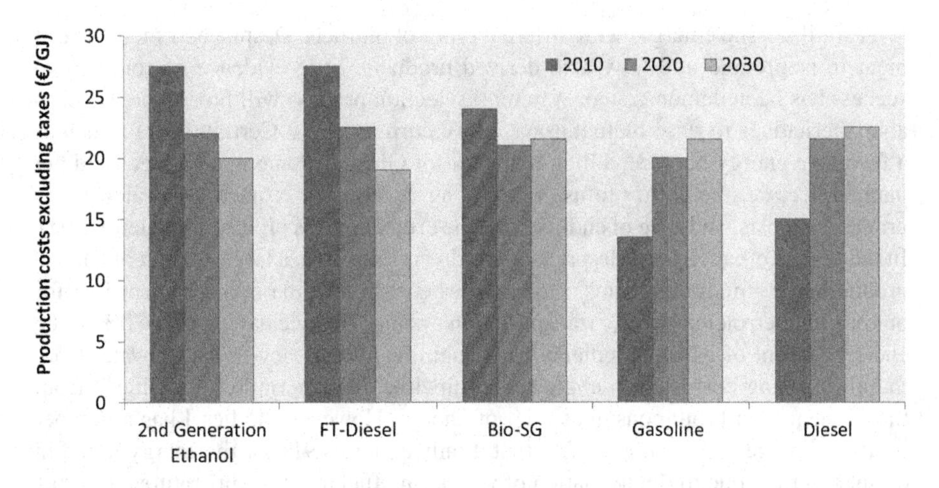

FIGURE 3.1 Cost of different biofuels for the year (2010) and the predicted developmental cost for the year 2030.

technology must all be considered to maximize biofuel production's total economics. The location of biofuel production and upgrading plants must also be considered. Plants that have been upgraded can be put separately or as part of a separate biorefinery.

Furthermore, biofuel production is a great opportunity to be linked with petroleum-refining activities. When comparing the entire cost of alternative fuels to traditional fuels, the conversion efficiency and accompanying emissions must be included throughout the cycle. It is worth emphasizing that these challenges have received insufficient attention in the context of biofuel production (Furimsky 2013).

To make lignocellulosic ethanol commercially feasible, a number of difficulties must be addressed first. The contribution of biomass costs to the total cost of lignocellulosic bioethanol production demonstrates that they are a substantial component of the total cost (Gnansounou & Dauriat 2011). Additionally, the pretreatment stage consumes a lot of energy, which increases the cost of ethanol production. Additionally, the contribution of enzyme costs to the economics of lignocellulosic biofuel production continues to be a point of contention among scientists. While some authors highlight the cost of enzymes as a significant impediment to biofuel production, others tacitly imply that it is not, either because they feel the cost is low, or because they expect the cost will decrease due to technological advancement or other developments. Other studies simply indicate that there is a dearth of publicly available information on the cost of enzymes.

The literature estimates the cost contribution of enzymes to lignocellulosic ethanol production at \$0.3–\$0.11/L – a significant discrepancy (Mussatto et al. 2010). However, using saccharification and fermentation yields already reported in the scientific literature, the cost contribution of enzymes was calculated to be \$0.39/gal. While sugarcane juice, corn, and cassava juice are all cheaper than ethanol. As a result, reducing the contribution of enzymes to biofuel production costs is critical (Mussatto et al. 2010).

A similar issue emerges with different types of biofuels. Despite enormous investment in isoprenoid and fatty acid-derived products, little evidence of commercial success has been demonstrated. A natural selection process will favor more sustainable alternatives to raise biofuel usage above current levels. Certain factors, such as a favorable energy balance, a high potential for GHG abatement, a greater yield per planted hectare, the ability to use rain-fed agriculture, or require less water, lower production costs, and ease of end-use, will be crucial in this process (Nogueira 2011). In the case of ethanol, its widespread use is due primarily to a large amount of natural production by microorganisms, rather than because it is the most efficient fuel for our current petroleum-centric transportation system. Despite having only 70% of the energy content of gasoline, ethanol has a significant tendency to gather water from the air, causing corrosion in engines and pipelines. Furthermore, distilling it from the fermentation broth consumes a lot of energy. However, biodiesel has a number of disadvantages, including the fact that it only contains 91% of the energy found in diesel and that, due to the formation of wax in the fuel at low temperatures, it is difficult to transport using current distribution infrastructure, resulting in geographic restrictions on its use.

3.10 CONCLUSIONS

This chapter discusses the possibilities for biofuel production and the constraints associated with diverse feedstocks and technological advancements in the process. Biomass is an energy-dense feedstock for biofuels production; however, the conversion process is hindered by various operating circumstances, the kind of conversion method employed, and the cost of the advantages derived from biomass. Achieving cost-effective biofuel generation from biomass depends on the effectiveness of cellulolytic fungal enzymes and the availability of biofuel fermenting strains capable of fermenting a wide range of sugars and starches. According to the findings of this study, the social and economic implications of biofuels are quite significant. It has the potential to contribute to local and national rural economy diversification, economic growth, job creation, and energy security through the production and consumption of biofuels, among other things. Additionally, it is feasible to substitute imported commodities, which has an effect on the trade balance both directly and indirectly, to increase energy supply and to diversify the economy by developing new industries that utilize biofuels as a fuel source.

REFERENCES

Agbor, V. B., Cicek, N., Sparling, R., Berlin, A., & Levin, D. B. (2011). Biomass pretreatment: Fundamentals toward application. *Biotechnology Advances* Vol. 29, Issue 6, pp. 675–685. Doi: 10.1016/j.biotechadv.2011.05.005.

Ahmad, F. B., Zhang, Z., Doherty, W. O. S., & O'Hara, I. M. (2019). The prospect of microbial oil production and applications from oil palm biomass. *Biochemical Engineering Journal* Vol. 143, pp. 9–23. Doi: 10.1016/j.bej.2018.12.003.

Alaswad, A., Dassisti, M., Prescott, T., & Olabi, A. G. (2015). Technologies and developments of third generation biofuel production. *Renewable and Sustainable Energy Reviews* Vol. 51, pp. 1446–1460. Doi: 10.1016/j.rser.2015.07.058.

Alayoubi, R., Mehmood, N., Husson, E., Kouzayha, A., Tabcheh, M., Chaveriat, L., Sarazin, C., & Gosselin, I. (2020). Low temperature ionic liquid pretreatment of lignocellulosic biomass to enhance bioethanol yield. *Renewable Energy* Vol. 145, pp. 1808–1816. Doi: 10.1016/j.renene.2019.07.091.

Ambaye, T. G., Vaccari, M., Bonilla-Petriciolet, A., Prasad, S., van Hullebusch, E. D., & Rtimi, S. (2021). Emerging technologies for biofuel production: A critical review on recent progress, challenges and perspectives. *Journal of Environmental Management* Vol. 290, Issue March. Doi: 10.1016/j.jenvman.2021.112627.

Babu, V., Thapliyal, A., & Patel, G. K. (2013). *Biofuels Production* (Vol. 148). Wiley-Scrivener, USA.

Bailis, R., Solomon, B. D., Moser, C., & Hildebrandt, T. (2014). Biofuel sustainability in Latin America and the Caribbean - A review of recent experiences and future prospects. *Biofuels* Vol. 5, Issue 5, pp. 469–485. Doi: 10.1080/17597269.2014.992001.

Bak, J. S., Ko, J. K., Han, Y. H., Lee, B. C., Choi, I. G., & Kim, K. H. (2009). Improved enzymatic hydrolysis yield of rice straw using electron beam irradiation pretreatment. *Bioresource Technology* Vol. 100, Issue 3, pp. 1285–1290. Doi: 10.1016/j.biortech.2008.09.010.

Behera, S., Arora, R., Nandhagopal, N., & Kumar, S. (2014). Importance of chemical pretreatment for bioconversion of lignocellulosic biomass. *Renewable and Sustainable Energy Reviews* Vol. 36, pp. 91–106. Doi: 10.1016/j.rser.2014.04.047.

Berni, M. D., & Manduca, P. C. (2014). Bioethanol program in Brazil: Production and utilization of trade-offs for CO_2 abatement. *Journal of Clean Energy Technologies* Doi: 10.7763/jocet.2014.v2.91.

Binod, P., Satyanagalakshmi, K., Sindhu, R., Janu, K. U., Sukumaran, R. K., & Pandey, A. (2012). Short duration microwave assisted pretreatment enhances the enzymatic saccharification and fermentable sugar yield from sugarcane bagasse. *Renewable Energy* Vol. 37, Issue 1, pp. 109–116. Doi: 10.1016/j.renene.2011.06.007.

Bjornsson, W. J., MacDougall, K. M., Melanson, J. E., O'Leary, S. J. B., & McGinn, P. J. (2012). Pilot-scale supercritical carbon dioxide extractions for the recovery of triacylglycerols from microalgae: A practical tool for algal biofuels research. *Journal of Applied Phycology* Vol. 24, Issue 3, pp. 547–555. Doi: 10.1007/s10811-011-9756-2.

Brachet, A., Christen, P., & Veuthey, J. L. (2002). Focused microwave-assisted extraction of cocaine and benzoylecgonine from coca leaves. *Phytochemical Analysis* Vol. 13, Issue 3, pp. 162–169. Doi: 10.1002/pca.637.

Chandra, A. (2021). *Biofuels Annual.* United States Department of Agriculture. Report Number: IN2021-0072, New Delhi (India).

Chang, K. L., Thitikorn-amorn, J., Hsieh, J. F., Ou, B. M., Chen, S. H., Ratanakhanokchai, K., Huang, P. J., & Chen, S. T. (2011). Enhanced enzymatic conversion with freeze pretreatment of rice straw. *Biomass and Bioenergy* Vol. 35, Issue 1, pp. 90–95. Doi: 10.1016/j.biombioe.2010.08.027.

Chen, H., Xiao, Q., Weng, H., Zhang, Y., Yang, Q., & Xiao, A. (2020). Extraction of sulfated agar from *Gracilaria lemaneiformis* using hydrogen peroxide-assisted enzymatic method. *Carbohydrate Polymers* Vol. 232. Doi: 10.1016/j.carbpol.2019.115790.

Chen, K. S., Lin, Y. C., Hsu, K. H., & Wang, H. K. (2012). Improving biodiesel yields from waste cooking oil by using sodium methoxide and a microwave heating system. *Energy* Vol. 38, Issue 1, pp. 151–156. Doi: 10.1016/j.energy.2011.12.020.

Da Silva, A. S. A., Inoue, H., Endo, T., Yano, S., & Bon, E. P. S. (2010). Milling pretreatment of sugarcane bagasse and straw for enzymatic hydrolysis and ethanol fermentation. *Bioresource Technology* Vol. 101, Issue 19, pp. 7402–7409. Doi: 10.1016/j.biortech.2010.05.008.

Daza Serna, L. V., Orrego Alzate, C. E., & Cardona Alzate, C. A. (2016). Supercritical fluids as a green technology for the pretreatment of lignocellulosic biomass. *Bioresource Technology* Vol. 199, pp. 113–120. Doi: 10.1016/j.biortech.2015.09.078.

De la Rubia, M. A., Fernández-Cegrí, V., Raposo, F., & Borja, R. (2011). Influence of particle size and chemical composition on the performance and kinetics of anaerobic digestion process of sunflower oil cake in batch mode. *Biochemical Engineering Journal* Vol. 58–59, Issue 1, pp. 162–167. Doi: 10.1016/j.bej.2011.09.010.

De Vrije, T., Mars, A. E., Budde, M. A. W., Lai, M. H., Dijkema, C., De Waard, P., & Claassen, P. A. M. (2007). Glycolytic pathway and hydrogen yield studies of the extreme thermophile *Caldicellulosiruptor saccharolyticus*. *Applied Microbiology and Biotechnology* Vol. 74, Issue 6, pp. 1358–1367.

Duncan, S., Jing, Q., Katona, A., Kazlauskas, R. J., Schilling, J., Tschirner, U., & Wafa Aldajani, W. (2010). Increased saccharification yields from aspen biomass upon treatment with enzymatically generated peracetic acid. *Applied Biochemistry and Biotechnology* Vol. 160, Issue 6, pp. 1637–1652. Doi: 10.1007/s12010-009-8639-3.

Duque, A., Manzanares, P., & Ballesteros, M. (2017). Extrusion as a pretreatment for lignocellulosic biomass: Fundamentals and applications. *Renewable Energy* Vol. 114, pp. 1427–1441. Doi: 10.1016/j.renene.2017.06.050.

Ebadian, M., van Dyk, S., McMillan, J. D., & Saddler, J. (2020). Biofuels policies that have encouraged their production and use: An international perspective. *Energy Policy* Vol. 147. Doi: 10.1016/j.enpol.2020.111906.

Fadel, H., Marx, F., El-Sawy, A., & El-Ghorab, A. (1999). Effect of extraction techniques on the chemical composition and antioxidant activity of *Eucalyptus camaldulensis* var. brevirostris leaf oils. *European Food Research and Technology* Vol. 208, Issue 3, pp. 212–216. Doi: 10.1007/s002170050405.

Falsafi, Z., Raofie, F., Kazemi, H., & Ariya, P. A. (2020). Simultaneous extraction and fractionation of petroleum biomarkers from tar balls and crude oils using a two-step sequential supercritical fluid extraction. *Marine Pollution Bulletin* Vol. 159, Issue June. Doi: 10.1016/j.marpolbul.2020.111484.

Food and Agriculture Organization (FAO). (2013). *The State of Food and Agriculture - Executive Summary*. FAO. http://www.fao.org/icatalog/inter-e.htm

Furimsky, E. (2013). Hydroprocessing challenges in biofuels production. *Catalysis Today* Vol. 217. Doi: 10.1016/j.cattod.2012.11.008.

Gabhane, J., Prince William, S. P. M., Vaidya, A. N., Mahapatra, K., & Chakrabarti, T. (2011). Influence of heating source on the efficacy of lignocellulosic pretreatment - A cellulosic ethanol perspective. *Biomass and Bioenergy* Vol. 35, Issue 1, pp. 96–102. Doi: 10.1016/j.biombioe.2010.08.026.

Gáspár, M., Kálmán, G., & Réczey, K. (2007). Corn fiber as a raw material for hemicellulose and ethanol production. *Process Biochemistry* Vol. 42, Issue 7, pp. 1135–1139. Doi: 10.1016/j.procbio.2007.04.003.

Gaurav, N., Sivasankari, S., Kiran, G. S., Ninawe, A., & Selvin, J. (2017). Utilization of bioresources for sustainable biofuels: A review. *Renewable and Sustainable Energy Reviews* Vol. 73, Issue November 2016, pp. 205–214. Doi: 10.1016/j.rser.2017.01.070.

Gnansounou, E., & Dauriat, A. (2011). Technoeconomic analysis of lignocellulosic ethanol. *Biofuels*. Doi: 10.1016/B978-0-12-385099-7.00006-1.

Gonçalves, F. A., Ruiz, H. A., Nogueira, C. D. C., Santos, E. S. Dos, T. J. A., & De Macedo, G. R. (2014). Comparison of delignified coconuts waste and cactus for fuel-ethanol production by the simultaneous and semi-simultaneous saccharification and fermentation strategies. *Fuel* Vol. 131, pp. 66–76. Doi: 10.1016/j.fuel.2014.04.021.

Gu, B. J., Wang, J., Wolcott, M. P., & Ganjyal, G. M. (2018). Increased sugar yield from pre-milled Douglas-fir forest residuals with lower energy consumption by using planetary ball milling. *Bioresource Technology* Vol. 251, pp. 93–98. Doi: 10.1016/j.biortech.2017.11.103.

Gupta, R., & Lee, Y. Y. (2009). Pretreatment of hybrid poplar by aqueous ammonia. *American Institute of Chemical Engineers Biotechnology Program* Vol. 25, pp. 357–364. Doi: 10.1021/bp.133.

He, Z., Wang, Z., Zhao, Z., Yi, S., Mu, J., & Wang, X. (2017). Influence of ultrasound pretreatment on wood physiochemical structure. *Ultrasonics Sonochemistry* Vol. 34, pp. 136–141. Doi: 10.1016/j.ultsonch.2016.05.035.

Hendriks, A. T. W. M., & Zeeman, G. (2009). Pretreatments to enhance the digestibility of lignocellulosic biomass. *Bioresource Technology* Vol. 100, Issue 1, pp. 10–18. Doi: 10.1016/j.biortech.2008.05.027.

Hermiati, E., Laksana, R. P. B., Fatriasari, W., Kholida, L. N., Thontowi, A., Yopi, A. D. R., Champreda, V., & Watanabe, T. (2020). Microwave-assisted acid pretreatment for enhancing enzymatic saccharification of sugarcane trash. *Biomass Conversion and Biorefinery* Doi: 10.1007/s13399-020-00971-z.

Hernández-Salas, J. M., Villa-Ramírez, M. S., Veloz-Rendón, J. S., Rivera-Hernández, K. N., González-César, R. A., Plascencia-Espinosa, M. A., & Trejo-Estrada, S. R. (2009). Comparative hydrolysis and fermentation of sugarcane and agave bagasse. *Bioresource Technology* Vol. 100, Issue 3, pp. 1238–1245. Doi: 10.1016/j.biortech.2006.09.062.

Hideno, A., Inoue, H., Tsukahara, K., Fujimoto, S., Minowa, T., Inoue, S., Endo, T., & Sawayama, S. (2009). Wet disk milling pretreatment without sulfuric acid for enzymatic hydrolysis of rice straw. *Bioresource Technology* Vol. 100, Issue 10, pp. 2706–2711. Doi: 10.1016/j.biortech.2008.12.057.

Ho, D. P., Ngo, H. H., & Guo, W. (2014). A mini review on renewable sources for biofuel. *Bioresource Technology* Vol. 169, pp. 742–749. Doi: 10.1016/j.biortech.2014.07.022.

Huber, G. W., & Dumesic, J. A. (2006). An overview of aqueous-phase catalytic processes for production of hydrogen and alkanes in a biorefinery. *Catalysis Today* Vol. 111, Issue 1–2, pp. 119–132. Doi: 10.1016/j.cattod.2005.10.010.

IEA. (2011). *Technology Roadmap Biofuels for Transport*. International Energy Agency, Paris, France, 56.

pretreatment of wheat straw for methane production. *Bioresource Technology* Vol. 102, Issue 12, pp. 6750–6756. Doi: 10.1016/j.biortech.2011.03.107.

Joshi, G., Pandey, J. K., Rana, S., & Rawat, D. S. (2017). Challenges and opportunities for the application of biofuel. *Renewable and Sustainable Energy Reviews* Vol. 79, pp. 850–866. Doi: 10.1016/j.rser.2017.05.185.

Kádár, Z., Vrije, T. D., Budde, M. A., Szengyel, Z., Réczey, K., & Claassen, P. A. (2003). Hydrogen production from paper sludge hydrolysate. In *Biotechnology for Fuels and Chemicals* (pp. 557–566). Humana Press, Totowa, NJ.

Kalyani, D., Lee, K. M., Kim, T. S., Li, J., Dhiman, S. S., Kang, Y. C., & Lee, J. K. (2013). Microbial consortia for saccharification of woody biomass and ethanol fermentation. *Fuel* Vol. 107, pp. 815–822. Doi: 10.1016/j.fuel.2013.01.037.

Kanitkar, A., Sabliov, C. M., Balasubramanian, S., Lima, M., & Boldor, D. (2011). Microwave-assisted extraction of soybean and rice bran oil: Yield and extraction kinetics. *Transactions Of The Asabe* Vol. 54, Issue 4, pp. 1387–1394.

Klein-Marcuschamer, D., Oleskowicz-Popiel, P., Simmons, B. A., & Blanch, H. W. (2012). The challenge of enzyme cost in the production of lignocellulosic biofuels. *Biotechnology and Bioengineering* Vol. 109, Issue 4. Doi: 10.1002/bit.24370.

Kumar, A., Kumar, N., Baredar, P., & Shukla, A. (2015a). A review on biomass energy resources, potential, conversion and policy in India. *Renewable and Sustainable Energy Reviews* Vol. 45, pp. 530–539. Doi: 10.1016/j.rser.2015.02.007.

Kumar, R. R., Rao, P. H., & Arumugam, M. (2015b). Lipid extraction methods from microalgae: A comprehensive review. *Frontiers in Energy Research* Vol. 3, Issue, Jan. Doi: 10.3389/fenrg.2014.00061.

Kumari, D., & Singh, R. (2018). Pretreatment of lignocellulosic wastes for biofuel production: A critical review. *Renewable and Sustainable Energy Reviews* Vol. 90, pp. 877–891. Doi: 10.1016/j.rser.2018.03.111.

Kuster Moro, M., Sposina Sobral Teixeira, R., Sant'Ana da Silva, A., Duarte Fujimoto, M., Albuquerque Melo, P., Resende Secchi, A., & Pinto da Silva Bon, E. (2017). Continuous pretreatment of sugarcane biomass using a twin-screw extruder. *Industrial Crops and Products* Vol. 97, pp. 509–517. Doi: 10.1016/j.indcrop.2016.12.051.

Li, P., Sakuragi, K., & Makino, H. (2019a). Extraction techniques in sustainable biofuel production: A concise review. *Fuel Processing Technology* Vol. 193, Issue February, pp. 295–303. Doi: 10.1016/j.fuproc.2019.05.009.

Liu, H., Grot, S., & Logan, B. E. (2005). Electrochemically assisted microbial production of hydrogen from acetate. *Environmental Science & Technology* Vol. 39, Issue 11, pp. 4317–4320.

Liu, Y., Zhou, H., Wang, S., Wang, K., & Su, X. (2015). Comparison of γ-irradiation with other pretreatments followed with simultaneous saccharification and fermentation on bioconversion of microcrystalline cellulose for bioethanol production. *Bioresource Technology* Vol. 182, pp. 289–295. Doi: 10.1016/j.biortech.2015.02.009.

Luque-García, J. L., & Luque De Castro, M. D. (2004). Focused microwave-assisted Soxhlet extraction: Devices and applications. *Talanta* Vol. 64, Issue 3, pp. 571–577. Doi: 10.1016/j.talanta.2004.03.054.

Ma, H., Liu, W. W., Chen, X., Wu, Y. J., & Yu, Z. L. (2009). Enhanced enzymatic saccharification of rice straw by microwave pretreatment. *Bioresource Technology* Vol. 100, Issue 3, pp. 1279–1284. Doi: 10.1016/j.biortech.2008.08.045.

Malode, S. J., Prabhu, K. K., Mascarenhas, R. J., Shetti, N. P., & Aminabhavi, T. M. (2021). Recent advances and viability in biofuel production. *Energy Conversion and Management: X* Vol. 10, Issue September 2020, p. 100070. Doi: 10.1016/j.ecmx.2020.100070.

Malhotra, R (2007). Road to emerging alternatives—biofuels and hydrogen. *Journal of the Petrotech Society* Vol. 4, pp. 34–40.

Maran, J. P., & Priya, B. (2015). Supercritical fluid extraction of oil from muskmelon (Cucumis melo) seeds. *Journal of the Taiwan Institute of Chemical Engineers* Vol. 47, pp. 71–78. Doi: 10.1016/j.jtice.2014.10.007.

Marshall, C. W., Ross, D. E., Fichot, E. B., Norman, R. S., & May, H. D. (2012). Electrosynthesis of commodity chemicals by an autotrophic microbial community. *Applied and Environmental Microbiology* Vol. 78, Issue 23, pp. 8412–8420.

McIntosh, S., & Vancov, T. (2010). Enhanced enzyme saccharification of Sorghum bicolor straw using dilute alkali pretreatment. *Bioresource Technology* Vol. 101, Issue 17, pp. 6718–6727. Doi: 10.1016/j.biortech.2010.03.116.

Mikulski, D., & Kłosowski, G. (2020). Microwave-assisted dilute acid pretreatment in bioethanol production from wheat and rye stillages. *Biomass and Bioenergy* Vol. 136. Doi: 10.1016/j.biombioe.2020.105528.

Mohan, S. V., Agarwal, L., Mohanakrishna, G., Srikanth, S., Kapley, A., Purohit, H. J., & Sarma, P. N. (2011). Firmicutes with iron dependent hydrogenase drive hydrogen production in anaerobic bioreactor using distillery wastewater. *International Journal of Hydrogen Energy* Vol. 36, Issue 14, pp. 8234–8242.

Mohan, S. V., Mohanakrishna, G., Reddy, B. P., Saravanan, R., & Sarma, P. N. (2008). Bioelectricity generation from chemical wastewater treatment in mediatorless (anode) microbial fuel cell (MFC) using selectively enriched hydrogen producing mixed culture under acidophilic microenvironment. *Biochemical Engineering Journal* Vol. 39, Issue 1, pp. 121–130.

Mohan, S. V., Srikanth, S., Babu, M. L., & Sarma, P. N. (2010). Insight into the dehydrogenase catalyzed redox reactions and electron discharge pattern during fermentative hydrogen production. *Bioresource Technology* Vol. 101, Issue 6, pp. 1826–1833.Mussatto, S. I., Dragone, G., Guimarães, P. M. R., Silva, J. P. A., Carneiro, L. M., Roberto, I. C., Vicente, A., Domingues, L., & Teixeira, J. A. (2010). Technological trends, global market, and challenges of bio-ethanol production. *Biotechnology Advances* Vol. 28, Issue 6. Doi: 10.1016/j.biotechadv.2010.07.001.

Mussatto, S. I., & Roberto, I. C. (2004). Alternatives for detoxification of diluted-acid lignocellulosic hydrolyzates for use in fermentative processes: A review. *Bioresource Technology* Vol. 93, Issue 1, pp. 1–10. Doi: 10.1016/j.biortech.2003.10.005.

Ni, H., Hsu, C. S., Lee, P., Wright, J., Chen, R., Xu, C., & Shi, Q. (2015). Supercritical carbon dioxide extraction of petroleum on kieselguhr. *Fuel* Vol. 141, pp. 74–81. Doi: 10.1016/j.fuel.2014.09.126.

Nikolić, S., Pejin, J., & Mojović, L. (2016). Challenges in bioethanol production: Utilization of cotton fabrics as a feedstock. *Chemical Industry and Chemical Engineering Quarterly* Vol. 22, Issue 4, pp. 375–390. Doi: 10.2298/CICEQ151030001N.

Nogueira, L. A. H. (2011). Does biodiesel make sense? *Energy* Vol. 36, Issue 6. Doi: 10.1016/j.energy.2010.08.035.

Peng, L., Ye, Q., Liu, X., Liu, S., & Meng, X. (2019). Optimization of aqueous enzymatic method for *Camellia sinensis* oil extraction and reuse of enzymes in the process. *Journal of Bioscience and Bioengineering* Vol. 128, Issue 6, pp. 716–722. Doi: 10.1016/j.jbiosc.2019.05.010.

Potumarthi, R., Baadhe, R. R., Nayak, P., & Jetty, A. (2013). Simultaneous pretreatment and sacchariffication of rice husk by *Phanerochete chrysosporium* for improved production of reducing sugars. *Bioresource Technology* Vol. 128, pp. 113–117. Doi: 10.1016/j.biortech.2012.10.030.

Prashanth Kumar, C., Rena, M. A., Khapre, A. S., Kumar, S., Anshul, A., Singh, L., Kim, S. H., Lee, B. D., & Kumar, R. (2019). Bio-Hythane production from organic fraction of municipal solid waste in single and two stage anaerobic digestion processes. *Bioresource Technology* Vol. 294, Issue August, p. 122220. Doi: 10.1016/j.biortech.2019.122220.

Rastegari, A. A., Yadav, A. N., & Gupta, A. (2019). *Prospects of Renewable Bioprocessing in Future Energy Systems: Production by Cyanobacteria* (Issue February 2020). Doi: 10.1007/978-3-030-14463-0.

Rastogi, P., Kumar, N., Gupta, B. L., & Fuskele, V. (2021). Advances in biofuel technology: A review. *SSRN Electronic Journal* Vol. 11, Issue 4, pp. 32–41. Doi: 10.2139/ssrn.3822333.

Rena, Zacharia, K. M. B., Yadav, S., Machhirake, N. P., Kim, S. H., Lee, B. D., Jeong, H., Singh, L., Kumar, S., & Kumar, R. (2020). Bio-hydrogen and bio-methane potential analysis for production of bio-hythane using various agricultural residues. *Bioresource Technology* Vol. 309, Issue February), p. 123297. Doi: 10.1016/j.biortech.2020.123297.

Rooni, V., Raud, M., & Kikas, T. (2017). The freezing pre-treatment of lignocellulosic material: A cheap alternative for Nordic countries. *Energy* Vol. 139, pp. 1–7. Doi: 10.1016/j.energy.2017.07.146.

Saha, B. C., & Cotta, M. A. (2006). Ethanol production from alkaline peroxide pretreated enzymatically saccharified wheat straw. *Biotechnology Progress* Vol. 22, Issue 2, pp. 449–453. Doi: 10.1021/bp050310r

Sakuragi, K., Li, P., Otaka, M., & Makino, H. (2016). Recovery of bio-oil from industrial food waste by liquefied dimethyl ether for biodiesel production. *Energies* Vol. 9, Issue 2, pp. 1–8. Doi: 10.3390/en9020106.

Segneanu, A.-E., Sziple, F., Vlazan, P., Sfarloaga, P., Grozesku, I., & Daniel, V. (2013). Biomass extraction methods. *Biomass Now - Sustainable Growth and Use*. Doi: 10.5772/55338.

Sen, B., Chou, Y. P., Wu, S. Y., & Liu, C. M. (2016). Pretreatment conditions of rice straw for simultaneous hydrogen and ethanol fermentation by mixed culture. *International Journal of Hydrogen Energy* Vol. 41, Issue 7, pp. 4421–4428. Doi: 10.1016/j.ijhydene.2015.10.147.

Sheikh, M. M. I., Kim, C. H., Park, H. H., Nam, H. G., Lee, G. S., Jo, H. S., Lee, J. Y., & Kim, J. W. (2015). A synergistic effect of pretreatment on cell wall structural changes in barley straw (*Hordeum vulgare* L.) for efficient bioethanol production. *Journal of the Science of Food and Agriculture* Vol. 95, Issue 4, pp. 843–850. Doi: 10.1002/jsfa.7004.

Sindhu, R., Binod, P., & Pandey, A. (2016). Biological pretreatment of lignocellulosic biomass - An overview. *Bioresource Technology* Vol. 199, pp. 76–82. Doi: 10.1016/j.biortech.2015.08.030.

Sipponen, M. H., & Österberg, M. (2019). Aqueous ammonia pre-treatment of wheat straw: Process optimization and broad spectrum dye adsorption on nitrogen-containing lignin. *Frontiers in Chemistry* Vol. 7. Doi: 10.3389/fchem.2019.00545.

Song, Y., Wi, S. G., Kim, H. M., & Bae, H. J. (2016). Cellulosic bioethanol production from Jerusalem artichoke (*Helianthus tuberosus* L.) using hydrogen peroxide-acetic acid (HPAC) pretreatment. *Bioresource Technology* Vol. 214, pp. 30–36. Doi: 10.1016/j.biortech.2016.04.065.

Srikanth, S., Mohan, S. V., Devi, M. P., Peri, D., & Sarma, P. N. (2009). Acetate and butyrate as substrates for hydrogen production through photo-fermentation: Process optimization and combined performance evaluation. *International Journal of Hydrogen Energy* Vol. 34, Issue 17, pp. 7513–7522.

Taha, M., Shahsavari, E., Al-Hothaly, K., Mouradov, A., Smith, A. T., Ball, A. S., & Adetutu, E. M. (2015). Enhanced biological straw saccharification through coculturing of lignocellulose-degrading microorganisms. *Applied Biochemistry and Biotechnology* Vol. 175, Issue 8, pp. 3709–3728. Doi: 10.1007/s12010-015-1539-9.

Xiang, Y., Xiang, Y., & Wang, L. (2016). Cobalt-60 gamma-ray irradiation pretreatment and sludge protein for enhancing enzymatic saccharification of hybrid poplar sawdust. *Bioresource Technology* Vol. 221, pp. 9–14. Doi: 10.1016/j.biortech.2016.09.032.

Yan, X., Wang, Z., Zhang, K., Si, M., Liu, M., Chai, L., Liu, X., & Shi, Y. (2017). Bacteria-enhanced dilute acid pretreatment of lignocellulosic biomass. *Bioresource Technology* Vol. 245, pp. 419–425. Doi: 10.1016/j.biortech.2017.08.037.

Yang, J., (Sophia)He, Q., & Yang, L. (2019). A review on hydrothermal co-liquefaction of biomass. *Applied Energy* Vol. 250, pp. 926–945. Doi: 10.1016/j.apenergy.2019.05.033.

Yin, D. L. (Tyler), Jing, Q., AlDajani, W. W., Duncan, S., Tschirner, U., Schilling, J., & Kazlauskas, R. J. (2011). Improved pretreatment of lignocellulosic biomass using enzymatically-generated peracetic acid. *Bioresource Technology* Vol. 102, Issue 8, pp. 5183–5192. Doi: 10.1016/j.biortech.2011.01.079.

Yoo, J., Alavi, S., Vadlani, P., & Amanor-Boadu, V. (2011). Thermo-mechanical extrusion pretreatment for conversion of soybean hulls to fermentable sugars. *Bioresource Technology* Vol. 102, Issue 16, pp. 7583–7590. Doi: 10.1016/j.biortech.2011.04.092.

Youngs, H., & Somerville, C. (2012). Development of feedstocks for cellulosic biofuels. *F1000 Biology Reports* Vol. 4, Issue 1. Doi: 10.3410/B4-10.

Yue, L., Cheng, J., Tang, S., An, X., Hua, J., Dong, H., & Zhou, J. (2021). Ultrasound and microwave pretreatments promote methane production potential and energy conversion during anaerobic digestion of lipid and food wastes. *Energy* Vol. 228. Doi: 10.1016/j.energy.2021.120525.

Zhang, Q., He, J., Tian, M., Mao, Z., Tang, L., Zhang, J., & Zhang, H. (2011). Enhancement of methane production from cassava residues by biological pretreatment using a constructed microbial consortium. *Bioresource Technology* Vol. 102, Issue 19, pp. 8899–8906. Doi: 10.1016/j.biortech.2011.06.061.

Zhao, X., Zhou, Y., Zheng, G., & Liu, D. (2010). Microwave pretreatment of substrates for Cellulase production by solid-state fermentation. *Applied Biochemistry and Biotechnology* Vol. 160, Issue 5, pp. 1557–1571. Doi: 10.1007/s12010-009-8640-x

Zhao, X. B., Wang, L., & Liu, D. H. (2007). Effect of several factors on peracetic acid pretreatment of sugarcane bagasse for enzymatic hydrolysis. *Journal of Chemical Technology and Biotechnology* Vol. 82, Issue 12, pp. 1115–1121. Doi: 10.1002/jctb.1775.

Zheng, Y., Pan, Z., & Zhang, R. (2009). Overview of biomass pretreatment for cellulosic ethanol production. *International Journal of Agricultural and Biological Engineering* Vol. 2, Issue 3, pp. 51–68. Doi: 10.3965/j.issn.1934-6344.2009.03.051-068.

Zhu, J. Y., Pan, X. J., Wang, G. S., & Gleisner, R. (2009). Sulfite pretreatment (SPORL) for robust enzymatic saccharification of spruce and red pine. *Bioresource Technology* Vol. 100, Issue 8, pp. 2411–2418. Doi: 10.1016/j.biortech.2008.10.057.

4 Paradigm Shift from Biofuel to Biorefinery
Prospects and Roadmap

Ramalingam Kayalvizhi, Ramachandran Devasena Umai, and Samuel Jacob
SRM Institute of Science and Technology

CONTENTS

DOI: 10.1201/9781003197737-6

4.1 INTRODUCTION

For decades, scientists and researchers have been working toward an ecofriendly alternative to fossil fuels without compromising the future. To meet the present demand and for a sustainable future, there is an uttermost need to move toward biorefinery. Sustainability is the basic goal for the existence of an effective economy and society. Along with other renewable resources like photovoltaic, wind power, solar, and hydropower, biofuels also help to overcome the present energy demand. Nonrenewable fossil fuel leads to environmental pollution and cannot meet the growing demand. To overcome the demand for raw materials due to the population explosion, there needs to be a shift from fossil fuels to bioresources as feedstock. Biorefinery aims at the production of high-value-added products, in addition to biofuels from biowaste, at the cost of less energy consumption. To compete with other growing countries, there is a need for moving toward sustainable production of green chemicals. The current utilization of feedstocks was changed from food crops to nonfood crops and biowaste/organic residues from various industries (Balan 2014).

4.1.1 Biorefinery

Biorefinery is a confinement that converts the biowaste into a high-value-added product with little or no release of pollutants to the environment compared to fossil fuels. Biorefinery aims at the production of high-value-added products along with reducing biowaste accumulation. Biorefinery is the sole key to unlocking the treasures of bioresources and to an unlimited supply of biofuels and high-value-added chemicals to mankind. These high-value-added chemicals have high potential and unrealized application in industrial, pharmaceutical, cosmetics, food, and many other industries.

4.1.2 Biorefinery vs. Biofuel

The idea of biofuel as a renewable alternative to fossil fuel started many decades ago. Many feedstocks have been analyzed for their efficiency to produce biofuels. The feedstock varies from food crops to nonfood crops and even nonconventional biomass. As researchers work toward green chemistry, the idea of biofuel production is now combined with the production of many value-added platform chemicals.

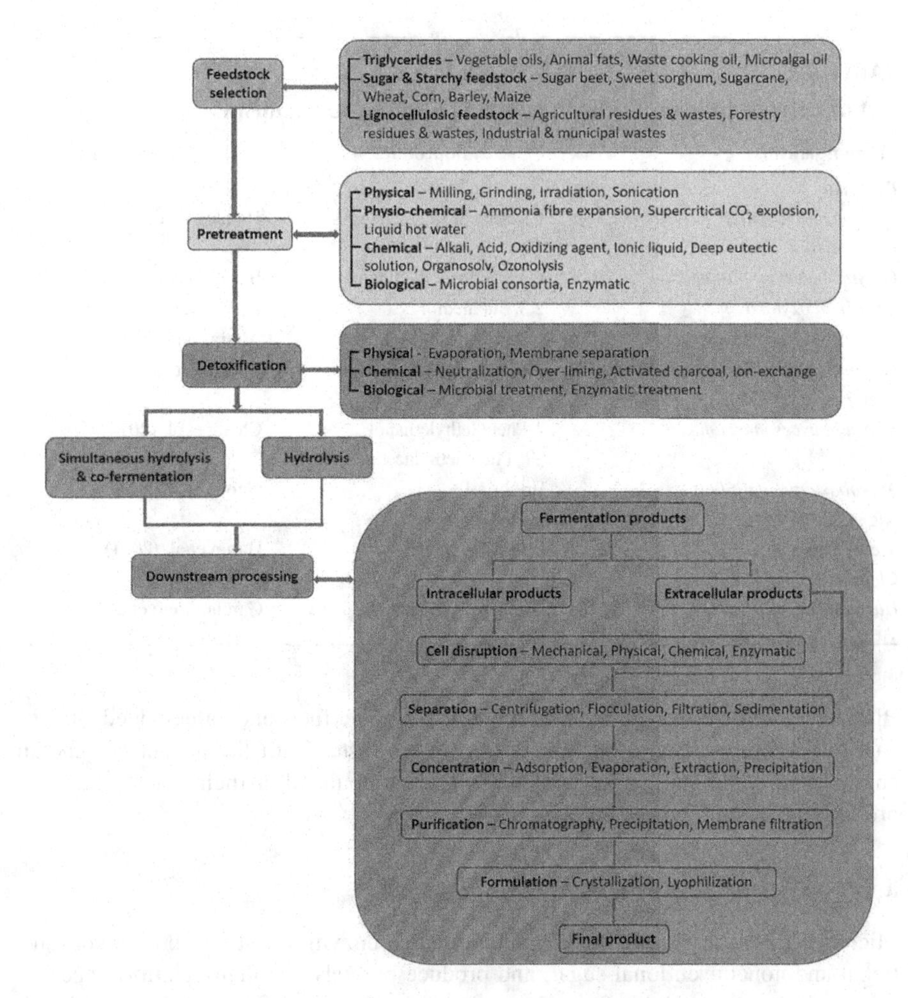

FIGURE 4.1 Overview of process stages in a biorefinery approach.

Figure 4.1 shows the overall summary of steps in a biorefinery approach involved in the production of value-added products.

Biorefinery—an integrated approach shows the following benefits compared to other methods of production of fuel and chemicals:

- It is a renewable, sustainable, and alternative approach to the production of biofuels and chemicals
- It is both economical and ecofriendly
- It has lesser emission of greenhouse gases during the process
- It shows prevention and effective utilization of organic waste accumulation
- It is cheaper and has safer feedstocks and the process strategy
- Availability of feedstock is throughout the year, and there is easy accessibility
- There is scalability.

TABLE 4.1
List of Selected Bioproducts Produced from Microorganisms

Microorganisms	Bioproduct	References
Bacteria		
Zymomonas mobilis	Bioethanol	Hidalgo et al. (2021)
Engineered *E. coli*		
Corynebacterium glutamicum	L-Lysine and L-glutamate	Baritugo et al. (2018)
Klebsiella oxytoca	2,3-Butanediol	Ji et al. (2008)
Engineered *Escherichia coli*	1,2-Propanediol	Altaras and Cameron (2000)
Engineered *Bacillus methanolicus* sp.	Methanol-based cadaverine	Naerdal et al. (2015)
Fungi		
Kluyveromyces marxianus	2-Phenylethylethanol, 2- Phenylethylacetate	Chang et al. (2014)
Aureobasidium pullulans	Erythritol	Guo et al. (2016)
Algae		
Scenedesmus sp.	Ethanol	Harun et al. (2011)
Chlorococcum infusionum		
Haematococcus pluvialis	Biodiesel, Astaxanthin, PHB	García Prieto et al. (2017)

Biorefinery is a facility that aims at manufacturing biofuels and value-added chemicals from bioresources, especially targeting biowaste. With the advent of genetic engineering techniques, microorganisms are manipulated at their genetic level to produce the desired product at the maximum level.

4.1.3 Selection of Potent Microorganisms

Microorganisms are loaded with many valuable enzymes that can utilize conventional and nonconventional sugar, and produce biofuels and many platform chemicals as shown in Table 4.1. Bacteria, fungi, and yeast have been extensively utilized for the production of various bioproducts, and now the attention is moving toward microalgae, which is a promising third-generation biofuel production machinery.

4.1.4 Microbial Biotechnology

Microbial biotechnology is a challenging branch of biotechnology with high hopes and a promising future. With the recent advancement in the genetic engineering field, manipulation of microbial genes is becoming more and more reliable and economical. Microbes with their rich and unique enzyme system, metabolic pathway, and ability to adapt to extreme conditions make them highly explorable in the area of production of high-market-value products. With the advent of genetic engineering tools, it has become easy to manipulate the genetic makeup and receive a soaring level of attention due to the high level of desired product production. This can be achieved by the introduction of new novel genes or the elimination of existing undesirable genes.

4.1.5 Metabolic Engineering

Metabolic engineering is another important tool to manipulate the metabolic pathway within an organism by altering biochemical pathways and enzyme systems. It aims at the effective utilization of conventional and nonconventional carbon sources, optimizing the production of desired products and reduction in by-product formation. This can be achieved by following strategies like (1) increased production of the product by overexpression of the key enzyme, (2) blocking the competing pathways by knocking out their enzyme expression, (3) eliminating feedback inhibition, (4) maximizing the product production by reducing product toxicity, (5) overexpression of desired product by blocking the diversion of substrate utilization, and (6) utilization of unconventional carbon source. The efficiency of microorganisms can be elevated by random or site-specific mutagenesis. In some cases, the metabolic engineering of microorganisms may result in the production of the desired product in large amounts due to the beneficial alteration in the genome of microorganisms. But, in some cases, it will end up in some random mutation without any benefits (Liao 2016).

4.1.6 Choice of Feedstocks

The utilization of food crops as a biomass source for fuel production ended up in scarcity and the increased price of food crops. To overcome this problem, nonfood crops are utilized as carbon sources. Now, for the production of bioproducts, a wide variety of potent lignocellulosic and organic waste materials are available without competing with the food crops. This includes agricultural waste, forest waste, industrial organic waste, domestic waste, etc. The value of these biowastes depends on their carbon content and their bioavailability. The carbon content of the bioresources depends on the location and climatic condition. Cellulose and hemicellulose (sugar polymers) are the major components of lignocellulosic biomass. Upon hydrolysis hemicellulose yields five-carbon- and six-carbon-containing sugars, and cellulose yields only glucose—a six-carbon-containing sugar molecule. These sugars are converted to biofuels and value-added compounds by the enzymatic action of microorganisms. The chemical composition of feedstocks plays a vital role in the type and amount of production of value-added chemicals.

4.2 BIOFUELS

Biofuels refer to a liquid or gaseous fuel produced by the enzymatic action of microorganisms on various biomasses. Biofuel is a promising sustainable, renewable fuel substitute or alternative to fossil fuels. These biofuels are a promising fuel choice for future generations to meet their energy needs without compromising the health of Mother Earth. These biofuels are renewable and cheaper due to abundant feedstock availability and lead to minimal environmental pollution and climatic change. These biofuels and chemicals derived through the biorefinery approach are extensively used in many industries, including pharmaceutical, food and feed, cosmetics, textiles, chemical, and automobile industries. First-generation biofuels refer to the fuels produced from food crops. Second- and third-generation biofuels refer to the fuel produced from nonfood crops (Naik et al. 2010). Lignocellulosic biofuels are the fuels produced from the cellulose and hemicellulose portions of the biomass by the action of microorganisms. Table 4.2 shows selected feedstocks and the biofuels produced by the action of microorganisms. This lignocellulosic biomass represents all agricultural waste, by-products, and industrial waste. Globally, about half the biomass produced is nonedible to mankind. These biowastes are the fundamental source of biofuel production (Demirbas 2006).

4.2.1 Biodiesel

Biodiesel is a clear yellow-colored liquid renewable fuel. It is a promising substitute for fossil fuel-derived diesel. Biodiesel is an ethyl ester of long-chain fatty acid. It is derived from lipids by the process of transesterification. It is produced by the action of methanol or ethanol on vegetable/animal oil. Alcohol reacts with oil under alkaline conditions and produces biodiesel and glycerol. Biodiesel is obtained from different types of bioresources. Scientists and researchers are working toward utilizing more and more lignocellulosic sources that are renewable, cost effective, affordable, transportable, and do not compete with human or animal feeds. Different types of bioresources are analyzed for their efficiency toward an efficient feedstock for biodiesel production. Biodiesel being a nonflammable and nonexplosive fuel is safe on roads. Emissions from the biodiesel engine are less toxic compared to commercial diesel.

4.2.2 Bioalcohols

4.2.2.1 Bioethanol

Bioethanol produced from lignocellulosic biomass is a clear, colorless liquid and highly flammable (Taherzadeh et al. 2013). In 1897, Nikolas otto developed an internal combustion engine that utilizes ethanol as motor fuel. Later, in 1900, Dr. Rudolph diesel presented a peanut oil-based diesel engine at the Paris exposition. Brazil and the USA are the leading producers of bioethanol globally (Saini et al. 2014). Bioethanol is produced by the process of fermentation where sugar is converted to ethanol by the enzymes produced by the microorganism. The major steps involved in bioethanol production are pretreatment of biomass, removal of lignin residue, fermentation of sugar molecule, and finally recovery and concentration of ethanol. The major feedstocks of bioethanol can be categorized into two types (Saini et al. 2014).

1. **Sugar based**: sugar cane juice, sweet molasses, sweet sorghum
2. **Starch based**: corn, maize, cassava, barley, rye, potatoes

Initially, food crops were used as feedstocks for bioethanol production, and, later on, nonconventional bioresources such as forest residues, straws, and other agricultural by-products, grass, and waste paper from chemical pulps were utilized as feedstocks (Taherzadeh et al. 2013).

Ethanol produced by the fermentation process is distilled and concentrated. The cost of conversion of ethanol to biofuel is high. Researchers are focused on the pretreatment of biomass and metabolic engineering of microorganisms, which are key to reducing the cost of production of the final product—ethanol.

TABLE 4.2
List of Selected Feedstocks and the Biofuel Produced

Biofuel	Feedstock	References
Biodiesel	Rapeseed oil	Mazanov et al. (2016)
	Canola oil	Issariyakul and Dalai (2010)
	Soybean oil	Du et al. (2003)
	Sunflower oil	Granados et al. (2007)
	Palm oil	Kalam and Masjuki (2002)
	Beef Tallow	Da Cunha et al. (2009)
	Mutton Tallow	Faleh et al. (2018)
	Poultry fat	Venkat Reddy et al. (2006)
	Pork lard	Janchiv et al. (2012)
	Yellow grease	Panchal et al. (2017)
Bioethanol	Sweet sorghum bagasse	Goshadrou et al. (2011)
	Miscanthus	Lee and Kuan (2015)
	Softwood biomass	Nitsos et al. (2012)
	Sugar bagasse	Hernawan et al. (2017)
	Cassava starch	Pradyawong et al. (2018)
	Starch by-products	Gronchi et al. (2019)
Biobutanol	Potatoes	Nimcevic and Gapes (2000)
	Corn fiber	Qureshi et al. (2006)
	Liquefied corn starch	Ezeji et al. (2007)
	Cassava bagasse	Lu et al. (2012)
	Sweet sorghum bagasse	Cai et al. (2013)
	Sucrose and sugarcane juice	Jiang et al. (2014)
	Corn cob bagasse	Cai et al. (2016)
Biogas	Food waste	Mookherjee et al. (2017)
	Kitchen waste	Pasha et al. (2015)
	Pig manure	Cabeza et al. (2017)
	Municipal solid waste	Alibardi and Cossu (2015)
	Agro-industrial residues	Corneli et al. (2016)
Biohydrogen	Sweet potato residues	Yokoi et al. (2001)
	Cornstalk waste	Zhang et al. (2007)
	Dairy wastewater	Mohan et al. (2007)
	palm oil mill effluent	Ismail et al. (2010)
	Textile wastewater	Li et al. (2012)
	Cassava wastewater	Amorim et al. (2014)
	Sugarcane vinasse	Júnior et al. (2014)
	Organic waste	Alibardi and Cossu (2016)
Syngas	Dairy waste	Gordillo et al. (2009)
	Food waste	Ahmed and Gupta (2010)
	Sewage sludge	Calvo et al. (2013)
	Woody biomass	Ong et al. (2015)
	Recycled food waste and acidic soil from the forest	Yang et al. (2016)
	Blended biomass and coal	Yan et al. (2018)
	Wood chip and coconut waste	Sulaiman et al. (2018)

4.2.2.2 Biobutanol

After ethanol production, the ABE (Acetone–Butanol–Ethanol) fermentation process is considered the world's second-largest alcohol production. Butanol—primary alcohol—is a flammable colorless liquid with high flash points and low vapor pressure. Due to its less volatile and less explosive nature, it is safer to handle compared to ethanol (García et al. 2011). In 1861, the microbial fermentation of butanol was first reported by Louis Pasteur (Gabriel and Crawford 1930). Several species of *Clostridium* prove to be potent industrial microorganisms for butanol fermentation. Many lignocellulosic biomasses act as a promising feedstock for the commercial production of butanol. Initially, sugar cane juices, molasses, and sweet beets were utilized for butanol production, but, later on, with the advent of many biowaste disposal strategies, the focus has now moved on to the utilization of agricultural by-products. Being a promising fuel next to ethanol, butanol is an industrial solvent significantly used in cosmetics (lipsticks, eye makeup, nail care products, perfumes) and in hydraulic and brake fluids. In the pharmaceutical industry, it is highly used in the production of many drugs along with vitamins, antibiotics, therapeutic hormones, etc. It is used in industries as a platform chemical in the production of several chemicals like butyl acetate, butyl acrylate, methacrylate, acrylate esters, glycol ethers, butylamines, and amino resins. Other industrial applications include the manufacturing of detergents, paints, paint thinners, super absorbents, and surface coatings (García et al. 2011; Dürre 2007).

4.2.3 Biogas

Biogas is a clean alternative renewable fuel produced by the process of anaerobic digestion of organic waste material. Chemically, biogas is composed of 55%–60% methane and 30%–40% carbon dioxide. Anaerobic digestion comprises four steps, namely hydrolysis, acidogenesis, acetogenesis, and methanogenesis, leading to the conversion of biomass into biogas (Mookherjee et al. 2017). Many microorganisms are involved in various stages of biogas production. Some of the most significant microbes are *Clostridium* sp., *Corynebacterium* sp., *Pseudomonas* sp., *Staphylococcus* sp., *Propionicbacterium* sp., *Actinomyces* sp., *Methanobacterium* sp., and *Methanobrevibacter* sp. (Bhatia et al. 2020). Animal dung was initially used as the prime source of biogas production. Later, the idea of utilization of organic waste such as agricultural waste, food waste, and kitchen waste for its biogas production efficiency is highly noted (Mookherjee et al. 2017). Biogas is a potent clean fuel, and a great source of power and heat. In many countries like India and China, biogas is extensively utilized as cooking gas. Being an efficient vehicle fuel, biogas is used to run IC engines. Other applications include fuel for street lights and fuel to run small engines for irrigation purposes.

4.2.4 Biohydrogen

Biohydrogen is a clean renewable fuel with innumerable applications other than transportation fuel. Biohydrogen generation by photo and/or dark fermentation of organic materials replaced conventional methods (oxidation of fossil fuels, electrolysis of water, and biophotolysis) of hydrogen production (Hassan et al. 2018). Microorganisms such as *Enterobacter* sp., *Bacillus* sp., *Thermoanaerobacterium* sp., and *Clostridium* sp. are commonly utilized for dark fermentation, and genera like *Rhodobacter* sp., *Rhodopseudomonas* sp., and *Rhodovulum* sp. are used for photo-fermentation, respectively.

The idea of utilization of energy crops for biofuel production is replaced by many agricultural and industrial biowaste. Numerous carbohydrate-rich substrates/biowastes, which are modest and plentiful, are used for biohydrogen generation. Biohydrogen generation utilizes organic waste as feedstock, and thus reduces environmental pollution (reducing CO_2 emission) and waste accumulation (Limongi et al. 2021). Biohydrogen holds greater promise as a transport fuel. Currently, many researchers are working on increasing the long life of cars running on biohydrogen as fuel. Biohydrogen is utilized as a fuel cell for electricity generation. The combination of biogas and biohydrogen as transport fuel reduces the emission of carbon dioxide and the consumption of fuel. Other than as transport fuel, biohydrogen is a key feedstock for many industrial chemical productions.

4.2.5 Syngas

Syngas (synthesis gas) is a combustible gas and a mixture of carbon dioxide, hydrogen, methane, carbon monoxide, smaller amounts of water vapor, sulfur compounds (hydrogen sulfide, carbonyl sulfide), and ammonia (NETL 2021). Syngas is produced by the technique of gasification of biomass at high temperature in the presence of oxidants (Heidenreich and Foscolo 2015). A wide variety of feedstocks such as petroleum waste, municipal sewage sludge, refinery waste and residue, hydrocarbon residues, and carbonaceous biowaste and biomass have been utilized for syngas production (Speight 2015).

Syngas is a highly potent combustible gas and is used in the production of transport fuel (Nielsen 2002). As shown in Figure 4.2, syngas is used to generate power for the turbines, IC engines, and fuel cells. Syngas is a chemical platform to produce a variety of chemicals like ethylene, acetic acid, formaldehyde, DME, methyl acetate, and fertilizers like ammonia, urea, and fuels like petrol, gasoline, diesel, and kerosene.

4.3 VALUE-ADDED PRODUCTS

By the enzymatic action of various microorganisms, lignocellulosic biomasses are converted into many commercially important chemicals and chemical intermediates. Some of the valuable chemicals produced are glycerol, furfural, organic acids, lipids, sugar alcohols, amino acids, etc. These industrially important chemicals have a wide variety of applications. These are used as anti-freeze agents, biosurfactants, animal feeds, fertilizers, cosmetics, sweeteners, additives, biomaterials, polymers, resins, antioxidants, and pharmaceutical/therapeutic products.

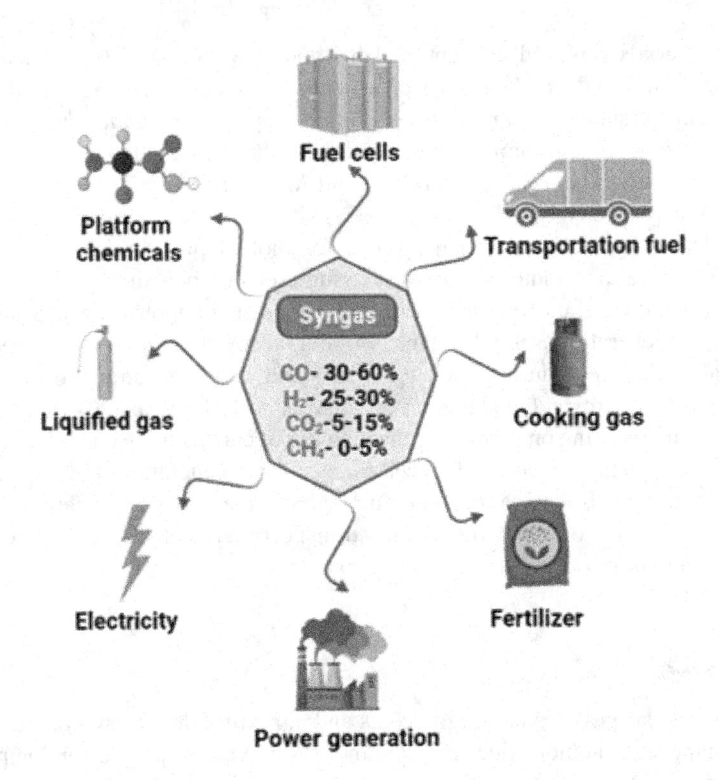

FIGURE 4.2 Applications of syngas.

4.3.1 Furfural and Its Derivatives

Furfural (2-furaldehyde) is a colorless organic liquid. It is produced from xylose—a 5-carbon-containing sugar present in lignocellulosic biomass. In 1922, the leading manufacturer of oats, Quaker oats company initiated production of furfural on a large scale from its by-product, oats hull, and encouraged other agricultural and industrial wastes like sugarcane bagasse and corn cobs for furfural production (Takkellapati et al. 2018). Furfural is produced by hydrolysis of lignocellulosic biomass (to release pentose sugar) and cyclodehydration of pentose sugar. Some of the commercially important derivatives of furfural are furoic acid, tetrahydrofuran, tetrahydrofurfuryl alcohol, dihydropyran, and methyltetrahydrofuran. As illustrated in Figure 4.3, furfural and its derivatives are widely used in pharmaceutical industries and many agrochemical industries as insecticides, nematicides, fungicides, flavor enhancers, lubricants, transportation fuel, feedstock for bioplastic, and resin production (Kabbour and Luque 2020).

4.3.2 Glycerol and Its Derivatives

Glycerol, commonly known as glycerin, is chemically 1,2,3-propanetriol. It is an important feedstock for the production of many value-added products. Glycerol—a

FIGURE 4.3 Applications of furfural and its derivatives.

by-product produced during biodiesel production—plays a crucial role in the oleochemical industry. Common feedstocks for glycerol production are waste cooking oil, waste vegetable oil, nonedible oil, oil wasted during processing, animal fats, various biomass, and microalgae. Glycerol is produced at a rate of 10wt.% of the total biodiesel produced by the transesterification process. Thus, the biodiesel industry contributes about 66% of global glycerol production (Sathianachiyar and Devaraj 2013). Glycerol is a commercially valuable feedstock in many pharmaceuticals, agrochemicals, cosmetics, and textile industries. As illustrated in Figure 4.4, glycerol is widely used for the production of surfactants, cosmetics, personal care products, pharmaceuticals, beverages, lubricants, and textiles (Takkellapati et al. 2018).

4.3.3 Isoprene

Isoprene (C_5H_8) is a volatile monomer with the chemical formula, 2-methyl-1,3-butadiene. It is a monomer with five-carbon building blocks, utilized for the production

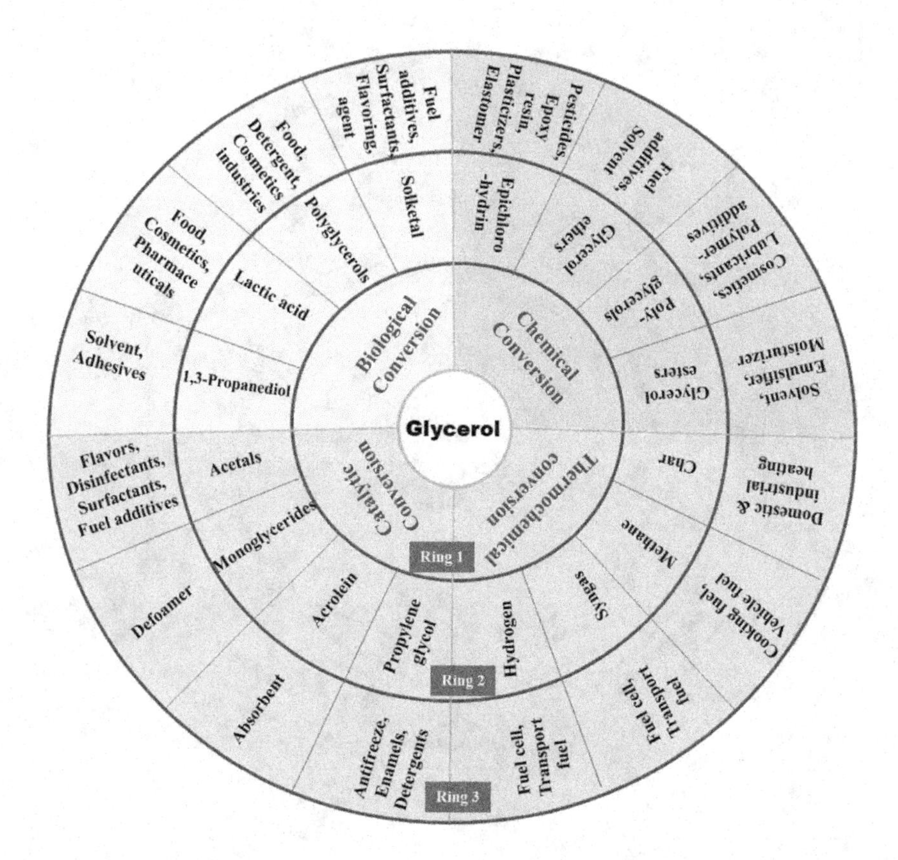

FIGURE 4.4 Commercial application of glycerol and its derivatives.

of polyisoprene polymer. This polymer is used to produce synthetic rubber, which is used to make footwear, rubber tires, sports goods, mechanical instruments, medical appliances, adhesives, cosmetics, and flavorings. It is also used as a fuel additive for jet fuel and gasoline. Companies like Amyris, Braskem, and Michelin have been jointly working on the production and commercialization of polyisoprene polymers from renewable resources since 2014. Ajinomoto Co., Inc. and Bridgestone Corp. have been working together to develop synthetic rubber (isoprene polymer) from biomass since 2012. Common feedstocks for isoprene production are glucose and glycerol. Microorganisms such as *Bacillus subtilis*, *Saccharomyces cerevisiae*, *Escherichia coli*, engineered *Methanosarcina acetivorans* (Carr et al. 2021), *Ralstonia eutropha* (Lee et al. 2019), and some microalgae and cyanobacteria are used extensively for isoprene production from biomass.

4.3.4 Organic Acids

Microbial fermentation of biowaste yields a wide variety of organic acids. These microbes break down starch, cellulose, hemicellulose, and hydrocarbon residues

TABLE 4.3
List of Organic Acids Produced by the Biorefinery Approach

Organic Acid	Microorganisms	Applications	References
Glyceric acid	*Gluconobacter oxydans* *Glucaonobactetr frateurii* *Acetobacter tropicalis*	Used as a reagent, surfactant, fabric conditioning agent	Habe et al. (2009)
Lactic acid	*Lactobacilli* sp. *Candida* sp. *Rhizopus oryzae*	Used as cleaning, moisturizing, flavoring, antimicrobial agents	Rahman et al. (2011)
Propionic acid	*Propionibacterium acidipropionici* *Clostridium propionicum*	Used as flavoring and preservative agent	Chen et al. (2013)
Levulinic acid	*Clostridium thermoaceticum* *Rhodobacter sphaeroides*	Used in food additives, solvents, fuel additives, and pharmaceuticals	Morone et al. (2015)
Muconic acid	*Arthrobacter* sp. *Pseudomonas* sp.	Derivatives include adipic acid, Caprolactam, Terephthalic acid	Pleissner et al. (2017)
Lactobionic acid	*Pseudomonas graveolens* *Penicillium chrysogenum*	Used as antioxidant, emulsifying, humectant, and chelating agent	Pleissner et al. (2017)
Malic acid	*Aspergillus flavus* *Saccharomyces cerevisiae*	Used as a taste enhancer, chelating agent, and acidulant	Kövilein et al. (2019)
Succinic acid	*E. coli* *Zymomonas mobilis* *Saccharomyces cerevisiae*	Used in pharmaceuticals, paint industry, detergents, polyester	Jiang et al. (2019)
Acrylic acid	*Propionibacterium shermanii* *Alcaligenes faecalis*	Acrylate esters are used in the manufacturing of dispersant, detergent, polymers	Lebeau et al. (2020)
Muconic acid	*Arthrobacter* sp. *Pseudomonas* sp.	Precursor for the synthesis of adipic acid, caprolactam, terephthalic acid, and nylon-6,6	Chen et al. (2021)

present in the organic waste into organic acids such as acetic acid, citric acid, glycolic acid, lactic acid, fumaric acid, malic acid, adipic acid, itaconic acid, propionic acid, succinic acid, levulinic acid, 3-hydroxy propionic acid, glucaric acid, butyric acid, xylonic acid, glyceric acid, and many more (Alonso et al. 2014). These organic acids are used in the cosmetics, pharmaceuticals, food, and chemical industries, as shown in Table 4.3.

4.3.5 Microbial Lipids

Microorganisms having the capacity to store lipids from 20% to 70% in their biomass are called oleaginous microorganisms. These oleaginous microorganisms are utilized for the production of microbial lipids or single cell oils (SCOs).

Cryptococcus curvatus was found to convert 61.3% of corncob hydrolysate to lipids (Chang et al. 2013). The bacterium *Botryococcus braunii* was proven to store 75% lipid of its biomass (Meng et al. 2009). Commonly used oleaginous microorganisms are bacteria (*Bacillus alcalophilus, Acinetobacter calcoaceticus, Arthrobacter* sp., *Rhodococcus opacus*), fungi (*Mortierella isabelline, Aspergillus oryzae, Mortierella vinacea, Humicola lanuginosa*), yeast (*Rhodosporidium, Candida, Rhodotorula, Cryptococcus, Lipomyces, Yarrowia, Trichosporon*), and microalgae (*Cylindrotheca* sp., *Crypthecodinium cohnii, Isochrysis* sp., *Monallanthus salina, Nannochloris* sp., *Nannochloropsis* sp., *Neochloris oleoabundans, Nitzschia* sp., *Schizochytrium* sp.). The feedstock for lipid production varies from agricultural residues, forest residues, lignocellulosic biomass, starch and sugary residues, industrial waste, glycerol, and waste and nonedible oils. The type and concentration of fatty acids produced from these biomasses vary according to the type of biomass and organism used for the production process. The most common fatty acids produced by these oleaginous microorganisms are palmitic acid, stearic acid, oleic acid, and linoleic acid (Ruan et al. 2013). Microbial lipids are considered an efficient high-value feedstock for the oleochemical industries in the production of biodiesel, soaps, detergents, wax esters, and oleo-gels. It is also used as fuel additives, lubricants, water repellents, flotation agents, corrosion inhibitors, and cosmetic additives. It is used in the preparation of many pharmaceutical formulations like laxatives, emollients, demulcents, and liniments, and is also used as a dental cement setting retardant.

4.3.6 Sugar Alcohols

Sugar alcohol is a water-soluble, non-cyclic hydrogenated carbohydrate produced by the process of reduction of sugar. It contains one or more hydroxyl groups. Common sugar alcohols produced by microbes are sorbitol, arabitol, mannitol, xylitol, erythritol, maltitol, lactitol, and isomalt (a mixture of gluco-mannitol and gluco-sorbitol). A wide variety of bacteria (*Weisella* sp., *Lactobacillus* sp., *Oenococcus* sp., *Leuconostoc* sp.), fungi (*Yarrowia lipolytica, Penicillium* sp., *Moniella* sp., *Aureobasidium* sp., *Pseudozyma tsukubaensis*), and yeast (*Moniliella* sp., *Trichosporonoides* sp., *Candida* sp., *Debaromyces* sp., *Kluveromyces* sp.) are used in sugar alcohols production (Rice et al. 2019). Sugar alcohols (polyols/polyhydric alcohols) are extensively used as a sweetener in the food industry. Due to its anti-cariogenic properties, it is widely utilized in the formulation of toothpaste for patients with gum problems. It is used in the preparation of hard candy, chewing gum, chocolates, spreads, ice creams, beverages, baked goods, frozen desserts, and confectionary items. It is highly used as a sweetener/sugar substitute, especially by diabetic patients. It is also used in restaurants as a table-top sweetener and as a thickener, flavor enhancer, humectant, cooling agent, and the dusting powder for chewing gums/candies (Park et al. 2016).

4.4 CHALLENGES IN BIOREFINERY

Global consumption of energy and organic chemicals comes from fossil fuels, and the consumption rate is increasing by approximately 7% annually (Maity 2015). To meet the demand the government emphasizes blending bioethanol or biodiesel with petrol, which requires slight modification in automobile engines. There are many hurdles and challenges in the production and utilization of bioproducts produced from biorefinery. Throughout the world, researchers and governments are working to a great extent toward maximum utilization of biomass for the production of many value-added products (Cesaro and Belgiorno 2015).

4.4.1 Feedstock Diversity

There is a large array of organic feedstock available throughout the globe. These biomasses are known for their abundance in availability and inexpensiveness in nature. The expenses are directly linked to land usage and the cost of transportation. To further reduce the cost, the Government of India launched the National Biodiesel Mission (NBM) in December 2009 and encouraged the cultivation of *Jatropha curcas* (for oilseeds) in unused less fertile lands in several states for biofuel production (NBM 2021).

The source and the type of biomass determine its physical and chemical characteristics (Maity 2015).

- **Agricultural Residues**: Sugar cane and sweet sorghum bagasse, corn stover (leaves, stalks, cobs, and husks), wheat, rice and barley straw, rice hulls, nut hull, fruit stones, branches, bark, and leaves
- **Livestock Residues**: Animal manure (from chicken, cattle, and pigs), dead livestock, and its nonedible parts
- **Forestry Residues and Wastes**: hardwood and softwood post-harvest waste, unusable thin branches, and removed dead and dying trees
- **Industrial Wastes**: post-production waste – waste paper, waste cardboard, wood waste, yard waste, and sawdust
- **Domestic and Municipal Wastes**: Sewage sludge, municipal solid waste (MSW), residential organic waste, and garden waste
- Food waste and food processing waste

The chemical composition of the biomass varies based on its type, source, geographical location, seasonal variation, and quality. The overall composition of biomass used for the biorefinery process includes 75% of cellulose, starch, and sucrose, 20% of lignin, and 5% of oil and other components. These diversities directly attribute immense research in this area to standardize and optimize the biorefinery process at each level.

4.4.2 Sugar Utilization

The genetic makeup of the microorganisms determines their enzyme production and combination and ultimately determines their metabolic pathway. The major components of hydrolysate of the biomass are C5 and C6 sugars, sugar alcohols, sugar acids, and lignin. Genetic and metabolic engineering aims at widening the substrate utilization spectrum and utilization efficiency of the microorganism. To increase the efficiency of saccharification and fermentation, consortia of potent microorganisms are used in the biorefinery processes. Consortia of C5 and C6 carbon metabolizing microorganisms are co-fermented for complete utilization of carbon residue in the hydrolysate (Francois et al. 2020).

4.4.3 Lignin Detoxification

During the biorefinery process, lignin acts as an important inhibitor in sugar utilization. The main aim of biomass pretreatment is to degrade lignin and expose sugar moieties. Lignin makes up to 40% of biomass composition, and it is comprised of aromatic alcohols, namely, coniferyl, sinapyl, and coumaryl alcohols. Enzymes such as manganese peroxidase, lignin peroxidase, versatile peroxidase, dye-decolorizing peroxidase, and laccase derived from *Phanerochaete chrysosporium*, *Pleurotus* sp., *Phlebia radiata*, and *Sphingomonas paucimobilis* are commonly used for lignin detoxification (Zhang et al. 2015).

4.5 INNOVATIONS IN BIOREFINERY

4.5.1 Integrated Bioprocessing in the Production of Value-Added Chemicals

With the advancing research in microbial processes, industries are looking forward to adopting strategies for the valorization of biomass and waste, which has led to the emergence of the biorefinery sector (Fava et al. 2015). To maintain sustainability, an integrated biorefinery concept is essential. These biorefineries target zero waste production by employing recycling/reusing of resources and integrating different technologies into one platform (Nizami et al. 2017). Integrated biorefineries exploit multiple components of a feedstock and intermediate products to produce multiple products. A biorefinery model will be more viable if integrated strategies are adopted. Integrated technology focuses not just on the primary refining of raw materials but also goes further to secondary refining, thereby utilizing the raw materials in an optimized way. In this way, waste generation can be minimized, and the maximum profit can be achieved through the production of a variety of products, and hence a sustainable supply chain will be ensured (Schieb et al. 2015). For instance, Yu et al. (2012) have used sweet sorghum stems to produce multiple products like butanol, ethanol, and wood–plastic composites targeting the maximum utilization of biomass. The biorefinery sector is emerging as an advantageous competitor to oil refineries, as products with similar applications are being produced, employing multiple conversion technologies that target to minimize environmental impact, production cost, and energy consumption.

4.5.2 Immobilization and Recycling of Cells and Enzymes

Cell immobilization is gaining interest in biorefineries, as it has several benefits over traditional cell systems. With the possibility of cells being reused in prolonged fermentation, cell immobilization techniques give more yield, since the substrate uptake is greater and there is minimal cell washout (Wirawan et al. 2020; Sekoai et al. 2018; Khammee et al. 2021). It also decreases the fermentation time, as the initial cell concentration used is high. Immobilization also makes the recovery and downstream processing easier, and hence will be favorable in reducing the cost of production. It should be noted that the carriers used for immobilizing cells are not toxic to the cells. The development of cheaper carriers with high mechanical strength and stability is necessary. Carriers based on chitosan, cellulose, and calcium alginate are widely used for cell immobilization (Sebayang et al. 2019). Recently, Kyriakou et al. (2020) have reported enhanced ethanol production using Saccharomyces cerevisiae cells immobilized on biochar-based carriers.

Biorefineries utilizing lignocellulosic biomass require enzymes such as cellulase and hemicellulase to get monosaccharide sugars, which can further be fermented to produce various products. These enzymes can be recycled from different process streams such as the hydrolysate, the broth after fermentation is carried out, and the stillage (Jørgensen and Pinelo 2017). Before enzymatic hydrolysis, to break the linkage between lignin and other structural constituents of the plant cell wall, viz., cellulose and hemicellulose, a pretreatment step is needed. A pretreatment step would ensure that the enzymes have better access to the cellulose/hemicellulose fraction, thereby achieving maximum conversion into products (Bhatia et al. 2020).

Though several developments are being made on enzyme recycling for the hydrolysis of lignocellulosic biomass, there are some challenges pertinent to its industrial applications, which include the presence of inhibitors and insoluble lignin in the hydrolysate, stability of the enzymes, and the requirement of different enzymes (Jørgensen and Pinelo 2017). Enzyme immobilization is an advancing technique in lignocellulosic biomass hydrolysis that offers lower cost, higher stability, and easier reusability of enzymes that can potentially boost the production of biofuels and other value-added products in biorefineries (Zanuso et al. 2021).

4.5.3 Yeast Cell Surface Display Systems

Cell surface display is a technique that involves the expression of a protein of interest on a cell surface by fusing the protein genetically to an anchor protein on the surface of the cell. Yeast cell surface display systems employ the fusion of lignocellulose degrading enzymes on *Saccharomyces cerevisiae* cell membrane through a metabolic engineering approach. Hence, through yeast cell surface display systems, consolidated bioprocessing can be accomplished as hydrolysis and fermentation can be done in a single step (Tabañag et al. 2018). Co-displaying different enzymes that are required for cellulose or hemicellulose hydrolysis has been found to enhance fermentation (Kotaka et al. 2008). Yeast cell surface display systems are not only applicable for cellulose hydrolysis but also for degrading xylan by expressing xylanase enzyme on the cell surface (Fujita et al. 2002; Katahira et al. 2004). Yeast cell surface display engineering finds its applications in waste biorefinery as well (Liu et al. 2016). More focus on identifying new anchor proteins and advances in the production of heterologous proteins are required for the success of these whole-cell biocatalysts.

4.6 BIOECONOMY

4.6.1 Biorefineries for a Sustainable Future

Many industries are looking forward to utilizing biomass for the production of biofuels and other chemicals owing to its carbon neutrality. Moreover, biomass is an advantageous raw material, as it is a renewable resource (Ubando et al. 2020). India is a country that has the potential to play a key role in the bio-based economy owing to its large geographical area and high biomass availability. The bioeconomy of India has had steady growth through the years, and it reached 70.2 billion dollars in 2020. By 2025, it is projected to grow to 150 billion dollars (BIRAC 2021). Sustainable production of biofuels and other value-added products can be achieved if the biomass resources are utilized appropriately. In the production of chemicals and fuels, there is a necessity to shift to a circular economy to reduce the environmental impacts caused by the existing industries, especially the developing countries. Through a circular economic approach, biorefineries can produce bioproducts by maximizing the utilization of resources through recycling/reusing and minimizing waste generation (Liu et al. 2021). This can be applied throughout the supply chain starting from raw materials, products, processes, and energy. New biorefineries can create a boost in the economy of a country by establishing direct and indirect job opportunities, thereby leading to a sustainable society (Domac et al. 2005).

4.6.2 Biorefineries—The Future of Bioeconomy

Among the bio-based products, more attention is being paid to bioethanol production. The global demand for bioethanol, which was 33.7 billion USD in 2020, is expected to reach 64.8 billion USD by 2025 (Markets and Markets 2021). The increasing demand and interest in bioethanol production can be attributed to the fuel blending policies adopted by many countries. For instance, the ethanol blending program in India aims at achieving 20% ethanol blending by 2025. India's import of petroleum in 2020 was 185MMT. Ethanol blending would minimize such fuel imports. Furthermore, ethanol has less carbon emission, thereby reducing its harmful impact on the environment (Ministry of Petroleum and Natural Gas, Government of India 2021).

The growth of the bioeconomy is not just marked by biofuels but also by other platform chemicals. According to a report by the Department of Energy of the United States, biomass can be used to produce 12 important building blocks that are employed to produce a variety of bio-based products that can replace petroleum-based chemicals (Werpy and Peterson 2004). Though it took more than a decade for these innovations to be recognized, these molecules have the potential to contribute significantly to the growth of the bio-based economy shortly (American Chemical Society 2017). Sustainability is the key to any bio-based economy. Government, policymakers, and stakeholders should consider implementing policies that will have a less negative effect on the economy, society, and environment, thereby benefitting the growth of biorefineries.

4.7 SUMMARY

The utilization of biomass by biorefineries is highly advantageous as biofuels, and other value-added chemicals can be produced sustainably and can potentially replace fossil fuel-based products. In this chapter, we have discussed various major products that can be produced in biorefineries such as biofuels, organic acids, glycerol, isoprene, microbial lipids, and sugar alcohols and their applications. The main aim of biorefineries is to produce products with a reduction in environmental impacts and with zero waste products so that sustainable development can be achieved. However, there are some obstacles and challenges faced by the biorefining industries that need to be addressed to help valorize biomass efficiently. This chapter also outlines some new developments and strategies applicable to biorefineries and how biorefineries can contribute to the bioeconomy. Further advancement in research to overcome the challenges faced in the valorization of biomass and waste and the promotion of appropriate policies that promote sustainability will lead to a successful bioeconomy through biorefineries.

BIBLIOGRAPHY

Ahmed, I. I., and Gupta, A. K. 2010. Pyrolysis and gasification of food waste: Syngas characteristics and char gasification kinetics. *Appl. Energy* 87:1. doi: 10.1016/j.apenergy.2009.08.032.

Alibardi, L., and Cossu, R. 2015. Composition variability of the organic fraction of municipal solid waste and effects on hydrogen and methane production potentials. *J. Waste Manag.* 36:147–155 doi: 10.1016/j.wasman.2014.11.019.

Alibardi, L., and Cossu, R. 2016. Effects of carbohydrate, protein and lipid content of organic waste on hydrogen production and fermentation products. *Waste Manag.* 47:69–77 doi: 10.1016/j.wasman.2015.07.049.

Alonso, S., Rendueles, M., and Dı´az, M. 2014. Microbial production of specialty organic acids from renewable and waste materials. *Crit. Rev. Biotechnol.* 1–17. doi: 10.3109/07388551.2014.904269.

Altaras, N. E., and Cameron, D. C. 2000. Enhanced production of (R)-1,2-propanediol by metabolically engineered *Escherichia coli. Biotechnol. Prog.* 16:940–946 doi: 10.1021/bp000076z.

Amorim, N. C. S., Alves, I., Martins, J. S., and Amorim, E. L. C. 2014. Biohydrogen production from cassava wastewater in an anaerobic fluidized bed reactor. *Braz. J. Chem. Eng.* 31:603–612 doi: 10.1590/0104-6632.20140313s00002458.

Balan, V. 2014. Current challenges in commercially producing biofuels from lignocellulosic biomass. *ISRN Biotechnol.* 2014:1–31 doi:10.1155/2014/463074.

Baritugo, K. A. G., Kim, H. T., David, Y. C., et al. 2018. Recent advances in metabolic engineering of *Corynebacterium glutamicum* as a potential platform microorganism for biorefinery. *Biofuels, Bioprod. Bioref.* 12:5. doi: 10.1002/bbb.1895.

Bhatia, S. K., Jagtap, S. S., Bedekar, A. A., et al. 2020. Recent developments in pretreatment technologies on lignocellulosic biomass: Effect of key parameters, technological improvements, and challenges. *Bioresour. Technol.* 300:122724. doi: 10.1016/j.biortech.2019.122724.

Bioethanol Market. 2021. Markets and markets. Accessed 23 September 2021 https://www.marketsandmarkets.com/Market-Reports/bioethanol-market-131222570.html.

BIRAC (Biotechnology Industry Research Assistance Council). 2018. Assessing the regional competitiveness of the Indian bioeconomy. BIRAC (Biotechnology Industry Research Assistance Council). Accessed 23 September 2021. https://birac.nic.in/webcontent/1594624859_Regional_Competitiveness_of_Indian_Bioeconomy.pdf.

Cabeza, I., Acosta, M., and Hernandez, M. 2017. Evaluation of the biochemical methane potential of pig manure, organic fraction of municipal solid waste and cocoa industry residues in Colombia. *Chem. Eng. Trans.* 57:55–60 doi: 10.3303/CET1757010.

Cai, D., Dong, Z., Wang, Y., et al. 2016. Co-generation of microbial lipid and bio-butanol from corn cob bagasse in an environmentally friendly biorefinery process. *Bioresour. Technol.* 216:345–351 doi: 10.1016/j.biortech.2016.05.073.

Cai, D., Zhang, T., Zheng, J., et al. 2013. Biobutanol from sweet sorghum bagasse hydrolysate by a hybrid pervaporation process. *Bioresour. Technol.* 145:97–102 doi: 10.1016/j.biortech.2013.02.094.

Calvo, L.F., García, A.I., and Otero, M. 2013. An experimental investigation of sewage sludge gasification in a fluidized bed reactor. *Sci. World J.* 2013:1–8 doi: 10.1155/2013/479403.

Carr, S., Aldridge, J., and Buan, N. R. 2021. Isoprene production from municipal wastewater biosolids by engineered archaeon *Methanosarcina acetivorans*. *Appl. Sci.* 11:3342. doi: 10.3390/app11083342.

Cesaro, A., and Belgiorno, V. 2015. Combined biogas and bioethanol production: Opportunities and challenges for industrial application. *Energies* 8:8121–8144 doi:10.3390/en8088121.

Chang, J. J., Ho, C. Y., Mao, C. T., et al. 2014. A thermo- and toxin-tolerant kefir yeast for biorefinery and biofuel production. *Appl. Energy* 132:465–474 doi: 10.1016/j.apenergy.2014.06.08.

Chang, Y. H., Chang, K. S., Hsu, C. L., 2013. A comparative study on batch and fed-batch cultures of oleaginous yeast *Cryptococcus* sp. In glucose-based media and corncob hydrolysate for microbial oil production. *Fuel* 105:711–717 doi: 10.1016/j.fuel.2012.10.033.

Chen, Y., Fu, B., Xiao, G., et al. 2021. Bioconversion of lignin-derived feedstocks to muconic acid by whole-cell biocatalysis. *ACS Food Sci. Technol.* 1:382–387. doi: 10.1021/acsfoodscitech.1c00023.

Chen, Y., Li, X., Zheng, X., and Wang, D. 2013. Enhancement of propionic acid fraction in volatile fatty acids produced from sludge fermentation by the use of food waste and *Propionibacterium acidipropionici*. *Water Res.* 47:615–22 doi: 10.1016/j.watres.2012.10.035.

Corneli, E., Dragoni, F., Adessi, A., Philippis, R. D., Bonari, E., and Ragaglini, G. 2016. Energy conversion of biomass crops and agro-industrial residues by combined biohydrogen/biomethane system and anaerobic digestion. *Bioresour. Technol.* 211:509–518 doi: 10.1016/j.biortech.2016.03.134.

Da Cunha, M. E., Krause, L. C., Moraes, M. S. A., et al. 2009. Beef tallow biodiesel produced in a pilot scale. *Fuel Process. Technol.* 90:570–575 doi:10.1016/j.fuproc.2009.01.001.

Demirbas, A. 2006. Alternative fuels for transportation. *Energy Explor. Exploit.* 24:45–54. doi:10.1260/014459806779387985.

Domac, K., Richards, S., and Risovic, S. 2005. Socio-economic drivers in implementing bioenergy projects. *Biomass Bioenergy*. 28:97–106 doi: 10.1016/j.biombioe.2004.08.002.

Du, W., Xu, Y., and Liu, D. 2003. Lipase-catalysed transesterification of soya bean oil for biodiesel production during continuous batch operation. *Biotechnol. Appl. Biochem.* 38:103. doi:10.1042/ba20030032.

Dürre, P. 2007. Biobutanol: An attractive biofuel. *Biotechnol. J.* 2:1525–1534 doi:10.1002/biot.200700168.

Ezeji, T.C., Qureshi, N., and Blaschek, H.P. 2007. Production of acetone butanol (AB) from liquefied corn starch, a commercial substrate, using *Clostridium beijerinckii* coupled with product recovery by gas stripping. *J. Ind. Microbiol. Biotechnol.* 34:771–777 doi: 10.1007/s10295-007-0253-1.

Faleh, N., Khila, Z., Wahada, Z., Pons, M.-N., Houas, A., and Hajjaji, N. 2018. Exergoenvironmental life cycle assessment of biodiesel production from mutton tallow transesterification. *Renew. Energy.* 127:74–83. doi:10.1016/j.renene.2018.04.046.

Fava, F., Totaro, G., Diels, L., et al. 2015. Biowaste biorefinery in Europe: Opportunities and research & development needs. *New Biotechnol.* 32:100–108. doi: 10.1016/j.nbt.2013.11.003.

Francois, J. M., Alkim, C., and Morin, N. 2020. Engineering microbial pathways for production of bio-based chemicals from lignocellulosic sugars: Current status and perspectives. *Biotechnol. Biofuels* 13:118. doi: 10.1186/s13068-020-01744-6.

Fujita, Y., Katahira, S., Ueda, M., 2002. Construction of whole-cell biocatalyst for xylan degradation through cell-surface xylanase display in *Saccharomyces cerevisiae. J. Mol. Catal. B Enzym.* 17:189–195. doi: 10.1016/S1381-1177(02)00027-9.

Gabriel, C. L., and Crawford, F. M. 1930. Development of the butyl-acetonic fermentation industry. *Ind. Eng. Chem. Res.* 22:1163–1165 doi:10.1021/ie50251a014.

García, V., Päkkilä, J., Ojamo, H., Muurinen, E., and Keiski, R. L. 2011. Challenges in biobutanol production: How to improve the efficiency? *Renew. Sustain. Energy Rev.* 15:964–980 doi:10.1016/j.rser.2010.11.008.

García Prieto, C. V., Ramos, F. D., Estrada, V., Villar, M. A. and Diaz, M. S. 2017. Optimization of an integrated algae-based biorefinery for the production of biodiesel, astaxanthin and PHB. *Energy.* 139:1159–1172 doi: 10.1016/j.energy.2017.08.036.

Gordillo, G., Annamalai, K., and Carlin, N. 2009. Adiabatic fixed-bed gasification of coal, dairy biomass, and feedlot biomass using an air steam mixture as an oxidizing agent. *Renew. Energy.* 34:2789e2797. doi: 10.1016/j.renene.2009.06.004.

Goshadrou, A., Karimi, K., and Taherzadeh, M. J. 2011. Bioethanol production from sweet sorghum bagasse by *Mucor hiemalis. Ind. Crops Prod.* 34:1219–1225 doi:10.1016/j.indcrop.2011.04.018.

Granados, M. L., Poves, M. D. Z., Alonso, D. M., et al. 2007. Biodiesel from sunflower oil by using activated calcium oxide. *Appl. Catal. B.* 73:317–326 doi:10.1016/j.apcatb.2006.12.017.

Gronchi, N., Favaro, L., Cagnin, L., 2019. Novel yeast strains for the efficient saccharification and fermentation of starchy by-products to bioethanol. *Energies* 12:714. doi: 10.3390/en12040714.

Guo, J., Li, J., Chen, Y., Guo, X., and Xiao, D. 2016. Improving erythritol production of *Aureobasidium pullulans* from xylose by mutagenesis and medium optimization. *Appl Biochem. Biotechnol.* 180:717–727 doi: 10.1007/s12010-016-2127-3.

Habe, H., Fukuoka, T., Kitamoto, D., and Sakaki, K. 2009. Biotechnological production of D-glyceric acid and its application. *Appl. Microbiol. Biotechnol.* 84:445–452 doi: 10.1007/s00253-009-2124-3.

Harun, R., Jason, W. S. Y., Cherrington, T., and Danquah, M. K. 2011. Exploring alkaline pre-treatment of microalgal biomass for bioethanol production. *Appl. Energy.* 88:3464–3467. doi:10.1016/j.apenergy.2010.10.048.

Hassan, A. H. S., Mietzel, T., Brunstermann, R., et al. 2018. Fermentative hydrogen production from low-value substrates. *World J. Microbiol. Biotechnol.* 34:176 doi:10.1007/s11274-018-2558-9.

Heidenreich, S., and Foscolo, P. U. 2015. New concepts in biomass gasification. *Prog. Energy Combust. Sci.* 46:72–95 doi: 10.1016/j.pecs.2014.06.002.

Hernawan, M., Maryana, R., Pratiwi, D., et al. 2017. Bioethanol production from sugarcane bagasse by simultaneous saccharification and fermentation using *Saccharomyces cerevisiae. AIP Conference Proceedings.* 1823:020026. doi: 10.1063/1.4978099.

Hidalgo, A. M. L., Magana, G., Rodriguez, F., Rodriguez, A. D. L., and Sanchez, A. 2021. Co-production of ethanol-hydrogen by genetically engineered *Escherichia coli* in sustainable biorefineries for lignocellulosic ethanol production. *Chem. Eng. J.* 406:126829 doi: 10.1016/j.cej.2020.126829.

INDIA: National Mission on Biodiesel. 2013. The Asian and pacific energy forum 2013. Accessed 15 September 2021. https://policy.asiapacificenergy.org/node/2777.

Ismail, I., Hassan, M. A., Rahman, N. A., and Chen, S. S. 2010. Thermophilic biohydrogen production from palm oil mill effluent (POME) using suspended mixed culture. *Biomass Bioenerg*. 34:42–47 doi: 10.1016/j.biombioe.2009.09.009.

Issariyakul, T., and Dalai, A. K. 2010. Biodiesel production from green seed canola oil. *Energy Fuels* 24:4652–4658 doi:10.1021/ef901202b.

Janchiv, A., Oh, Y., and Choi, S. 2012. High quality biodiesel production from pork lard by high solvent additive. *Sci. Asia*. 38:95–101. doi:10.2306/scienceasia1513–1874.2012.38.095.

Ji, X. J., Huang, H., Li, S., Du. J., and Lian, M. 2008. Enhanced 2,3-butanediol production by altering the mixed acid fermentation pathway in *Klebsiella oxytoca. Biotechnol Lett*. 30:731–734 doi: 10.1007/s10529-007-9599-8.

Jiang, M., Zhang, W., Yang, Q., et al. 2019. Metabolic regulation of organic acid biosynthesis in *Actinobacillus succinogenes. Front. Bioeng. Biotechnol*. 7:216. doi: 10.3389/fbioe.2019.00216.

Jiang, W., Zhao, J., Wang, Z., and Yang, S.T. 2014. Stable high-titer n-butanol production from sucrose and sugarcane juice by *Clostridium acetobutylicum* JB 200 in repeated batch fermentations. *Bioresour. Technol*. 163:172–179 doi: 10.1016/j.biortech.2014.04.047.

Jørgensen, H. and Pinelo, M. 2017. Enzyme recycling in lignocellulosic biorefineries. *Biofpr*. 11:150–167 doi: 10.1002/bbb.1724.

Júnior, A. D. N. F., Wenzel, J., Etchebehere, C., and Zaiat, M. 2014. Effect of organic loading rate on hydrogen production from sugarcane vinasse in thermophilic acidogenic packed bed reactors. *Int. J. Hydrog. Energy*. 39:16852–16862 doi: 10.1016/j.ijhydene.2014.08.017.

Kabbour, M., and Luque, R. 2020. Furfural as a platform chemical: From production to applications. *Recent Adv. Develop. Plat. Chem*. 283–297 doi: 10.1016/B978-0-444-64307-0.00010-X.

Kalam, M., and Masjuki, H. 2002. Biodiesel from palm oil- an analysis of its properties and potential. *Biomass Bioenergy*. 23:471–479. doi:10.1016/s0961-9534(02)00085-5.

Katahira, S., Fujita, Y., Mizuike, A., Fukuda, H., and Kondo, A. 2004. Construction of a xylan-fermenting yeast strain through codisplay of xylanolytic enzymes on the surface of xylose-utilizing *Saccharomyces cerevisiae* cells. *Appl. Environ. Microbiol*. 70:5407–5414 doi: 10.1128/AEM.70.9.5407-5414.2004.

Khammee, P., Ramaraj, R., Whangchai, N., Bhuyar, P., and Unpaprom, Y. 2021. The immobilization of yeast for fermentation of macroalgae Rhizoclonium sp. for efficient conversion into bioethanol. *Biomass Convers. Biorefin*. 11:827–835 doi:10.1007/s13399-020-00786-y.

Kotaka, A., Bando, H, Kaya M, et al. 2008. Direct ethanol production from barley beta-glucan by sake yeast displaying *Aspergillus oryzae* beta-glucosidase and endoglucanase. *J. Biosci. Bioeng*. 105:622–627 doi: 10.1263/jbb.105.622.

Kövilein, A., Kubisch, C., Cai, L., and Ochsenreither, K. 2019. Malic acid production from renewables: A review. *J. Chem. Technol. Biotechnol*. 95:513–526 doi: 10.1002/jctb.6269.

Kyriakou, M., Patsalou, M., Xiaris, N., et al. 2020. Enhancing bioproduction and thermotolerance in *Saccharomyces cerevisiae* via cell immobilization on biochar: Application in a citrus peel waste biorefinery. *Renew. Energy*. 155:53–64. doi: 10.1016/j.renene.2020.03.087.

Lebeau, J., Efromson, J. P., and Lynch, M. D. 2020. A review of the biotechnological production of methacrylic acid. *Front. Bioeng. Biotechnol*. 8:207. doi: 10.3389/fbioe.2020.00207.

Lee, H. W., Park, J. H., Lee, H. S., et al. 2019. Production of bio-based isoprene by the mevalonate pathway cassette in *Ralstonia eutropha. J. Microbiol. Biotechnol*. 29:1656–1664 doi: 10.4014/jmb.1909.09002.

Lee, W. C., and Kuan, W. C. 2015. *Miscanthus* as cellulosic biomass for bioethanol production. *Biotechnol. J*. 10:840–854. doi:10.1002/biot.201400704.

Li, Y. C., Chu, C. Y., Wu, S. Y., et al. 2012. Feasible pretreatment of textile wastewater for dark fermentative hydrogen production. *Int. J. Hydrog. Energy*. 37:15511–15517 doi: 10.1016/j.ijhydene.2012.03.131.

Liao, J. C., Mi, L., Pontrelli, S., and Luo, S. 2016. Fuelling the future: Microbial engineering for the production of sustainable biofuels. *Nat. Rev. Microbiol.* 14:288–304. doi:10.1038/nrmicro.2016.32.

Limongi, A.R., Viviano, E., De Luca, M., Radice, R.P., Bianco, G., and Martelli, G. 2021. Biohydrogen from microalgae: Production and applications. *Appl. Sci.* 11:1616. doi: 10.3390/ app11041616.

Liu, Y., Lyu, Y., Tian, J., Zhao, J., Ye, N., Zhang, Y., and Chen, L. 2021. Review of waste biorefinery development towards a circular economy: From the perspective of a life cycle assessment. *Renew. Sustain. Energy Rev.* 139:110716. doi: 10.1016/j.rser.2021.110716.

Liu, Z., Ho, S. H., Hasunuma, T., Chang, J. S., Ren, N. Q. and Kondo, A. 2016. Recent advances in yeast cell-surface display technologies for waste biorefineries. *Bioresour. Technol.* 215:324–333 doi: 10.1016/j.biortech.2016.03.132.

Lu, C.C., Zhao, J.B., Yang, S.T., and Wei, D. 2012. Fed-batch fermentation for n-butanol production from cassava bagasse hydrolysate in a fibrous bed bioreactor with continuous gas stripping. *Bioresour. Technol.* 104:380–387 doi: 10.1016/j.biortech.2011.10.089.

lwinstel. 2017. "Top value added chemicals: The biobased economy 12 years later" American Chemical Society, 16 March. Accessed 24 September 2021. https://communities.acs.org/t5/GCI-Nexus-Blog/Top-Value-Added-Chemicals-The-Biobased-Economy-12-Years-Later/ba-p/15759.

Maity, S. K. 2015. Opportunities, recent trends and challenges of integrated biorefinery: Part I. *Renew. Sustain. Energy Rev.* 43:427–1445 doi: 10.1016/j.rser.2014.11.092.

Mazanov, S. V., Gabitova, A. R., Usmanov, R. A., et al. 2016. Continuous production of biodiesel from rapeseed oil by ultrasonic assist transesterification in supercritical ethanol. *J. Supercrit. Fluids.* 118:107–118 doi:10.1016/j.supflu.2016.07.009.

Ministry of Petroleum and Natural Gas, Government of India. 2021. Roadmap for ethanol blending in India 2020–25. Accessed 24 September 2021. https://mopng.gov.in/files/uploads/Ethanol_blending_in_India_15072021bnew.pdf.

Meng, X., Yang, J., Xu, X., Zhang, L., Nie, Q. and Xian, M. 2009. Biodiesel production from oleaginous microorganisms. *Renew. Energy.* 34:1–5 doi: 10.1016/j.renene.2008.04.014.

Mohan, S. V., Babu, V. L., and Sarma, P. N. 2007. Anaerobic biohydrogen production from dairy wastewater treatment in sequencing batch reactor (AnSBR): Effect of organic loading rate. *Enzyme Microb. Technol.* 41:506–515 doi: 10.1016/j.enzmictec.2007.04.007.

Mookherjee, I., Maity, A., Hasan, M., Ahmed, F., and Sinha, P. K. 2017. Investigation of biogas generation from food waste. *Int. J. Innov. Res. Sci. Eng. Technol.* 6:6. doi:10.15680/IJIRSET.2017.0606012.

Morone, A., Apte, M., and Pandey, R. A. 2015. Levulinic acid production from renewable waste resources: Bottlenecks, potential remedies, advancements and applications. *Renew. Sustain. Energy Rev.* 51:548–565 doi: 10.1016/j.rser.2015.06.032.

Naerdal, I., Pfeifenschneider, J., Brautaset, T., and Wendisch, V.F. 2015. Methanol-based cadaverine production by genetically engineered *Bacillus methanolicus* strains. *Microb. Biotechnol.* 8:342–350. doi:10.1111/1751-7915.12257.

Naik, S. N., Goud, V. V., Rout, P. K., and Dalai, A. K. 2010. Production of first and second generation biofuels: A comprehensive review. *Renew. Sustain. Energy Rev.* 14:578–597 doi:10.1016/j.rser.2009.10.003.

National Energy Technology Laboratory. 2021. Syngas composition. Accessed 15 August 2021. https://netl.doe.gov/research/coal/energy-systems/gasification/gasifipedia/syngas-composition#:~:text=This%20can%20vary%20significantly%20depending,amounts%20of%20the%20sulfur%20compounds.

Nielsen, J. R. R. 2002. Syngas in perspective. *Catal. Today.* 71:243–247 doi:10.1016/s0920-5861(01)00454-0.

Nimcevic, D., and Gapes, J. R. 2000. The acetone–butanol fermentation in pilot plant and pre-industrial scale. *J. Mol. Microbiol. Biotechnol.* 2:15–20.

Nitsos, C. K., Matis, K. A., and Triantafyllidis, K. S. 2012. Optimization of hydrothermal pretreatment of lignocellulosic biomass in the bioethanol production process. *ChemSusChem*. 0000:1–14 doi: 10.1002/cssc.201200546.

Nizami, A. S., Rehan, M., Waqas, M., et al. 2017. Waste biorefineries: Enabling circular economies in developing countries. *Bioresour. Technol*. 241:1101–1117. doi: 10.1016/j.biortech.2017.05.097.

Ong, Z., Cheng, Y., Maneerung, T., et al. 2015. Cogasification of woody biomass and sewage sludge in a fixed-bed downdraft gasifier. *AIChE J*. 61:8. doi: 10.1002/aic.14836.

Panchal, B. M., Deshmukh, S. A., and Sharma, M. R. 2017. Production and kinetic transesterification of biodiesel from yellow grease with dimethyl carbonate using methanesulfonic acid as a catalyst. *Environ. Prog. Sustain. Energy*. 36:802–807. doi:10.1002/ep.12559.

Park, Y. C., Oh, E. J., Jo, J. H, Jin, Y. S., and Seo, J. H. 2016. Recent advances in biological production of sugar alcohols. *Curr. Opin. Biotechnol*. 37:105–113 doi: 10.1016/j.copbio.2015.11.006.

Pasha, K. M., Akram, A., Narasimhulu, K., and Kodandaramaiah, G. N. 2015. Waste management by anaerobic digestion of kitchen waste- A review. *Int. J. Innov. Res. Sci. Eng. Technol*. 4:3. doi: 10.15680/IJIRSET.2015.0403029.

Pleissner, D., Dietz, D., Duuren, J. B.J.H.V., et al. 2017. Biotechnological production of organic acids from renewable resources. *Adv. Biochem. Eng. Biotechnol*. 166:373–410 doi: 10.1007/10_2016_73.

Pradyawong, S., Juneja, A., Sadiq, M. B., Noomhorm, A., and Singh, V. 2018. Comparison of cassava starch with corn as a feedstock for bioethanol production. *Energies*. 11:3476. doi:10.3390/en11123476.

Qureshi, N., Li, X. L., Hughes, S., Saha, B. C. and Cotta, M. A. 2006. Butanol production from corn fiber xylan using *Clostridium acetobutylicum*. *Biotechnol. Prog*. 22:673–680 doi: 10.1021/bp050360w.

Rahman, M. A. A., Tashiro, Y., and Sonomoto, K. 2011. Lactic acid production from lignocellulose-derived sugars using lactic acid bacteria: Overview and limits. *J. Biotechnol*. 156:286– 301. doi:10.1016/j.jbiotec.2011.06.017.

Rice, T., Zannini, E., Arendt, E. K., and Coffey, A. 2019. A review of polyols – biotechnological production, food applications, regulation, labelling and health effects. *Crit. Rev. Food Sci. Nutr*. 60:2034–2051. doi:10.1080/10408398.2019.1625859.

Ruan, Z., Zanotti, M., Zhong, Y., Liao, W., Ducey, C., and Liu, Y. 2013. Cohydrolysis of lignocellulosic biomass for microbial lipid accumulation. *Biotechnol. Bioeng*. 110:1039–1049 doi: 10.1002/bit.24773.

Saini, J. K., Saini, R., and Tewari, L. 2014. Lignocellulosic agriculture wastes as biomass feedstocks for second-generation bioethanol production: Concepts and recent developments. *3 Biotech*. 5:337–353. doi:10.1007/s13205-014-0246-5.

Sathianachiyar, S., and Devaraj, A. 2013. Biopolymer production by bacterial species using glycerol, a byproduct of biodiesel. *Int. J. Sci. Res*. 3:1–5.

Schieb, P. A., Lescieux-Katir, H., Thénot, M. and Clément-Larosière, B. 2015. An original business model: The integrated biorefinery. In *Biorefinery 2030*, edited by Schieb, P.A, 25–56. Berlin: Springer. doi: 10.1007/978-3-662-47374-0_2.

Sebayang, F., Bulan, R., Hartanto, A. and Huda, A. 2019. Enhancing the efficiency of ethanol production from molasses using immobilized commercial *Saccharomyces cerevisiae* in two-layer alginate-chitosan beads, in *IOP Conference Series: Earth and Environmental Science*, (IOP Publishing), 012014.

Sekoai, P. T., Awosusi, A. A., Yoro, K. O., et al. 2018. Microbial cell immobilization in biohydrogen production: A short overview. *Crit. Rev. Biotechnol*. 38:157–171 doi: 10.1080/07388551.2017.1312274.

Speight, J. G. 2015. *Gasification Processes for Syngas and Hydrogen Production*. Gasification for Synthetic Fuel Production, Woodshield publishing, 119–146. doi:10.1016/b978-0-85709-802-3.00006-0.

Sulaiman, S. A., Roslan, R., Inayat, M., and Yasin Naz, M. 2018. Effect of blending ratio and catalyst loading on co-gasification of wood chips and coconut waste. *J. Energy Inst.* 91:779–785 doi: 10.1016/j.joei.2017.05.003.

Tabañag, I. D. F., Chu, I., Wei, Y. H. and Tsai, S. L. 2018. The role of yeast-surface-display techniques in creating biocatalysts for consolidated bioprocessing. *Catalysts* 8:94. doi: 10.3390/catal8030094.

Taherzadeh, M. J., Lennartsson, P. R., Teichert, O., and Nordholm, H. 2013. *Bioethanol Production Processes*. Biofuels Production, Scrivener Publishing LLC, 211–253. doi:10.1002/9781118835913.ch8.

Takkellapati, S., Li, T., and Gonzalez, M. A. 2018. An overview of biorefinery derived platform chemicals from a cellulose and hemicellulose biorefinery. *Clean Technol Environ Policy*. 20:1615–1630. doi:10.1007/s10098-018-1568-5.

Ubando, A. T., Felix, C. B. and Chen, W. H. 2020. Biorefineries in circular bioeconomy: A comprehensive review. *Bioresour. Technol.* 299:122585. doi: 10.1016/j.biortech.2019.122585.

Venkat Reddy, C. R., Oshel, R., and Verkade, J. G. 2006. Room-temperature conversion of soybean oil and poultry fat to biodiesel catalyzed by nanocrystalline calcium oxides. *Energy Fuels*. 20:1310–1314 doi:10.1021/ef050435d.

Werpy, T. and Petersen, G. 2004. *Top Value Added Chemicals from Biomass: Volume I: Results of Screening for Potential Candidates from Sugars and Synthesis Gas* (No. DOE/GO-102004-1992). National Renewable Energy Lab., Golden, CO. doi:10.2172/15008859. https://www.osti.gov/biblio/15008859.

Wirawan, F., Cheng, C. L., Lo, Y. C., et al. 2020. Continuous cellulosic bioethanol co-fermentation by immobilized *Zymomonas mobilis* and suspended *Pichia stipitis* in a two-stage process. *Appl. Energy*. 266:114871. doi: 10.1016/j.apenergy.2020.114871.

Yan, L., Cao, Y., Li, X., and He, B. 2018. Characterization of a dual fluidized bed gasifier with blended biomass/coal as feedstock. *Bioresour. Technol.* 254:97–106 doi: 10.1016/j.biortech.2018.01.067.

Yang, Z., Koh, S. K., Ng, W. C. et al. 2016. Potential application of gasification to recycle food waste and rehabilitate acidic soil from secondary forests on degraded land in Southeast Asia. *J. Environ. Manag.* 172:40–48 doi: 10.1016/j.jenvman.2016.02.020.

Yokoi, H., Saitsu, A., Uchida, H., Hirose, J., Hayashi, S., and Takasaki, Y. 2001. Microbial hydrogen production from sweet potato starch residue. *J. Biosci. Bioeng.* 91:58–63 doi: 10.1016/S1389–1723(01)80112-2.

Yu, J., Zhang, T., Zhong, J., Zhang, X. and Tan, T. 2012. Biorefinery of sweet sorghum stem. *Biotechnol. Adv.* 30:811–816. doi: 10.1016/j.biotechadv.2012.01.014.

Zanuso, E., Gomes, D. G., Ruiz, H. A., Teixeira, J. A. and Domingues, L. 2021. Enzyme immobilization as a strategy towards efficient and sustainable lignocellulosic biomass conversion into chemicals and biofuels: Current status and perspectives. *Sustain. Energy Fuels.* 5:4233–4247 doi: 10.1039/D1SE00747E.

Zhang, G. C., Liu, J. J., Kong, I. I., Kwak, S., and Jin, Y. S. 2015. Combining C6 and C5 sugar metabolism for enhancing microbial bioconversion. *Curr. Opin. Chem. Biol.* 29:49–57 doi: 10.1016/j.cbpa.2015.09.008.

Zhang, M. L., Fan, Y. T., Xing, Y., et al. (2007). Enhanced biohydrogen production from cornstalk wastes with acidification pretreatment by mixed anaerobic cultures. *Biomass Bioenerg.* 31:250–254 doi: 10.1016/j.biombioe.2006.08.004.

5 Challenges and Opportunities Associated with Second-Generation Biofuel

Saurabh Singh, Gowardhan Kumar Chouhan, Akhilesh Kumar, and Jay Prakash Verma
Banaras Hindu University

CONTENTS

5.1 INTRODUCTION

Second-generation biofuel production involves the use of biomass that is not in competition with the food crops for the production of fuels (Havlík et al., 2011). In general, it involves the use of lignocellulosic biomass for the production of fuels that are considered waste in agriculture and need to be disposed. Though with regard to carbon emissions, second-generation biofuels have a net positive carbon emission, but overall it is friendly compared to the emissions when burned (Mat Aron et al., 2020).

Second-generation biofuel production refers to the use of nonfood lignocellulosic biomass utilization for the production of biofuels (Naik et al., 2010). Agro-residues, food wastes and other lignocellulosic materials that are unfit for consumption directly or after processing can be considered under second-generation biofuel production. The rise in the production of second-generation biofuel production gained pace after the realization of the fact that first-generation biofuel feedstocks competed against the food materials that pose threat to food security with respect to the use of agricultural land as well as direct food demand (Gheewala et al., 2013; Singh et al., 2020b).

DOI: 10.1201/9781003197737-7

The ever-increasing population and their associated food demand continuously led to increase in the agriculture residue production. Thus, agricultural residue production increased consequently. Among the major crops that are grown as staple crops, such as wheat paddy and wheat, produce huge amounts of agricultural waste (Kumar et al., 2019; Rena et al., 2020). This agricultural waste is lignocellulosic in nature and can be harnessed to produce utilizable forms of fuel. The lignocellulosic biomass is rich in molecules with monomers of glucose units (cellulose) and pentose sugars (hemicellulose) (Zabed et al., 2016). However, the biomass is recalcitrant in nature and cannot be easily degraded by the microbes (Sharma et al., 2019). This recalcitrance of biomass is due to the presence of lignin in the biomass matrix, which is recalcitrant to most of the enzymes secreted by the microbes usually present in nature (Bomble et al., 2017; Singh et al., 2018b).

5.2 BIOMASS PRODUCTION

Though it has been estimated that second-generation biofuel production won't be able to achieve the total transportation fuel level production due to the biomass limit, the biomass limit makes it only 20–30 realistically achievable if sustainable biomass supply is considered for biofuel production (Inderwildi and King, 2009). The current second-generation biofuel production is compelled upon by the fact that huge amounts of lignocellulosic biomass is production on farmlands as a result of food production. Furthermore, the issues such as the burning of agro waste (agro biomass) to clear off the land implies that certain sustainable solutions need to be explored to handle the biomass produced in such large quantities. There are many studies indicating the health and environmental effects of stubble burning in India (Abdurrahman et al., 2020; Beig et al., 2020; Mittal et al., 2009; Satpathy and Pradhan, 2020; Singh et al., 2018a), United States (Dhammapala et al., 2006, 2007; Jimenez et al., 2007), Australia (Reisen et al., 2011), and Canada (Malhi and Kutcher, 2007). Further, the potential of biomass in energy production is described by many in the past for specific crops; Nigeria (Agbro and Ogie, 2012; Sokan-Adeaga and Ana, 2015), Italy (Gómez et al., 2017), Iran (Taghizadeh-Alisaraei et al., 2017), USA (Voigt et al., 2013), and Tanzania (Mwamila et al., 2009). Overall biomass potential studies have been described in the following reviews by (Chen et al., 2009; Gupta and Verma, 2015; Kumar et al., 2020; Simmons et al., 2008). The biomass can also be used for the production of different chemicals, which has been described in the reviews by (Bisaria, 1991; Ramesh et al., 2019; Rishikesh et al., 2021).

5.3 SECOND-GENERATION BIOFUEL CONVERSION TECHNOLOGIES

The lignocellulosic biomass conversion can be categorized into two basic types: thermochemical and enzymatic pathway. The thermochemical pathway involves the conversion of the biomass at various temperatures ranging between 150°C and 1,000°C, with or without the application of catalysts. In the enzymatic pathway, a broad three-step conversion takes place: pretreatment, cellulosic hydrolysis, and fermentation

(Singh and Verma, 2019). The pretreatment stage consists of the removal of lignin from the lignocellulosic matrix, followed by hydrolysis of the resultant biomass, and finally fermentation of the hexose and pentose sugars. The technologies employed in the deconstruction process of the lignocellulosic matrix result in emissions of varying proportions. The thermochemical pathways produce more emissions compared to enzymatic pathways. The biochemical or the enzymatic pathways have a slightly better performance than the thermochemical pathways in terms of greenhouse gas emissions and fossil fuel consumption, but thermochemical pathways are known for less water consumption, and better performance in environmental impact categories when operated as a biorefinery with mixed alcohol co-products (Singh et al., 2021). This is because of the credits associated with the co-products during the process (Mu et al., 2010). The biochemical pathways consume a lot of water, which supersedes the co-products credit produced during the process (Singh et al., 2021). Thus, thermochemical pathways have much less water consumption than biochemical pathways. Thermochemical conversions include processes such as gasification and pyrolysis of various types depending on the temperatures used.

In a study by Iribarren et al. (2013), Fischer Tropsch (FT) synthesis products produced less environmental emissions compared to fossil diesel, rapeseed biodiesel, soybean biodiesel, and other fossil alternatives. The abiotic depletion potential, global warming potential, ozone depletion potentials were significantly less in Fischer Tropsch synthesis when compared to their alternative counterparts' potentials. Mixed alcohol synthesis is a type of gasification that involves the compression of syngas before it combines with methanol. Another type of gasification is methanol to gasoline, in which methanol is produced with dimethyl ether and olefins as the intermediate products. In a study by Reno et al. (2011), it was observed that the abiotic depletion potential is little expressive, which is due to the fossil fuel consumption at the harvesting stage and transportation stage of the biomass. Human toxicity value increases due to the gasification of sugarcane bagasse, which emits many particulates. Eutrophication impact in biodiesel production from the same is ten times higher than methanol production due to the consumption of a high level of fertilizers during the cultivation of soybean. An improved version of methanol to gasoline is called syngas to distillates in which methanol dehydration and hydrocarbon formation takes place in a single reactor. In a study by Phillips et al. (2011), direct conversion of methanol to gasoline was achieved through a fluidized bed method in which methanol was firstly converted into dimethyl ether and then gasoline in a single reactor. This directly reduces the cost of another reactor required at the time of capital investment of the setup. Syngas fermentation is yet another type of gasification in which the syngas is cleaned from the contaminants and then allowed to ferment in the presence of microbes. The product obtained during the production of syngas fermentation directly depends upon the pretreatment of the biomass feedstock, gasifier used, gas cleaning process, and the microbe used for the fermentation. The other factors also play a key role in the product formation, such as the environment of the process, fuel synthesis. Syngas fermentation technology not only helps in the production of biofuels but also reduces greenhouse gas emissions by capturing the carbon content from the gasifier. Though this has been studied for quite a while now, its realization at the commercial level is yet to be achieved.

Pyrolysis can be categorized into three types: conventional pyrolysis, long duration of 5–30 minutes and temperatures of ranging between 400°C and 500°C; fast pyrolysis with comparatively lesser duration of 0.5–2.0 seconds and temperatures of 400°C–650°C; and flash pyrolysis or ultrafast pyrolysis that has a duration of less than 0.5 seconds and temperatures range between 700°C and 1,000°C. Pyrolysis since long has provided fuel energy and is considered to provide transportation fuels at the industrial level. A study by Hsu (2012) shows how pyrolysis fuels are better than fossil fuels. Greenhouse gas (GHG) emissions of 117 and 98 g/km and net energy values (NEVs) of 1.09 and 0.92 MJ/km for pyrolysis-derived gasoline and pyrolysis-derived diesel, respectively. Flash pyrolysis is a technique in which the biomass is subjected to high temperatures for a very short period. In another study, the biomass was converted into gases, solid biochar, and condensable pyrolytic vapor at high temperatures in the absence of oxygen (Chen et al., 2018). It was observed that the normalized values of the impact category showed meaningful results, in which GWP and effect of photochemical smog were significant, while other impacts were insignificant. This result suggests that fuel generated from flash pyrolysis is more ecofriendly than its fossil counterpart.

Biochemical pathways include pretreatment processes that can be physical, chemical, biological, or a combination of these. The physical pretreatment consists of temperature, pressure, and moisture variations, while chemical pretreatment involves the use of various acids and alkalis, and in biological pretreatment lignin-degrading microbes, usually fungus, are used for the deconstruction of lignocellulosic biomass. In a study by Prasad et al. (2016), which suggested liquid hot water to be the best pretreatment method, the CO_2 eq emissions from different treatments were 14.30, 385.00, 0.94, and 9.23 kg for the steam explosion (SE), dilute acid (DA), liquid hot water (LHW), and organosolv (OS) pretreatment, respectively. The terrestrial eutrophication potentials followed DA > LHW > SE > OS, while the aquatic eutrophication potential followed OS > SE > DA > LHW. But taking into account the values, terrestrial eutrophication values are quite less; thus, SE and OS are the ones with the highest emissions. Pretreatment technologies also require water for washing the biomass at multiple stages. Acid pretreatment (DA) method usually involving sulfuric acid consumes the highest amount of water for dilution, which is required at multiple stages during the process. The acidification potential of the DA method is the highest with LHW, SE and OS following next in order. The hydrolysis phase consists of the conversion of cellulosic and hemicellulosic content to hydrolyze into fermentable sugars. It can be of two basic types: separate hydrolysis and fermentation, and combined hydrolysis and fermentation. Separate hydrolysis and fermentation consist of enzymatic saccharification and fermentation in two separate containers, while the cultivation of the enzyme is done separately. Combined hydrolysis and fermentation can be of two types: simultaneous saccharification and fermentation, which involves saccharification or hydrolysis and fermentation together but enzyme production separately, while consolidate bioprocessing involves all the things in a single reactor. Combined hydrolysis and fermentation reduce the costs involved significantly at the capital stage. In biochemical pathways, water requirement is usually high, which increases the cost as well as its life cycle assessment attribute.

5.4 CONSTRAINTS AND OPPORTUNITIES ASSOCIATED WITH SECOND-GENERATION BIOFUEL CONVERSION

Type of feedstock and its availability has been a major problem in the production of biofuel from lignocellulosic biomass (Singh et al., 2020a; Singh and Verma, 2019). The type of feedstock also classifies the generation of biofuel production (Singh et al., 2019b). The change in the feedstock can be due to many factors, such the change in species, variety, part of the plant, and many other factors. This leads to the variation in the amount of lignin, hemicellulose, and cellulose content variation in the feedstock, which creates a problem at the processing step, if a process is standardized for a particular type of feedstock. Minor compounds associated with the lignocellulosic biomass also leads to the formation of by products that may inhibit the standardized processing of the biomass into simpler sugars. Variation in terms of moisture is also a major factor associated with the type of feedstock. During the operating procedures the moisture most of the time reduces the calorific value followed by variation in heating rate, hydration times, stability of the digestate, pH of the process, and also the thermal efficiency of the process. Another feedstock-associated factor is the particle size and shape that greatly influences the rate of process and its properties. Thermochemical processes are more affected by the particle shape and size, while biochemical processes are less sensitive to it. The biochemical properties of the feedstock also play a major role in the determination of the by product and the major product derivative. Other properties such as compressibility, surface area, bulk density of the biomass is other feedstock-associated factors determining the process efficiency. Surface area of the material at the microstructural level helps in the determination of the available surface area for the enzymatic reactions to take place during a biochemical conversion process. Thus, mixed type of feedstock presents a major challenge, and a process that standardized the processing of mixed biomass at the biorefinery level would certainly revolutionize the second-generation biofuel production field. This is followed by another major challenge, i.e., the availability of the feedstock. Regular supply of feedstock to the biorefinery is a major challenge in the smooth functioning of second-generation biofuel production plants. If we take into consideration the amount of biomass produced annually that can be available for biofuel production, it is immense, but its regular supply to the biorefineries is limited. It is because most of the agro waste from the farmlands is generated in the form of bulk at the time of harvest of the crop. This is followed by a period when little to no biomass is produced from farmlands for biorefineries. Alternate options can be the use of technology that can use wood residue from tree pruning in the biorefinery, while there is little to no biomass available from the farmlands for the biorefineries.

While the constraints associated with the feedstock and its types seem to thwart the smooth functioning of the biorefinery, it also presents us with a huge opportunity to utilize it. Across the globe, millions of tons of agricultural waste is produced annually that can be utilized for second-generation biofuel production. According to survey, Europe—one of the most developed continents—produced 0.7 billion tons of agricultural and forestry waste during the period of 1998–2006 from the western region alone (Guo et al., 2010; Weiland, 2000). China tops the list among the major producers of agricultural waste in Asian continent. Apart from this, wood residue production in China has risen ten times to almost 100 million m^3 in the last two decades. If one takes a close look at the data available at FAOstat, it can be seen that Russia has always maintained its wood residue production to be around 10 million m^3.

A number of important factors are present that show the advancement of biofuel over conventional fuel such as lowering GHG emissions in the environment, but before considering it as an environmental and economically viable alternate many other challenges have to be resolved. A number of researches have been carried on biofuel around the world to increase the economics for sustainable development. Due to conventional fuel transportation, a huge amount of GHG emission in the environment is released which is one of the major causes of global warming. At present, due to geographical, societal, environmental, and technical constraints, the high production cost of biofuel does not favor biofuel use as an alternative to conventional fuel. Many developed and underdeveloped countries started using ethanol/bioethanol as an alternative source as compared to conventional fuel for transportation due to good compatibility with gasoline and high-octane number. However, a blended form of ethanol and gasoline is a good alternative that releases less amount of GHG in the environment, but due to low energy value of ethanol as compared to gasoline makes it an alternative with low fuel economy (Masum et al., 2015; Moka et al., 2014). Moreover, most of the bioethanol is produced by a fermentation process that includes initial pretreatment cost, and, for better results, expensive enzymes have been used, which make the process costlier than the conventional gasoline (Khoo, 2015; Malhi and Kutcher, 2007; Moka et al., 2014). Other than that biodiesel has also attained good attention as an alternative to conventional diesel due to its compatibility with conventional diesel. Biodiesel can be used directly or indirectly in combustion engines, but their large-scale commercialization has to be attained. At present, vegetable oil is one of the most prominent sources of biodiesel, and more than 90% of total biodiesel has been produced only from vegetable oils all around the world (Meher et al., 2006). The production of biodiesel is expected to reduce by 2.53% in the upcoming 10 years (OECD 2020). Besides, the economics of biofuel production is affected by limitations of feedstock as well as lacking good production technology.

Nowadays, the use of biofuels is essential and environment friendly, but the popularity of their use around the world has highlighted many environmental concerns for their use. Biofuel is derived from bioproducts, and it has the potential to be "carbon neutral" over their life cycles. Although, its combustion releases CO_2 in the environment, which is used by the plant for the preparation of their food and ultimately help in overcoming the GHG emission, biofuel is a useful alternative to petroleum fuels due to low GHG emissions in the environment. Many countries try to encourage the use of biofuel, and cut the use of petroleum fuels and set standards worldwide for lowering GHG emissions in the future. Many developed and developing countries, including India, have set their GHG emission reduction target and development program to generate energy alternative to petroleum fuels in the future. This initiative encourages the biofuel industry to grow and develop, but simultaneously, despite many advantages of biofuel over petroleum fuel, extensive development of the biofuel industry with considering sustainability issues may cause negative effects on the environment directly and indirectly. Moreover, certain issues have also been reported with ethanol-blended fuel such as after oxidation may produce various toxic aldehydes, such as formaldehyde, acetaldehyde, etc. This fuel after burning emits some toxic aldehydes and some aromatic compounds in the environment that are hazardous to the environment, which is not regulated in emissions laws (Dhillon et al., 2013).

The end use constraints refer to the use of biofuel in the existing machinery setup. For instance, the replacement of bioethanol with the conventional gasoline can present a constraint with the existing automobile technology. Currently, there have been provisions to use the bioethanol with gasoline in the blended form usually ranging between 5% and 20% (v/v). This allows little to no changes in the automobile technology to work with bioethanol as the fuel.

5.5 MICROBES FOR SECOND-GENERATION BIOFUEL PRODUCTION

With advances in the biochemical conversion technologies of the biomass, there have been many techniques applied for the enhancement for the rate of enzymatic hydrolysis of lignocellulosic biomass by cellulose-degrading microbes. Many microbes that have potential application in the biomass degradation are mentioned in Table 5.1.

With advances in microbial technology and the increasing need of more and more advanced biofuels, it becomes imperative to explore the microbial options that have high stability with respect to temperature. This is where thermophilic microbes have a major role to play in cellulose's enzymatic conversion (Singh et al., 2019a). With respect to advances in enzymes for biomass conversion, gene-level studies have explored many new options. *pBC*elR gene isolated from *Bacillus subtilis* CD4 enhanced the activity of two specific enzymes. It enhanced the activity of endoglucanase as well as β-glucosidase when observed in the transformed strain of *Escherichia coli* (JM83) (Srivastava et al., 1999). Same way another such gene *cel*S was found to code for cellulase and exoglucanase, which was isolated from *Bacillus subtilis*. This gene was inserted in the *Escherichia coli* strain to express its activity (Jung et al., 1996). *pNB*6 and *pNA*1 genes also showed endoglucanase as well as exoglucanase and endoglucanase activity, respectively, when inserted in *E. coli* RI (the transformed strain). Same way, we have already isolated the cellulose-degrading bacterial strains from the soil samples of hot springs for the isolation of genes coding for such enzymes. Some other genes such as *cbh*1, *cbh*2, and *egl* were found to get induced to a moderate level in the presence of cellobiose. The problem with these started in the later stages of cultivation when glucose started accumulating in the culture medium (Ilmen et al., 1997). This gives an insight to use different inducers for enhancing the enzyme production. Another gene *XYR*1 (Xylanase regulator) was found to activate cellulase gene expression in *Aspergillus niger* (van Peij et al., 1998). The expression of *egl*1 and *egl*2 (endoglucanases), *cbh*1 and *cbh*2 (cellobiohydrolases), and several other genes encoding the side chain-cleaving hemicellulases were found to be coordinately regulated by this xylanolytic activator. Another mechanism through which enzyme activity can be enhance is through synergism. The genes such as *egl1/cel*7b and *eg2/cel*5a, *eg3/cel*12a (Okada et al., 1998; Saloheimo et al., 1988), *eg4/cel*61a, and

TABLE 5.1
Microbes with Their Enzymes Used for Cellulose Degradation

Microbes	Isolation Source	Enzymes Evaluated	Major Finding	Reference
Aspergillus niger MK543209	Egyptian soils	Exo-β-glucanase activity, Endo-β-1,4-glucanase activity, β-glucosidase activity (cellobiase)	Improvement of paper wastes conversion	Darwesh et al. (2020)
Klebsiella sp. MD21	Gut of Helicoverpa armigera	Endoglucanase, exoglucanase, β-glucosidase and xylanase	Deconstruction of the saw dust and filter paper	Dar et al. (2018)
Bacillus amyloliquefaciens subsp. *Plantarum*	Ward poultry	Total cellulase	Optimum conditions for each one for cellulase production were determined, useful for biofuel production	Hussain et al. (2017)
Bacillus megaterium, Bacillus subtilis subsp. Subtilis, Anoxybacillus flavithermus WK1	Soil	Total cellulase	As above	Hussain et al. (2017)
Escherichia coli	Bovine rumen	Exo-β-glucanase activity, Endo-β-1,4-glucanase activity, β-glucosidase activity	Ethanol and hydrogen production from corn straw	Pang et al. (2017)
Bacillus sp. BMP01 and *Ochrobactrum oryzae* BMP03	Gut of Cryptotermes brevis	Endoglucanase and xylanase	Termite's gut potential sources of lignocellulose-degrading bacteria	Tsegaye et al. (2019)
Pseudomonas mendocina, Burkholderia pseudomallei, Chryseobacterium luteola, Klebsiella oxytoca, and *Klebsiella terrigena*	Gut of Coptotermes formosanus	NA	Significantly degrade filter paper	Egwuatu and Appeh (2018)

(Continued)

TABLE 5.1 (*Continued*)
Microbes with Their Enzymes Used for Cellulose Degradation

Microbes	Isolation Source	Enzymes Evaluated	Major Finding	Reference
Ochrobactrum oryzae BMP03 and *Bacillus* sp. BMP01	Wood-feeding termite's guts: Isoptera: *Cryptotermes brevis*	NA	Enhance bioethanol production	Tsegaye et al. (2018)
Aspergillus fumigatus	Cellulosic waste-contaminated soil	Endoglucanase, xylanase	Process optimization of endoglucanase improves saccharification yields of *Pennisetum* sp.	Mohapatra et al. (2018)
Clostridium sp. DBT-IOC-C19	Enrichment cultures of Himalayan hot spring	NA	Ethanol production through consolidated bioprocessing	Singh et al. (2017)
T2-D2 (*Bacillus* sp.), E1-PT (*Pseudomonas* sp.), and D1-PT (*Pseudomonas* sp.), C1-BT (*Bacillus* sp.), E1-BT (*Bacillus* sp.)	Mangrove forests, Sundarbans in Bangladesh	Endoglucanase (CMC/cellulose) and exoglucanase (microcrystalline/avicellase) activities	High cellulolytic activity	Biswas et al. (2020)
Bacillus methylotrophicus 1EJ7	Rotten wood of Qinling (China)	Cellulase β-glucosidase and endoglucanase	Effective cellulose-degrading strains	Ma et al. (2020)
Streptomyces fulvissimus CKS7	Soil sample	Cellulases, amylase, xylanase, and pectinase	Promising candidate for biodegradation cellulosic wastes	Mihajlovski et al. (2021)

*eg5/cel*45a (Saloheimo et al., 1997), and *cbh1/cel7*a and *cbh2/cel6*a (Teeri et al., 1987) were found to act synergistically to convert cellulose into cellobiose, which are acted upon by β-glucosidases *bgl1/cel3*a and *bgl2/cel1*1a to convert it into glucose (Barnett et al., 1991).

5.6 CONCLUSIONS AND FUTURE PERSPECTIVES

The increasing world population corresponds with increasing food production, which subsequently generates more and more agro waste. The conversion of this into useable form is a task that heavily relies on the exploration of efficient technologies for the conversion of lignocellulosic biomass. The current technologies either lack in terms of sustainability or in terms of efficiency. Thermochemical conversions are driven by unsustainable heating options, whereas the biochemical conversions lack in terms of efficiency. Therefore, the biochemical conversion technologies need to be explored more in order to come up with a more advanced and efficient conversion of lignocellulosic waste. The thermophilic enzymes seem to be viable option for such a case and would certainly help in boosting the production of second-generation bioethanol production. Other options that should be explored are the determination of the inhibitors that are produced during the process of conversion of lignocellulosic biomass in simpler sugars.

ACKNOWLEDGMENT

Saurabh Singh is highly grateful to the University Grants Commission (UGC Ref. No. 3819/NET-JULY-2018) for providing the Junior Research Fellowship and Senior Research Fellowship to carry out the research work.

REFERENCES

Abdurrahman, M.I., Chaki, S., Saini, G. (2020) Stubble burning: Effects on health & environment, regulations and management practices. *Environmental Advances* 2, 100011.

Agbro, E.B., Ogie, N.A. (2012) A comprehensive review of biomass resources and biofuel production potential in Nigeria. *Research Journal in Engineering and Applied Sciences* 1, 149–155.

Barnett, C.C., Berka, R.M., Fowler, T. (1991) Cloning and amplification of the gene encoding an extracellular β-glucosidase from *Trichoderma reesei*: Evidence for improved rates of saccharification of cellulosic substrates. *Bio/technology* 9, 562–567.

Beig, G., Sahu, S.K., Singh, V., Tikle, S., Sobhana, S.B., Gargeva, P., Ramakrishna, K., Rathod, A., Murthy, B. (2020) Objective evaluation of stubble emission of North India and quantifying its impact on air quality of Delhi. *Science of The Total Environment* 709, 136126.

Bisaria, V. (1991) Bioprocessing of agro-residues to glucose and chemicals. *Bioconversion of Waste Materials to Industrial Products* 187–223.

Biswas, S., Saber, M.A., Tripty, I.A., Karim, M.A., Islam, M.A., Hasan, M.S., Alam, A.S.M.R.U., Jahid, M.I.K., Hasan, M.N. (2020) Molecular characterization of cellulolytic (endo- and exoglucanase) bacteria from the largest mangrove forest (Sundarbans), Bangladesh. *Annals of Microbiology* 70, 68.

Bomble, Y.J., Lin, C.-Y., Amore, A., Wei, H., Holwerda, E.K., Ciesielski, P.N., Donohoe, B.S., Decker, S.R., Lynd, L.R., Himmel, M.E. (2017) Lignocellulose deconstruction in the biosphere. *Current Opinion in Chemical Biology* 41, 61–70.

Chen, L., Xing, L., Han, L. (2009) Renewable energy from agro-residues in China: Solid biofuels and biomass briquetting technology. *Renewable and Sustainable Energy Reviews* 13, 2689–2695.

Chen, Z., Wang, M., Jiang, E., Wang, D., Zhang, K., Ren, Y., Jiang, Y. (2018) Pyrolysis of torrefied biomass. *Trends in Biotechnology* 36, 1287–1298.

Dar, M.A., Shaikh, A.A., Pawar, K.D., Pandit, R.S. (2018) Exploring the gut of *Helicoverpa armigera* for cellulose degrading bacteria and evaluation of a potential strain for lignocellulosic biomass deconstruction. *Process Biochemistry* 73, 142–153.

Darwesh, O.M., El-Maraghy, S.H., Abdel-Rahman, H.M., Zaghloul, R.A. (2020) Improvement of paper wastes conversion to bioethanol using novel cellulose degrading fungal isolate. *Fuel* 262, 116518.

Dhammapala, R., Claiborn, C., Corkill, J., Gullett, B. (2006) Particulate emissions from wheat and Kentucky bluegrass stubble burning in eastern Washington and northern Idaho. *Atmospheric Environment* 40, 1007–1015.

Dhammapala, R., Claiborn, C., Simpson, C., Jimenez, J. (2007) Emission factors from wheat and Kentucky bluegrass stubble burning: Comparison of field and simulated burn experiments. *Atmospheric Environment* 41, 1512–1520.

Dhillon, G.S., Kaur, S., Brar, S.K. (2013) Perspective of apple processing wastes as low-cost substrates for bioproduction of high value products: A review. *Renewable and Sustainable Energy Reviews* 27, 789–805.

Egwuatu, T.F., Appeh, O.G. (2018) Isolation and characterization of filter paper degrading bacteria from the guts of *Coptotermes formosanus*. *Journal of Bioremediation & Biodegradation* 9.

Gheewala, S.H., Damen, B., Shi, X. (2013) Biofuels: Economic, environmental and social benefits and costs for developing countries in Asia. *Wiley Interdisciplinary Reviews: Climate Change* 4, 497–511.

Gómez, L.D., Amalfitano, C., Andolfi, A., Simister, R., Somma, S., Ercolano, M.R., Borrelli, C., McQueen-Mason, S.J., Frusciante, L., Cuciniello, A. (2017) Valorising faba bean residual biomass: Effect of farming system and planting time on the potential for biofuel production. *Biomass and Bioenergy* 107, 227–232.

Guo, X.M., Trably, E., Latrille, E., Carrere, H., Steyer, J.-P. (2010) Hydrogen production from agricultural waste by dark fermentation: A review. *International Journal of Hydrogen Energy* 35, 10660–10673.

Gupta, A., Verma, J.P. (2015) Sustainable bio-ethanol production from agro-residues: A review. *Renewable and Sustainable Energy Reviews* 41, 550–567.

Havlík, P., Schneider, U.A., Schmid, E., Böttcher, H., Fritz, S., Skalský, R., Aoki, K., De Cara, S., Kindermann, G., Kraxner, F. (2011) Global land-use implications of first and second generation biofuel targets. *Energy Policy* 39, 5690–5702.

Hsu, D.D. (2012) Life cycle assessment of gasoline and diesel produced via fast pyrolysis and hydroprocessing. *Biomass and Bioenergy* 45, 41–47.

Hussain, A.A., Abdel-Salam, M.S., Abo-Ghalia, H.H., Hegazy, W.K., Hafez, S.S. (2017) Optimization and molecular identification of novel cellulose degrading bacteria isolated from Egyptian environment. *Journal of Genetic Engineering and Biotechnology* 15, 77–85.

Ilmen, M., Saloheimo, A., Onnela, M.-L., Penttilä, M.E. (1997) Regulation of cellulase gene expression in the filamentous fungus *Trichoderma reesei*. *Applied and Environmental Microbiology* 63, 1298–1306.

Inderwildi, O.R., King, D.A. (2009) Quo vadis biofuels? *Energy & Environmental Science* 2, 343–346.

Iribarren, D., Susmozas, A., Dufour, J. (2013) Life-cycle assessment of Fischer–Tropsch products from biosyngas. *Renewable Energy* 59, 229–236.

Jimenez, J.R., Claiborn, C.S., Dhammapala, R.S., Simpson, C.D. (2007) Methoxyphenols and levoglucosan ratios in PM2. 5 from wheat and Kentucky bluegrass stubble burning in eastern Washington and northern Idaho. *Environmental Science & Technology* 41, 7824–7829.

Jung, K.H., Chun, Y.C., Lee, J.-C., Kim, J.H., Yoon, K.-H. (1996) Cloning and expression of a *Bacillus* sp. 79-23 cellulase gene. *Biotechnology Letters* 18, 1077–1082.

Khoo, H.H. (2015) Review of bio-conversion pathways of lignocellulose-to-ethanol: Sustainability assessment based on land footprint projections. *Renewable and Sustainable Energy Reviews* 46, 100–119.

Kumar, C.P., Rena, M.A., Khapre, A.S., Kumar, S., Anshul, A., Singh, L., Kim, S.H., Lee, B. D., Kumar, R. (2019) Bio-Hythane production from organic fraction of municipal solid waste in single and two stage anaerobic digestion processes. *Bioresource Technology* 294, 122220. Doi: 10.1016/j.biortech.2019.122220.

Kumar, S.J., Kumar, N.S., Chintagunta, A.D. (2020) Bioethanol production from cereal crops and lignocelluloses rich agro-residues: Prospects and challenges. SN Applied Sciences 2, 1–11.

Ma, L., Lu, Y., Yan, H., Wang, X., Yi, Y., Shan, Y., Liu, B., Zhou, Y., Lü, X. (2020) Screening of cellulolytic bacteria from rotten wood of Qinling (China) for biomass degradation and cloning of cellulases from *Bacillus methylotrophicus. BMC Biotechnology* 20, 2.

Malhi, S., Kutcher, H. (2007) Small grains stubble burning and tillage effects on soil organic C and N, and aggregation in northeastern Saskatchewan. *Soil and Tillage Research* 94, 353–361.

Masum, B., Masjuki, H.H., Kalam, M., Palash, S., Habibullah, M. (2015) Effect of alcohol–gasoline blends optimization on fuel properties, performance and emissions of a SI engine. *Journal of Cleaner Production* 86, 230–237.

Mat Aron, N.S., Khoo, K.S., Chew, K.W., Show, P.L., Chen, W.H., Nguyen, T.H.P. (2020) Sustainability of the four generations of biofuels–A review. *International Journal of Energy Research* 44, 9266–9282.

Meher, L.C., Sagar, D.V., Naik, S. (2006) Technical aspects of biodiesel production by transesterification—A review. *Renewable and Sustainable Energy Reviews* 10, 248–268.

Mihajlovski, K., Buntić, A., Milić, M., Rajilić-Stojanović, M., Dimitrijević-Branković, S. (2021) From agricultural waste to biofuel: Enzymatic potential of a bacterial isolate *Streptomyces fulvissimus* CKS7 for bioethanol production. *Waste and Biomass Valorization* 12, 165–174.

Mittal, S.K., Singh, N., Agarwal, R., Awasthi, A., Gupta, P.K. (2009) Ambient air quality during wheat and rice crop stubble burning episodes in Patiala. *Atmospheric Environment* 43, 238–244.

Mohapatra, S., Padhy, S., Das Mohapatra, P.K., Thatoi, H.N. (2018) Enhanced reducing sugar production by saccharification of lignocellulosic biomass, Pennisetum species through cellulase from a newly isolated *Aspergillus fumigatus. Bioresource Technology* 253, 262–272.

Moka, S., Pande, M., Rani, M., Gakhar, R., Sharma, M., Rani, J., Bhaskarwar, A.N. (2014) Alternative fuels: An overview of current trends and scope for future. *Renewable and Sustainable Energy Reviews* 32, 697–712.

Mu, D., Seager, T., Rao, P.S., Zhao, F. (2010) Comparative life cycle assessment of lignocellulosic ethanol production: Biochemical versus thermochemical conversion. *Environmental management* 46, 565–578.

Mwamila, B., Temu, R., Oscar, K., John, G., Majamba, H., Maliondo, S., Kulindwa, K., Temu, A., Lupala, J., Chijoriga, M. (2009) Feasibility of large-scale biofuel production in Tanzania.

Naik, S.N., Goud, V.V., Rout, P.K., Dalai, A.K. (2010) Production of first and second generation biofuels: A comprehensive review. *Renewable and Sustainable Energy Reviews* 14, 578–597.
OECD-FAO Agricultural Outlook. (2020) *OECD-FAO Agricultural Outlook, 2020–2029.* OECD Publishing, Paris/FAO, Rome.
Okada, H., Tada, K., Sekiya, T., Yokoyama, K., Takahashi, A., Tohda, H., Kumagai, H., Morikawa, Y. (1998) Molecular characterization and heterologous expression of the gene encoding a low-molecular-mass endoglucanase from *Trichoderma reesei* QM9414. *Applied and Environmental Microbiology* 64, 555–563.
Pang, J., Liu, Z.-Y., Hao, M., Zhang, Y.-F., Qi, Q.-S. (2017) An isolated cellulolytic Escherichia coli from bovine rumen produces ethanol and hydrogen from corn straw. *Biotechnology for Biofuels* 10, 165.
Phillips, O. (1997) The changing ecology of tropical forests. *Biodiversity & Conservation* 6, 291–311.
Phillips, S.D., Tarud, J.K., Biddy, M.J., Dutta, A. (2011) Gasoline from woody biomass via thermochemical gasification, methanol synthesis, and methanol-to-gasoline technologies: A technoeconomic analysis. *Industrial & Engineering Chemistry Research* 50, 11734–11745.
Prasad, A., Sotenko, M., Blenkinsopp, T., Coles, S.R. (2016) Life cycle assessment of lignocellulosic biomass pretreatment methods in biofuel production. *The International Journal of Life Cycle Assessment* 21, 44–50.
Ramesh, D., Muniraj, I.K., Thangavelu, K., Karthikeyan, S., (2019) Chemicals and fuels production from agro residues: A biorefinery approach. In *Sustainable Approaches for Biofuels Production Technologies*, N. Srivastava, M. Srivastava, P.K. Mishra, S.N. Upadhyay P.W. Ramteke, V.K. Gupta (Eds.), Springer, Cham, pp. 47–71.
Reisen, F., Meyer, C.M., McCaw, L., Powell, J.C., Tolhurst, K., Keywood, M.D., Gras, J.L. (2011) Impact of smoke from biomass burning on air quality in rural communities in southern Australia. *Atmospheric Environment* 45, 3944–3953.
Rena, Zacharia, K.M.B., Yadav, S., Machhirake, N.P., Kim, S.H., Lee, B.D., Jeong, H., Singh, L., Kumar, S., Kumar, R. (2020) Bio-hydrogen and bio-methane potential analysis for production of bio-hythane using various agricultural residues. *Bioresource Technology* 309, 123297. Doi: 10.1016/j.biortech.
Renó, M.L.G., Lora, E.E.S., Palacio, J.C.E., Venturini, O.J., Buchgeister, J., Almazan, O. (2011) A LCA (life cycle assessment) of the methanol production from sugarcane bagasse. *Energy* 36, 3716–3726.
Rishikesh, M.S., Harish, S., Mahendran Prasanth, S., Gnana Prakash, D. (2021) A comprehensive review on lignin obtained from agro-residues: Potential source of useful chemicals. *Biomass Conversion and Biorefinery*, 1–24.
Saloheimo, M., Lehtovaara, P., Penttilä, M., Teeri, T.T., Ståhlberg, J., Johansson, G., Pettersson, G., Claeyssens, M., Tomme, P., Knowles, J.K. (1988) EGIII, a new endoglucanase from *Trichoderma reesei*: The characterization of both gene and enzyme. *Gene* 63, 11–21.
Saloheimo, M., Nakari-SetäLä, T., Tenkanen, M., Penttilä, M. (1997) cDNA cloning of a *Trichoderma reesei* cellulase and demonstration of endoglucanase activity by expression in yeast. *European Journal of Biochemistry* 249, 584–591.
Satpathy, P., Pradhan, C. (2020) Biogas as an alternative to stubble burning in India. *Biomass Conversion and Biorefinery*, 1–12.
Sharma, H.K., Xu, C., Qin, W. (2019) Biological pretreatment of lignocellulosic biomass for biofuels and bioproducts: An overview. *Waste and Biomass Valorization* 10, 235–251.
Simmons, B.A., Loque, D., Blanch, H.W. (2008) Next-generation biomass feedstocks for biofuel production. *Genome Biology* 9, 1–6.
Singh, J., Singhal, N., Singhal, S., Sharma, M., Agarwal, S., Arora, S., (2018a) Environmental implications of rice and wheat stubble burning in north-western states of India. *Advances in Health and Environment Safety*. Springer, pp. 47–55.

Singh, N., Mathur, A.S., Tuli, D.K., Gupta, R.P., Barrow, C.J., Puri, M. (2017) Cellulosic ethanol production via consolidated bioprocessing by a novel thermophilic anaerobic bacterium isolated from a Himalayan hot spring. *Biotechnology for Biofuels* 10, 73.

Singh, S., Gaurav, A.K., Verma, J.P., (2020a) Genetically modified microbes for second-generation bioethanol production. In *Fungal Biotechnology and Bioengineering*, A.E.-L. Hesham, R.S. Upadhyay, G.D. Sharma, C. Manoharachary, V.K. Gupta (Eds.), Springer Cham, pp. 187–198.

Singh, S., Jaiswal, D.K., Krishna, R., Mukherjee, A., Verma, J.P. (2020b) Restoration of degraded lands through bioenergy plantations. *Restoration Ecology* 28, 263–266.

Singh, S., Jaiswal, D.K., Sivakumar, N., Verma, J.P. (2019a) Developing efficient thermophilic cellulose degrading consortium for glucose production from different agro-residues. *Frontiers in Energy Research* 7, 61.

Singh, S., Kumar, A., Verma, J.P., (2021) Techno-economic analysis of second-generation biofuel technologies. In *Bioprocessing for Biofuel Production*, N. Srivastava, M. Srivastava, P.K. Mishra, V.K. Gupta (Eds.), Springer, Cham, pp. 157–181.

Singh, S., Pereira, A.P., Verma, J.P. (2019b) Research and production of third-generation biofuels. In *Bioprocessing for Biomolecules Production*, G. Molina, V.K. Gupta, Brahma N. Singh, N. Gathergood (Eds.), Wiley, 401–416.

Singh, S., Verma, J., de Araujo Pereira, A.P., Sivakumar, N., (2018b) Production of biofuels and chemicals from lignin. In *Production of Biofuels and Chemicals from Lignin*, Z. Fang, R.L. Smith Jr. (Eds.), Springer (2016), p. 442. Doi: 10.1007/978-981-10-1965-4, ISBN: 978-981-10-1964-7. Elsevier.

Singh, S., Verma, J.P., (2019) Bioethanol production from different lignocellulosic biomass. In *Sustainable Biofuel and Biomass*. Arindam Kuila (Eds.) Apple Academic Press, New York, 281–300.

Sokan-Adeaga, A.A., Ana, G.R. (2015) A comprehensive review of biomass resources and biofuel production in Nigeria: Potential and prospects. *Reviews on Environmental Health* 30, 143–162.

Srivastava, K., Verma, P., Srivastava, R. (1999) A recombinant cellulolytic *Escherichia coli*: Cloning of the cellulase gene and characterization of a bifunctional cellulase. *Biotechnology Letters* 21, 293–297.

Taghizadeh-Alisaraei, A., Assar, H.A., Ghobadian, B., Motevali, A. (2017) Potential of biofuel production from pistachio waste in Iran. *Renewable and Sustainable Energy Reviews* 72, 510–522.

Teeri, T.T., Lehtovaara, P., Kauppinen, S., Salovuori, I., Knowles, J. (1987) Homologous domains in *Trichoderma reesei* cellulolytic enzymes: Gene sequence and expression of cellobiohydrolase II. *Gene* 51, 43–52.

Tsegaye, B., Balomajumder, C., Roy, P. (2018) Biodegradation of wheat straw by *Ochrobactrum oryzae* BMP03 and *Bacillus* sp. BMP01 bacteria to enhance biofuel production by increasing total reducing sugars yield. *Environmental Science and Pollution Research* 25, 30585–30596.

Tsegaye, B., Balomajumder, C., Roy, P. (2019) Isolation and characterization of novel lignolytic, cellulolytic, and hemicellulolytic bacteria from wood-feeding termite *Cryptotermes brevis*. *International Microbiology* 22, 29–39.

van Peij, N.N., Gielkens, M.M., de Vries, R.P., Visser, J., de Graaff, L.H. (1998) The transcriptional activator XlnR regulates both xylanolytic and endoglucanase gene expression in *Aspergillus niger*. *Applied and Environmental Microbiology* 64, 3615–3619.

Voigt, T., Lee, D., Kling, G. (2013) *Perennial Herbaceous Crops with Potential for Biofuel Production in the Temperate Regions of the USA*. CABI Publishing Wallingford, England.

Weiland, P. (2000) Anaerobic waste digestion in Germany–Status and recent developments. *Biodegradation* 11, 415–421.

Zabed, H., Sahu, J., Boyce, A.N., Faruq, G. (2016) Fuel ethanol production from lignocellulosic biomass: An overview on feedstocks and technological approaches. *Renewable and Sustainable Energy Reviews* 66, 751–774.

6 Advanced Oxidation-Based Pretreatment of Wastewater for Enhanced Biofuel Production

Krutarth H. Pandit, Parth M. Khandagale, Adwait T. Sawant, and Sameena N. Malik
Institute of Chemical Technology

CONTENTS

6.1 INTRODUCTION

In the recent times, water pollution has become a major issue for human and environmental health due to the wastewater (industrial and home) being directly discharged into the water bodies. Everyday different kinds of toxic pollutants from fertilizers, pesticides, pharmaceuticals, dyes, etc. are released into the water. The pollutant molecules are biorefractory (cannot breakdown easily) to the microorganisms. These compounds are also recalcitrant to conventional biological treatment methods

DOI: 10.1201/9781003197737-8

like aerobic, anaerobic digestion, and hence they find their way in rivers and seas. Additionally, the depletion of fossil fuels has led to the search for alternative greener and cleaner sources of fuels.

AOPs are the processes that break down different substrates by oxidative radicals produced from various chemical reactions (M'Arimi et al. 2020). AOPs are classified based on the oxidative radical used for the oxidation mechanism namely hydroxyl, sulfate, ozone, etc. These radicals act as powerful oxidizing agents (have high oxidation potential) and destroy the wastewater constituents by transforming them into nontoxic products by breaking them down, thereby, making this technology viable and useful for treatment applications (Deng and Zhao 2015). Ozonation, Wet Air Oxidation, Fenton Oxidation, Photocatalysis, Cavitation, Hydrogen Peroxidation, Ultraviolet Radiation, etc. are some of the commonly used AOPs. Typically, AOPs have been extensively used in wastewater treatment applications, detoxification of hazardous substances, as well as remediation of soil, etc. (Paździor et al. 2019; Zhou et al. 2019; Jallouli et al. 2018). Recent trends have shown the increased use of advanced oxidation processes for different fields like organic synthesis, medicinal treatment, and production of energy (Zeng and Lu 2018; Michelin and Hoffmann 2018; Gellé and Moores 2019). Due to the increase in the demand for clean and green energy, these advanced oxidation processes have found their way into applications related to the bioenergy sector. The major activities where AOPs are used are pretreatment and enhancement of raw material (wastewater). They have a variety of applications like pretreatment of bioethanol and biohydrogen substrates, pretreatment of biomass that is of lignocellulosic origin for the production of biogas, bioenergy effluent post-treatment, enhancing transesterification process for biodiesel production, etc. (Tamilarasan et al. 2019; Yang and Wang 2019; Kumar et al. 2019; Guimarães et al. 2019; Malani et al. 2019).

Biogas, biodiesel, and biomethane are the most common sources of bioenergy. Out of the total renewable energy production, around 80% comes from biomass (Strzalka, Schneider, and Eicker 2017). Due to the increase in global energy demand, it is estimated that bioenergy will contribute to a one-third of the global energy supply by 2050, as opposed to the current 6%–10% of global energy demand (Roopnarain and Adeleke 2017; Guo et al. 2015). The increased production of bioenergy will have several benefits like reduction in carbon footprint and emissions along with reduced dependence on fossil fuels (Hoang et al. 2021).

Biogas is one of the highest produced forms of bioenergy worldwide. Raw materials are required to produce biogas from various wastes like food waste, municipal waste, biomass, plant crops, anaerobic sludge, etc. It has been extensively studied that biogas can be effectively and efficiently produced from raw substrates. However, for other organic substrates like organic or industrial-based ones, necessary pretreatment is needed to enhance the efficacy, yield, conversion, and productivity (Sambusiti et al. 2013; Rodriguez et al. 2015). The conventional pretreatment methods that may either be of biological or physiochemical origin proved to be less suitable for treatment with certain recalcitrant compounds, e.g., the ability of these methods to break down the crystalline structure of hemicellulose leading to lower yield and process efficiency. AOPs have an efficiency that offers significant improvement that led to enhance the breakdown of such compounds.

The next bioenergy source in liquid form is bioethanol, which is an important biofuel. The worldwide demand for biofuel has been increasing due to the need to reduce dependency on conventional crude oil, thereby reducing greenhouse gasses and carbon emissions. The worldwide bioethanol production in 2019 was around 110 billion liters, and this number is expected to rapidly increase in near future ("Alternative Fuels Data Center: Maps and Data – Global Ethanol Production by Country or Region" 2021). Fermentation acts as the primary process used for the production of bioethanol from different raw substrates. The first step of the process is pretreatment of the raw substrate, where size reduction takes place and the complex structure of hemicellulose is broken down leading to substrate solubilization. The major disadvantage of pretreatment processes is its low solubility. Hence, high solubility is needed to harness the high sugar concentration from substrates leading to improved bioethanol production. To make the pretreatment process efficient, AOPs like Fenton oxidation, ultrasound oxidation, and microwave-enhanced AOPs have been applied for pretreatment of the lignocellulosic biomass, thus increasing the cellulose and hemicellulose composition, and consequently increasing bioethanol production (Rodriguez et al. 2017).

Recent trends have shown that the generation of biodiesel has gained much interest due to rising fuel prices worldwide. The primary limitation of biodiesel production is the low yield from transesterification from organic substrates, which makes the process less cost effective. The traditional methods used for enhancing these transesterification processes are acid, alkali, and enzyme-catalyzed processes (Li et al. 2008). The application of AOPs can also be used to augment the yield of biodiesel and to improve the overall process. Lysing the bio algae cells and extractions of oils can be done to produce biodiesel. AOPs can be applied to these lysing processes to enhance these processes before biodiesel extraction (Li et al. 2008).

This book chapter focuses on the application of AOPs for pretreatment of wastewater to increase biofuel production through wastewater treatment system. The introduction section mainly focuses on the problem of efficient wastewater treatment, followed by types of commonly used AOPs and the recent trends with examples of these processes that are commonly observed at an industrial level. There are different forms of bioenergy, namely biogas, bioethanol, and biodiesel, which are focused upon with some preliminary insights into how AOPs can be incorporated into their production process, thereby enhancing the production. The applications section focuses on the direct application of AOPs for the treatment of bioenergy substrates. AOPs like Ozonation, Fenton Oxidation, Wet Air Oxidation, Photocatalysis, Ultraviolet Radiation, Hydrogen Peroxide, Cavitation Methods, and Electrochemical Oxidation are discussed. On-site treatment systems will be highlighted in detail for each oxidation technology, including the treatment steps and processes parameters based on the results obtained along with the enhancement in the production that has been observed. To summarize the sections, each oxidation technology with possible applications and significant advantages and disadvantages with some annotations on each technology has been discussed. The conclusion and future scope section will discuss oxidation technologies and the future scope for improvement and applications of these processes and its possibilities for the integration of AOPs that can be implemented into wastewater treatment for the production of bioenergy.

6.2 TREATMENT OF WASTEWATER WITH ADVANCED OXIDATION TECHNOLOGIES

6.2.1 Introduction to AOPs

AOPs are classified by the oxidative radical used in the mechanism of oxidation for that AOP. These oxidation radicals may be either generated *in-situ* from oxidants or substrates. The oxidative agents are powerful and effective for the oxidation of recalcitrant and inert compounds, which increases the biodegradability of the wastewater, thus resulting in increasing the production of bioenergy as well. It has been observed that unstable compounds with double and triple bonds have preferential reactions with most AOPs. This reactivity depends on the oxidation potential of the oxidative radical observed in the oxidation technology. The oxidation potential of different common oxidants is shown in Table 6.1.

The different oxidation technologies that are commonly used in the bioenergy sector such as ozonation, Fenton oxidation, wet air oxidation, electrochemical oxidation, ultraviolet radiation, photocatalysis, cavitation methods, and hydrogen peroxide. The pictographic representation of the same can be seen in Figure 6.1 (M'Arimi et al. 2020).

6.2.2 Motivation for the Use of AOPs in the Bioenergy Sector

Almost all biomass substrates, including lignocellulosic biomass (Zheng et al. 2014; Taherzadeh and Karimi 2008) or organic effluent (Apollo, Onyango, and Ochieng 2014), have to undergo some kind of pretreatment step that will increase hydrolysis followed by the overall fermentation for the production of bioenergy. The primary goal of pretreatment steps is to break down or release the confined hemicellulose structure (commonly seen in natural substrates), thus, releasing the fermentable organic chemicals and consequently increasing the yield of bioenergy. A similar schematic can be seen in Figure 6.2 (M'Arimi et al. 2020).

TABLE 6.1
Oxidants and Oxidation Potential

Name	Oxidative Agents	Oxidation Potential (V)
Hydroxyl radical	OH^-	2.80
Ozone	O_3	2.07
Sulfate	SO_4^{2-}	2.01
Manganate	MnO_4^{2-}	1.77
Hydrogen peroxide	H_2O_2	1.77
Chlorine dioxide	ClO_2	1.57
Chlorine	Cl_2	1.36
Chromate	$Cr_2O_7^{2-}$	1.23
Superoxide radical	O_2^-	1.23

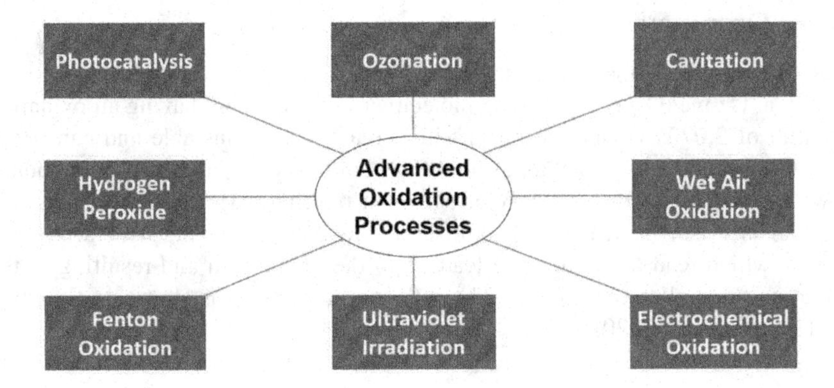

FIGURE 6.1 Commonly used advanced oxidation processes.

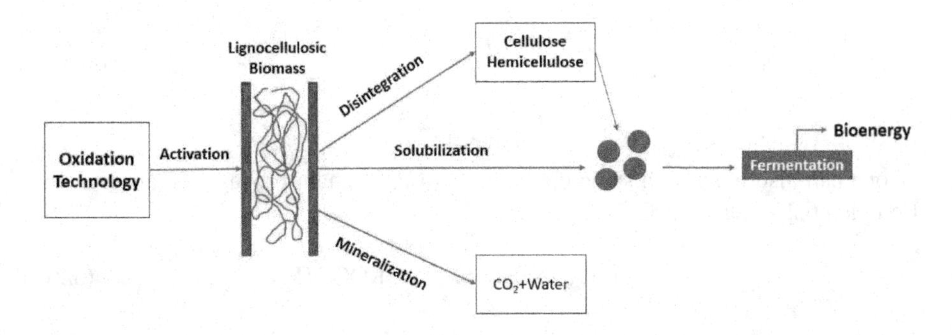

FIGURE 6.2 Schematic of production of bioenergy from lignocellulosic biomass.

There are two types of methods i.e., physical and chemical, whereas physical methods include communion, steam explosion, extraction, and liquid hot water (Zheng et al. 2014; Taherzadeh and Karimi 2008), chemical methods consists of acid or alkali pretreatment. The primary limitation of individual physical methods is their low productivity, substrate breakdown, and consequently low bioenergy yield. For acid-based chemical methods, acids like sulfuric acid, nitric acid, acetic acid, and hydrochloric acid are used (Monlau et al. 2012). The alkali-based methods used chemicals such as lime and sodium hydroxide. Lower bioenergy yields have been observed for acid-based chemicals than alkali-based ones. The use of chemical pre-treatment is very limited, and is not feasible due to high recurring costs and microbial poisoning due to certain chemicals. Biological pretreatment is an alternative where biological organisms like bacteria (Zhang et al. 2011), fungi (Muthangya, Manoni, and Kajumulo 2009) or enzymes (Sarkar et al. 2012) are used for treatment. Although these methods prove to be ineffective but have applications based on a lab-scale level. Hence, it can be seen that there is a need for innovative approach that comprises advanced, effective, less energy-intensive, fast, and cost-effective processes.

Hence, we will look at the different advanced oxidation technologies and their applications in the wastewater treatment sector.

6.2.3 Ozonation

Ozonation, as the name suggests, is based on ozone as the oxidative radical for oxidation. Ozone (O_3) is a triatomic molecule of oxygen atom having an oxidation potential of 2.07 V. Ozone, despite its high reactivity, is unstable and can break down C–C double and triple bonds, and C–N double bonds, and N–N triple bonds as well. This proves to be highly effective in treating toxic recalcitrant organic compounds. Also, ozone reacts with different olefins forming ring structured compounds, which tend to be unstable leading to the breakdown and resulting in the formation of smaller compounds. The reaction mechanism is shown in Equation 6.1 (M'Arimi et al. 2020).

Olefin Group + Ozone → Intermidiate $\xrightarrow{H+}$ Breakdown of Olefin Chain + HO^- (6.1)

Ozone can also react with some compounds to form chain propagation, as seen in Equation 6.2.

$$RH + O_3 \rightarrow HR^+O_3^- \rightarrow ROOOH \quad (6.2)$$

Electrophilic addition reactions of ozone are also seen to form reactive radicals seen in Equations 6.3 and 6.4.

$$R + O_3 \rightarrow R^+OOO^- \quad (6.3)$$

$$R{+}OOO^- \rightarrow R + O^* + O_2^* \quad (6.4)$$

A combination of O_3 and H_2O_2 is also used to accelerate the reaction and to aid the formation of more oxidative radicals. Addition of H_2O_2 reduces the ozone requirement and improve the kinetics of the overall reaction. The peroxide group and ozone react together in three different ways. Equation 6.5 shows a direct reaction with ozone to form hydroxyl radicals. Equation 6.6 shows the disassociation of peroxide into hydroperoxide, thus generating hydroxyl radical, while Equation 6.7 shows indirect formation of hydroxyl radical (Wang et al. 2002).

$$2O_3 + H_2O_2 \rightarrow 2OH^* + 3O_2 \quad (6.5)$$

$$H_2O_2 \leftrightarrow H^+ + HO_2^- \quad (6.6)$$

$$HO_2^- + O_3 \rightarrow OH^* + O_2^* + O_{2-} \quad (6.7)$$

Hence, it can be observed that the oxidation and breakdown of compounds through ozonation occurs through two major routes.

1. Direct ozone attack on organic compounds (Equations 6.1–6.4)
2. Addition of H_2O_2 to form hydroxyl radical
   2.1. Direct formation of hydroxyl radical (Equation 6.5)
   2.2. Indirect formation of hydroxyl radical (Equation 6.6)

Generally, a combination of ozonation with other technologies like H_2O_2 or UV is used for rapid and economically viable treatment (Gadipelly et al. 2014). Ozonation finds its application as a pretreatment step for producing bioenergy or can be used as the final effluent treatment polishing step, prior to the effluent is released into the environment. Toxic chemicals like phenols can be degraded by ozonation, thus detoxifying the water (Gilbert 1987). This proves to be efficient for removing color, while most of the organic colorants contain phenols or have unsaturated double bonds. Due to this property of the ozonation process, the overall biodegradability of the wastewater is increased, making it more efficient for biological treatment. Ozonation can be used as a pretreatment step for the degradation of the complex recalcitrant compounds (Battimelli et al. 2010).

The application of ozonation for final treatment or final polishing step has not proved to be feasible, since low COD (chemical oxygen demand) removal for organic effluents has been observed (Coca, Peña, and González 2005). This low COD removal is caused due to low mineralization of organic substrates. Hence, higher levels of ozone doses are needed for high mineral-based substrates, which increases the process cost, reduce biodegradability, and can aid the formation of toxic products (Ikehata and Li 2018). Another disadvantage of the process is the inability of ozone to break down larger molecule chemicals as compared to other treatment methods like Sonolysis (Cesaro and Belgiorno 2013).

Malik et al. (2020) studied the enhancement in biogas production by pretreatment of yard waste with ozonation, Fenton, and peroxone treatment. For ozonation, the reaction was performed in a 1.5L glass reactor by batch mode. Different tests were carried out to optimize the ozone dosage and the operating pH of the process. The ozone dosage was varied between 0.2 and 2.5 g/h, and the pH was varied between 3 and 9. For these batch studies, 10g of yard waste was mixed with 0.5L distilled water for a batch time of 15 minutes to 1 hour. The results show that the highest TRS (total reducing sugar) concentration of 70.5 mg/g biomass was seen at a pH value of 3. TRS values of 47.2, 53.0, and 60.5 mg/g biomass were for pH values of 9, 7, and 5, respectively. Hence, it can be seen that the degradation mechanism is favorable at acidic pH range. At higher pH values, the reduced degradation of cellulose, hemicellulose, and lignin was observed. The reason behind this phenomenon is scavenging of hydroxyl radical by lignin degradation products or some associated intermediate compounds; for ozone dosage optimization it was found out that the TRS concentration increases with increased ozone concentration up to an optimum value of 1.8 g/h at a value of 155 mg/g biomass, after which it reduces. It is observed that the highest TRS value of 425 mg/g was obtained at pH of 3, ozone dosage of 1.8 g/h, and treatment time of 30 minutes.

Yeom et al. (2002) analyzed the effect of pretreatment using ozone on the biodegradability of municipal wastewater sludge. A cylindrical-shape reactor was used for ozonation reaction having a flow rate of 50 mg-O_3/L gas with an increment of up to an ozone dosage of 0.2 gO_3/g-SS. There is a 200%–300% rise in methane production due to ozonation pretreatment, as the biodegradability is increased.

Malik et al. (2019) evaluated the application of ozone pretreatment on molasses-based biomethanated distillery wastewater on the composting process. The distillery effluent is characterized by COD levels of 10^5 mg/L, dark brown color, high BOD (biochemical oxygen demand) levels, and high organic carbon content. The ozone feeding was at a flow rate of 4.6 g/h into a glass column reactor for 1 L quantity of distillery wastewater. Post-treatment, the gas chromatography–mass spectrometry (GS–MS) and Fourier transform infrared (FTIR) analysis was done to understand the degradation of organic compounds. It was observed that the biodegradation rate was higher for ozone pretreated wastewater. This, in turn, reduced the carbon/nitrogen ratio to values of 13.4, 14.6, and 17.6, respectively, for press mud 1:3, 1:4, and 1:5. The germination index value of 96% was seen with the highest toxicity reduction, as opposed to only 21% for untreated wastewater. This study shows the effect of ozone pretreatment for distillery wastewater and, how it is effective for bio composting and producing good quality compost.

6.2.4 Fenton Oxidation

Fenton oxidation is an effective process used for the treatment of the recalcitrant compounds in wastewater. Fenton reagents, i.e., Fe^{2+} and H_2O_2, are used in the process to generate hydroxyl radicals. Equation 6.8 is the primary Fenton reaction.

$$Fe^{2+} + H_2O_2 \rightarrow Fe^{3+} + OH + OH^* \tag{6.8}$$

Ferrous ions initiate the breakdown of H_2O_2 to produce OH radicals. This OH radical oxidizes the organic products by removal of protons, resulting in the production of organic radicals, as seen in Equation 6.9.

$$RH + OH^* \rightarrow R^* + H_2O \tag{6.9}$$

The chain of substrate may be elongated by the radical that is reacting with other organic substrates. Catalyst in excess may result in scavenging of OH radicals, thereby, decreasing the oxidation potential as seen in Equation 6.10.

$$Fe^{2+} + OH^* \rightarrow Fe^{3+} + OH \tag{6.10}$$

The utilization of photo-Fenton in the process leads to an increase in the oxidation efficiency and lowers the usage of chemicals. In this process, solar energy/UV radiation is used as a medium to produce radicals from the reactants. Heterogeneous Fenton catalysts have become a promising treatment technology. The catalyst is fixed on a solid catalyst bed, thereby oxidizing the refractory organics effectively (M'Arimi et al. 2020).

UV-H_2O_2 along with Fenton reagents (Fe^{2+}, Fe^{3+}), called the *photo-Fenton process*, generates a larger quantity of OH radicals than the standard Fenton treatment.

Photoreduction of Fe^{3+} for the reproduction of Fe^{2+} takes place in this process (Xu et al. 2020). The produced Fe^{2+} combines with hydrogen peroxide to give Fe^{3+} and OH^* seen in Equation 6.11.

$$Fe\left(OH\right)_2^+ + hv \rightarrow Fe^{2+} + OH^* \tag{6.11}$$

Further, H_2O_2 undergoes direct photolysis to generate OH^*. This accelerates the rate of deterioration of toxic contaminants seen in Equation 6.12.

$$H_2O_2 + hv \rightarrow 2 \cdot OH \tag{6.12}$$

To increase the effectiveness of the photo-Fenton process, ligands like oxalate, EDTA, and various organic carboxylic acids are added and reacted with Fe^{3+} (photocatalysis). Utilization of these organic ligands accelerates the reduction of Fe^{3+} to Fe^{2+}, resulting in large production of OH radicals and low ferric-sludge generation.

Bhange et al. (2010) investigated the pretreatment of garden biomass (GB) with the help of Fenton's reagent and the impact of concentrations of H_2O_2 and Fe^{2+} on biodegradation of recalcitrant compounds. The aim was to extract a considerable amount of cellulose, lignin, and hemicellulose present in GB. Initially, garden biomass containing twigs, clippings, roots, flowers, etc. were screened, followed by air drying for about 24 hours and sun drying for about 3 days. The dried product was then crushed with the help of a pulverizer. Every experiment was performed using 100 mL of Fenton's reagent ($FeSO_4.7H_2O$) and 5 g of garden biomass Fenton's reagent. The concentration of the reagent varied from 250 to 1,000 ppm, while that of H_2O_2 varied from 1,000 to 10,000 ppm. Investigation of Fenton's pretreatment on cellulose and hemicellulose degradation proved to be efficacious. H_2O_2 and Fe^{2+} play an important role in estimating the efficacy of the degradation. Increasing the concentration of H_2O_2 and higher reaction temperatures led to effective extraction of lignin and cellulose from GB, thereby increasing rates of degradation of recalcitrant compounds. Increasing concentrations of Fe^{2+} result in scavenging of OH^*, thereby reducing the degradation efficiency. Results indicated that garden biomass comprised of 94.10% organic matter, 25.68% lignin, 26.2% hemicellulose, 38.5% cellulose, and 49.12% organic carbon. The total nitrogen content of garden biomass was estimated to be around 1.65%. During the pretreatment, Fe^{2+} ion concentration was limited to catalytic amounts, because high dosage may result in adsorption on the substrate being treated, thus, affecting the activity of the treatment.

Zawieja and Brzeska (2019) studied the generation of biogas in methane fermentation of excess sludge with the help of Fenton's reagent (Dose of Fe^{2+} ions −0.08 gFe/g; Fe^{2+}:H_2O_2 of 1:5 was the preferred condition). Excess sludge undergoing deep oxidation of Fenton's reagent led to a rise in SCOD (soluble chemical oxidation demand) value. Moreover, the efficacy of the methanation process increased. The 28-day fermentation process of sludge resulted in 0.53 L/g VSS of biogas yield, 59% of degradation degree, and a 7-fold rise in the SCOD value.

6.2.5 Cavitation Methods

The process of formation of cavities or microbubbles and subsequent collapse that lead to the release of energy is known as cavitation. Usually, there are two major types of cavitation routes that are commonly used such as ultrasound cavitation and hydrodynamic cavitation. The ultrasound cavitation technique has been broadly used in industries. Ultrasound frequencies are used for treatment in the range from 16 kHz to 100 MHz. Very high temperatures greater than 5,000 K and more than 100 atm are generated at localized areas in microbubbles, which causes break down of the water molecule to form hydroxyl radicals that can be seen in Equation 6.13.

$$H_2O + 0.5O_2 \rightarrow OH^- + OH^* \quad (6.13)$$

This produced hydroxyl radical reacts with some substrate that produces the radical of the substrate. The reaction is seen in Equation 6.14.

$$R + OH^* \rightarrow R^* + H_2O \quad (6.14)$$

Further substrate chain reactions can occur with another substrate or with oxygen molecules as well. The reactions are seen in Equations 6.15 and 6.16.

$$R^* + O_2 \rightarrow ROO^* \quad (6.15)$$

$$ROO^* + R1H \rightarrow ROOH + R1^* \quad (6.16)$$

A significant advantage of using cavitation is that it does not require any addition of chemicals, and this cuts down the recurring costs of the process. Powerful oxidative radicals are produced by this process, which leads to the breakdown of substrates into small particles, thus increasing biodegradability. Hence, sonication is an effective method for pretreatment of bioenergy substrates, since the biodegradability and bioenergy yields are increased (Bougrier et al. 2006; Braguglia et al. 2012). However, the primary limitation of sonication is the process that is not optimized fully for industrial application due to the large energy requirement associated with it.

Patil et al. (2016) investigated the pretreatment of waste straw with HC (hydrodynamic cavitation) to enhance biogas production. Different wheat straw to water concentrations ranging from 0.5% to 1.5% have been tested in HC stator and rotor assembly with different rotor speeds ranging from 2,300 to 2,700 RPM. The treatment times have also been varied ranging from 2 to 6 minutes. Post-treatment, the methane yield obtained was 77.9 mL, which is almost a 100% increase from the untreated value of 31.8 mL. This confirmed the feasible application of HC in the pretreatment of wheat straw wastewater. An experimental approach that used a combination of KOH along with HC was tested, which further increased the methane yield to 172.3 mL, which is 440% more than the initial value. The research has concluded that significant biogas and bioenergy production has been obtained with the application of HC. This pretreatment method can be efficiently used for increasing the anaerobic digestion efficiency of wheat straw, which is a lignocellulosic-based biomass. The work has highlighted the scale-up abilities of hydrodynamic cavitation treatment devices. Hence, it was proved that hydrodynamic cavitation is a very efficient pretreatment mechanism to enhance the production of biogas of wheat straw using anaerobic digestion.

Fardinpoor et al. (2021) studied the production of biofuel using cyanobacteria using HC-assisted sodium hydroxide pretreatment. These cyanobacteria cultures have been known to grow in different waste and wastewater types producing landless biomass. Biofuels obtained by processing this biomass are known as third-generation biofuels. The production of methane was analysed by HC-assisted NaOH treatment of cyanobacteria. The study highlighted that HC is a low-cost method for cell disruption. The results showed an increased methane production by around 20% with an increase of 2%–35% of soluble COD. Further optimization of the process can be done with the help of batch and continuous testing data.

Zieliński et al. (2019) reported the cavitation methods as pretreatment for the production of biogas from an agricultural biogas plant. The agricultural lignocellulosic waste was tested by two cavitation methods, namely, ultrasound and HC. Due to hydrodynamic cavitation pretreatment, biogas production was enhanced by 16.6%. Similarly, the biogas production was enhanced by 24.6% in ultrasound pretreatment. The results have emphasized that the consumption of energy from HC was less as compared to ultrasound treatment. Hence, it has been observed that even though both cavitation technologies were effective, hydrodynamic cavitation was found to be cost effective with a scope of scale-up and industrial application.

6.2.6 Photocatalysis

Photocatalysis is the type of technique in which free radicals are formed when it interacts with photons. When the fine slurry of semiconductor particles was introduced in water with an irradiation of energy, photon interactions took place from the semiconductor surface and molecules of water. Semiconductors used for photocatalysis are ZnO, CdS, ZnS, and TiO_2. TiO_2 is widely used semiconductor in photocatalysis, where it also offers various advantages of being environment friendly, cheap, chemically stable, and less toxic in nature (Maryam et al. 2021).

Mechanism of photocatalysis:

Initially the metal surface is subjected to energy (either solar or UV), which results in the formation of electron hole pairs (M'Arimi et al. 2020), as seen in Equation 6.17.

$$TiO_2 + \text{Energy} \rightarrow TiO_2(\text{hole}) + e^{-1} \tag{6.17}$$

The electron formed reduces the oxygen-forming superoxide at the metal electrode, as seen in Equation 6.18.

$$O_2 + e^{-1} \rightarrow (O_2)^{*-} \tag{6.18}$$

The TiO_2 (hole) formed oxidizes hydroxyl ions (OH^-) or water molecules to generate hydroxyl radicals (OH^*), as seen in Equations 6.19 and 6.20.

$$TiO_2\,(\text{hole}) + H_2O \rightarrow TiO_2 + H^{+1} + OH^* \tag{6.19}$$

$$TiO_2\,(\text{hole}) + OH^{-1} \rightarrow TiO_2 + OH^* \tag{6.20}$$

It is also observed that the TiO_2 (hole) formed can directly oxidize the substrate, generating a substrate radical. This can be seen in Equation 6.21.

$$TiO_2\,(\text{hole}) + R-H \rightarrow R^* + TiO_2 + H^+ \tag{6.21}$$

Figure 6.3 shows the mechanism of photocatalysis for treating organic pollutants (M'Arimi et al. 2020).

Photocatalysis is a well-known technique that can be used for the treatment of wastes containing high COD concentration. In biological treatment system, photocatalysis acts as a pretreatment, ensuring biodegradability of highly biorefractory pollutants present in the wastewater. Waste-activated sludge is subjected to photocatalytic alkaline treatment using NaOH and TiO_2 dosage of 0.4% and 0.5 g/L, respectively. Alkaline-photocatalytic pretreatment of waste shows methane production up to 450 NmL/g, which is observed to be 70%–72% higher compared to the raw sludge. Thus, the methane production has significantly increased as compared to the individual methods of alkaline and photocatalytic treatment (Maryam et al. 2021).

Treatment was done using $C_3N_{4+x}H_y$ and TiO_2 nanotubes as catalysts in visible light photocatalytic treatment to wastewater, where it was observed that around 95%–97% of 2-chlorophenol was removed with enhanced biogas production. The TiO_2 nanotubes have melamine around 2%, which increases the degradation of 2-chlorophenol, where the initial concentration of 2-chlorophenol was approx. 30–32 mg/L.

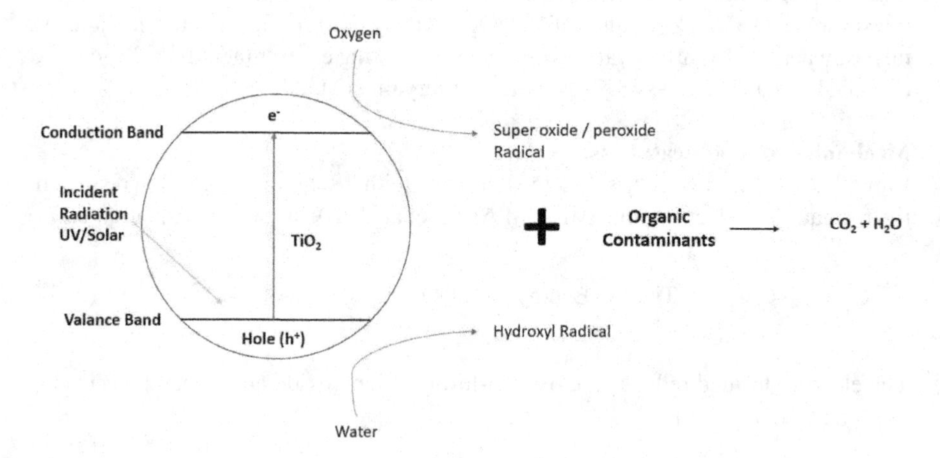

FIGURE 6.3 Mechanism of photocatalysis for treating organic pollutants.

The pH of the waste was maintained at near 7. The catalysts (TiO_2/C_3N_4) help in solubilizing the rigid structure of the wastewater sludge, thereby releasing soluble organics and improving the soluble chemical oxygen demand to 4,550–4,600 mg/L. The maximum removal of 2-chlorophenol was achieved after 3 hours (96.6%). After the photocatalytic pretreatment, anaerobic digestion of sludge shows that the methane production increases to 720–725 mL/kg, where volatile solids play a vital role in proving it as useful technique for treating wastewater and biogas generation (Anjum et al. 2018a).

Cassava starch wastewater has a high polluting potential due to very high organic loads and high cyanide concentration. It is treated by maintaining the pH and degrading the cyanide present in it. pH is adjusted using oyster shells, which are a source of $CaCO_3$, and cyanide is degraded using photocatalysis where the Degussa P25 TiO_2 is used as a catalyst. Stabilization of pH near 6 was achieved after 6 hours of reaction. This was effective in controlling the effluent acidity. After the stabilization of pH, photocatalysis using P25 TiO_2 catalyst is performed resulting in degradation of cyanide concentration to 73%–75%. Thus, photocatalytic pretreatment of this starch wastewater reduces the toxic content, cyanide concentration, and 25%–30% increase in the amount of biogas was observed when subjected to anaerobic digestion (Andrade et al. 2020).

On-site photocatalytic treatment of WAS using TiO_2 catalyst in anaerobic digestion enhances the degradation of WAS and increases methane production in photocatalytic anaerobic fermenter. Reactor consists of photocatalytic and digestion unit in which the function of photocatalytic unit is to supply soluble organics constantly, which does not have a negative impact on the activity of methanogens where the optimum photocatalytic time was 4 hours/day. Anaerobic digestion was carried out for 35 days, which has shown methane production up to 1,260–1,270 mL methane/L sludge. The VS and total COD removal were around 66%–68% and 60%–62%, respectively. When photocatalysis is not used as a pretreatment, the methane production ranges below 950 mL/L in sludge. Also, total COD and VS removal are around 42%–44% and 48%–50%, respectively. Hence photocatalysis as a pretreatment enhances methane production and efficacious wastewater treatment (Liu et al. 2014).

It was observed that the visible light photocatalysis of wastewater using ZnO/ZnS polyaniline nanocomposite catalyst increases the degradability of contaminants, accelerates hydrolysis of WAS, and increases the formation of soluble COD by 6–7 times. The pH of the solution was maintained around 3–5, and pretreatment was carried out for a duration of 360 minutes. The catalyst dosage was kept constant at 0.25 g/mL. Photocatalysis as a pretreatment (ZnO–ZnS polyaniline) of wastewater followed by anaerobic digestion results in removal of 68%–70% total COD and 60%–62% organic matter. Furthermore, photocatalytic pretreated sludge improves the biogas production by 1.5–1.7 times (1,600–1,650 mL/L VS) as compared to raw sludge (1,020–1,030 mL/L volatile solids) (Anjum et al. 2018b).

Photocatalysis can be coupled with other oxidation processes to achieve higher efficiency in treating the wastewater. UV (TiO_2) along with H_2O_2 treatment has resulted in higher removal efficiency of COD and phenols in pharmaceutical wastewaters (Gadipelly et al. 2014). Therefore, photocatalysis is an efficacious technique for pretreatment of wastewater containing high COD content and large refractory pollutants.

6.2.7 Ultraviolet (UV) Radiation

Advanced oxidation processes (AOPs) have been widely used for the treatment of recalcitrant compounds present in wastewater. UV-based AOPs are widely preferred in the field of wastewater treatment. Photolysis ($UV + H_2O_2$) is one of the promising technologies used in AOPs. UV radiation is used in combination with other AOPs like photocatalysts, hydrogen peroxide, Fenton oxidation, or ozonation. UV, being used as a stand-alone method, increases the process costs, which is the main shortcoming of most of the oxidation processes (M'Arimi et al. 2020). UV radiation enhances oxidizing agents such as H_2O_2 and ozone to form free radicals. UV radiation is combined with a wavelength of 280 mm for the spontaneous formation of OH radicals (Equation 6.1):

$$H_2O_2 \xrightarrow{hv} 2OH^* \quad (6.22)$$

UV is treated with ozone to form O1 (D), which is further reacted with water to form hydrogen peroxide. This H_2O_2 in the presence of UV produces two OH radicals (Equations 6.23 and 6.24).

$$O_3 \xrightarrow{hv} O^1\,(D) + O_2 \quad (6.23)$$

$$O1\,(D) + H_2O \xrightarrow{hv} H_2O_2\ 2OH^* \quad (6.24)$$

6.2.7.1 UV/H_2O_2 System

Utilization of UV radiation along with H_2O_2 reagent in wastewater treatment is one of the efficient methods for effective removal of organic and recalcitrant compounds in wastewaters (Cuerda-Correa et al. 2019). In addition, H_2O_2 being a costly reagent increases the production costs of the process. The initial concentration of H_2O_2 must be regulated in order to obtain the maximum efficiency of the process. H_2O_2 in excess may lead to scavenging of OH^- radicals. This may have a negative impact on the overall degradation efficiency.

6.2.7.2 UV/TiO_2 system

Alvarado-Morales et al. (2017) evaluated the application of an oxidation process merging UV with TiO_2 for photocatalytic pretreatment of wheat straw for an increased biogas generation. TiO_2 being highly active, cost-effective, and a chemically stable catalyst (semiconductor) is preferred over other catalysts like CdS and zinc oxide. Photocatalytic oxidation of biomass could be an effective choice rather than the usage of chemicals (pressure and temperature conditions), which may lead to contamination of the degradation process. The main objective of pre-treatments is to degrade the valuables and sugars present in lignocellulosic biomass. Photocatalytic pretreatment led to an improved anaerobic decomposition of wheat straw. With the help of UV/TiO_2 system [1.5% (w/w) TiO_2 and UV radiation (wavelength varying from 200 to 400 mm) for 3 hours], ferulic and vanillic acids were identified in the range of 1.67 ± 0.01 and 91.18 ± 2.00, respectively. This treatment method has increased the methane yield by 37%. Overall, the lignin-rich biomass was effectively treated under mild conditions to produce valuable products like vanillic acid, ferulic acid, and vanillin. The long duration of irradiation had a positive impact on the methanation process leading to the generation of products like carboxylic acids and aldehydes. Irradiation time was increased from 2 to 3 hours to check the degradation efficiency. Catalyst concentration being constant (1.5% (w/w) TiO_2), longer irradiation time raised the concentration of vanillic acid by 57.7%. A high dosage of TiO_2 (2.0% (w/w)) with an irradiation time of 3 hours led to an increase in vanillic acid by 21.6%. UV radiation as a stand-alone process resulted in negligible decomposition of lignocellulose.

6.2.8 Electrochemical Oxidation

The basic principle of electrochemical oxidation is using redox reactions for the removal of pollutants in the wastewater. Pollutants are oxidized at anode, while reduction of heavy metals is done at cathode. The two types of electrochemical oxidation are direct and indirect oxidation (Garcia-Segura et al. 2018). Direct oxidation takes place on anode involving charge transfer between the anode surface and pollutant. However, the electrolysis must be conducted at a potential higher than that of water oxidation potential. If this is not followed, the electrode surface may get poisoned, thereby stopping the process. On the other hand, in indirect oxidation, high-level oxidant species are generated at the electrode surface. Reactive oxygen species and chlorine active species are further classified in indirect electrochemical oxidation method. In reactive oxygen, *in situ*-generated hydroxyl radicals are main oxidants, while in chlorine active species *in-situ*-generated chlorine activated species are used for pollutant removal as main oxidants (Gadipelly et al. 2014; Garcia-Segura et al. 2018).

In electrochemical oxidation, effective pollutant degradation occurs when the electrode (anode) is diamond coated. It is observed that total organic carbon (TOC) removal is 96%–98% when boron doped diamond (BDD) is used and is preferred due to its high corrosion stability. This facilitates generation of hydroxyl radicals at anode, which serves as an oxidant for treatment of pollutants (Loos et al. 2018). Due to the high standard reduction potential (2.8 V), it nonselectively reacts with the organic contaminants present in the wastewater until total mineralization occurs (Sirés and Brillas 2012; Gadipelly et al. 2014).

Babu et al. (2009) reported that sulfamethoxazole, diclofenac, and 17-alpha-ethinylestradiol readily degrades in sewage wastewater and real effluent wastewater (Loos et al. 2018). Wastewater containing 1,365 mg/L COD, 22.1 mg/L TOC and 3,550 mg/L ppm TDS (total dissolved solids) is subjected to electrochemical oxidation at varying current densities. Major contaminants present in the waste are dexamethasone and gentamicin. Flow rate of water was kept at 10 L/hr. pH was adjusted around 7.5–7.7, and a constant dosage of NaCl was maintained at 3 g/L. When current density was varied from 2 to 4 A/dm^2, COD removal was found to be increased. 85%–87% COD removal was achieved when current density was kept constant at 4 A/dm^2. Thus, electrochemical oxidation employed for treating wastewater at high current densities increases the COD removal (Babu et al. 2009).

TABLE 6.2
Contaminant Removal by Electrochemical Oxidation Using Various Anodes

Major Contaminants	Anode	Result and Inference	References
Phenolic wastewater	Ti/RuO_2	Complete removal of COD and phenol is achieved	Fajardo et al. (2017)
Perfluorooctanoic acid	PbO_2	92%–93% removal of perfluorooctanoic acid	Zhuo et al. (2017)
Dexamethasone and gentamicin	NaCl	85%–87% COD removal is observed at current density 4 A/dm	Babu et al. (2009)
Perfluoro-octane	$Ti/Sb–SnO_2$	More than 99% organic removal is observed	Yang et al. (2017)
Tetracycline	$Ti/Sb–SnO_2$	80%–82% (TOC) removal	Zhi et al. (2017)
TDW	Pt/Ti	84%–86%, 70%–72%, and 35% removal of COD, BOD_5, and TKN	Vlyssides et al. (1999)
Ferulic acid	BDD	Complete COD removal (9,500 Wh/m^3 of energy consumed)	Ellouze et al. (2016)

Textile dye wastewater was treated using electrochemical oxidation using Pt/Ti and stainless steel 304 as anode and cathode, respectively. After TDW is passed through the electrolytic cell, the organic contaminants in the wastewater were oxidized to water and CO_2. Post 18 minutes of electrolysis at a current density of 0.89 A/cm^2, 36% HCl (2 mL) was added, which resulted in 84%–86% and 70%–72% COD and BOD_5 removal, respectively. 35% total Kjeldhal's nitrogen (TKN) removal was achieved. Biodegradability has shown an increase from 0.4629 to 0.6578 (Vlyssides et al. 1999).

Table 6.2 summarizes the removal of contaminants by electrochemical oxidation using various anode (Martínez-Huitle and Panizza 2018).

Hybrid technologies involving electrochemical oxidation can enhance the treatment of wastewater by efficient removal of contaminants. The combination of electrochemical oxidation with physical and chemical processes (coagulation/flocculation) improves the efficiency of COD removal and increases biodegradability of the waste (M'Arimi et al. 2020). These hybrid techniques not only improve the removal efficiency of contaminants but also reduce the total cost of operation. The addition of processes can improve the COD reduction, since the AOP will target the recalcitrant molecules, thus reducing the load on other treatment methods and thereby reducing process cost. Further experimentation can be done in this area to study the impact of electrochemical oxidation as a pretreatment for treating wastewaters to generate the bioenergy.

6.2.9 Hydrogen Peroxide

H_2O_2 is an oxidizing agent and is merged with different advanced oxidation processes, thereby acting as a medium for the generation of OH radicals. Hydrogen peroxide oxidation, as a stand-alone technique, is comparatively a less-efficient process, as high dosage requirements lead to reduced energy yields and increased costs. H_2O_2 plays a crucial role in the abatement of toxic contaminants (Oturan and Aaron 2014).

Song et al. (2013) investigated the pretreatment of rice straw using H_2O_2 so as to obtain an effective biogas yield. The pretreatment led to a significant reduction in the hemicellulose (3.7%–60.3%), cellulose (0.9%–22%), and lignin (0.4%–12.5%) content. Increasing degradation rates in the case of hemicellulose reported that H_2O_2 could be an efficacious agent for the degradation of rice straw cellulose, thus enhancing the methane yield. Under optimal conditions, i.e., 2.68% H_2O_2 concentration, 6.18-d pretreatment time, and 1.08 S/I ratio, results showed a higher methane yield of around 290 mL/g VS (volatile solids).

Siciliano et al. (2016) studied the pretreatment of olive mill waste using H_2O_2 (in alkaline conditions, room temperature) to enhance the anaerobic biodegradability. This is a simplified treatment, as any additional catalysts are not being used. Methane yield of around 0.328 LCH_4/g COD removed was reported. Polyphenols were reduced by around 72%. Many experiments were conducted to analyse the effect of pH and peroxide dosage on process performance. The kinetic analysis of experiments investigated the reaction mechanisms and optimized the operating procedures. It was reported that significant biogas production was not detected on raw olive mill waste, whereas a COD reduction of about 77% was observed.

6.2.10 Wet Air Oxidation (WAO)

Wet air oxidation (WAO) is an established technique used for treating highly toxic and organic industrial waste water efficaciously. It belongs to a class of thermal chemical oxidation methods (M'Arimi et al. 2020). WAO is advantageous when the waste to be treated has high concentration of bio refractory contaminants (M'Arimi et al. 2020). The operating pressures and temperatures for WAO process are in the range of 500–20,000 kPa and 398–593 K, respectively. The inorganic and organic components are oxidized in liquid phase at high pressures and temperatures. As the solubility of oxygen in the waste solution increases thereby providing a large driving force for oxidation (Mishra et al. 1995). High pressure conditions are required during WAO operation to ensure that the water remains in liquid state (Mishra et al. 1995). The oxidizing conditions of wastewater depend upon the types of industrial effluents. Oxidation mainly depends on the operating temperature, RTD (residence time distribution), partial pressure of O_2, and oxidizability of contaminant (Mishra et al. 1995).

The end products resulting after WAO treatment include carbon dioxide and other nonharmful products. The organic contaminants present in the waste are converted as seen in Table 6.3.

TABLE 6.3
Organic Contaminants Present in Wastewater

Organic Contaminants	End Products After Conversion
Carbon	Carbon dioxide
Sulfur	Sulfates
Nitrogen	Ammonia, nitrogen trioxide, or N_2
Halogens	Halides

6.2.11 Mechanism of WAO

Equations 6.25 and 6.26 describe the radical formation, whereas Equations 6.27 and 6.28 show prolonged substrate chains or formation of catalyst species (M'Arimi et al. 2020):

$$\text{R-H} + O_2 \rightarrow R^* + (HO_2)^- \quad (6.25)$$

$$\text{R-H} + (HO_2)^- \rightarrow H_2O_2 + R^* \quad (6.26)$$

$$R^* + O_2 \rightarrow ROO^* \quad (6.27)$$

$$R_1OO^* + R2H \rightarrow R_1OOH + R_2^* \quad (6.28)$$

WAO is very useful in treating excess amount of sludge, biomass production, and lignocellulosic pretreatment of biomass (M'Arimi et al. 2020; Bertanza et al. 2015).

WAO is environment friendly, as it does not produce sulfur or nitrogen oxides. If cellulose present in the biomass is economically and efficiently converted to fuel, then WAO can be effectively used to produce useful chemicals and energy from biomass residue (Mishra et al. 1995). The noncatalytic WAO study of biomass resulted in lignin formation along with carbohydrates. The parameters maintained were 120°C–240°C and 0.8–3.5 MPa (O_2 partial pressure). The solid fraction obtained after pretreating biomass with WAO contained cellulose with some amounts of lignin, while the liquid fraction contained lignin and hemicellulose. If ethanol is desired as the end product, the cellulose obtained is treated to give glucose by making it susceptible toward acid hydrolysis. At elevated temperatures, oxidation of biomass takes place resulting in the formation of glucose, methanol, and organic acids.

WAO is used as a pretreatment for treating distillery effluent having composition of COD and BOD around 100–150 and 35–50 kg/L, respectively. TDS makes a contribution of 80–100 kg/L (Malik et al. 2014). The experiment was carried out on a distillery effluent (molasses distillery) having pH around 7 in 1% biomass, where the BOD and COD compositions were around 45,000–50,000 mg/L and 6,000–8,000 mg/L, respectively (Malik et al. 2014). The results have shown that WAO-pretreated waste has reduced COD to 65%–68% and 85%–90% during aerobic and anaerobic coupled biological treatment processes, respectively. WAO-pretreated wastewater has higher BI (in the range of 0.4–0.8), which enhances the biogas production (Malik et al. 2014).

During the treatment of industrial waste, sulfides and chromium present in the wastewater limit the production of biogas at low costs. Thus, pretreatment of the effluent helps in maximizing biogas production in short period of time with lower costs (Prabakar et al. 2018; Kumar et al. 2017).

A laboratory scale study conducted by Padoley et al. (2012) demonstrated the generation of biogas from a distillery effluent (biomethanated) using WAO as a pretreatment. The experimental study was performed at temperature and pressure around 423–473 K and 6–12 bar, respectively, with reaction time varying from 15 to 120 minutes. The distillery effluent had a pH and COD around 7–8 and 40,000–42,000 mg/L, respectively. The biodegradability index of the distillery effluent ranges from 0.15 to 0.2. The experimental results show that maximum BI (0.85–0.88) was obtained at 473 K, 1.2×10^6 Pascal with the reaction time around 120 minutes. The efficiency of WAO was found to be increased when initial COD loading was high. However, BI of 0.4 was good enough to yield a considerable amount of biogas. Hence, the operating parameters for experiment were kept at 448 K, 6 bar, and 30 minutes. The WAO-pretreated distillery effluent have shown COD reduction in the range of 30%–35% and methane generation around 50%, whereas the untreated waste had shown 5%–10% COD reduction and 1%–2% methane generation. It was observed that the post-treatment by-products were also formed such as volatile fatty acids, which indicates a feasible potential for the formation of anaerobic biogas (Padoley et al. 2012).

As compared to the conventional WAO, catalytic wet air oxidation (CWAO) requires less energy. Catalyst rises the oxidation rates of pollutants, thereby increasing the efficiency of the process, whereas CWAO operates at lower temperature and pressure than conventional WAO. Thus, same degree of COD removal can be achieved at lower operating conditions using CWAO (Levec and Pintar 2007).

Treatment of distillery wastewater using homogeneous CWAO as a pretreatment is also studied. The pH and COD of the waste was around 8–9 and 40,000 mg/L. After pretreating with CWAO, activated carbon adsorption followed by anaerobic and aerobic digestions was conducted. The catalyst used was $FeSO_4$. Oxidation is controlled (0.65–0.7 MPa partial pressure of O_2) and catalyst loading was around 30–35 mg/L. CWAO followed by carbon adsorption has shown to increase the BI from 0.2 to 0.4 and COD reduction up to 70%–75%. BI was further increased to 0.5–0.55 when anaerobic digestion was performed along with biogas generation. It is estimated that 68%–70% methane is formed for 1 m^3 of water (1–1.2 Nm^3 of biogas generation). Furthermore, aerobic digestion was performed as a final step, which enhanced the COD removal to 90%–92% and increases the BI to 0.58 (Bhoite and Vaidya 2018).

WAO is effective for treating wastewater having high COD content. It can be coupled with other AOPs to increase the treatment efficiency. The high operating conditions required for WAO will be an expensive process. Studies conducted on wet air oxidation indicated that COD removal was found to be more efficient when pure oxygen flow was used instead of air. Flow COD removal was 32% at 0.5 L/min when air was used, but it was increased drastically to 56% when oxygen was used (Lin, Ho, and Wu 1996). However, using pure oxygen is not as economical as compared to air. If pure oxygen is preferred over air for WAO then the capacity of the plant has to be increased, which is not always feasible. Hence wet air oxidation is used for treating or pretreating industrial waste efficaciously to enhance biodegradability and biogas formation.

6.3 SUMMARY OF ADVANCED OXIDATION TECHNOLOGIES

After the treatment section for analyzing the treatment results of different oxidation technologies on bioenergy production, a general summary of different technologies has shown. The summary highlights the advantages, disadvantages, along with the possible application scope and the specific comments. It is elaborated in detail in Table 6.4.

TABLE 6.4
Summary of AOPs

AOP	Advantages	Disadvantages	Application Scope	Remarks	Reference
Ozonation	• Sludge formation is low. • Increased biodegradability of wastewater. • Efficient Breakdown of lignin is seen.	• Toxicity due to excess dosage. • High cost. • Low mineralization of substrates.	• Biofuel Production by pretreatment of biomass. • Sludge Minimization • Post treatment of final effluent.	Dosage should be well maintained, low solubilization seen, and complicated application due to complex reactions.	Chandra et al. (2012), Wu, Ein-Mozaffari, and Upreti (2013), M'Arimi et al. (2020)
Cavitation	• Low COD loss. • Breakdown of phenolic compounds and lignin. • Increased biodegradability • HC has scope of scale-up.	• High energy costs. • HC not used at commercial scale. • Complex reactor for UC. • Scale-up problems observed.	• Pretreatment of lignocellulosic biomass. • Recalcitrant effluent pretreatment before anaerobic digestion. • Algae biomass disruption for biofuel production.	Process optimization needed with simple and scalable reactors.	Subhedar, Ray, and Gogate (2018), Fardinpoor et al. (2021), Patil et al. (2016), Zieliński et al. (2019)
Fenton	• Simple reactor setup. • Increased biodegradability. • High yield of biogas.	• pH-sensitive process. • Loss of COD. • High chemical costs. • Sludge formation is more.	• Production of biomass from biofuel. • Leachate treatment. • Final effluent treatment.	Photo-Fenton process can be worked upon and can be used effectively.	Michalska et al. (2012), Kato et al. (2014)

(Continued)

TABLE 6.4 (*Continued*)
Summary of AOPs

AOP	Advantages	Disadvantages	Application Scope	Remarks	Reference
Wet Air Oxidation	• Low sludge formation and COD loss. • Increased Biodegradability	• High cost. • Low solubilization. • High energy requirement. • Corrosion	• Production of bioenergy from biomass. • Sludge Minimization. • Final Effluent Treatment.	Heterogeneous catalysts for preventing microbial poisoning. Scale-up and process optimization needed.	Banerjee et al. (2009), Arvaniti, Bjerre, and Schmidt (2012)
Photocatalysis	• Low COD loss and sludge formation. • Fast process. • Increase in biodegradability	• Complex reactor. • Optimized in small scale only.	• Upgrading biofuel products. • Sludge Minimization	Detailed research into different catalysts needed. Process scale-up and optimization needed as well.	Zhu, Zhu, and Wu (2012), Alvarado-Morales et al. (2017)
UV/H_2O_2	• Low sludge formation and COD loss. • Simple reactor.	• High chemical cost. • Low solubilization and biodegradability.	• Biofuel production from lignocellulosic biomass. • Final Effluent Treatment.	Combination and integration of other processes with this process is needed.	Hamid et al. (2017); Amorós-Pérez et al. (2017)
Electrochemical Oxidation	• High COD removal. • Effective on recalcitrant compounds	• Scale-up issues • High energy requirement. • Low selectivity.	• Microbial fuel cell. • Biofuel effluent treatment.	Process optimization and scale-up of processes is needed.	Jadhav et al. (2017), Rózsenberszki et al. (2017), Rathour et al. (2019)

6.4 CONCLUSION AND FUTURE SCOPE

This book chapter evaluates the incorporation of AOPs for the production of bioenergy through wastewater. In the current scenario, due to the depletion of fossil fuels, bioenergy has gained interest, and extensive research has been carried out in this field for the search of efficient treatment methods. The study has shown that AOPs are efficient for the treatment of lignocellulosic biomass like wheat/rice straw, bagasse, etc. The advanced oxidation pretreatment breaks down the lignin structure, making the wastewater susceptible to further treatment (hydrolysis) and subsequently producing higher amounts of bioenergy. The bioenergy yield is dependent on different factors like source of raw material, type of AOP used, and the operating conditions. More comparative studies are needed for analyzing different raw materials with various advanced oxidation technologies and a need to observe the bioenergy yield from the same. The dosage and exposure time of the AOPs are critical parameters that needed to be understood clearly before the application of AOP; in the treatment and production step, further optimization studies can be done as well to understand the impact of dosage and exposure time on production. The summary table gives future and possible insights into various commonly used advanced oxidation technologies.

For high COD effluent containing recalcitrant organic compounds, advanced oxidation technologies are used before biological treatment to eliminate the nonbiodegradable and recalcitrant compounds. The application of AOPs from this angle has two primary advantages. The first one being the enhanced recovery of biofuels, whereas the second one being the possibility of final post-treatment of effluent before disposal. WAO can be used to increase the production of other products like volatile acids from wastewater. However, the application of these AOPs should be done in a judicious manner, since an optimum dosage would be needed. For example, for H_2O_2 pretreatment, the hydrogen peroxide should be consumed in the pretreatment stage itself, or else it can hamper the action of biological processes further downstream. It is also necessary to understand the constituents of the wastewater in detail, as direct application of AOP can lead to the formation of even more toxic and recalcitrant compounds. The primary limitations of WAO are that the system has tendency of getting corroded along with very high energy requirements due to the high temperature and pressure requirements. Further research on CWAO is being done and new catalysts are being developed, which will lead to milder conditions for the setup. Cavitation methods successfully disrupt the bio algae cells, thus producing high biodiesel.

For the production of bioenergy by fermentation through excess sludge treatment, ultrasound treatment has proved to be better than other technologies, since it has the ability to break down large-sized molecules. This breakdown leads to an increase in biodigestion, thus enhancing production. Complex reactor setup and scalability of the process are the two primary hurdles in large-scale industrial applications. Additionally, a combination of ultrasound treatment with other technologies can prove to be a feasible and cost-effective technique.

Analysis of cost efficiency versus technical feasibility on different AOPs can be done for a wider scope of applications. The bioenergy substrates can also be used to provide other value-added bulk chemicals other than biofuels. Some cost-benefit analysis studies can also be done to highlight the cost aspect of these processes, thus making them industrially attractive. The major disadvantages, however, of AOPs in the bioenergy sector are associated with the cost of chemicals and the high energy requirement. Studies on reducing the cost impact of the processes through catalyzed AOPs can be done. A combination of different AOPs with each other or with the conventional processes can be done as a means of cost reduction. Generally, a combination of AOPs like UV/hydrogen peroxide, UV/O_3, ultrasound/ozone, or hydrogen peroxide, etc. are some commonly used combinations leading to enhanced hydroxyl radical generation, resulting in higher oxidation rates and leading to better treatment. A combination of AOPs has proved to be a promising way for the enhancement in production of biofuels where systematic research into the existing and newer techniques can prove to be important in the field of bio and renewable energy, wastewater treatment, and biofuels.

REFERENCES

"Alternative fuels data center: Maps and data – global ethanol production by country or region." 2021. Accessed July 15. https://afdc.energy.gov/data/10331.

Alvarado-Morales, M., P. Tsapekos, M. Awais, M. Gulfraz, and I. Angelidaki. 2017b. "TiO_2/UV based photocatalytic pretreatment of wheat straw for biogas production." *Anaerobe* 46 (August). doi:10.1016/j.anaerobe.2016.11.002.

Andrade, L.R.S., I.A. Cruz, L. de Melo, D. da Silva Vilar, L.T. Fuess, G.R. e Silva, V.M.S. Manhães, N.H. Torres, R.N. Soriano, R.N. Bharagava, and L.F.R. Ferreira. 2020. "Oyster shell-based alkalinization and photocatalytic removal of cyanide as low-cost stabilization approaches for enhanced biogas production from cassava starch wastewater." *Process Safety and Environmental Protection* 139: 47–59. doi:10.1016/j.psep.2020.04.008.

Anjum, M., H.A. Al-Talhi, S.A. Mohamed, R. Kumar, and M.A. Barakat. 2018a. "Visible light photocatalytic disintegration of waste activated sludge for enhancing biogas production." *Journal of Environmental Management* 216: 120–27. doi:10.1016/j.jenvman.2017.07.064.

Anjum, M., R. Kumar, S.M. Abdelbasir, and M.A. Barakat. 2018b. "Carbon nitride/titania nanotubes composite for photocatalytic degradation of organics in water and sludge: Pre-treatment of sludge, anaerobic digestion and biogas production." *Journal of Environmental Management* 223 (May): 495–502. doi:10.1016/j.jenvman.2018.06.043.

Apollo, S., Ma.S. Onyango, and A. Ochieng. 2014. "Integrated UV photodegradation and anaerobic digestion of textile dye for efficient biogas production using zeolite." *Chemical Engineering Journal* 245 (June). doi:10.1016/j.cej.2014.02.027.

Arvaniti, E., A.B. Bjerre, and J.E. Schmidt. 2012. "Wet oxidation pretreatment of rape straw for ethanol production." *Biomass and Bioenergy* 39 (April). doi:10.1016/j.biombioe.2011.12.040.

Babu, B.R., P. Venkatesan, R. Kanimozhi, and C. Ahmed Basha. 2009. "Removal of pharmaceuticals from wastewater by electrochemical oxidation using cylindrical flow reactor and optimization of treatment conditions." *Journal of Environmental Science and Health - Part A Toxic/Hazardous Substances and Environmental Engineering* 44 (10): 985–94. doi:10.1080/10934520902996880.

Banerjee, S., R. Sen, R.A. Pandey, T. Chakrabarti, D. Satpute, B.S. Giri, and S. Mudliar. 2009. "Evaluation of wet air oxidation as a pretreatment strategy for bioethanol production from rice husk and process optimization." *Biomass and Bioenergy* 33 (12). doi:10.1016/j.biombioe.2009.09.001.

Battimelli, A., D. Loisel, D. Garcia-Bernet, H. Carrere, and J.-P. Delgenes. 2010. "Combined ozone pretreatment and biological processes for removal of colored and biorefractory compounds in wastewater from molasses fermentation industries." *Journal of Chemical Technology & Biotechnology* 85 (7). doi:10.1002/jctb.2388.

Bertanza, G., R. Galessi, L. Menoni, R. Salvetti, E. Slavik, and S. Zanaboni. 2015. "Wet oxidation of sewage sludge: Full-scale experience and process modeling." *Environmental Science and Pollution Research* 22 (10): 7306–16. doi:10.1007/s11356-014-3144-9.

Bhange, V.P, S.P.M. Prince William, A. Sharma, J. Gabhane, A.N Vaidya, and S.R Wate. 2015. "Pretreatment of garden biomass using Fenton's reagent: Influence of Fe^{2+} and H_2O_2 concentrations on lignocellulose degradation." *Journal of Environmental Health Science and Engineering* 13 (1). doi:10.1186/s40201-015-0167-1.

Bhoite, G.M., and P.D. Vaidya. 2018. "Improved biogas generation from biomethanated distillery wastewater by pretreatment with catalytic wet air oxidation." *Industrial and Engineering Chemistry Research* 57 (7): 2698–704. doi:10.1021/acs.iecr.7b04281.

Bougrier, C., C. Albasi, J.P. Delgenès, and H. Carrère. 2006. "Effect of ultrasonic, thermal and ozone pre-treatments on waste activated sludge solubilisation and anaerobic biodegradability." *Chemical Engineering and Processing: Process Intensification* 45 (8). doi:10.1016/j.cep.2006.02.005.

Braguglia, C.M., A. Gianico, and G. Mininni. 2012. "Comparison between ozone and ultrasound disintegration on sludge anaerobic digestion." *Journal of Environmental Management* 95 (March). doi:10.1016/j.jenvman.2010.07.030.

Cesaro, A., and V. Belgiorno. 2013. "Sonolysis and ozonation as pretreatment for anaerobic digestion of solid organic waste." *Ultrasonics Sonochemistry* 20 (3). doi:10.1016/j.ultsonch.2012.10.017.

Chandra, R., H. Takeuchi, and T. Hasegawa. 2012. "Hydrothermal pretreatment of rice straw biomass: A potential and promising method for enhanced methane production." *Applied Energy* 94 (June). doi:10.1016/j.apenergy.2012.01.027.

Coca, M., M. Peña, and G. González. 2005. "Variables affecting efficiency of molasses fermentation wastewater ozonation." *Chemosphere* 60 (10). doi:10.1016/j.chemosphere.2005.01.090.

Cuerda-Correa, E.M., M.F. Alexandre-Franco, and C. Fernández-González. 2019. "Advanced oxidation processes for the removal of antibiotics from water. An overview." *Water* 12 (1). doi:10.3390/w12010102.

Deng, Y., and R. Zhao. 2015. "Advanced oxidation processes (AOPs) in wastewater treatment." *Current Pollution Reports* 1 (3). doi:10.1007/s40726-015-0015-z.

Ellouze, S., Ma. Panizza, A. Barbucci, G. Cerisola, T. Mhiri, and S.C. Elaoud. 2016. "Ferulic acid treatment by electrochemical oxidation using a BDD anode." *Journal of the Taiwan Institute of Chemical Engineers* 59: 132–37. doi:10.1016/j.jtice.2015.09.008.

Fajardo, A.S., H.F. Seca, R.C. Martins, V.N. Corceiro, I.F. Freitas, M. Emília Quinta-Ferreira, and R.M. Quinta-Ferreira. 2017. "Electrochemical oxidation of phenolic wastewaters using a batch-stirred reactor with NaCl electrolyte and Ti/RuO_2 anodes." *Journal of Electroanalytical Chemistry* 785: 180–89. doi:10.1016/j.jelechem.2016.12.033.

Fardinpoor, M., N.A. Perendeci, V. Yılmaz, B.E. Taştan, and F. Yılmaz. 2021. "Effects of Hydrodynamic cavitation-assisted NaOH pretreatment on biofuel production from cyanobacteria: Promising approach." *Bioenergy Research.* doi:10.1007/s12155-021-10286-0.

Gadipelly, C., A. Pérez-González, G.D. Yadav, I. Ortiz, R. Ibáñez, V.K. Rathod, and K.V. Marathe. 2014a. "Pharmaceutical industry wastewater: Review of the technologies for water treatment and reuse." *Industrial and Engineering Chemistry Research* 53 (29): 11571–92. doi:10.1021/ie501210j.

Garcia-Segura, S., J.D. Ocon, and M.N. Chong. 2018. "Electrochemical oxidation remediation of real wastewater effluents — A review." *Process Safety and Environmental Protection* 113: 48–67. doi:10.1016/j.psep.2017.09.014.

Gellé, A., and A. Moores. 2019. "Plasmonic nanoparticles: Photocatalysts with a bright future." *Current Opinion in Green and Sustainable Chemistry* 15 (February). doi:10.1016/j.cogsc.2018.10.002.

Gilbert, E. 1987. "Biodegradability of ozonation products as a function of COD and DOC elimination by example of substituted aromatic substances." *Water Research* 21 (10). doi:10.1016/0043-1354(87)90180-1.

Guimarães, V., M.S. Lucas, and J.A. Peres. 2019. "Combination of adsorption and heterogeneous photo-Fenton processes for the treatment of winery wastewater." *Environmental Science and Pollution Research* 26 (30). doi:10.1007/s11356-019-06207-6.

Guo, M., W. Song, and J. Buhain. 2015. "Bioenergy and biofuels: History, status, and perspective." *Renewable and Sustainable Energy Reviews* 42 (February). doi:10.1016/j.rser.2014.10.013.

Hamid, S., I. Ivanova, T.H. Jeon, R. Dillert, W. Choi, and D.W. Bahnemann. 2017. "Photocatalytic conversion of acetate into molecular hydrogen and hydrocarbons over Pt/TiO_2: PH dependent formation of Kolbe and Hofer-Moest products." *Journal of Catalysis* 349 (May). doi:10.1016/j.jcat.2017.02.033.

Hoang, A.T., S. Nizetic, H.C. Ong, C.T. Chong, A.E. Atabani, and V.V. Pham. 2021. "Acid-based lignocellulosic biomass biorefinery for bioenergy production: Advantages, application constraints, and perspectives." *Journal of Environmental Management* 296 (October): 113194. doi:10.1016/j.jenvman.2021.113194.

Ikehata, K., and Y. Li. 2018. "Ozone-based processes." In *Advanced Oxidation Processes for Waste Water Treatment*. Elsevier. doi:10.1016/B978-0-12-810499-6.00005-X.

Jadhav, D.A., S.G. Ray, and M.M. Ghangrekar. 2017. "Third generation in bio-electrochemical system research – A systematic review on mechanisms for recovery of valuable by-products from wastewater." *Renewable and Sustainable Energy Reviews* 76 (September). doi:10.1016/j.rser.2017.03.096.

Jallouli, N., L.M. Pastrana-Martínez, A.R. Ribeiro, N.F.F. Moreira, J.L. Faria, O. Hentati, A.M.T. Silva, and M. Ksibi. 2018. "Heterogeneous photocatalytic degradation of ibuprofen in ultrapure water, municipal and pharmaceutical industry wastewaters using a TiO_2/UV-LED system." *Chemical Engineering Journal* 334 (February). doi:10.1016/j.cej.2017.10.045.

Kato, D.M., N. Elía, M. Flythe, and B.C. Lynn. 2014. "Pretreatment of lignocellulosic biomass using Fenton chemistry." *Bioresource Technology* 162 (June). doi:10.1016/j.biortech.2014.03.151.

Kumar, B., N. Bhardwaj, and P. Verma. 2019. "Pretreatment of Rice straw using microwave assisted $FeCl_3$-H_3PO_4 system for ethanol and oligosaccharides generation." *Bioresource Technology Reports* 7 (September). doi:10.1016/j.biteb.2019.100295.

Kumar, G., P. Sivagurunathan, A. Pugazhendhi, N.B.D. Thi, G. Zhen, K. Chandrasekhar, and A. Kadier. 2017. "A comprehensive overview on light independent fermentative hydrogen production from wastewater feedstock and possible integrative options." *Energy Conversion and Management* 141: 390–402. doi:10.1016/j.enconman.2016.09.087.

Levec, J., and A. Pintar. 2007. "Catalytic wet-air oxidation processes: A review." *Catalysis Today* 124 (3–4): 172–84. doi:10.1016/j.cattod.2007.03.035.

Li, Y., M. Horsman, N. Wu, C.Q. Lan, and N. Dubois-Calero. 2008. "Biofuels from microalgae." *Biotechnology Progress*. doi:10.1021/bp070371k.

Lin, S. H., S.J. Ho, and C.L. Wu. 1996. "Kinetic and performance characteristics of wet air oxidation of high-concentration wastewater." *Industrial and Engineering Chemistry Research* 35 (1): 307–14. doi:10.1021/ie950251u.

Liu, C., W. Shi, H. Li, Z. Lei, L. He, and Z. Zhang. 2014. "Improvement of methane production from waste activated sludge by on-site photocatalytic pretreatment in a photocatalytic anaerobic fermenter." *Bioresource Technology* 155: 198–203. doi:10.1016/j.biortech.2013.12.041.

Loos, G., T. Scheers, K. Van Eyck, A. Van Schepdael, E. Adams, B. Van der Bruggen, D. Cabooter, and R. Dewil. 2018. "Electrochemical oxidation of key pharmaceuticals using a boron doped diamond electrode." *Separation and Purification Technology* 195 (October 2017): 184–91. doi:10.1016/j.seppur.2017.12.009.

Malani, R.S., V. Shinde, S. Ayachit, A. Goyal, and V.S. Moholkar. 2019. "Ultrasound–assisted biodiesel production using heterogeneous base catalyst and mixed non–edible oils." *Ultrasonics Sonochemistry* 52 (April). doi:10.1016/j.ultsonch.2018.11.021.

Malik, S.N., K. Madhu, V.A. Mhaisalkar, A.N. Vaidya, and S.N. Mudliar. 2020. "Pretreatment of yard waste using advanced oxidation processes for enhanced biogas production." *Biomass and Bioenergy* 142 (November) doi:10.1016/j.biombioe.2020.105780.

Malik, S.N., P.C. Ghosh, A.N. Vaidya, and S.N. Mudliar. 2019. "Ozone pre-treatment of molasses-based biomethanated distillery wastewater for enhanced bio-composting." *Journal of Environmental Management* 246 (September): 42–50. doi:10.1016/j.jenvman.2019.05.087.

Malik, S.N., T. Saratchandra, P.D. Tembhekar, K.V. Padoley, S.L. Mudliar, and S.N. Mudliar. 2014. "Wet air oxidation induced enhanced biodegradability of distillery effluent." *Journal of Environmental Management* 136 (March): 132–38. doi:10.1016/j.jenvman.2014.01.026.

M'Arimi, M.M., C.A. Mecha, A.K. Kiprop, and R. Ramkat. 2020. "Recent trends in applications of advanced oxidation processes (AOPs) in bioenergy production: Review." *Renewable and Sustainable Energy Reviews*. doi:10.1016/j.rser.2019.109669.

Martínez-Huitle, C.A., and M. Panizza. 2018. "Electrochemical oxidation of organic pollutants for wastewater treatment." *Current Opinion in Electrochemistry* 11: 62–71. doi:10.1016/j.coelec.2018.07.010.

Maryam, A.Z., M. Badshah, M. Sabeeh, and S.J. Khan. 2021. "Enhancing methane production from dewatered waste activated sludge through alkaline and photocatalytic pretreatment." *Bioresource Technology* 325 (January): 124677. doi:10.1016/j.biortech.2021.124677.

Michalska, K., K. Miazek, L. Krzystek, and S. Ledakowicz. 2012. "Influence of pretreatment with Fenton's reagent on biogas production and methane yield from lignocellulosic biomass." *Bioresource Technology* 119 (September). doi:10.1016/j.biortech.2012.05.105.

Michelin, C., and N. Hoffmann. 2018. "Photocatalysis applied to organic synthesis – a green chemistry approach." *Current Opinion in Green and Sustainable Chemistry* 10 (April). doi:10.1016/j.cogsc.2018.02.009.

Monlau, F., A. Barakat, J.P. Steyer, and H. Carrere. 2012. "Comparison of seven types of thermo-chemical pretreatments on the structural features and anaerobic digestion of sunflower stalks." *Bioresource Technology* 120 (September). doi:10.1016/j.biortech.2012.06.040.

Muthangya, M., A.M. Mshandete, and A.K. Kivaisi. 2009. "Two-stage fungal pre-treatment for improved biogas production from sisal leaf decortication residues." *International Journal of Molecular Sciences* 10 (11). doi:10.3390/ijms10114805.

Oturan, M.A., and J.-J. Aaron. 2014. "Advanced oxidation processes in water/wastewater treatment: Principles and applications. A review." *Critical Reviews in Environmental Science and Technology* 44 (23). doi:10.1080/10643389.2013.829765.

Padoley, K.V., P.D. Tembhekar, T. Saratchandra, A.B. Pandit, R.A. Pandey, and S.N. Mudliar. 2012. "Wet air oxidation as a pretreatment option for selective biodegradability enhancement and biogas generation potential from complex effluent." *Bioresource Technology* 120: 157–64. doi:10.1016/j.biortech.2012.06.051.

Patil, P.N., P.R. Gogate, L. Csoka, A. Dregelyi-Kiss, and M. Horvath. 2016. "Intensification of biogas production using pretreatment based on hydrodynamic cavitation." *Ultrasonics Sonochemistry* 30 (May): 79–86. doi:10.1016/j.ultsonch.2015.11.009.

Paździor, K., L. Bilińska, and S. Ledakowicz. 2019. "A review of the existing and emerging technologies in the combination of AOPs and biological processes in industrial textile wastewater treatment." *Chemical Engineering Journal* 376 (November). doi:10.1016/j.cej.2018.12.057.

Prabakar, D., V.T. Manimudi, T. Mathimani, G. Kumar, E.R. Rene, and A. Pugazhendhi. 2018. "Pretreatment technologies for industrial effluents: Critical review on bioenergy production and environmental concerns." *Journal of Environmental Management* 218: 165–80. doi:10.1016/j.jenvman.2018.03.136.

Rathour, R., V. Kalola, J. Johnson, K. Jain, D. Madamwar, and C. Desai. 2019. "Treatment of various types of wastewaters using microbial fuel cell systems." *Microbial Electrochemical Technology*. doi:10.1016/B978-0-444-64052-9.00027-3.

Rodriguez, C., A. Alaswad, J. Mooney, T. Prescott, and A.G. Olabi. 2015. "Pre-treatment techniques used for anaerobic digestion of algae." *Fuel Processing Technology* 138 (October). doi:10.1016/j.fuproc.2015.06.027.

Rodriguez, C., A. Alaswad, K.Y. Benyounis, and A.G. Olabi. 2017. "Pretreatment techniques used in biogas production from grass." *Renewable and Sustainable Energy Reviews* 68 (February). doi:10.1016/j.rser.2016.02.022.

Roopnarain, A., and R. Adeleke. 2017. "Current status, hurdles and future prospects of biogas digestion technology in Africa." *Renewable and Sustainable Energy Reviews* 67 (January). doi:10.1016/j.rser.2016.09.087.

Rózsenberszki, T., L. Koók, P. Bakonyi, N. Nemestóthy, W. Logroño, M. Pérez, G. Urquizo, C. Recalde, R. Kurdi, and A. Sarkady. 2017. "Municipal waste liquor treatment via bioelectrochemical and fermentation (H_2+CH_4) processes: Assessment of various technological sequences." *Chemosphere* 171 (March). doi:10.1016/j.chemosphere.2016.12.114.

Sambusiti, C., F. Monlau, E. Ficara, H. Carrère, and F. Malpei. 2013. "A comparison of different pre-treatments to increase methane production from two agricultural substrates." *Applied Energy* 104 (April). doi:10.1016/j.apenergy.2012.10.060.

Sarkar, N., S.K. Ghosh, S. Bannerjee, and K. Aikat. 2012. "Bioethanol production from agricultural wastes: An overview." *Renewable Energy* 37 (1). doi:10.1016/j.renene.2011.06.045.

Siciliano, A., M.A. Stillitano, and S. de Rosa. 2016. "Biogas production from wet olive mill wastes pretreated with hydrogen peroxide in alkaline conditions." *Renewable Energy* 85 (January). doi:10.1016/j.renene.2015.07.029.

Sirés, I., and E. Brillas. 2012. "Remediation of water pollution caused by pharmaceutical residues based on electrochemical separation and degradation technologies: A review." *Environment International* 40 (1): 212–29. doi:10.1016/j.envint.2011.07.012.

Song, Z.-l., G.-h. Yag, Y.-z. Feng, G.-x. Ren, and X.-h. Han. 2013. "Pretreatment of rice straw by hydrogen peroxide for enhanced methane yield." *Journal of Integrative Agriculture* 12 (7). doi:10.1016/S2095-3119(13)60355-X.

Strzalka, R., D. Schneider, and U. Eicker. 2017. "Current status of bioenergy technologies in Germany." *Renewable and Sustainable Energy Reviews* 72 (May). doi:10.1016/j.rser.2017.01.091.

Subhedar, P.B., P. Ray, and P.R. Gogate. 2018. "Intensification of delignification and subsequent hydrolysis for the fermentable sugar production from lignocellulosic biomass using ultrasonic irradiation." *Ultrasonics Sonochemistry* 40 (January). doi:10.1016/j.ultsonch.2017.01.030.

Taherzadeh, M., and K. Karimi. 2008. "Pretreatment of lignocellulosic wastes to improve ethanol and biogas production: A review." *International Journal of Molecular Sciences* 9 (9). doi:10.3390/ijms9091621.

Tamilarasan, K., J.R. Banu, M.D. Kumar, G. Sakthinathan, and J.-H. Park. 2019. "Influence of mild-ozone assisted disperser pretreatment on the enhanced biogas generation and biodegradability of green marine macroalgae." *Frontiers in Energy Research* 7 (September). doi:10.3389/fenrg.2019.00089.

Vlyssides, A.G., M. Loizidou, P.K. Karlis, A.A. Zorpas, and D. Papaioannou. 1999. "Electrochemical oxidation of a textile dye wastewater using a Pt/Ti electrode." *Journal of Hazardous Materials* 70 (1–2): 41–52. doi:10.1016/S0304-3894(99)00130-2.

Wang, S., F. Shiraishi, and K. Nakano. 2002. "A synergistic effect of photocatalysis and ozonation on decomposition of formic acid in an aqueous solution." *Chemical Engineering Journal* 87 (2). doi:10.1016/S1385-8947(02)00016-5.

Wu, J., F. Ein-Mozaffari, and S. Upreti. 2013. "Effect of ozone pretreatment on hydrogen production from barley straw." *Bioresource Technology* 144 (September). doi:10.1016/j.biortech.2013.07.001.

Xu, M., C. Wu, and Y. Zhou. 2020. "Advancements in the fenton process for wastewater treatment." In *Advanced Oxidation Processes - Applications, Trends, and Prospects*. IntechOpen. doi:10.5772/intechopen.90256.

Yang, B., J. Wang, C. Jiang, J. Li, G. Yu, S. Deng, S. Lu, P. Zhang, C. Zhu, and Q. Zhuo. 2017. "Electrochemical mineralization of perfluorooctane sulfonate by novel F and Sb Co-doped Ti/SnO_2 electrode containing Sn-Sb interlayer." *Chemical Engineering Journal* 316: 296–304. doi:10.1016/j.cej.2017.01.105.

Yang, G., and J. Wang. 2019. "Ultrasound combined with dilute acid pretreatment of grass for improvement of fermentative hydrogen production." *Bioresource Technology* 275 (March). doi:10.1016/j.biortech.2018.12.013.

Yeom, I.T., K.R. Lee, Y.H. Lee, K.H. Ahn, and S.H. Lee. 2002. "Effects of ozone treatment on the biodegradability of sludge from municipal wastewater treatment plants." *Water Science and Technology* 46 (4–5). doi:10.2166/wst.2002.0641.

Zawieja, I., and K. Brzeska. 2019. "Biogas production in the methane fermentation of excess sludge oxidized with Fenton's reagent." *E3S Web of Conferences* 116 (September). doi:10.1051/e3sconf/201911600104.

Zeng, J., and J. Lu. 2018. "Mechanisms of action involved in ozone-therapy in skin diseases." *International Immunopharmacology* 56 (March). doi:10.1016/j.intimp.2018.01.040.

Zhang, Q., J. He, M. Tian, Z. Mao, L. Tang, J. Zhang, and H. Zhang. 2011. "Enhancement of methane production from cassava residues by biological pretreatment using a constructed microbial consortium." *Bioresource Technology* 102 (19). doi:10.1016/j.biortech.2011.06.061.

Zheng, Y., J. Zhao, F. Xu, and Y. Li. 2014. "Pretreatment of lignocellulosic biomass for enhanced biogas production." *Progress in Energy and Combustion Science* 42 (June). doi:10.1016/j.pecs.2014.01.001.

Zhi, D., J. Qin, H. Zhou, J. Wang, and S. Yang. 2017. "Removal of tetracycline by electrochemical oxidation using a Ti/SnO_2–Sb anode: Characterization, kinetics, and degradation pathway." *Journal of Applied Electrochemistry* 47 (12): 1313–22. doi:10.1007/s10800-017-1125-7.

Zhou, Z., X. Liu, K. Sun, C. Lin, J. Ma, M. He, and W. Ouyang. 2019. "Persulfate-based advanced oxidation processes (AOPs) for organic-contaminated soil remediation: A review." *Chemical Engineering Journal* 372 (September). doi:10.1016/j.cej.2019.04.213.

Zhu, Z., M. Zhu, and Z. Wu. 2012. "Pretreatment of sugarcane bagasse with NH_4OH–H_2O_2 and ionic liquid for efficient hydrolysis and bioethanol production." *Bioresource Technology* 119 (September). doi:10.1016/j.biortech.2012.05.111.

Zhuo, Q., Q. Xiang, H. Yi, Z. Zhang, B. Yang, Kai Cui, X. Bing, Z. Xu, X. Liang, Q. Guo, and R. Yang. 2017. "Electrochemical oxidation of PFOA in aqueous solution using highly hydrophobic modified PbO_2 electrodes." *Journal of Electroanalytical Chemistry*. 801. doi:10.1016/j.jelechem.2017.07.018.

Zieliński, M., M. Dębowski, M. Kisielewska, A. Nowicka, M. Rokicka, and K. Szwarc. 2019. "Cavitation-based pretreatment strategies to enhance biogas production in a small-scale agricultural biogas plant." *Energy for Sustainable Development* 49 (April): 21–26. doi:10.1016/j.esd.2018.12.007.

7 Nanocatalyzed Pretreatment of Wastewater for Biofuel Production

Shreyansh P. Deshmukh, Aman N. Patni, Ameya S. Mantri, and Sameena N. Malik
Institute of Chemical Technology

CONTENTS

ABBREVIATIONS

ABR Anaerobic Batch Reactor
AFBR Anaerobic Fluidized Bed Reactor
AGS Anaerobic Granular Sludge
AP Amino Phenol
BOD Biological Oxygen Demand
COD Chemical Oxygen Demand
CR Congo Red
CSTR Continuous Stirring Tank Reactor
GAC Granulated Activated Carbon
HRT Hydraulic Retention Time
MA Methanogenic Archaea

DOI: 10.1201/9781003197737-9

MB	Methylene Blue
MBR	Membrane Bioreactor4
	NA - Nitro-Aniline
MO	Methyl Orange
MWCNT	Multiwalled Carbon Nanotubes
NP	Nitro-Phenol
OLR	Organic Loading Rate
ORP	Oxidation Reduction Potential
PBR	Packed Bed Reactor
PWW	Petroleum wastewater
SHY	Specific Hydrogen Yield
SRB	Sulfate Reducing Bacteria
SRT	Sludge Retention Time
TDS	Total Dissolved Solutes
UASBR	Up flow Anaerobic Sludge Blanket Reactor
VFA	Volatile Fatty Acids
WAS	Waste-Activated Sludge
ZVI	Zero-Valent Iron

7.1 INTRODUCTION

Fossil fuels are the primary sources of global energy requirement, and are consumed for heat and electricity generation, transportation, industrial sector, as well as for residential purposes. Burning of coal and fossil fuels emits CO_x, SO_x, and NO_x into the atmosphere. These emissions contribute in global warming, acid rains, smog, and respiratory disorders (Kumar et al. 2019). The combustion of coal emits large amount of fly ash into the atmosphere, which promotes air pollution and cause respiratory illness. The statistics show that the limited natural resources for fossil fuels such as crude oil reservoirs and coal will get exhausted in the near future (Patel et al. 2018). Past studies suggest that the biofuel can be a potential alternative to fossil fuel. Use of waste in biofuel generation helps reduce the environmental pollution—the major concern of future (Levin and Chahine 2010; Patel and Kalia 2013). Also, biofuels containing high energy content can be seen as a promising approach to combat against the increasing global energy demand with decreasing fossil fuel resources. Hydrogen can be considered as a convenient source of energy, as it is ecofriendly, efficient, has no by-products (Elreedy et al. 2017; Patni et al. 2021). About 90% of the global hydrogen production is used in food preparation, chemical manufacturing (mainly CH_3OH and NH_3), and industrial processes. The requirement of hydrogen is expected to rise exponentially, as it will penetrate into transportation industry as an alternative to fuels to meet the demand of refineries (Levin and Chahine 2010).

Physiochemical processes such as steam reforming and coal gasification that are operated at high temperatures and pressures are used for hydrogen production in industries. However, the production of hydrogen is highly expensive, as the operating conditions of the various processes are energy intensive (Rambabu et al. 2021). Also, as these processes are dependent on the burning of fossil fuels, they consume nearly 1%–3% of energy demand globally (Wünschiers and Lindblad 2002). Therefore, a balance between renewable and sustainable energy resources needs to be maintained, as it is necessary to develop efficient and clean method for hydrogen production. Biological production is economically viable, effective, and clean route for hydrogen production (Budzianowski 2012). Simultaneously, the use biological feedstock helps improving the bio-based economy (Redwood et al. 2009). Bio-based feedstock such as biomass containing high organic content, precisely industrial and agricultural, are used for fermentation process (Hallenbeck 2005).

7.1.1 Wastewater as Substrate

Wastewater with high COD concentration consists of high carbohydrates and solid wastes rich with organic compounds content, and shows satisfactory results when used as feedstock for generating biofuels (Levin and Chahine 2010; do Nascimento Junior et al. 2021). Researchers have studied wastewater from pharmaceutical industry (Malik et al. 2019; Krishna 2013), palm oil mill (Ahmad et al. 2016), dairy and food processing industry (Ferreira et al. 2018), cheese whey (do Nascimento Junior et al. 2021), cattle (Ferreira et al. 2018), poultry (Ferreira et al. 2018), textile industry (Lin et al. 2017; Malik et al. 2018), processing industries (Zhu et al. 2010), petroleum industry (Ahmad 2020), and brewery (Ferreira et al. 2018). Processes such as dark fermentation, anaerobic digestion, and photosynthetic fermentation are the commonly used biological methods for hydrogen production, where these processes require ambient process conditions and require less energy (Ahmad 2020). Dark fermentation can be applied at large scale by optimizing the parameters such as substrate concentration, pH, and temperature (Kumar et al. 2019; Usman et al. 2019), studied the effect of different process parameters such as temperature, pH, HRT, reactor configuration, and OLR on biohydrogen production. They observed that the major issues faced for biohydrogen production are low substrate conversion and low productivity. Bacterial growth, hydrogen production, degradation of substrate, and by-product formation are temperature dependent (Perna et al. 2013). Increase in temperature shows better substrate degradation and hydrogen production due to increase in entropy but lacks energy recovery, which further increases the cost (Mu et al. 2006). On the other hand, ambient temperature enhances the enzyme activity, which simultaneously promotes the solubility of biofuel (Hernández-Mendoza et al. 2014; Silva et al. 2019). Reduction in pH limits the bacterial growth and affects the metabolic bacteria production (Stavropoulos et al. 2016). The dormant form of bacteria that promotes hydrogen production can survive extreme pH conditions, but, on the other hand, methanogens get eliminated (Shida et al. 2009). HRT is an important operating parameter that influences production of hydrogen and limits the activity of methanogens during fermentation (Hawkes et al. 2007). Hydrogen yield increases with decrease in HRT, as it promotes washout of microorganisms that reduces the size of reactor and enhances the availability of substrate (Ueno et al. 2001). OLR shows major impact on the performance of reactor due to overload of organic content, this further lowers the hydrogen yield (Kargi et al. 2012). Lin and Lay (2004) studied the effect of carbon, nitrogen, and phosphate on hydrogen production. They observed that optimum C/N ratio enhanced the stability of the process, while optimum phosphate concentration enhanced the overall hydrogen yield. The configuration of bioreactor plays an important role in determining the yield of hydrogen. Mainly

CSTR, UASBR, PBR, and AFBR are used for biohydrogen production using wastewater (Usman et al. 2019). CSTR increases contact between the microorganisms and their substrates, which results in higher mass transfer. However, due to its vigorous mixing pattern, microbes get washed out at lower HRT (Show et al. 2011). PBR allows substrates to pass through column where bed has granules attached for reaction. Low HRT in this case works productively without washing out biomass. Its simple process operations and anaerobic environment makes the hydrogen production more successful. Only problem while working on PBR is that the side reactions can occur (Show et al. 2011). AFBR is a combination of CSTR and PBR. The induced and upward flow of liquid wastewater enhances catalytic activity and hence boosts hydrogen production. Also, there is very less biomass being washed out. The only disadvantage in AFBR is that the energy desired for fluidization is more (Anand et al. 2016; Zhang et al. 2008). MBR, as the name suggests, is a bioreactor involving ultrafiltration membranes, which allows separation as well as activated sludge treatment. Low SRT and HRT conditions prove helpful in hydrogen production (Oh et al. 2004). UASBR has the blanket of granulated sludge present in the reactor, which is helpful in treating the wastewater abundant in microbes. Despite having lower HRT, this bioreactor enhances hydrogen production (Intanoo et al. 2016).

7.1.2 Impact of Nanoparticles on Biological Hydrogen Production

Researchers have reported the potential application of nanomaterials in pretreating the contaminated water by maintaining these optimum parameters. The results of study have shown excellent efficiency in removal of contaminants at low cost and ability to be recycled (Baruah et al. 2016). These nanoengineered materials show properties like strong solute mobility, dispersibility, nano size, hydrophobicity, hydrophilicity, high reactivity, and large surface area with porous characteristic (Daer et al. 2015; Wu et al. 2019). The recent advancements in the application of nanocatalysts for water treatment show adaptability of the process (Mahmoodi and Arami 2009). Introducing nanomaterials in the dark fermentation, supported pathway for the production of biohydrogen has shown attractive results and easy adaptability of process. As seen in Figure 7.1, Malik et al. (2021) proposed the mechanism for nanomaterial-based dark fermentative pathway supporting biohydrogen production. The suggested mechanism mainly follows two pathways:

1. NADH-dependent biohydrogen production

$$NADH + H^+ + 2Fd^{2+} \rightarrow 2H^+ + NAD^+ + 2Fd^{2+} \quad (7.1)$$

$$2H^+ + 2Fd^{2+} \rightarrow 2Fd^+ + H_2 \quad (7.2)$$

2. Biohydrogen production by decomposition of pyruvate

$$CH_3COCOO^- + CoA + Fd_{ox} \rightarrow \text{Acetyl CoA} + Fd_{red} + CO_2 \quad (7.3)$$

$$Fd_{red} + 2H^+ \rightarrow Fd_{ox} + 2H_2 \quad (7.4)$$

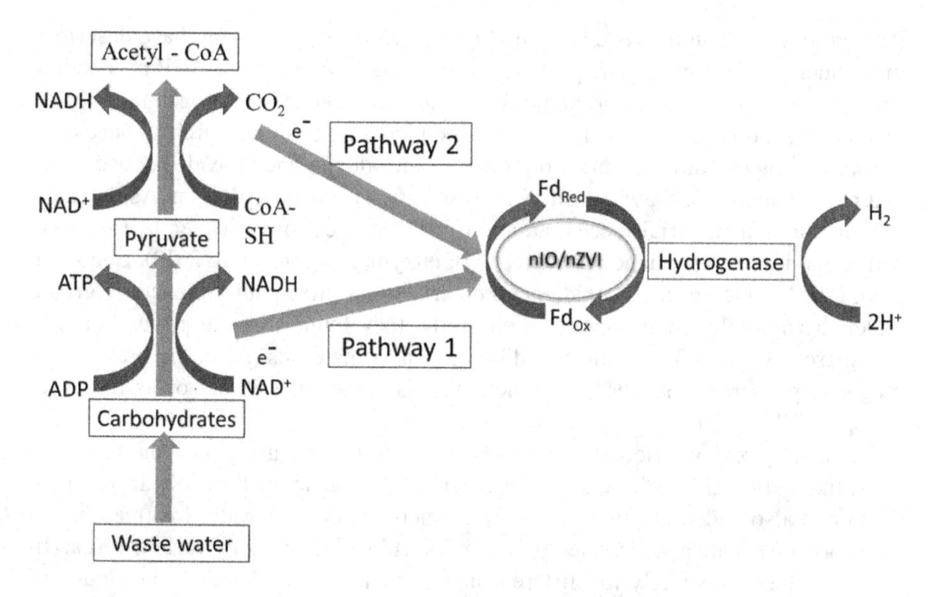

FIGURE 7.1 Mechanism for production of biological hydrogen using dark fermentation.

According to the mechanism, hydrogen generation increases with the feeding of *n*ZVI and *n*IO, as it enhances the ferredoxin oxidoreductase activity. Optimal concentration of *n*ZVI and *n*IO catalyzes the activity of ferredoxin, hydrogenase enzymes, and pyruvate oxidoreductase ferredoxin and results into enhanced hydrogen production. Similarly, Dutta (2020) explained the mechanism for photocatalysis using TiO_2. According to the proposed mechanism, an electron (e^-) – hole (h^+) pair is formed at the photocatalyst particles during the presence of solar radiation. The holes are strong oxidizing agents that oxidize the electron donor organic compounds and reduce itself to form H_2O and CO_2. The photoexcitation (7.5) and electron scavenging due to presence of oxygen (7.6), using TiO_2 photocatalyst, is given as

$$TiO_2 \xrightarrow{hv \gg E_g} h^+ + e^- \tag{7.5}$$

$$e^- + O_2 \rightarrow O_2^- \tag{7.6}$$

The suggested mechanism of the photocatalysis reaction can be summarized as follows:

i. Reactant transfer from fluid to the surface of photocatalyst
ii. Adsorption of reactant on the surface
iii. Reaction of photocatalyst with the adsorbed reactant
iv. Desorption of mineralized product
v. Removal of product

Patel et al. (2018), in their study on biological hydrogen production, have described the influence of different nanoparticles in yield and efficiency of BHP production. The drawback noticed for scaling up the process of producing hydrogen in dark fermentation conditions was the high cost of the incoming feed. Researchers have tried for optimizing the antimicrobial properties by introducing metal oxide nanoparticles and metal nanoparticles, which show high selectivity toward hydrogen. As the addition of iron nanomaterials accelerate the transfer of electrons between hydrogenase and ferreodoxin and enhance the activity of enzymes (Leong et al. 2019). Engliman et al. (2017) added iron (II) oxide nanomaterials to a mixed culture bacteria grown under thermophilic conditions. In their study, they found that the production rate was increased up to 34% with the addition of nanomaterials. Also, the presence of magnetic properties into the iron nanomaterials show a possibility of its recycling (Yang and Wang 2018).

This study reviews biofuel production from wastewater using various nanocatalysts, like ZrO_2, ZnO, Pd, Fe^0, and Fe_2O_3. The mechanism of the biohydrogen production is also added in this work. Various methods of wastewater treatment using microorganisms are also depicted in this work. Individual work by various researchers is put forth separately for different nanoparticles to get detailed knowledge of the method. The mechanism and the enhancement in selectivity of nanoparticles for production of biofuel is also highlighted in this work.

7.2 NANOPARTICLES FOR BIOLOGICAL HYDROGEN PRODUCTION

7.2.1 Zero-Valent Iron (Fe^0)

Feng et al. (2014) observed an increase in fatty acid yield and protein degradation after the addition of ZVI due to increase in acidification and hydrolysis. It has also been observed that an increase in methane production occurs by nearly 40%–45% with change in sludge reduction ratio by 12%–13%. Activity of methanogens were increased with the addition of ZVI. The methane production was increased due to the generation of acetate, which acts as a substrate for the methanogens, and ZVI's role as electron donor enables it to reduce CO_2. The reduction of carbon dioxide can be studied through the following mechanism (7.7 and 7.8):

$$CO_2 + 4Fe^0 + 8H^+ \rightarrow CH_4 + 4Fe^{2+} + 2H_2O \quad (7.7)$$

$$CO_2 + 4H_2 \rightarrow CH_4 + 2H_2O \quad (7.8)$$

After adding ZVI, the activities of main enzymes involved in hydrolysis and acidification were increased. It improved the conversion of organic materials present in the waste water to VFAs by catalyzing the digestion process. ZVI could influence the concentration of hydrogen-utilizing microorganisms like homoacetogens and hydrogenotrophic methanogens to utilize H_2 (Feng et al. 2014).

Eaktasang (2017) studied impact of *n*ZVI dose on biohydrogen generation at 25°C in anaerobic condition at lab scale. The seed sludge was taken from an anaerobic digester at a brewing industry using glucose. *n*ZVI dosage ranged from 0 to 500 mg/L at constant pH (5.5) and COD concentration of 2,000 ppm. The generation of hydrogen increased when the *n*ZVI dosage was increased, which was greater than that of the reactor without the *n*ZVI addition. They theorized the reduction of glucose with the following mechanism (7.9):

$$4Fe^0 + C_6H_{12}O_6 + 12H_2O \rightarrow 2Fe_2O_3 + 6CO_2 + 18H_2 \tag{7.9}$$

Glucose was broken down in the presence of *n*ZVI to CO_2 and H_2. Other advantages were pollution management by organic matter reduction via Fe^0-reducing processes and biodegradation via hydrogen-producing microorganisms (Numfon 2017).

ZVI simultaneously enhances methane production and sulfate reduction in AGS reactors. Undissociated H_2S inhibits acetogens, methanogens, and sulfate, although ZVI lowers this inhibition. By buffering pH and causing iron sulfide precipitation, SRB kills bacteria; this improves sulfate reduction capacity, especially in worsening conditions. When used in an ABR, ZVI can lower the oxidation reduction potential (ORP) and act as an acid buffer, both of which are necessary for MA to grow (Zhang et al. 2011). ZVI improves $CH_3CH_2CO_2$ degradation and methanogenesis, while also reducing the inhibitory effects of undissociated H_2S on acetogens, MA, and SRB in the reactor, resulting in improved methane synthesis and sulfate reduction (Liu et al. 2015).

Zero-valent iron-activated carbon (ZVI-AC) and ZVI have been used in bio-H_2 generation, as it was easily oxidized to Fe^{2+} by chemical and electrochemical reactions. The microelectrolysis system was formed when ZVI and AC, which acts as anode and cathode, respectively, comes in contact with each other. It results in synchronous incidences of redox reactions on the area of a large number of electrodes which results in substantial electron flow. Microelectrolysis with ZVI-AC resulted in a substantial increase in hydrogen generation rates. Moreover, iron acts as the cofactor for hydrogenase and sulfur and iron protein (Das et al. 2006). The sludge was performed with ZVI powder added in the range of 0–500 mg/L to explore the impacts of ZVI-AC and ZVI. Heavy metals in excessive amounts have a tendency to cause inhibition or toxicity. It might be the cause of the lower hydrogen production at doses higher than 400 mg/L. The highest hydrogen yield was obtained for ZVI-AC among the catalysts used. In this batch experiment, it was observed that the greater acetate concentrations were invariably associated by increased hydrogen generation. The dark fermentative hydrogen generation was successfully increased by ZVI-AC microelectrolysis in the study. Nearly 50% enhancement of hydrogen yield was observed at a concentration of 300 mg/L of ZVI-AC. The enhanced results were attributed to the microelectrolysis system's capacity to better optimize environmental variables and microbial community structure for process. The approach proposed in this study has a great potential for improving dark fermentative hydrogen generation, and it should be investigated further (Zhang et al. 2015).

Methane production rose in all *n*ZVI-added bioreactors, with the addition of 2.5 and 5.0 g/L *c*NZVI achieving the maximum increase of 28%. The addition of bulk ZVI nanoparticles only resulted in a 5% rise in CH_4, demonstrating the benefit of nanoscale particles. NZVI amendments enhanced the biogas generation and reduced the CO_2 emissions from the bioreactor by around 58%. Methanogens are known for their sluggish metabolism and great sensitivity to system changes (such as temperature and pH). The use of ZVI has been shown to enhance the anaerobic degradation conditions. Because of its low cost and lack of by-products, iron as an amendment is appealing. CO_2 may be eliminated biotically when hydrogenotrophic methanogens use hydrogen as a reducing agent to convert CO_2 to CH_4. According to the process, NZVI may also conduct a redox reaction with CO_2 and H_2O to create iron carbonate and hydrogen (7.10).

$$Fe^0 + CO_2 + H_2O \rightarrow FeCO_3 + H_2 \tag{7.10}$$

As compared to unmodified control reactors, adding *n*ZVI to ABR resulted in increase in CH_4 generation and reduced the COD. The influence of *n*ZVI on pH and ORP had positive effect on enhancing methanogenesis (Carpenter et al. 2015).

An UASBR was used to store scrap iron. The ZVI has increased the methanogenesis, as evidenced by a jump in CH_4 content from 47.9% to 64.8% in the biogas. The azo dye's unsaturated link was also cleaved, making the wastewater more biodegradable. Low ORP and near-neutral pH favored the growth of methanogens, which aided the biological conversion of organic acids to CH_4. The coagulation caused by the Fe^{2+} dissolved in the ZVI provides a cost-effective and efficient way to treat the azo dye wastewater (Zhang et al. 2011).

7.2.2 Zirconium Dioxide

Sandoval et al. (2011) in their studies synthesized a ZrO_2 nanocatalyst impregnated with GAC to remove organic and arsenic contaminants from waste water. Arsenic, being toxic and carcinogenic, is one of the most common components of wastewater contaminants. GAC mediums like bituminous and lignite were used for synthesis of water samples. Water samples used were 5 mM $NaHCO_3$ buffered with water containing 120 μg/L of Ar. Water is either accommodated with no other competing ions known as ultrapure water and with other competing ions known as model groundwater. Bituminous-based Zr-GAC medium showed better adsorption than lignite-based Zr-GAC medium for Ar removal from wastewater due to its higher surface area and generation of new pores on its surface (John et al. 2017). Freundlich adsorption isotherm model was used, and the value of $1/\eta$ was found to be less than 1, which proves as a successful adsorption. However, the extent of adsorption by these catalysts (2.4 mg As/g Zr used) was not as good as from the commercially available Zr nanopowders (3.6 mg As/g Zr used). Use of oxo-anions proved a bit useful, and it increased the adsorption capacity of bituminous Zr-GAC from 2.4 to 3.0 mg As/g Zr used (Sandoval et al. 2011). Vijayalakshmi et al. (2020) in their studies reported photolytic decomposition of pharmaceutical wastewater by using polyaniline-based ZrO_2 nanocatalyst (PANI/ZrO_2). UV light was irradiated for 120 minutes on CIP molecules, and it was found that their decomposition rate was 96.6%. (Vijayalakshmi et al. 2020). Qiu et al. (2011) synthesized a ZrO_2 solid base nanocatalyst packed with $C_4H_4O_6HK$. This catalyst was characterized and used for biofuel production from transesterification reaction of soyabean oil and methanol. During the reaction the parameters optimized were CH_4OH to oil molar ratio, quantity of nanocatalyst used, reaction time, and temperature of reaction. Soyabean oil to methanol ratio was kept to be 1:16 for producing higher yield and shifting equilibrium to product's side (Liang et al. 2010). This reaction is a three-phase reaction, so to overcome the diffusion problem we use only 6% catalyst and only for 120 minutes. Researchers found that rise in temperature will increase the biofuel yield, and hence higher temperature was preferred, but experimentally methanol starts forming bubbles and begins to vaporize above 60°C (Qiu et al. 2011). So, the reaction with 1:16 oil to methanol ratio using 6% nanocatalyst at 60°C for 120 minutes gave the best biofuel yield of 98.03%. After reaction, glycerol, being denser, was removed by centrifugation; then methanol, being volatile, was evaporated by rotary evaporator; and, hence, the catalyst was regenerated. The nanocatalyst used in the study proved to be effective, and it was found that even after its use in five cycles, its activity and lifetime was mostly unchanged (Qiu et al. 2011).

7.2.3 Zinc Oxide

The metal ion toxicity released into the stream due to usage of metal oxide nanoparticles was reported by (Brunner et al. 2006). Further, Jiang et al. (2007) reported that due to presence of microorganisms the toxicity was released by ZnO nanoparticle and not by Zn^{2+} metal ion. Otero-González, Field, and Sierra-Alvarez (2014) analyzed the effect of ZnO and Zn^{2+} concentration on the methanogenic activity for two feedings of substrate (first feeding, 30% inhibition; second feeding, 70% inhibition). They reported that the toxicity of ZnO nanoparticles differed toward acetate- and hydrogen-based methanogens. The hydrogenotrophic methanogenesis inhibition values in an AGS were observed to be higher compared to the acetoclastic methanogenesis for both feeding. This observation suggest that the ZnO toxicity severely affects the acetoclastic methanogens than hydrogenotrophic methanogens. They concluded that the high ZnO nanoparticle concentration can increase the toxicity and simultaneously disrupt the activity of methanogens. The inhibition can occur at low concentration due to the accumulation of biosolids by ZnO nanoparticles, which may increase the actual concentration in contact with the microorganism and can further enhance the UASBR with a high SRT.

CH_4 formation was observed to be influenced by the concentrations of sodium dodecyl-benzene sulfonate (SDBS) (Jiang et al. 2007). Mu et al. (2011) studied the influence of *n*ZnO, *n*TiO_2, *n*SiO_2, and *n*Al_2O_3 on methane formation in WAS. They observed that the *n*ZnO had a great influence on methane formation in the WAS fermentation system compared to other nanoparticles. The data that was reported after 18 days of fermentation for different concentrations of *n*ZnO shows that the methane generated in the presence of 0.150 g/g-TSS was 0.0245 L/g-VSS, while 0.03 g/g-TSS produced 0.0995 L/g-VSS of methane. The control test for methane formation reported the inhibition rates to be 81.1% and 22.8% with 0.129.1 L/g-VSS CH_4 produced. The lower concentration of *n*ZnO gives no impact on methane generation, while overall CH_4 generation was dependent on the dosage of *n*ZnO. The researchers predicted that the addition nanoparticles does not influence the solubilized carbohydrates and proteins. This solubilized material further hydrolyzed to form monosaccharaides and amino acids, and this hydrolysis inhibits effect on methane formation during higher dosage of *n*ZnO (Mu et al. 2011).

Ahmad (2020) studied the bioprocess involved in formation of methane using ZnO nanoparticles for PWW. It has analyzed the production process using modified first-order model (7.11), Gomptez model (7.12), and Cone model (7.13). It was reported 7.65 L of methane formation with 4.5 g-VS L of PWW. They observed that the lag time (λ) increased with increasing ZnO concentration; however, no lag time was reported at 3.5 and 4.5 ratios of ZnO. Also, Gompertz model was observed to be the best fitting, and the result was supported by RMSPE and SEE data.

$$CH_4 = Y\left[1-\exp\left(k_{\text{hyd}}t\right)\right] \tag{7.11}$$

$$CH_4 = Y\exp\left\{-\exp\left[(\lambda - t)R_m^e / (Y+1)\right]\right\} \tag{7.12}$$

$$CH_4 = \frac{Y}{1+\left(k_{\text{hyd}}t\right)^{-n}} \tag{7.13}$$

Xia et al. (2021) discussed the effect of different parameters in sustainable production of biofuel using *Chlorella vulgaris* bacteria and ZnO and Fe_3O_3 nanoparticles in synthetic saline water medium. Likelihood of producing high-quality biofuel was reported using the enhanced lipid content of the algal biomass. It was observed that the pH levels were reduced with the addition of nanoparticles, precisely the rate of reduction was higher with ZnO compared to Fe_2O_3 particles. pH level was seen to be directly proportional to the ammonia content in the stream with increasing time. The observations had shown that the rate of reduction of pH level was higher in the initial 20 days; then it linearly reduced till 80th day. A sudden rise was observed in pH level post the 80th day. This rise in pH levels may be due to higher consumption of CO_2 due photosynthesis at a faster rate (Lam et al. 2017). Also, greater phosphorous recovery was observed using ZnO nanoparticles compared to Fe_2O_3 particles.

7.2.4 Palladium

Huff et al. (2018) studied the formation of hydrogen gas by carrying out hydrolysis of $NaBH_4$ in presence of palladium nanocatalyst supported by carbon nanotubes. The main purpose of using nanocatalyst is its high surface area to volume ratio. Due to this, active sites of the material increases (Li et al. 2005). Palladium nanocatalyst was formed by impregnating (MWCNTs) by adding palladium (II) chloride to aq. beta-cyclodextrin after 2 hours of stirring. This addition was done because nano catalyst alone forms agglomerate that are difficult to reproduce, but impregnation of MWCNTs solves this problem. The reaction was proceeding by using 10 mg of synthesized catalyst at 7 pH and 295 K, which was added with deionized water (100 mL) and 835 μmoles of $NaBH_4$. Researchers tried three types of catalyst Pd/MWCNT, Pd NP alone, and MWCNT alone (Huff et al. 2018). The results showed the highest hydrogen yield with Pd/MWCNT catalyst, which was 25% better than Pd NP alone and 53% better than primary MWCNT. The best hydrogen production results were obtained at 295K and 7 pH, which produced 835 μmoles of hydrogen at 21.7 mL/min/gram of catalyst used (mL/min/g_{cat}) (Huff et al. 2018). Singhania and Bhaskarwar (2018a) studied hydrogen production by self-synthesized carbon nanotubes (CNT) impregnated palladium catalyst by using decomposition reaction of hydrogen iodide under thermochemical water splitting sulfur–iodine (SI) cycle. They worked out various different concentration of Pd along with CNT. SI cycle was used, because the conventional method for water decomposition requires 3,000°C, but this SI cycle decomposes water (which is used as a precursor here) only at around 550°C Singhania and Bhaskarwar (2018b). 1%, 3%, and 5% by weight of Pd was added to CNTs and its conversion capacity to convert hydrogen iodide to hydrogen was measured. Results showed that 3% Pd gave the best results of conversion as 23.7% at 550°C. Further, the increase in Pd% resulted in lesser conversion yield. This was justified by the reason that the quantity of deformity existing in the catalyst were the highest at 3% Pd/CNT, and this high defect helps to increase decomposition reaction. A stability test for 3% Pd/CNT catalyst was also performed after 100 hours of its decomposition, but the conversion rate remained constant, which makes this catalyst one of the best, highly active, and mainly a stable catalyst for HI decomposition (Singhania and Bhaskarwar 2018a). Garole et al. (2019) reported a very straightforward method for biosynthesis of Pd nanocatalyst. They performed some experiments to investigate the activity of so formed nanocatalyst to reduce organic pollutants such as MO, MB, and 4-NP (Kojima et al. 2002). Three different experiments were carried out. In first and second, dye solution of MO and MB (3 mL of 1.0 mM), respectively, and sodium borohydride (1 mL of 0.3 mM) along with pure water (6 mL) were mixed with 20 mg Pd nanocatalyst and were left for reaction to occur. In third experiment, 4-NP (0.3 mL of 2 mM) and sodium borohydride (1 mL of 0.3 mM) along with pure water were added with 20 mg Pd nanocatalyst. The reaction was recorded by using UV-visible spectrometer, until the reduction reaction completes (that was indicated by formation of colorless products). All three solutions of MB, MO, 4-NP showed the maximum absorbance wavelength peak (λ_{max}) at 664, 465, and 402 nm, respectively. The remarked rate of reaction was 0.20 min^{-1} for MB, 0.70 min^{-1} for MO, and 0.30 min^{-1} for 4-NP, and all the three reactions followed pseudo-first-order reaction (Garole et al. 2019). Baran and Menteş (2020) developed a new palladium catalyst, which was obtained from gum arabic catalyst that was named as GA–Sch–Pd. This catalyst was tested and was analyzed by using TGA, FE-SEM, TEM, XRD, etc. The catalyst was used to treat the environmental contaminants like NA, NP, CR, MB, and MO. All these contaminants are hazardous to human health as well as aquatic life, which need to be detoxified. GA–Sch–Pd denotes Pd catalyst with Schiff-based modified gum arabic stabilizer. Catalyst was tested on some organic dyes and some aromatic compounds. $NaBH_4$ was used as reducing agent. o-NP produced bands at 283 and 410 nm, which on reduction changed to single band at 290 nm that showed the formation of o-phenylenediamine only in 45 seconds. Another aromatic compound studied was p-NP where it showed peak at 400 nm that got changed to 300 nm just in 90 seconds and manifested the presence of *p*-aminophenol. Organic dyes like CR, MO, and MB showed UV-visible spectrum peaks at 496, 465, and 663 nm, respectively. All these reactions shown very fast timing for reduction reaction, which is very less as compared to many other catalysts. Also, the major plus point of this nanocatalyst is that it is easily recyclable as it easily separates and hence is reused (Baran and Menteş 2020). Kempasiddaiah et al. (2021) harmonized a magnetic catalyst from Pd that can be recycled. Many analytical techniques were used to characterize this newly formed catalyst called *N*-heterocyclic carbine palladium (II) $\left((CH_3)_3 - NHC - Pd - Fe_3O_4\right)$ tethered by magnetic nanoparticles. Its performance check was done by introducing catalyst in reduction reaction of organic dyes with $NaBH_4$ at room temperature (Sajjadi et al. 2019). Organic dyes include 4-NP (which was reduced to 4-AP), MB, and MO. HCOOH was also used as hydrogen source to reduce Cr(IV) to Cr(III), which even after 2 hours of conventional reaction remained the same. As the nanocatalyst was added to this reaction, the reduction enhanced and the reaction was completed in 25 minutes. Leaching actions are more prominent in these kinds of catalysts, but this nanocatalyst showed very less amount of leaching (<0.0099 μg/L) as compared to others (Kempasiddaiah et al. 2021). De Corte et al. (2012) mentioned the use of bio-Pd catalyst with the use of Pd-reducing agents like *Escherichia coli*, *Desulfovibrio desulfuricans*, *Clostridium pasteurianum*, and *Desulfovibrio vulgaris* to reduce Cr (IV) to Cr (III) from polluted industrial wastewater (De Corte et al. 2012).

7.2.5 Iron Oxide

Malik et al. (2014) used Iron Oxide NP ranging from 0 to 200 mg/L and pH ranging from 5 to 7. They optimized the conditions to be initial concentration of substrate to be 110 g/L and concentration of Iron Oxide NP 50 mg/L for pH 6. Mixed microbial culture was used to treat the distillery wastewater. It was observed that the cumulative hydrogen production was 380 mL, where the Andrew's model was used to describe the kinetics of hydrogen production and the effect of iron oxide (7.14).

$$\mathrm{SHY} = \frac{(\mathrm{SHY}_m \times S)}{\left(K_s + S + \dfrac{S^2}{K_i}\right)} \tag{7.14}$$

where SHY_m = Maximum specific hydrogen yield, K_i = Inhibition constant, K_s = Half saturation constant, and S = Concentration Iron Oxide NP. It was observed that maximum SHY was 44.28 mL H_2/g COD. The authors theorized that high substrate concentrations become hostile to microorganisms as a result of a reduction in pH or a rise in hydrogen partial pressure, lowering the SHY. Taherdanak et al. (2016) investigated the impact of FeNPs and Fe^{2+} ions on H_2 generation over range of 50 mg/L in anaerobic sludge using glucose. 15% and 37% increase in the H_2 yield was observed with the addition of Fe^{2+} ions and FeNPs. Increase in H_2 yield is related to the reduction in CH_3CH_2OH and $CH_3CH_2CO_2H$ production that utilized the H_2 resulting in less H_2 used. It was revealed that the creation of CH_3CH_2OH had a greater impact on hydrogen output than the generation of other by-products.

Nath et al. (2015) synthesized the FeNP using different concentrations of leaf extract of *Sygygium cumini* and $FeSO_4$ solution. The bacteria strain of *Enterobacter cloacae* DH-89 was used for dark fermentation. Increase in H_2 production was observed after introducing both $FeSO_4$ and FeNP. It was also reported the H_2 production of 950 mL/L at controlled environment. After the addition of $FeSO_4$ the rate was increased by 650 at 25 mg/L. At higher concentration, the H_2 production declined. It was hypothesized that the high quantity of $FeSO_4$ has inhibited the bacterial metabolism resulting in minimal H_2 generation. The addition of FeNP increased the H_2 production by twofold (2,100 mL/L) at concentration of 100 mg/L of FeNP. Further, it was found that iron is essential in ferredoxin. It serves as an electron carrier in the process of hydrogenases for H_2 production. Therefore, at particular dosage of FeNP or iron, improved ferredoxin activity is observed (Nath et al. 2015; Wang and Wan 2008). Mohanraj et al. (2014) used Murrayakoenigii leaf extract for the synthesis of FeNP. Higher efficiency was achieved at 6–7 pH and 10 g/L of glucose and 7.5 g/L of sucrose, respectively, in the range of 25–125 mg/L. It was observed that an increase in the production of H_2 after fermentation of substrate (glucose and sucrose) using *E. cloacae* and adding FeNP and $FeSO_4$, where the sugar type also affects the H_2 production when in contact with FeNP. The study suggested that an increase in H_2 production to be more due FeNP than $FeSO_4$. It was hypothesized that the appropriate dose of iron oxide nanoparticles was crucial for enhanced hydrogen with acetate and butyrate, in which it had observed a decreasing trend in production of C_2H_5OH (Ethanol) and $C_3H_5O_2$ (Propionate) after adding the FeNP. Hence, the hypothesized metabolic pathway shown in Figure 7.2 depicts that supplementing with the right amount of iron oxide nanoparticles may boost ferredoxin oxidoreductase and ferredoxin activity, resulting in a lot of hydrogen generation.

Beckers et al. (2013) in their study reported the biohydrogen production and the use of Pd, Au, Cu, and metallic FeNP. The strain used as hydrogen-producing microorganism was *C. butyricum*, and the concentration of FeNP was 10^{-6} mol/L. The highest hydrogen production rates (HPR) using Fe/SiO_2 catalysts were 58% greater

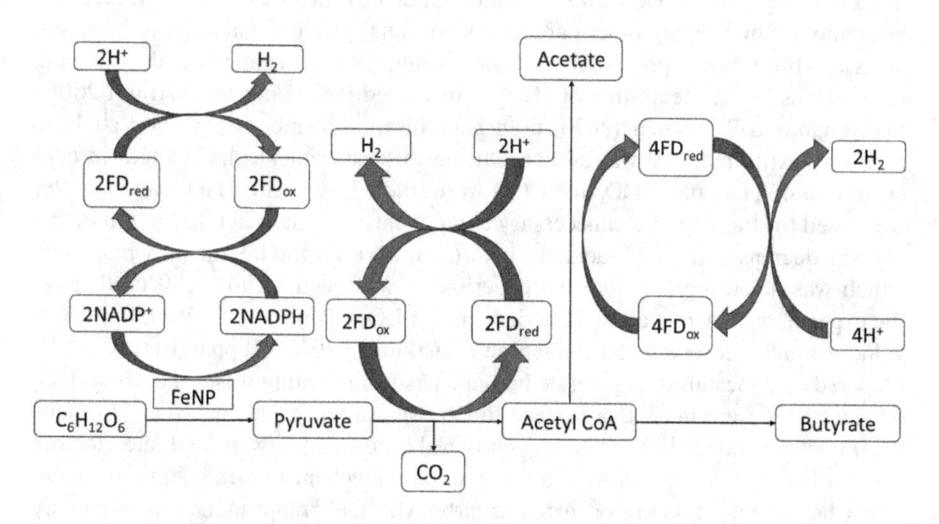

FIGURE 7.2 Metabolic pathway of fermentative hydrogen production process by FeNP.

than the 1.97 0.2 mL/h without catalyst addition. The bacteria had utilized the immobilized iron oxide active sites for oxidation/reduction chemical reactions, allowing them to move electrons quicker without consuming or metabolizing the iron in a dissolved form supplied to the medium.

Reddy et al. (2017) synthesized H_2 by dark fermentation of wastewater using sugarcane bagasse as substrate where the effect of use of both Fe^{2+} ions and $FeSO_4$ NP was studied. It was observed that increase in H_2 production (1.2 mol H_2/mol glucose) when $FeSO_4$ NP was added till 200 mg/L. After 200 mg/L, decrease in H_2 production was observed. It was inferred that the higher iron concentration can cause poisoning and the production of reactive oxygen species, which has a detrimental impact on microorganism growth. The FeNP increases the rate of e^- transfer between NADPH and hydrogenases, which increases the activity of ferredoxin oxidoreductase. Gadhe et al. (2015) explored the use of Fe_2O_3 and NiO NP for H_2 production. The dairy wastewater was used with Fe_2O_3 ranging from 0.5 to 100 mg/L. About 24% in H_2 production at 50 mg/L was achieved. After that the sudden decrease was observed in H_2 production due to toxicity. It was hypothesized that rate of e– transfer was improved by the addition of Fe_2O_3 NP between NADPH to [Fe–Fe] hydrogenase. Based on the several reviews on the literature, it was discovered that adding FeNP to the mix boosts the activity of ferredoxin oxidoreductase, which is responsible for H_2 production.

7.2.6 Titanium Dioxide

Mahmoodi and Arami (2009) studied a pilot scale reaction in order to degrade and reduce toxicity of textile wastewater. For this reaction, titanium dioxide was used to photocatalytically degrade textile dye known as Acid Blue 25. The parameters that need to be optimized were pH, H_2O_2, and concentration of Acid Blue 25 dye. TiO_2 nanoparticles were used due to its nontoxic nature, being chemically inert, and majorly being photoactive nature. All the parameters mentioned were checked one after another by keeping other parameters constant. Daphnia magna was employed to assess the toxicity present in the wastewater. Decolorization (i.e. degradation) increases as the concentration of H_2O_2 is increased (Mahmoodi and Arami 2009). Saravanan and Sasikumar (2020) used nanofiltration membranes in their study to treat the textile industry wastewater. During this experiment, the changes in concentration of TDS, pH, BOD, and COD were studied. A filter of TiO_2 nanoparticles were used to strain out the unnecessary components of wastewater. It was found that pH was decreased from 11 (acidic) to 6.5 (neutral) after the treatment, while TDS, which was 4,100 mg/L before the experiment, was declined to 2,000 mg/L after the experiment. There was a 33.3% decrease in BOD value, and 54.5% fall in COD value was also observed (Saravanan and Sasikumar 2020). Hippargi et al. (2021) prepared a nanocatalyst to generate hythane gas from treating wastewater. The TiO_2 nanoparticles were used as a surface to deposit Au and Pt nanoparticles to make Au–Pt/TiO_2 nanocatalyst. TiO_2 was chosen due to its ability to facilitate electron transfer between Au and Pt, which was the main mechanism idea behind this reaction. Chemical reaction that occurred in carboxylic acid groups in the wastewater are converted to hydrogen and methane gas, which together formed hythane gas. It was

TABLE 7.1
Comparative Study on Hydrogen Production Using Different Nanoparticles

Nanoparticle	Concentration	Substrate Feed	Organism	Reactor/Mode	Temperature	pH	Biofuel Yield		Reference
							CH_4	H_2	
ZVI	400 mg/L	Municipal wastewater treatment	*Clostridium* sp.	Anaerobic granular sludge reactor	30°C	7	–	1.22 mol H_2/mol glucose	Zhang et al. (2015)
ZVI-AC (Fe-activated carbon)	400 mg/L	Municipal wastewater treatment	*Clostridium* sp.	Anaerobic granular sludge reactor	30°C	7	–	1.33 mol H_2/mol glucose	Zhang et al. (2015)
ZVI (Fe^0)	500 g/L	Municipal wastewater treatment	Waste-activated sludge	UASB reactor	35°C	7	729.61 mL/VSS	–	Feng et al. (2014)
ZVI (Fe^0)	5 g/L	Brewery wastewater	Anaerobic membrane bioreactor	Anaerobic membrane bioreactor	30°C	-	183 mL	–	Carpenter et al. (2015)
ZVI (Fe^0)	25 mg/L	Wastewater treatment plant	Anaerobic digester		37°C	5.5	325 mL/g-VSS	–	Taherdanak et al. (2016)
Fe^0	100 mg/L		*Enterobacter Cloacae* DH-89	Batch	37°C	7	–	2,100 ml H_2/L	Nath et al. (2015)
ZnO	20 mg/L	Artificial wastewater	*Chlorella vulgaris*	Batch	Room Temp	6.5–8.0	2.08 g/L	–	Xia et al. (2021)
ZnO	6 mg/g-TSS	Municipal wastewater	Waste-activated sludge	Batch	35°C	6.7	125.5 mL/g-VSS	–	Mu et al. (2011)
ZnO	0.32–34.5 mg/L	Brewery Wastewater	Anaerobic granular sludge	Usab	30°C	7.2	–	–	Otero-González et al. (2014)

(Continued)

TABLE 7.1 (*Continued*)
Comparative Study on Hydrogen Production Using Different Nanoparticles

Nanoparticle	Concentration	Substrate Feed	Organism	Reactor/Mode	Temperature	pH	Biofuel Yield		Reference
							CH_4	H_2	
Pd impregnated with multiwalled carbon nanotubes (MWCNTs)	–	Hydrolysis Of $Nabh_4$	–	Batch	22°C	7	–	21.7 mL/min/g_{cat}	Huff et al. (2018)
CNT impregnated Pd nanocatalyst	–	Decomposition reaction of HI	–	Batch	550°C	–	–	23%	Singhania and Bhaskarwar (2018a)
Au–Pt/TiO_2 nanophotocatalyst	–	Waste water	–	Batch	–	-	4,230 µmol/h	9,390 µmol/h	Hippargi et al. (2021)
Fe^{3+}	50 mg/L	Dairy wastewater.	Anaerobic sludge	Batch	37°C	5.5	16.75 m mol/gcod	–	Gadheet al. (2015)
Fe^{2+}	200 mg/L	Wastewater treatment plant	Anaerobic holding tank	Batch	30°C		37.296 mL	–	Reddy et al. (2017)
Fe^{2+}	25 mg/L		*E. Cloacae* DH-89	Batch	37°C	7	–	1,600 mL H_2 /L	Nath et al. (2015)
Fe^{2+}	25 mg/L	Wastewater treatment plant	Anaerobic digester		37°C	5.5	325 mL/g-VSS	–	Taherdanak et al. (2016)
Fe	50 ppm	Synthetic Wastewater	*Clostridium pasteurianum*	Batch	35°C	7	-	1.8 mol H_2/mol xylase	Hsieh et al. (2016)
Fe_2O_3	50 mg/L	Artificial wastewater	*Chlorella vulgaris*	Batch	Room temp.	6.5–8.0	2.08 g/L	-	Xia et al. (2021)

(*Continued*)

TABLE 7.1 (*Continued*)
Comparative Study on Hydrogen Production Using Different Nanoparticles

Nanoparticle	Concentration	Substrate Feed	Organism	Reactor/Mode	Temperature	pH	Biofuel Yield		Reference
							CH_4	H_2	
TiO_2	800 ppm	Synthetic wastewater	*Clostridium pasteurianum*	Batch	35°C	7	-	2.1 mol H_2/mol xylase	Hsieh et al. (2016)
TiO_2	30 mg/g-TSS	Municipal wastewater	Waste-activated sludge	Batch	35°C	6.7	130.5 mL/g-VSS	-	Mu et al. (2011)
SiO_2	30 mg/g-TSS	Municipal wastewater	Waste-activated sludge	Batch	35°C	6.7	127.5 mL/g-VSS	-	Mu et al. (2011)
Al_2O_3	30 mg/g-TSS	Municipal wastewater	Waste-activated sludge	Batch	35°C	6.7	126.5 mL/g-VSS	-	Mu et al. (2011)

found that increase in deposition percentage of Au–Pt increased the hydrogen and methane yield. But the graph was again declined as the concentration goes higher due to poisoning of TiO_2. Therefore, 1% concentration of Au–Pt (0.5%w/w Au and 0.5%w/w Pt) gave the best results with evolution of methane around 9,390 μmol/h and hydrogen around 4,230 μmol/h. The presence of 5% CH_3COOH in the wastewater sample gave promising results as compared to the other concentrations (Hippargi et al. 2021).

Table 7.1 given below summarizes the effect of different Nanaoparticles on the biofuel yield.

7.3 CONCLUSION AND FUTURE ASPECTS

The work summarizes the effect of different nanocatalysts on the pretreatment of wastewater and simultaneous production of biological fuels. The comparison of biofuel yield by various nanocatalysts is discussed. The study focuses on the impact of nanoparticle characteristics and its concentration on biofuel production improvement. Simultaneously, the effect of nanoengineered particles on the growth of organism, efficiency, and feed degradation was reported. All catalysts produced higher yields as compared to other biofuel production technologies, making this an efficient, effective, and cost-effective technique for biofuel production. Furthermore, the effective wastewater treatment from diverse sectors such as the chemical, brewery, food processing, wine processing, and dairy industries was investigated. The varieties of microorganisms and organisms utilized in biofuel production, as well as the varied circumstances, were investigated.

Along with optimizing the existing technologies, researchers need to work on enhancing the efficiency and cost-effective application of unconventional water resources. Similarly, the effect of toxicity of nanocatalyst on environment and human health needs to be reduced. Green-synthesized nanocatalysts contain bioactive compounds that act as reducing agent or coating matrix, which may protect the microorganism cell from fermentation. These catalysts show interactive mechanism in reducing the pollutant from wastewater stream. Also, the low cost of production proves it to be a viable option for large-scale application.

REFERENCES

Ahmad, A. 2020. "Bioprocess evaluation of petroleum wastewater treatment with zinc oxide nanoparticle for the production of methane gas: Process assessment and modelling." *Applied Biochemistry and Biotechnology* 190 (3): 851–66. doi:10.1007/s12010-019-03137-4.

Ahmad, A., Azizul B., and A.H. Bhat. 2016. "Renewable and sustainable bioenergy production from microalgal co-cultivation with palm oil mill effluent (POME): A review." *Renewable and Sustainable Energy Reviews* 65: 214–34. doi:10.1016/j.rser.2016.06.084.

Anand, A.S., S. Adish Kumar, J. Rajesh Banu, and G. Ginni. 2016. "The performance of fluidized bed solar photo Fenton oxidation in the removal of COD from hospital wastewaters." *Desalination and Water Treatment* 57 (18): 8236–42. doi:10.1080/19443994.2015.1021843.

Baran, T., and A. Menteş. 2020. "Production of palladium nanocatalyst supported on modified gum Arabic and investigation of its potential against treatment of environmental contaminants." *International Journal of Biological Macromolecules* 161: 1559–67. doi:10.1016/j.ijbiomac.2020.07.321.

Baruah, S., M.N. Khan, and J. Dutta. 2016. "Perspectives and applications of nanotechnology in water treatment." *Environmental Chemistry Letters* 14 (1): 1–14. doi:10.1007/s10311-015-0542-2.

Beckers, L., S. Hiligsmann, S.D. Lambert, B. Heinrichs, and P. Thonart. 2013. "Improving effect of metal and oxide nanoparticles encapsulated in porous silica on fermentative biohydrogen production by clostridium butyricum." *Bioresource Technology* 133: 109–17. doi:10.1016/j.biortech.2012.12.168.

Brunner, T.J., P. Wick, P. Manser, P. Spohn, R.N. Grass, L.K. Limbach, A. Bruinink, and W.J. Stark. 2006. "In vitro cytotoxicity of oxide nanoparticles: Comparison to asbestos, silica, and the effect of particle solubility." *Environmental Science and Technology* 40 (14): 4374–81. doi:10.1021/es052069i.

Budzianowski, W.M. 2012. "Sustainable biogas energy in Poland: Prospects and challenges." *Renewable and Sustainable Energy Reviews* 16 (1): 342–49. doi:10.1016/j.rser.2011.07.161.

Carpenter, A.W., S.N. Laughton, and M.R. Wiesner. 2015. "Enhanced biogas production from nanoscale zero valent iron-amended anaerobic bioreactors." *Environmental Engineering Science* 32 (8): 647–55. doi:10.1089/ees.2014.0560.

Corte, S.D., T. Hennebel, B.D. Gusseme, W. Verstraete, and N. Boon. 2012. "Bio-Palladium: From metal recovery to catalytic applications." *Microbial Biotechnology* 5 (1): 5–17. doi:10.1111/j.1751-7915.2011.00265.x.

Daer, S., J. Kharraz, A. Giwa, and S.W. Hasan. 2015. "Recent applications of nanomaterials in water desalination: A critical review and future opportunities." *Desalination* 367: 37–48. doi:10.1016/j.desal.2015.03.030.

Das, D., T. Dutta, K. Nath, S.M. Kotay, A.K. Das, and T.N. Veziroglu. 2006. "Role of Fe-hydrogenase in biological hydrogen production." *Current Science* 90(12): 1627–37.

Dutta, S. 2020. *Wastewater Treatment Using TiO_2-Based Photocatalysts. Handbook of Smart Photocatalytic Materials.* INC. doi:10.1016/b978-0-12-819051-7.00010-5.

Eaktasang, N. 2017. "Effects of nanoscale zero-valent-iron (NZVI) particles on biohydrogen production from organic wastes." *International Proceedings of Chemical, Biological and Environmental Engineering* 100: 71–76. doi:10.7763/IPCBEE. 2017.V100.12.

Elreedy, A., E. Ibrahim, N. Hassan, A. El-Dissouky, M. Fujii, C. Yoshimura, and A. Tawfik. 2017. "Nickel-graphene nanocomposite as a novel supplement for enhancement of biohydrogen production from industrial wastewater containing mono-ethylene glycol." *Energy Conversion and Management* 140: 133–44. doi:10.1016/j.enconman.2017.02.080.

Engliman, N.S., P.M. Abdul, S.Y. Wu, and J.M. Jahim. 2017. "Influence of iron (II) oxide nanoparticle on biohydrogen production in thermophilic mixed fermentation." *International Journal of Hydrogen Energy* 42 (45): 27482–93. doi:10.1016/j.ijhydene.2017.05.224.

Feng, Y., Y. Zhang, X. Quan, and S. Chen. 2014. "Enhanced anaerobic digestion of waste activated sludge digestion by the addition of zero valent iron." *Water Research* 52: 242–50. doi:10.1016/j.watres.2013.10.072.

Ferreira, A., P. Marques, B. Ribeiro, P. Assemany, H.V. de Mendonça, A. Barata, A.C. Oliveira, A. Reis, H.M. Pinheiro, and L. Gouveia. 2018. "Combining biotechnology with circular bioeconomy: From poultry, swine, cattle, brewery, dairy and urban wastewaters to biohydrogen." *Environmental Research* 164 (December 2017): 32–38. doi:10.1016/j.envres.2018.02.007.

Gadhe, A., S.S. Sonawane, and M.N. Varma. 2015. "Enhancement effect of hematite and nickel nanoparticles on biohydrogen production from dairy wastewater." *International Journal of Hydrogen Energy* 40 (13): 4502–11. doi:10.1016/j.ijhydene.2015.02.046.

Garole, V.J., B.C. Choudhary, S.R. Tetgure, D.J. Garole, and A.U. Borse. 2019. "Palladium nanocatalyst: Green synthesis, characterization, and catalytic application." *International Journal of Environmental Science and Technology* 16 (12): 7885–92. doi:10.1007/s13762-018-2173-1.

Hallenbeck, P.C. 2005. "Fundamentals of the fermentative production of hydrogen." *Water Science and Technology* 52 (1–2): 21–29. doi:10.2166/wst.2005.0494.

Hawkes, F.R., I. Hussy, G. Kyazze, R. Dinsdale, and D.L. Hawkes. 2007. "Continuous dark fermentative hydrogen production by mesophilic microflora: Principles and progress." *International Journal of Hydrogen Energy* 32 (2): 172–84. doi:10.1016/j.ijhydene.2006.08.014.

Hernández-Mendoza, C.E., I. Moreno-Andrade, and G. Buitrón. 2014. "Comparison of hydrogen-producing bacterial communities adapted in continuous and discontinuous reactors." *International Journal of Hydrogen Energy* 39 (26): 14234–39. doi:10.1016/j.ijhydene.2014.01.014.

Hippargi, G., S. Anjankar, R.J. Krupadam, and S.S. Rayalu. 2021. "Simultaneous wastewater treatment and generation of blended fuel methane and hydrogen using Au-Pt/TiO_2 photoreforming catalytic material." *Fuel* 291 (December 2020). doi:10.1016/j.fuel.2020.120113.

Hsieh, P.H., Y.C. Lai, K.Y. Chen, and C.H. Hung. 2016. "Explore the possible effect of TiO_2 and magnetic hematite nanoparticle addition on biohydrogen production by clostridium pasteurianum based on gene expression measurements." *International Journal of Hydrogen Energy* 41 (46): 21685–91. doi:10.1016/j.ijhydene.2016.06.197.

Huff, C., J.M. Long, A. Heyman, and T.M. Abdel-Fattah. 2018. "Palladium nanoparticle multiwalled carbon nanotube composite as catalyst for hydrogen production by the hydrolysis of sodium borohydride." *ACS Applied Energy Materials* 1 (9): 4635–40. doi:10.1021/acsaem.8b00748.

Intanoo, P., P. Chaimongkol, and S. Chavadej. 2016. "Hydrogen and methane production from cassava wastewater using two-stage upflow anaerobic sludge blanket reactors (UASB) with an emphasis on maximum hydrogen production." *International Journal of Hydrogen Energy* 41 (14): 6107–14. doi:10.1016/j.ijhydene.2015.10.125.

Jiang, S., Y. Chen, Q. Zhou, and G. Gu. 2007. "Biological short-chain fatty acids (SCFAs) production from waste-activated sludge affected by surfactant." *Water Research* 41 (14): 3112–20. doi:10.1016/j.watres.2007.03.039.

John, C.C., R.R. Trussell, D.W. Hand, K.J. Howe and G. Tchobanoglous. 2017. "MWH's water treatment principles and design."

Kargi, F., N.S. Eren, and S. Ozmihci. 2012. "Bio-hydrogen production from cheese whey powder (CWP) solution: Comparison of thermophilic and mesophilic dark fermentations." *International Journal of Hydrogen Energy* 37 (10): 8338–42. doi:10.1016/j.ijhydene.2012.02.162.

Kempasiddaiah, M., V. Kandathil, R.B. Dateer, M. Baidya, S.A. Patil, and S.A. Patil. 2021. "Efficient and recyclable palladium enriched magnetic nanocatalyst for reduction of toxic environmental pollutants." *Journal of Environmental Sciences (China)* 101: 189–204. doi:10.1016/j.jes.2020.08.015.

Kojima, Y., K.I. Suzuki, K. Fukumoto, M. Sasaki, T. Yamamoto, Y. Kawai, and H. Hayashi. 2002. "Hydrogen generation using sodium borohydride solution and metal catalyst coated on metal oxide." *International Journal of Hydrogen Energy* 27 (10): 1029–34. doi:10.1016/S0360-3199(02)00014-9.

Krishna, R. 2013. "Bio hydrogen production from pharmaceutical waste water treatment by a suspended growth reactor using environmental anaerobic technology." *American Chemical Science Journal* 3 (2): 80–97. doi:10.9734/acsj/2013/2649.

Kumar, G., T. Mathimani, E.R. Rene, and A. Pugazhendhi. 2019. "Application of nanotechnology in dark fermentation for enhanced biohydrogen production using inorganic nanoparticles." *International Journal of Hydrogen Energy* 44 (26): 13106–13. doi:10.1016/j.ijhydene.2019.03.131.

Lam, M.K., M.I. Yusoff, Y. Uemura, J.W. Lim, C.G. Khoo, K.T. Lee, and H.C. Ong. 2017. "Cultivation of *Chlorella vulgaris* using nutrients source from domestic wastewater for biodiesel production: Growth condition and kinetic studies." *Renewable Energy* 103: 197–207. doi:10.1016/j.renene.2016.11.032.

Leong, Y.K., P.L. Show, J.C.W. Lan, R. Krishnamoorthy, D.T. Chu, D. Nagarajan, H.W. Yen, and J.S. Chang. 2019. "Application of thermo-separating aqueous two-phase system in extractive bioconversion of polyhydroxyalkanoates by cupriavidus necator H16." *Bioresource Technology* 287 (May): 121474. doi:10.1016/j.biortech.2019.121474.

Levin, D.B., and R. Chahine. 2010. "Challenges for renewable hydrogen production from biomass." *International Journal of Hydrogen Energy* 35 (10): 4962–69. doi:10.1016/j.ijhydene.2009.08.067.

Li, Y., X.M Hong, D.M Collard, and M.A El-sayed. 2005. "Metal-catalyzed cross-coupling reactions." *Choice Reviews Online* 43 (01): 43-0331. doi:10.5860/choice.43-0331.

Liang, J.H., X.Q. Ren, J.T. Wang, M. Jiang, and Z.J. Li. 2010. "Preparation of biodiesel by transesterification from cottonseed oil using the basic dication ionic liquids as catalysts." *Ranliao Huaxue Xuebao/Journal of Fuel Chemistry and Technology* 38 (3): 275–80. doi:10.1016/s1872-5813(10)60033-3.

Lin, C.Y., and C.H. Lay. 2004. "Carbon/nitrogen-ratio effect on fermentative hydrogen production by mixed microflora." *International Journal of Hydrogen Energy* 29 (1): 41–45. doi:10.1016/S0360-3199(03)00083-1.

Lin, C.Y., C.C. Chiang, M.L.T. Nguyen, and C.H. Lay. 2017. "Enhancement of fermentative biohydrogen production from textile desizing wastewater via coagulation-pretreatment." *International Journal of Hydrogen Energy* 42 (17): 12153–58. doi:10.1016/j.ijhydene.2017.03.184.

Liu, Y., Y. Zhang, and B.J. Ni. 2015. "Zero valent iron simultaneously enhances methane production and sulfate reduction in anaerobic granular sludge reactors." *Water Research* 75: 292–300. doi:10.1016/j.watres.2015.02.056.

Mahmoodi, N.M., and M. Arami. 2009. "Degradation and toxicity reduction of textile wastewater using immobilized titania nanophotocatalysis." *Journal of Photochemistry and Photobiology B: Biology* 94 (1): 20–24. doi:10.1016/j.jphotobiol.2008.09.004.

Malik, S.N., P.C. Ghosh, A.N. Vaidya, and S.N. Mudliar. 2018. "Catalytic ozone pretreatment of complex textile effluent using Fe^{2+} and zero valent iron nanoparticles." *Journal of Hazardous Materials* 357 (May): 363–75. doi:10.1016/j.jhazmat.2018.05.070.

Malik, S.N., S.M. Khan, P.C. Ghosh, A.N. Vaidya, G. Kanade, and S.N. Mudliar. 2019. "Treatment of pharmaceutical industrial wastewater by nano-catalyzed ozonation in a semi-batch reactor for improved biodegradability." *Science of the Total Environment* 678: 114–22. doi:10.1016/j.scitotenv.2019.04.097.

Malik, S.N., and S. Kumar. 2021. "Enhancement effect of zero-valent iron nanoparticle and iron oxide nanoparticles on dark fermentative hydrogen production from molasses-based distillery wastewater." *International Journal of Hydrogen Energy*. doi:10.1016/j.ijhydene.2021.06.125.

Malik, S.N., V. Pugalenthi, A.N. Vaidya, P.C. Ghosh, and S.N. Mudliar. 2014. "Kinetics of nano-catalysed dark fermentative hydrogen production from distillery wastewater." *Energy Procedia* 54: 417–30. doi:10.1016/j.egypro.2014.07.284.

Mohanraj, S., K. Anbalagan, S. Kodhaiyolii, and V. Pugalenthi. 2014. "Comparative evaluation of fermentative hydrogen production using enterobacter cloacae and mixed culture: Effect of Pd (II) ion and phytogenic palladium nanoparticles." *Journal of Biotechnology* 192 (Part A): 87–95. doi:10.1016/j.jbiotec.2014.10.012.

Mu, H., Y. Chen, and N. Xiao. 2011. "Effects of metal oxide nanoparticles (TiO_2, Al_2O_3, SiO_2 and ZnO) on waste activated sludge anaerobic digestion." *Bioresource Technology* 102 (22): 10305–11. doi:10.1016/j.biortech.2011.08.100.

Mu, Y., X.J. Zheng, H.Q. Yu, and R.F. Zhu. 2006. "Biological hydrogen production by anaerobic sludge at various temperatures." *International Journal of Hydrogen Energy* 31 (6): 780–85. doi:10.1016/j.ijhydene.2005.06.016.

Nascimento Junior, J.R.d., L.A.Z. Torres, A.B. Pedroni Medeiros, A.L. Woiciechowski, W.J. Martinez-Burgos, and C. Ricardo Soccol. 2021. "Enhancement of biohydrogen production in industrial wastewaters with vinasse pond consortium using lignin-mediated iron nanoparticles." *International Journal of Hydrogen Energy* 46 (54): 27431–43. doi:10.1016/j.ijhydene.2021.06.009.

Nath, D., A.K. Manhar, K. Gupta, D. Saikia, S.K. Das, and M. Mandal. 2015. "Phytosynthesized iron nanoparticles: Effects on fermentative hydrogen production by enterobacter cloacae DH-89." *Bulletin of Materials Science* 38 (6): 1533–38. doi:10.1007/s12034-015-0974-0.

Oh, S.E., P. Iyer, M.A. Bruns, and B.E. Logan. 2004. "Biological hydrogen production using a membrane bioreactor." *Biotechnology and Bioengineering* 87 (1): 119–27. doi:10.1002/bit.20127.

Otero-González, L., J.A. Field, and R. Sierra-Alvarez. 2014. "Fate and long-term inhibitory impact of ZnO nanoparticles during high-rate anaerobic wastewater treatment." *Journal of Environmental Management* 135: 110–17. doi:10.1016/j.jenvman.2014.01.025.

Patel, S.K.S., and V.C. Kalia. 2013. "Integrative biological hydrogen production: An overview." *Indian Journal of Microbiology* 53 (1): 3–10. doi:10.1007/s12088-012-0287-6.

Patel, S.K.S., J.K. Lee, and V.C. Kalia. 2018. "Nanoparticles in biological hydrogen production: An overview." *Indian Journal of Microbiology* 58 (1): 8–18. doi:10.1007/s12088-017-0678-9.

Patni, A.N., A.S. Mantri, and D. Kundu. 2021. "Ionic liquid promoted dehydrogenation of amine boranes: A review." *International Journal of Hydrogen Energy* 46 (21): 11761–81. doi:10.1016/j.ijhydene.2021.01.032.

Perna, V., E. Castelló, J. Wenzel, C. Zampol, D. M. Fontes Lima, L. Borzacconi, M. B. Varesche, M. Zaiat, and C. Etchebehere. 2013. "Hydrogen production in an upflow anaerobic packed bed reactor used to treat cheese whey." *International Journal of Hydrogen Energy* 38 (1): 54–62. doi:10.1016/j.ijhydene.2012.10.022.

Qiu, F., Y. Li, D. Yang, X. Li, and P. Sun. 2011. "Heterogeneous solid base nanocatalyst: Preparation, characterization and application in biodiesel production." *Bioresource Technology* 102 (5): 4150–56. doi:10.1016/j.biortech.2010.12.071.

Rambabu, K., G. Bharath, A. Thanigaivelan, D.B. Das, P.L. Show, and F. Banat. 2021. "Augmented biohydrogen production from rice mill wastewater through nano-metal oxides assisted dark fermentation." *Bioresource Technology* 319 (October 2020) doi:10.1016/j.biortech.2020.124243.

Reddy, K., M. Nasr, S. Kumari, S. Kumar, S.K. Gupta, A.M. Enitan, and F. Bux. 2017. "Biohydrogen production from sugarcane bagasse hydrolysate: Effects of PH, S/X, Fe^{2+}, and magnetite nanoparticles." *Environmental Science and Pollution Research* 24 (9): 8790–8804. doi:10.1007/s11356-017-8560-1.

Redwood, M.D., M. Paterson-Beedle, and L.E. MacAskie. 2009. "Integrating dark and light bio-hydrogen production strategies: Towards the hydrogen economy." *Reviews in Environmental Science and Biotechnology* 8 (2): 149–85. doi:10.1007/s11157-008-9144-9.

Sajjadi, M., M. Nasrollahzadeh, and M.R. Tahsili. 2019. "Catalytic and antimicrobial activities of magnetic nanoparticles supported N-heterocyclic palladium(II) complex: A magnetically recyclable catalyst for the treatment of environmental contaminants in aqueous media." *Separation and Purification Technology* 227 (June): 115716. doi:10.1016/j.seppur.2019.115716.

Sandoval, R., A.M. Cooper, K. Aymar, A. Jain, and K. Hristovski. 2011. "Removal of arsenic and methylene blue from water by granular activated carbon media impregnated with zirconium dioxide nanoparticles." *Journal of Hazardous Materials* 193: 296–303. doi:10.1016/j.jhazmat.2011.07.061.

Saravanan, N., and K.S.K. Sasikumar. 2020. "Waste water treatment process using nano TiO_2." *Materials Today: Proceedings* 33: 2570–72. doi:10.1016/j.matpr.2019.12.143.

Shida, G.M., A.R. Barros, C.M. dos Reis, E.L.C. de Amorim, M.H.R.Z. Damianovic, and E.L. Silva. 2009. "Long-term stability of hydrogen and organic acids production in an anaerobic fluidized-bed reactor using heat treated anaerobic sludge inoculum." *International Journal of Hydrogen Energy* 34 (9): 3679–88. doi:10.1016/j.ijhydene.2009.02.076.

Show, K.Y., D.J. Lee, and J.S. Chang. 2011. "Bioreactor and process design for biohydrogen production." *Bioresource Technology* 102 (18): 8524–33. doi:10.1016/j.biortech.2011.04.055.

Silva, A.N.d., W.V. Macêdo, I.K. Sakamoto, D.d.L.A.D. Pereyra, C.O. Mendes, S.I. Maintinguer, R.A.C. Filho, M.H.Z. Damianovic, M.B.A. Varesche, and E.L.C. de Amorim. 2019. "Biohydrogen production from dairy industry wastewater in an anaerobic fluidized-bed reactor." *Biomass and Bioenergy* 120 (November 2018): 257–64. doi:10.1016/j.biombioe.2018.11.025.

Singhania, A., and A.N. Bhaskarwar. 2018a. "Catalytic performance of carbon nanotubes supported palladium catalyst for hydrogen production from hydrogen iodide decomposition in thermochemical sulfur iodine cycle." *Renewable Energy* 127: 509–13. doi:10.1016/j.renene.2018.05.017.

Singhania, A., and A.N. Bhaskarwar. 2018b. "Platinum-titania catalysts for hydrogen-iodide decomposition in sulfur-iodine cycle for hydrogen production." *Chemistry Letters* 47 (12): 1482–85. doi:10.1246/cl.180770.

Stavropoulos, K.P., A. Kopsahelis, C. Zafiri, and M. Kornaros. 2016. "Effect of PH on continuous biohydrogen production from end-of-life dairy products (EoL-DPs) via dark fermentation." *Waste and Biomass Valorization* 7 (4): 753–64. doi:10.1007/s12649-016-9548-7.

Taherdanak, M., H. Zilouei, and K. Karimi. 2016. "The effects of FeO and NiO nanoparticles versus Fe^{2+} and Ni^{2+} ions on dark hydrogen fermentation." *International Journal of Hydrogen Energy* 41 (1): 167–73. doi:10.1016/j.ijhydene.2015.11.110.

Ueno, Y., S. Haruta, M. Ishii, and Y. Igarashi. 2001. "Microbial community in anaerobic hydrogen-producing microflora enriched from sludge compost." *Applied Microbiology and Biotechnology* 57 (4): 555–62. doi:10.1007/s002530100806.

Usman, T.M., J.R. Banu, M. Gunasekaran, and G. Kumar. 2019. "Biohydrogen production from industrial wastewater: An overview." *Bioresource Technology Reports* 7 (July): 100287. doi:10.1016/j.biteb.2019.100287.

Vijayalakshmi, S., E. Kumar, P. Sundara Venkatesh, and A. Raja. 2020. "Preparation of zirconium oxide with polyaniline nanocatalyst for the decomposition of pharmaceutical industrial wastewater." *Ionics* 26 (3): 1507–13. doi:10.1007/s11581-019-03323-8.

Wang, J., and W. Wan. 2008. "Effect of Fe^{2+} concentration on fermentative hydrogen production by mixed cultures." *International Journal of Hydrogen Energy* 33 (4): 1215–20. doi:10.1016/j.ijhydene.2007.12.044.

Wu, Y., H. Pang, Y. Liu, X. Wang, S. Yu, D. Fu, J. Chen, and X. Wang. 2019. "Environmental remediation of heavy metal ions by novel-nanomaterials: A review." *Environmental Pollution* 246: 608–20. doi:10.1016/j.envpol.2018.12.076.

Wünschiers, R., and P. Lindblad. 2002. "Hydrogen in education - A biological approach." *International Journal of Hydrogen Energy* 27 (11–12): 1131–40. doi:10.1016/S0360-3199(02)00098–8.

Xia, C., Q. Van Le, A. Chinnathambi, S.H. Salmen, S.A. Alharbi, and S. Tola. 2021. "Role of ZnO and Fe_2O_3 nanoparticle on synthetic saline wastewater on growth, nutrient removal and lipid content of *Chlorella vulgaris* for sustainable production of biofuel." *Fuel* 300 (May): 120924. doi:10.1016/j.fuel.2021.120924.

Yang, G., and J. Wang. 2018. "Improving mechanisms of biohydrogen production from grass using zero-valent iron nanoparticles." *Bioresource Technology* 266 (May): 413–20. doi:10.1016/j.biortech.2018.07.004.

Zhang, L., L. Zhang, and D. Li. 2015. "Enhanced dark fermentative hydrogen production by zero-valent iron activated carbon micro-electrolysis." *International Journal of Hydrogen Energy* 40 (36): 12201–8. doi:10.1016/j.ijhydene.2015.07.106.

Zhang, Y., Y. Jing, X. Quan, Y. Liu, and P. Onu. 2011. "A built-in zero valent iron anaerobic reactor to enhance treatment of azo dye wastewater." *Water Science and Technology* 63 (4): 741–46. doi:10.2166/wst.2011.301.

Zhang, Z.P., K.Y. Show, J.H. Tay, D.T. Liang, D.J. Lee, and A. Su. 2008. "The role of acid incubation in rapid immobilization of hydrogen-producing culture in anaerobic upflow column reactors." *International Journal of Hydrogen Energy* 33 (19): 5151–60. doi:10.1016/j.ijhydene.2008.05.016.

Zhu, G.F., P. Wu, Q.S. Wei, J.Y. Lin, Y.L. Gao, and H.N. Liu. 2010. "Biohydrogen production from purified terephthalic acid (PTA) processing wastewater by anaerobic fermentation using mixed microbial communities." *International Journal of Hydrogen Energy* 35 (15): 8350–56. doi:10.1016/j.ijhydene.2009.12.003.

8 Recent Status Potential Challenges and Future Perspectives of Biofuel Generated through Wastewater Treatment

Arpan Patra, Shruti Sharma, Krishna R. Nair, and Divya Gupta
Indian Institute of Technology

CONTENTS

DOI: 10.1201/9781003197737-10

8.1 INTRODUCTION

The rising consumption of fossil fuels carries with it the threat of increase in greenhouse gas emissions that can aggravate the global warming crisis. Thus, there has been extensive research to explore alternative sources of energy that are not only renewable but also carbon neutral. Biofuels are such potential sources that are being widely studied for applications in areas of energy generation and transportation. At present, the majorly used biofuels are bioethanol and biodiesel that are produced from crops like sugar cane, corn, soybean, etc., which raise concerns regarding the economic viability and their impact on food availability. One of the potential biofuel sources that have been proposed are algae, particularly the unicellular microalgae. These microalgae have higher biomass productivity per unit land area used than crops and generate substantial quantities of oil that can be converted to biodiesel (Pittman et al., 2011). They therefore reduce the greenhouse gas emissions associated with the utilization of fossil fuel and agriculture used for biofuel production. The algae produced can be converted to liquid or gas fuels by biochemical or thermochemical processes. Biochemical processes such as anaerobic digestion of algal biomass can produce biogas, while thermochemical processes such as gasification, pyrolysis, supercritical fluid extraction, hydrogenation and liquefaction can be used on algal biomass to produce syngas, hydrogen, ethanol, and gasoline. Despite the benefits associated with algal biofuels, there is a need for innovative, cost-effective technologies that can make biofuels compete with gasoline in terms of pricing. Isolation of high lipid-producing strains and genetically engineered commonly occurring strains to improve biofuel yield is one step to make it economically feasible. Microalgae provides another benefit that they can also be grown in salty water, such as sea water or wastewater. Another approach is the integration of algae production with wastewater treatment systems. These systems supply nutrients like nitrogen, phosphorus, sodium, potassium, etc., supporting algal growth and, in turn, providing oxygen to the heterotrophic microorganisms active in degrading the biodegradable organic matter in wastewater. Conventionally used wastewater treatment processes, such as activated sludge treatment process and sequential batch reactors, face challenges in removal of high concentrations of nutrients present in wastewater such as nitrogen and phosphorous. Specialized tertiary treatment needs to be carried out for removal of these materials. However, wastewater treatment with microalgae provides an added bonus of combining utilization of nutrients with treatment process (Pittman et al., 2011). Microalgal treatment performance is majorly affected by the wastewater composition. Three main nutrients required for proper growth of microalgae in wastewater are carbon, nitrogen, and phosphorous (Falkowski and Raven, 2007). If the nutrients are not available in sufficient quantity, the biomass growth can significantly reduce (Xin et al., 2010, Alketife et al., 2017). Therefore, enhanced nutrient removal can be achieved by an optimal ratio of nutrients. Moreover, micronutrients such as calcium, magnesium, potassium, manganese, silica, zinc, iron, and others are required for the biomass to grow (Mohsenpour et al., 2021).

Microalgae uses the nutrients available in wastewater for their cell growth. The algae produced can be converted to biofuel through various thermochemical processes, such as gasification, pyrolysis, and hydrothermal liquefaction. Other possible routes include biochemical methods such as fermentation, anaerobic digestion, and transesterification, which converts microalgae to biodiesel.

8.2 ALGAE TYPES, STRAINS FOR BIOFUEL PRODUCTION

Algae can consist of single-celled organisms or can have very complex forms. They are found in damp environments or water bodies (Singh et al., 2011). Depending on the size, algae are classified as microalgae or macroalgae (seaweed). Both the algae types have been explored for biofuel production. Based on the type of nutrition, algae growth can be photoautotrophic, mixotrophic, or heterotrophic. Photoautotrophic algae use light and inorganic carbon, i.e., carbon dioxide to produce glucose and oxygen. Heterotrophic algae use organic carbon as a source of energy; therefore, they can utilize organic compounds present in wastewater and can be used for wastewater treatment. Some algal strains can utilize both organic as well as inorganic forms of carbon, which are called mixotrophic. Mohsenpour et al. (2021) have reported the use of microalgal species for different kinds of wastewaters including municipal, agricultural, brewery, refinery, and industrial effluents. Microbes including *Scendesmus obliquus* and *Chlorella vulgaris* have been demonstrated to successfully remove nutrients (carbon, N and P) from piggery wastewater, dairy wastewater, and centrate.

There are over 40,000 algal species already identified that are classified in multiple major groupings, as reported by Dahiya (2015), which include cyanobacteria, green algae, diatoms, yellow-green algae, red algae, dinoflagellates, etc. For economic feasibility of the algae to biofuel conversion, the algae should contain at least 35% lipid content (Dahiya, 2015). *Botryococcus braunii* has been reported as the species showing highest lipid content (63%–86%), while the alga *Chlamydomonas reinhardtii* has been extensively studied among eukaryotic cells, and *Chlorella* sp. have been well studied for biofuel production. Despite many advantages of biofuel production from wastewater treatment using microalgae, the economics still remain a challenge. These are aerobic systems that require aeration and pumping operations and cultivation of biomass, all of which are energy and cost intensive. Stephenson et al. (2010) and Jorquera et al. (2010) have analyzed that the most energy-intensive step is the cultivation step. Therefore, many factors need to be considered while opting for algae-based wastewater treatment to improve the energy and economic feasibility of the process.

8.3 ALGAE IN WASTEWATER TREATMENT

Municipal wastewater treatment plants utilize large amounts of energy and money, which are further enhanced if nutrient recovery from wastewater is aimed at. Therefore, there has been an increasing interest in shifting from the classical activated sludge treatment plants to energy positive treatment options (Jeong and Jang, 2020). Use of microalgae for carbon and nutrients removal from wastewater is a potential option for recovering energy. Microalgae can also utilize nitrogen and phosphorous apart from carbon and can be converted to biofuel to be used a source of renewable energy.

Wastewater treatment is mainly carried out to reduce the oxygen demand, i.e., BOD or COD due to organic compounds present in wastewaters and, moreover, nutrients such as nitrogen and phosphorous need to be removed to prevent eutrophication of the receiving water bodies. The presence of such substances in large concentrations can have harmful effects on the marine life by reducing the dissolved oxygen (O_2) concentration levels in the water. The eukaryotic algae and cyanobacteria, therefore, serve as sustainable and environment friendly alternatives to the currently practiced technologies, as they serve as renewable sources of biomass (rather than specially grown crops for biofuel production) and enable biological fixation of CO_2 in an economically feasible manner (Almomani et al., 2019). Apart from utilization of organic and inorganic carbon, microalgae with mixotrophic nutrition also use the nitrogen and phosphorus for their growth. The ones with photosynthetic nutrition release oxygen into the water systems that get used up by the heterotrophic microorganisms, aiding the active removal of carbonaceous matter from the system. As algae uses carbon dioxide and generates oxygen gas through photosynthesis, it can also help in the bacterial growth, which require O_2 for consumption of organic compounds in the wastewater (Arun et al., 2020). A single-step process is enough for degradation of organic carbon and nutrients from wastewater, when a combination of algae and bacteria are used for wastewater treatment (Mohsenpour et al., 2021). Due to the assimilation of nitrogen into algal biomass, the greenhouse gas emissions are also reduced as compared to oxidative treatment. Moreover, algal photosynthesis increases the oxygen concentration in wastewater, which reduces the demand for external air supply. The influential factors affecting the selection and operation of wastewater treatment systems using microalgae are shown in Figure 8.1.

Wastewater generated from domestic applications have high concentrations of biodegradable organic matter and nutrients like nitrogen and phosphorus that can deplete the dissolved oxygen (DO) available in the aquatic bodies where the water is discharged. The operations of conventional biological treatment methods are energy intensive. Microalgae-based wastewater treatment process has shown many advantages that can meet the new demand for improved wastewater treatment. According

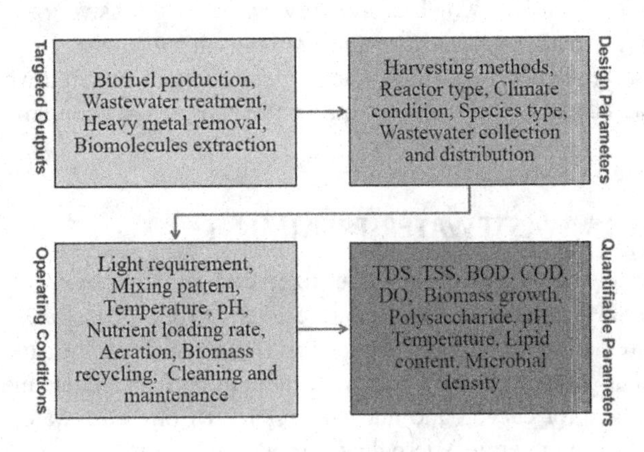

FIGURE 8.1 Factors affecting the selection and operation of microalgae-based wastewater treatment systems.

to Mohsenpour et al. (2021), two different wastewater streams can be identified in a municipal wastewater treatment plant as potential points where microalgae treatment process can be applied, i.e., primary settled wastewater or secondary treatment effluent. They suggest that treatment of primary effluent will prove to be more environment friendly, as they also contain more suitable nutrient concentration and suitable environment for microalgae growth.

However, considering the issues related to the complexity of wastewater characteristics and adaptability of microalgae species, and the challenges to the design and optimization of treatment processes in order to achieve higher removal efficiencies with lower costs, further exploration and research are still needed. Therefore, the following subsections explain in detail about the biotic and abiotic factors that influence the process of wastewater treatment by microalgae that will help in designing systems optimizing the process of biological matter removal and biofuel production.

8.3.1 Bacteria and Fungi

In a wastewater system, the populations of algae, bacteria, and fungi constitute a dynamic community supporting and hindering the growth of each member influencing the metabolic processes carried out by each organism. It, therefore, is impossible to operate a system with members of only one kingdom in an open environment. It has been studied that the microalgae can regulate the diversity in bacterial communities and can influence the composition of phyla depending on the source of inoculum introduced in the process and the source of wastewater. Generally, these can be the sludge from primary and secondary digestions or the digestate generated after the process. The steady-state composition of bacterial phyla will be different from the composition of the inoculum used in the beginning of the process.

The bacteria also support the growth of microalgae by providing CO_2 and inorganic matter in the form of ammonium and phosphates that are produced after mineralization of organic matter. In turn, the microalgae release photosynthetic oxygen that aids the bacteria to degrade the organic matter and also increase the dissolved oxygen concentration in the treated water. The oxygen is also utilized by the microalgae during their dark respiration.

Certain bacteria and fungi also enhance the growth and nutrient uptake by microalgae by excreting growth promoting factors, vitamins, etc. The phytohormones released by the bacteria and fungi also increase the lipid accumulation by microalgae, increasing the yield of biofuel produced. The microalgae, in turn, also releases compounds that enhance the mineralization of organic compounds by the heterotrophs. The community dynamics among bacteria, microalgae, and fungi also enable bioflocculation that reduces cell dilution and makes the process of cell harvesting easier, efficient, and less energy and cost intensive.

8.3.2 pH

Changes in pH and dissolved oxygen concentrations can affect the bacterial communities. An increase in pH has been studied to be detrimental to bacterial growth due to conformational changes in cell structure—respiratory chain damage increasing susceptibility to external factors such as chemicals and light. Rapid accumulation of inorganic nutrients by the microalgae and a slow rate of inorganic replenishment by the bacteria causes the pH of the system to suppress their growth. This can further reduce the growth of microalgae by decreasing the concentration of carbon dioxide required for photosynthesis. This asserts the need to set up an optimal environment for microalgal cultivation and organic/nutrient removal, and the best strategy will be to allow the communities to naturally acclimate the wastewater system and let it develop according to the processing conditions.

8.3.3 Temperature and Light

Temperature and light play an important role in checking that the population of algae is above that of bacteria and fungi. Light is fundamentally essential for normal functioning of microalgae, as it generates the energy required for oxygen evolution, ATP generation, and producing reducing compounds that fix CO_2 into organic matter. Maintaining availability of light below the saturation point is essential to balance the photosynthetic rate and energy utilized for illumination in cases when enough sunlight is not available. The illumination period and intensity of light can also influence the rate of removal of nitrogen and phosphorus, which decreases with the decrease in availability of these two factors.

The microalgal growth and treatment efficiency is also dependent on the environmental temperature. Temperature has been positively correlated with increase in cell number and associated negatively with biomass. The optimal temperature for microalgal growth is 15°C–25°C, and hence choosing the appropriate strain of microalgae that can thrive under the temperature conditions is important for optimum treatment of organic matter. Temperature also governs the photosynthetic assimilation process of carbon by the Calvin cycle, as the reactions are enzymatic, and the activities of enzymes are temperature dependent. The reaction rates are lower at lower temperatures. Operating at lower temperatures or the ambient conditions, however, reduces the power requirement for illumination for light availability and also increases solubility of oxygen. These conditions can enable the growth of microalgae over other competing indigenous organisms inhabiting the system.

8.4 PHOTOBIOREACTORS

In the last few decades, photobioreactors (PBRs) have emerged as a sustainable technology for the treatment of varieties of wastewater through bioremediation and biomass production, which can be further exploited for value-added applications. Typically, PBRs can be defined as a closed illuminated vessel containing phototrophic microorganisms, including micro and macroalgae, cyanobacteria, mosses, *etc*., designed to optimize biomass production through fine tuning of operating parameters (Znad, 2020). The PBRs can be operated in batch or continuous mode. For optimizing the algal production, different types of PBRs have been designed and could be classified on the basis of shape, mixing pattern, aeration facility, and the feeding habit of the microbes (Ashok et al., 2019). Based on shape, PBRs can be designated as tank/rectangular, flat panel, tubular, torus, and columnar.

TABLE 8.1
Comparison of Different Types of Photobioreactors Used for Microalgal Cultivation

PBR Configuration	Mixing Provision	S/V Ratio	Advantages	Disadvantages
Flat plate	Bubbling from bottom/ sides, airlift	High	High biomass yield Low oxygen build-up High photosynthesis efficiency	Dark zone formation Short light penetration
Stirred tank	Mechanical mixing	Low	Moderate biomass productivity Simple design High-value compounds	Low surface area-to-volume ratio Energy loss due to mechanical agitation
Tubular	Airlift/ bubbling		High mass transfer Good mixing Low power consumption	Fouling scale formation overheating
Column	Airlift/bubble	Low	Ease in operation Better mixing High mass transfer	Scaling reduces illumination region High mixing cost Sophisticated construction materials
Soft-frame			Less space requirement Foldable and replaceable	High construction cost dead zone formation
Hybrid			Low O&M cost reduced reactor size high carbohydrate content in microbial	Fouling negative energy balance

Source: Yadala and Cremaschi (2014), Gupta et al. (2015), Chang et al. (2017), Sheng et al. (2017), Hom-Diaz et al. (2017), and Vo et al. (2019).

Furthermore, bubble column, airlift, stirred tank, and rocking plate are the other types of PBRs categorized on the basis of mixing pattern. Besides, depending on the cultivation condition of microbes and oxygen availability in the reactor, the PBRs can be classified as phototrophic, heterotrophic, mixotrophic and aerobic, and anaerobic, respectively (Chen et al., 2012; Ashok et al., 2019). A comparison of various types of photobioreactors is presented in Table 8.1.

8.4.1 Classification of Photobioreactors

8.4.1.1 On the Basis of Shape

8.4.1.1.1 Tank/Rectangular

Closed tank reactors are often made up of polyethylene, where mixing can be carried out in a number of ways, including air bubbling, paddle mixing, and stirring.

For achieving the desired efficiency in the case of both the externally and internally illuminated systems, light distribution, path length, and intensity must be regulated to avoid the formation of dark regions (Kumar et al., 2013). High-volume tank PBRs suffer from nonuniform light distribution, and the use of several small units additionally poses the issue of cleaning and maintenance of the reactor. However, the challenges of light distribution and biomass accumulation in the tank-shaped PBRs could be addressed by adopting a circular design.

8.4.1.1.2 Flat Panel

The flat-panel reactors are cuboidal in shape and have a minimal light path (Singh and Sharma, 2012). The reactors can be made from glass, polycarbonates, *etc.* Due to its high surface area-to-volume ratio, a flat-panel reactor is widely employed to achieve higher light conversion efficiency. The major advantages of flat-panel reactors include high biomass yield, enhanced light penetration, easy regulation and monitoring of experimental parameters, and minimal risk of contamination. The agitation can be provided either by rotating the perforated tubes mechanically or bubbling air through them. Nevertheless, besides several advantages, higher land footprint, construction, operation, and maintenance costs of flat-panel systems limit its application at field scale (Huang et al., 2017).

8.4.1.1.3 Tubular

The tubular PBRs are currently one of the widely employed closed-typed reactors on an industrial scale. It allows better control on the conditioning of microbial culture and reduces the risk of contamination (Acién et al., 2017). In this system, the culture is subjected to flow through unidirectional tubings with high flow rates and is recirculated using mechanical pumps. The vertical tubing should have high transparency to ensure substantial light penetration. The illuminating sources can be either natural, or artificial, or a combination of both (Ashok et al., 2019). In terms of the economics involved, the tubular reactors are costlier to construct and require skilled maintenance, transportation, and handling (Ramírez-Mérida et al., 2015).

8.4.1.1.4 Columnar

As the name suggests, columnar PBRs are designed and constructed in vertical column configuration (Janoska et al., 2017), and the algal system is subjected to grow along the vertical axis. The microbial culture can be mixed in numerous ways, such as air/CO_2 bubbling, stirring, *etc.* Reportedly, columnar PBRs have a high rate of biomass growth, adequate gas–liquid mass transfer, and light/dark cycle control characteristics (Sierra et al., 2008; Jiménez-González et al., 2017). The major advantage of this system lies in its low land footprint due to its vertical configuration and simple design, which provides ease for upgradation and maintenance (Chang et al., 2016). Additionally, the circular shape of the reactor contributes to the uniform distribution of light and mixing attributes.

Further, a comparison of performance of various types of photobioreactors based on their configuration is summarized in Table 8.2.

TABLE 8.2
Performance Behavior of Various Configurations of Photobioreactors

PBR Type	Algal Species	Illumination Source	Light–Dark Cycle (h:h)	Biomass Concentration (kg/m^3)	Biomass Productivity Rate (g/m^2-d)	Reference
Vertical multicolumn airlift	*Chlorell pyrenoidosa*	Solar irradiation	NA	1.3–1.56	NA	Huang et al. (2016)
Horizontal tubular	*Arthrospira platensis*	Fluoroscent	NA	7.11	NA	Da Silva et al. (2016)
Bubble column	*Chaetoceros*	Artificial	NA	5.60	NA	Krichnavaruk et al. (2007)
Flat plate	*Dunaliella*	Solar irradiation	NA	7.3	NA	Barbosa et al. (2005)
Soft-frame biofilm based	*Pseudochoricystis ellipsoidea*	LED	12:12, 24:0	NA	8–10 g/m^2-d	Hamano et al. (2017)
Flat panel	*Chlamydomonas reinhardtii*	LED	24:0	4.5	29–54 g/m^2-d	de Mooij et al. (2016)
Closed column	Green algae	Metal halides	12:12	0.49–0.84	NA	Arias et al. (2017)
Soft-frame	*Ulva rigida, Cladophora sp.*, and *Ulva compressa*	Solar irradiation	NA	NA	NA	Chemodanov et al. (2017)
Hybrid-membrane based	*Chlorella vulgaris*			2.03		Bilad et al. (2014)
Flat Plate	*Botryococcus braunii*	Fluorescent	24:0	96.4	NA	Ozkan et al. (2012)

8.4.1.2 On the Basis of Mixing Pattern

8.4.1.2.1 Bubble Column

Bubble column PBRs constitute a cylindrical vessel where the height of the cylindrical column is greater than twice the diameter. The culture mixing and mass transfer are carried out by bubbling the gas mixture from the sparger. The illumination is provided externally to the reactor. Here, the photosynthetic efficiency primarily depends on the gas flow rate. The latter governs the light and dark cycle, as the liquid gets continuously circulated from the dark zone to the photic zone at a higher gas flow rate (Singh and Sharma, 2012). Low capital investment, easy handling, high surface area-to-volume ratio, and homogenous culture community are few major advantages of bubble column PBRs.

8.4.1.2.2 Airlift Photobioreactor

Airlift PBRs constitute two interconnecting zones where one of the tubes is subjected to sparging of a gas mixture, commonly referred to as riser. In contrast, the other tube that does not receive the gas mixture is called a downcomer (Pham et al., 2017). Airlift PBRs come in two different configurations: internal and external loop. In an internal loop reactor, the area is separated either by draft tubing or split cylinder, whereas in an external loop reactor, the riser and downcomer are physically separated by two different tubes. The mixing is achieved by sparging the bubble gas into the riser tube without any agitation, and here the working of rising could be viewed as similar to the bubble column where the sparged gas moves in the upward direction randomly. Due to the circular mixing patterns in the airlift PBRs, the algal cells are exposed to the flash light effect as the culture gets continuously circulated in the light and dark regions (Barbosa et al., 2003).

8.4.1.2.3 Stirred Tank Photobioreactor

Stirred tank reactors are the most conventional reactors in which the agitation is provided mechanically by impellers of different shapes and sizes (Mohan et al., 2019). The carbon source for algal growth is provided by bubbling the carbon dioxide-enriched air from the bottom, and baffles are used to minimize the formation of a vortex. Generally, fluorescent lamps and optical fibers are commonly used as an external illuminating source for such reactors. However, the application of stirred tank PBRs is limited due to the low surface area-to-volume ratio, which subsequently causes poor light-capturing efficiency (Sahle-Demessie et al., 2003).

8.4.2 Major Factors Affecting the Selection and Operation of Photobioreactors

Although there are numerous factors involved in the operation of PBRs, however, the set of desired output plays a pivotal role in the selection of PBRs. The few main parameters concerning the selection and operation of PBRs are discussed below.

8.4.2.1 Targeted Product Type and Quality

The product type and desired quality serve as the fundamentals for selecting PBRs for algae harvesting. A wide range of PBRs are commercially available, having their own strength and limitations.

8.4.2.2 Tolerance Limit of Microbial Culture

Another crucial parameter for the smooth functioning of PBR involves the selection of adaptable and tolerant algal strains under different reactor conditions. For instance, in open PBRs, the nutrient uptake and proliferation capacity of the selected strains must be sufficiently higher than that of other microbial communities in order to compete in case of nutrient deficient and cross-contamination reactor conditions (Chang et al., 2017). On the other hand, while selecting the algal strain for closed systems, one should consider the strong shear forces and oxygen toxicity generated due to sparging and building up of excessive molecular oxygen in the PBRs.

8.4.2.3 Resources Accessibility

Access to resources like land, nutrients, light, and carbon dioxide governs the selection of a PBR, its final size, and smooth operation. For algal growth, carbon, nitrogen, and phosphorus constitute the major nutrition, while other micronutrients required includes Ca, Mg, Cu, Mn, Zn, and S. The secondary wastewater treatment remains ineffective for the removal of nitrates and phosphates, and hence the wastewater can be used as a feed for algal-PBRs (Abdel-Raouf et al., 2012). In terms of land footprint, open systems require a large flat piece of land, and geological location is more crucial for open systems when compared to closed systems (Khan et al., 2018). The location of open systems should be chosen so that they should receive a sufficient amount of sunlight required to maintain a shallow culture medium for maximum photosynthetic efficiency. The other important parameters for algal cultivation are temperature, pH, light intensity, carbon source, and aeration. Usually, microalgae and mixed microbial cultures favor a wide range of pH values between 5 and 12 (Dubinsky and Rotem, 1974). The deviation of pH toward the lower value can cause acidification in the reactor. In comparison, a higher pH value (>9) may lead to the accumulation of dissolved oxygen in the reactor and can cause photooxidative damage to the microbial cells (Ashok et al., 2019). The availability of skilled professionals should also be taken into account for monitoring and maintenance of PBRs.

8.4.2.4 Economics

The targeted algal production rate and maximization of biomass yield determine the size and type of a PBR. The maximization of biomass yield with desired biomolecules can be obtained by employing closed-type PBRs. However, their scale-up proves to be expensive (Carvalho et al., 2006). Additionally, scaling up of closed types often involves high shear forces and build-up of excess oxygen, which might cause damage to microbial cells (Brennan et al., 2010). Open systems are generally preferred for achieving a higher productivity rate; nevertheless, such systems involve the risk of cross-contamination, evaporation loss of microbial culture, large land footprint, low mass-transfer efficiency, *etc.* Thus, it is clear that there are multiple factors involved that govern the economics of a PBR, and hence one has to establish a trade-off between the desired target and the cost of a PBR.

8.5 HIGH-RATE ALGAL PONDS

High-rate algae ponds (HRAP) can show good performance in places with sufficient solar radiation (Mohsenpour et al., 2021). Kohlheb et al. (2020) have identified that HRAPs provide environmental as well as economic benefits with increased carbon sequestration and reduced eutrophication potentials when compared to bacteria-based wastewater treatment processes. It has been reported that HRAPs treating wastewater can generate 800–1,400 GJ/ha/year in the form of harvested algal biomass for low-cost biofuel production (Mohsenpour et al., 2021). The mixing in PBRs usually through pumping as compared to paddlewheels used with HRAPs required 10 times more energy. The commercial production of algal biofuel in an HRAP is described in the following sections.

Algal production is frequently constrained by operating costs (for fertilizer and chemicals, and energy budgets of pumping water), increased energy needs (for biomass collection and dewatering operations), and technological compatibility with biomass-to-biofuel conversions (Craggs et al., 2011). HRAPs are a possible solution for some of the inadequacies. HRAP microalgae can be utilized as a feedstock, since they are cultivated as a by-product of wastewater treatment.

These ponds are shallow, open, and entirely aerated (by paddle wheel). These ponds' efficiency is dependent on symbiotic interactions between bacteria and microalgae. It's a low-cost wastewater treatment system that produces algal biomass as a by-product, which can then be utilized as a biofuel feedstock. Other mechanical and/or traditional wastewater treatment facilities are less effective in comparison to the HRAP facilities. Anaerobic digestions, thermochemical reactions, transesterification, and carbohydrate fermentation are among the conversion processes in HRAPs. Due to their ability to collect solar energy through algae photosynthesis, HRAPs can replace/reduce electricity usage by more than 50% (Arashiro et al., 2018). According to studies, utilizing HRAPs can minimize the amount of time required for nutrient removal per unit mass (Arashiro et al., 2018). A study reported that the HRAP can remove the same quantity of nutrients in four to eight days, while the water stabilization ponds (WSPs) take 30–60 days (Sutherland et al., 2020). HRAPs provide several other advantages. Because these ponds are mixed, nonmotile colonial algae-like as *Micratinium* and *Pediastrum* can be developed, unlike in traditional WSPs. Because these colonial algae are bigger, they may be collected passively with little effort and cost (Mehrabadi et al., 2015).

Freshwater algae and wastewater treatment facilities are two forms of biomass that may be harvested with HRAPs (Montemezzani et al., 2015). Freshwater algae require a constant supply of water as well as the availability of nutrients (Montemezzani et al., 2015). As a result, the freshwater algae's operational expenses in HRAPs are greater. The same in wastewater treatment, on the other hand, is more cost-effective. Water, fertilizers, and CO_2 are all given at no expense at the treatment facility. According to one study, 1 tonne of algal biomass with an average nitrogen content of 7% can be produced while treating around 2,500 m^3 of wastewater (Yen and Brune 2015). Due to the higher diameters of the algal flocs, which may be passively produced with regular gravity settings, the cost of wastewater treatment in HRAPs was further decreased. The wastewater treatment facilities are being combined with algal cultivation for biofuel production because of these beneficial components.

Carbohydrates, lipids, and proteins make up the majority of algal cells. The energy content of these algae cells is usually between 18 and 24 KJ/g (Craggs et al., 2011). The biological compositions, in combination with other factors, may be able to distinguish the energy recoveries from the algal flocs.

8.5.1 Parameters Affecting the Energy Production in HRAP

Researchers have long advocated for increased productivity in HRAPs by supporting algal cultivations that are high in energy and production, as well as efficient for wastewater treatment. As a result, investigations have shown some elements that might influence the algae species' dynamics and development. Except for bioethanol production, high lipid content, particularly high content of certain lipids (i.e., saturated and mono-saturated fatty acids with medium chain length, C^{16} to C^{18}, which do not require costly upgrading methods) might be helpful.

Abiotic variables, biotic factors, and design-related factors are the three primary types of factors in bioenergy production in WWT HRAPs. The following is a detailed discussion of these topics:

8.5.1.1 Abiotic Factors

The generation of algal biomass is influenced by both the light-dependent and light-independent or temperature-dependent phases of photosynthesis. As a result, both light and temperature affect the manufacturing efficiency. As a result, in a temperate environment, algal biomasses change with solar radiation. In summer, when there was three times more solar radiation than in winter, Sutherland et al. (2015) discovered a 250% increase in biomass content (measured as volatile suspended solids, VSS). The wavelength of solar radiation has an impact on productivity as well. Algal synthesis is catalyzed by wavelength ranges that are photosynthetically active radiation. As a result, the availability and exposure of sunlight to the pond water might boost biomass yield even further. The structure and other mechanical features of the pond can eventually be built so that productivity can be improved, according to studies. According to research, pond depth and hydraulic retention time are important factors in improving photoperiod and sunshine availability, and consequently biomass output.

Furthermore, the biomass output can be influenced by the combination of light availability and the chemical makeup of the microalgae. In general, strong light catalyzes lipid synthesis, resulting in the production of more mono- and di-saturated lipids. Excess electrons, on the other hand, can build up in the photosynthetic electron transport system, resulting in an excess of reactive oxygen, which hinders photosynthesis and destroys cell membrane lipids, proteins, and other macromolecules (Hu et al., 2008). Biomass productivity and nutrient removal efficiency both suffer, as a result. Because TAG synthesis costs double the quantity of the reductant NADPH (from the electron transport pathway), which is utilized for CO_2 absorption, algal cells produce more TAG instead of carbohydrate or protein under these conditions as a safety mechanism (Hu et al., 2008).

Temperature is an important factor in algal development. Many microalgal species can photosynthesize and thrive at temperatures ranging from 5°C to 40°C (Ras et al., 2013). Their ideal temperature, on the other hand, is usually between 28°C and 35°C (Park et al., 2011). As the temperature rises, it increases cell chlorophyll concentration, intracellular enzymatic activity, and biomass output. Until unfavorable temperatures are reached, biomass productivity will normally double for every 10°C increase in temperature (Ras et al., 2013).

The temperature has a significant impact on the biochemical makeup of microalgae, particularly the lipid profile. With the rising temperature, the ratio of unsaturated to saturated fatty acids falls in many species (Hu et al., 2008).

8.5.1.2 Biotic Factors

The biomass output and concentration in the wastewater treatment HRAP can be influenced greatly by the pond environment. While grazing on the zooplanktons, the rotifers can dramatically lower output by less than 10%. A ciliate eliminated an algae species after 5 days of an operational HRAP facility, according to research. As a result, regulating the environment in the HRAP is advantageous to biomass output. However, there is research underway to employ gazers to selectively promote the growth of the beneficial algal mass while controlling the growth of the poorly settleable, unicellular algae.

8.5.1.3 Design-Related Factors

8.5.1.3.1 CO_2 Availability

Another key parameter for increasing the C:N ratio of wastewater is CO_2 input. The ratio is required to stay within the 6:1 range, which is typical of biomass production. CO_2 injection during the summers has been shown to boost WWT HRAP biomass productivity by up to 30%, according to studies. Other research has found that rising CO_2 levels increases plant nitrogen availability by increasing the acidity of the pond water, moving the balance away from NH_3 and toward NO_4^+. As a result, CO_2 enhances nutrient recovery, algal productivity, and energy output.

Photosynthesis allows microalgae, like other plants, to collect CO_2 and convert it to use organic compounds such as lipids—a key component in biofuel generation.

CO_2 injection has also been shown to increase the fatty acid content of algal cells in other experiments. Muradyan et al. (2004) found that increasing the CO_2 concentration from 2% to 12% enhanced the fatty acid content of algal cells by more than 30%. The higher CO_2 concentration boosted de novo fatty acid synthesis in this range.

Furthermore, CO_2 injection has been proven to enhance lipid composition and boost algal energy content. Growth on a CO_2-enriched (2%–10% CO_2) culture medium, for example, altered the lipid profile of algal biomass and increased the degree of fatty acid saturation in various green algae (Nakanishi et al., 2014).

8.5.1.3.2 Hydraulic Retention Time (HRT)

Another key parameter that influences the culture concentration, dominant species, and algal/bacterial ratio is the HRT. The algal/bacterial ratio lowers, even more, when the pond's HRT rises and the nitrogen removal rate improves. According to Park et al. (2011), the percentage of algae in HRAP biomass rose from 56% to 80.5% when cultivated at a 4-day HRT. Longer retention durations reduce the number of algae in the biomass, which has an impact on the composition of the biomass. Furthermore, because bacteria have a greater ash content than microalgae, this biomass ash content will rise. Hydraulic retention time (HRT) has an impact on population dynamics owing to changes in algae species' particular growth rates, which can affect the biochemical composition and energy content of the biomass.

Aside from the above-listed parameters, there are a few more that can influence algal biomass production in the culture medium. These include mixing, culture mode, algae recycling, nutrients, and so forth. Mixing can lower the pond's boundary layer as well as its thermal stratification, which can affect nutrient assimilation and gas exchange efficiency.

In the wastewater treatment HRAP, the mode of cultivation can affect the dynamics and biomass productivity. Batch cultivation has been shown to boost the growth rates of smaller microalgae, whereas continuous cultivation promotes the growth of algae with greater growth rates, according to studies. The overall composition and energy content of wastewater treatment HRAP biomass would vary with the growing mode, since each algal species has a different biochemical makeup.

Microalgae cultivation in near-nutrient-limiting circumstances can boost lipid content, quality, and total energy content, but at the cost of lower production. Extending the HRAP HRT is one approach for achieving nutritional limitations. However, additional study is needed to determine the best HRTs for wastewater treatment HRAP in various seasons to create more useful biomass for biofuel generation while minimizing negative impacts on wastewater treatment rate, biomass output, and the algal/bacterial ratio.

Aeration is required to supply CO_2 in the cultivation of microalgae, which acts as a nutrient for growth and also aids in the uptake of inorganic N and P (Liu et al., 2020). However, the energy required to compress the air is an energy-intensive process and is one of the highest energy-intensive steps (Davis et al., 2016). Another option to supply carbon can be by adding organic carbon sources (which might be available in wastewater already) or inorganic salts containing carbonate or bicarbonate, which can be easily utilized by microbes as well as reduce the cost for compressing air and aeration (Evans et al., 2017; Perez-Garcia et al., 2011). Further, utilization of wastewater as the organic carbon source can itself reduce energy requirement and also give environmental benefits due to reduced resource use.

As a result, the amount of low-cost energy generated is determined by biomass productivity, chemical composition, and harvestability. The low-cost energy generation in the wastewater treatment HRAP may be enhanced by adjusting the HRT, adding CO_2, recycling biomass, and managing the HRAP ecosystem, according to the research. As a result, certain more innovations need to be handled.

- **Harvesting**: One of the fundamental problems that must be solved to improve combined wastewater treatment and resource recovery is cost-effective harvesting at full scale.
- **Effective CO_2 Addition**: CO_2 addition provides a lot of advantages for boosting HRAP performance, but it must be demonstrated to be efficient and cost-effective on a large scale.

- **Enhanced Phosphorus Removal**: Phosphorus removal in HRAPs must improve to meet stricter regulatory requirements on phosphorus discharge load limitations, if HRAPs are to remain a viable wastewater treatment technology.
- **Reducing Environmental Impacts**: While wastewater treatment HRAPs provide many environmental advantages over traditional facultative wastewater ponds, they also have certain disadvantages. Operational factors could be able to assist reduce them.

8.6 TECHNOLOGIES FOR BIOFUEL PRODUCTION

It has been estimated that 1.83 kg of CO_2 is required to produce 1 kg algal biomass, which can be sequestered in algal biomass (Khan et al., 2018). Microalgae have several advantages when being utilized as a source for biofuel. These include using water as electron donor for photosynthesis and high biomass generation as compared to crops used for biofuel generation, therefore no requirement of fertile land (Kumar et al., 2020). Algal biomass can be used to generate different products such as hydrogen, fatty acids by fermentation, gasoline, and bioethanol. Biofuels thus produced can be used for electricity generation. The extraction process should be specific for obtaining particular products. Biogas can be produced by anaerobic digestion, bioethanol through fermentation, and biodiesel by oil extraction. The cell wall in microalgae contains fatty acids and lipids. To obtain the oil, different methods can be used that can be classified as physical, chemical, mechanical, and enzymatic methods (Kumar et al., 2020). Liquid–liquid extraction is widely employed to extract oil, but it cannot be considered ecofriendly and also changes the product composition. Other well-known mechanical methods for oil extraction include bead milling and mechanical pressing. A chemical method, i.e., Soxhlet extraction is also popular. However, it also leads to use of toxic solvents and employs longer times (Kumar et al., 2020). Transesterification process converts lipids to biodiesel. It is carried out in the presence of alcohol, usually methanol or ethanol. Three moles of alcohol react with each mol of fatty acid to produce fatty acid methyl ester (FAME). The products are then distilled to retrieve biodiesel and separate the byproducts. Due to high water content in the feedstock, the transesterification process becomes less economically feasible.

Another process is saccharification for conversion of carbohydrates contained in algal biomass to bioethanol. Lignin needs to be removed from the biomass before saccharification. The enzymes then react with polysaccharides and convert them to monosaccharides (Kumar et al., 2020). It is reported that high starch containing microalgae can produce up to 140,000 L/ha/yr bioethanol. Two types of fermentation are used for conversion of sugars to bioethanol, separate hydrolysis and fermentation (SHF), and simultaneous saccharification and fermentation (SSF). In SSF, one pot synthesis is followed, whereas in SHF, first hydrolysis is carried out in one reactor and then the products are transferred to another reactor for fermentation. In SSF, the sugars produced during hydrolysis are simultaneously converted to ethanol and, therefore, their degradation is controlled.

It has been reported that to obtain high yield of glucose, lignin and hemicellulose have to be removed, as these compounds inhibit enzymes to access cellulose. In order to obtain bioethanol, lignocellulosic biomass has to be pretreated followed by enzymatic hydrolysis and fermentation (Mohapatra et al., 2017). Pretreatment can be used to weaken the cell walls and facilitate hydrolysis and fermentation. Pretreatment with alkali chemicals, e.g., sodium hydroxide, causes a swelling of the cell walls that increase the internal surface area and leads to disruption of lignin structure. Dilute acids like sulfuric or hydrochloric acid are also used to hydrolyze hemicellulose to its monomeric sugar units and make the cellulose more accessible to enzymatic action. Other than these, physicochemical processes such as use of steam and pressurized hot water, microwave, alkali–acid, and ozonolysis are also used to facilitate hemicellulose and lignin degradation (Raftery and Karim, 2017; Mohapatra et al., 2017). Dilute acid hydrolysis followed by steam treatment has been reported to be the most effective among different combinations (Raftery and Karim, 2017). A consortium of different organisms has to be used, as no single organism can utilize cellulose and hemicellulose, as well as facilitate fermentation. *Thermoanaerobacterium saccharolyticum* and *Clostridium thermocellum*, which are capable of hydrolyzing and fermenting hemicellulose and sugars, respectively, have been suggested as a consortium for ethanol production (Raftery and Karim, 2017). Cellulase enzyme, which helps in converting polysaccharides to monosaccharides, extracted from *Trichoderma reesei,* has been used to achieve glucose concentration of 0.13 g/L per g of pretreated (with acid followed by steam treatment at 120°C for 15 minutes) solid waste (Mtui and Nakamura, 2005). After cooling, it was subjected to saccharafication by adding *T. reesei* and incubated at 55°C for 8 hours. *Trichoderma viride* is another microbe that produces cellulase. *Saccharomyces cereviciae* and *Zymomonas mobilis* are used for ethanol production from reducing sugars. *S. cerevisiae* were then added, and the mixture was kept at 35°C for 72 hours for fermentation. Pretreatment helped in formation of sugars and further fermentation to alcohol by breaking the chemical bonds and hence increasing enzymatic hydrolysis and microbial reaction. Taherzadeh and Karimi (2007) suggested a drawback of the dilute acid hydrolysis process occurring in one stage. The sugars formed during the process also degrade and form undesirable byproducts such as furfural, 5-HMF, uronic acid, 4-hydroxybenzoic acid, formic acid, and levulinic acid, which can inhibit the formation of ethanol. Therefore, two-stage hydrolysis was suggested, where the first stage can be carried out at milder conditions (~170°C), converting hemicellulose to sugar monomers; and, in the second stage, cellulose can be hydrolyzed under more severe conditions (~220°C). Liquid is removed after the first stage, which will reduce the energy consumption and cause lesser sugar degradation. This two-step process leads to higher sugar yield with 80%–95% of the initial sugars being obtained in the first stage and 40%–60% in the second stage.

Biohydrogen is produced through dark fermentation. Pretreatment is important for dark fermentation to convert polysaccharides to their monomers. Different pretreatments reported in literature are enzymatic pretreatment, heating at 100°C–120°C, acidic and basic pretreatment, milling, and ultrasonication (Kumar et al., 2020).

Another treatment process for biofuel production from algae is hydrothermal liquefaction. This process is carried out at near-critical temperatures of water (250°C–400°C) in presence of water. The closed conditions and supercritical temperatures increase the pressure of the system, which leads to degradation of biomass.

The only solvent used in this process is water, which makes this process environment friendly, and it also reduces oxygen, sulfur, and hydrogen content in the residual biomass (Kumar et al., 2020). This is considered the most promising technology among others for biofuel production from algae. The properties of water at these conditions are different, and it acts as organic solvent due to lower dielectric constant as well as, under high pressure, the solubilized compounds can undergo polymerization to form biodiesel (Guo et al., 2015). It can lead to recovery of proteins and carbohydrates along with lipids. Also, in comparison to lipid extraction, it reduces resource utilization by decreasing the demand for energy, freshwater, and nutrients, such as N and P by 50%, 33% and 44%, respectively. Due to removal of oxygen during the HTL process, the energy density of bio-oil from HTL is higher. The main compounds present in bio-oil produced after HTL of algae include monoaromatics, aliphatic compounds, fatty acids, N-containing compounds, polyaromatics, and O-containing compounds (Guo et al., 2015). The carbohydrates first undergo dehydration to form furan compounds, which then undergo aromatization to form phenolic compounds.

The maximum yield of bio-oil from HTL has been reported to occur in the range of 250°C–375°C (Guo et al., 2015). Initially, the reaction takes the hydrolysis route, but with increase in temperature and also reaction duration, polymerization takes place. At lower temperatures, due to hydrolysis, more solubilization occurs, which can lead to reduced yield of bio-oil. After the temperature of 250°C, yield increases and oil content reduces from the water phase. Moreover, the structure of algae can be destroyed above this temperature, which can increase the oil yield. However, just below 250°C (i.e., between 225°C and 250°C), denaturing of proteins can lead to their aggregation forming a colloidal mass of algae (Garcia Alba et al., 2011).

Some challenges with HTL-produced bio-oil include high viscosity due to the presence of long-chain fatty acids, presence of heteroatoms, which can form NO_x and SO_x upon combustion. Therefore, upgrading of the biofuel is important for its use in transport applications. Vacuum distillation as a process for upgrading bio-oil used by Eboibi et al. (2014) led to an increase in yield, deoxygenation, and reduction in ash content.

8.7 SUMMARY

With the increasing consumption of fossil fuels and the energy demands, substantial research has been carried out to promote the carbon-neutral sources of energy. At present, the majorly used biofuels are bioethanol and biodiesel that are produced from crops like sugar cane, corn, soybean, etc., which raise concerns regarding the economic viability and their impact on food availability. Microalgae due to their unit biomass productivity per land area used has been promoted through several technological innovations and the development methodologies. The present chapter has summarized the biomass, algae that are primarily used for the production of biomasses, the biotic and abiotic factors related to the development of algal biomass, and the current state of the art methods to produce this biofuel—photobioreactor, HRAPS. These are simultaneously being used for the treatment and the uses of the biomass residues as biofuel feedstocks. The PBRs can be operated in batch or continuous mode. For optimizing the algal production, different types of PBRs have been designed and could be classified on the basis of shape, mixing pattern, aeration facility, and the feeding habit of the microbes. Furthermore, bubble column, airlift, stirred tank, and rocking plate are the other types of PBRs categorized on the basis of mixing pattern. HRAPs also are a possible solution for some of the inadequacies. HRAP microalgae can be utilized as a feedstock, since they are cultivated as a by-product of wastewater treatment. These can provide both economic as well as ecofriendly methods to convert the waste biomass to biofuel.

REFERENCES

Abdel-Raouf, N., Al-Homaidan, A. A., & Ibraheem, I. (2012). Microalgae and wastewater treatment. *Saudi Journal of Biological Sciences, 19*(3), 257–275.

Acién, F. G., Molina, E., Reis, A., Torzillo, G., Zittelli, G. C., Sepúlveda, C., & Masojídek, J. (2017). Photobioreactors for the production of microalgae. *Microalgae-based Biofuels and Bioproducts*, 1–44.

Alketife, A. M., Judd, S., & Znad, H. (2017). Synergistic effects and optimization of nitrogen and phosphorus concentrations on the growth and nutrient uptake of a freshwater *Chlorella vulgaris. Environmental Technology, 38*(1), 94–102.

Almomani, F. A. (2019). Assessment and modeling of microalgae growth considering the effects of CO_2, nutrients, dissolved organic carbon and solar irradiation. *Journal of Environmental Management, 247*, 738–748.

Arashiro, L. T., Montero, N., Ferrer, I., Acién, F. G., Gómez, C., & Garfí, M. (2018). Life cycle assessment of high rate algal ponds for wastewater treatment and resource recovery. *Science of the Total Environment, 622*, 1118–1130.

Arias, D. M., Uggetti, E., García-Galán, M. J., & García, J. (2017). Cultivation and selection of cyanobacteria in a closed photobioreactor used for secondary effluent and digestate treatment. *Science of the Total Environment, 587*, 157–167.

Arun, S., Sinharoy, A., Pakshirajan, K., & Lens, P. N. (2020). Algae based microbial fuel cells for wastewater treatment and recovery of value-added products. *Renewable and Sustainable Energy Reviews, 132*, 110041.

Ashok, V., Gupta, S. K., & Shriwastav, A. (2019). Photobioreactors for wastewater treatment. In *Application of Microalgae in Wastewater Treatment*, Gupta, S. K., & Bux, F. (Eds.), 383–409. Springer, Cham.

Barbosa, M. J., Janssen, M., Ham, N., Tramper, J., & Wijffels, R. H. (2003). Microalgae cultivation in air-lift reactors: Modeling biomass yield and growth rate as a function of mixing frequency. *Biotechnology and Bioengineering, 82*(2), 170–179.

Barbosa, M. J., Zijffers, J. W., Nisworo, A., Vaes, W., Van Schoonhoven, J., & Wijffels, R. H. (2005). Optimization of biomass, vitamins, and carotenoid yield on light energy in a flat-panel reactor using the A-stat technique. *Biotechnology and Bioengineering, 89*(2), 233–242.

Bilad, M. R., Discart, V., Vandamme, D., Foubert, I., Muylaert, K., & Vankelecom, I. F. (2014). Coupled cultivation and pre-harvesting of microalgae in a membrane photobioreactor (MPBR). *Bioresource Technology, 155*, 410–417.

Brennan, L., & Owende, P. (2010). Biofuels from microalgae—A review of technologies for production, processing, and extractions of biofuels and co-products. *Renewable and Sustainable Energy Reviews, 14*(2), 557–577.

Carvalho, A. P., Meireles, L. A., & Malcata, F. X. (2006). Microalgal reactors: A review of enclosed system designs and performances. *Biotechnology Progress, 22*(6), 1490–1506.

Chang, H. X., Fu, Q., Huang, Y., Xia, A., Liao, Q., Zhu, X., Zheng, Y. P. & Sun, C. H. (2016). An annular photobioreactor with ion-exchange-membrane for non-touch microalgae cultivation with wastewater. *Bioresource Technology, 219*, 668–676.

Chang, J. S., Show, P. L., Ling, T. C., Chen, C. Y., Ho, S. H., Tan, C. H., Nagarajan, D. and Phong, W. N. (2017). Photobioreactors. In *Current Developments in Biotechnology and Bioengineering*, Pandey, A., Lee, D., Chang, J., Sirohi, R., & Sim, S. J. (Eds.), 313–352. Elsevier, United States.

Chemodanov, A., Robin, A., & Golberg, A. (2017). Design of marine macroalgae photobioreactor integrated into building to support seagriculture for biorefinery and bioeconomy. *Bioresource Technology, 241*, 1084–1093.

Chen, H. W., Yang, T. S., Chen, M. J., Chang, Y. C., Lin, C. Y., Eugene, I., Wang, C., Ho, C. L., Huang, K. M., Yu, C. C., Yang, F. L. Wu, S. H., Lu, Y. C., & Chao, L. K. P. (2012). Application of power plant flue gas in a photobioreactor to grow Spirulina algae, and a bioactivity analysis of the algal water-soluble polysaccharides. *Bioresource Technology, 120*, 256–263.

Craggs, R. J., Heubeck, S., Lundquist, T. J., & Benemann, J. R. (2011). Algal biofuels from wastewater treatment high rate algal ponds. Algal biofuels from wastewater treatment high rate algal ponds. *Water Science and Technology, 63*(4), 660–665.

da Silva, M. F., Casazza, A. A., Ferrari, P. F., Perego, P., Bezerra, R. P., Converti, A., & Porto, A. L. F. (2016). A new bioenergetic and thermodynamic approach to batch photoautotrophic growth of Arthrospira (Spirulina) platensis in different photobioreactors and under different light conditions. *Bioresource Technology, 207*, 220–228.

Dahiya, A. (2015). Algae biomass cultivation for advanced biofuel production. In *Bioenergy*, 219–238. Academic Press, Jeffords Hall, Burlington.

Davis, R., Markham, J., Kinchin, C., Grundl, N., Tan, E. C., & Humbird, D. (2016). *Process Design and Economics for the Production of Algal Biomass: Algal Biomass Production in Open Pond Systems and Processing through Dewatering for Downstream Conversion* (No. NREL/TP-5100-64772). National Renewable Energy Lab (NREL), Golden, CO.

de Mooij, T., de Vries, G., Latsos, C., Wijffels, R. H., & Janssen, M. (2016). Impact of light color on photobioreactor productivity. *Algal Research, 15*, 32–42.

Dubinsky, Z., & Rotem, J. (1974). Relations between algal populations and the pH of their media. *Oecologia, 16*(1), 53–60.

Eboibi, B. E., Lewis, D. M., Ashman, P. J., & Chinnasamy, S. (2014). Effect of operating conditions on yield and quality of biocrude during hydrothermal liquefaction of halophytic microalga Tetraselmis sp. *Bioresource Technology, 170*, 20–29.

Evans, L., Hennige, S. J., Willoughby, N., Adeloye, A. J., Skroblin, M., & Gutierrez, T. (2017). Effect of organic carbon enrichment on the treatment efficiency of primary settled wastewater by Chlorella vulgaris. *Algal Research, 24*, 368–377.

Falkowski, P. G., & Raven, J. A. (2013). *Aquatic Photosynthesis*. Princeton University Press, Princeton, New Jersey.

Garcia Alba, L., Torri, C., Samorì, C., van der Spek, J., Fabbri, D., Kersten, S. R., & Brilman, D. W. (2012). Hydrothermal treatment (HTT) of microalgae: Evaluation of the process as conversion method in an algae biorefinery concept. *Energy & Fuels, 26*(1), 642–657.

Guo, Y., Yeh, T., Song, W., Xu, D., & Wang, S. (2015). A review of bio-oil production from hydrothermal liquefaction of algae. *Renewable and Sustainable Energy Reviews, 48*, 776–790.

Gupta, P. L., Lee, S. M., & Choi, H. J. (2015). A mini review: Photobioreactors for large scale algal cultivation. *World Journal of Microbiology and Biotechnology, 31*(9), 1409–1417.

Hamano, H., Nakamura, S., Hayakawa, J., Miyashita, H., & Harayama, S. (2017). Biofilm-based photobioreactor absorbing water and nutrients by capillary action. *Bioresource Technology, 223*, 307–311.

Hom-Diaz, A., Jaén-Gil, A., Bello-Laserna, I., Rodríguez-Mozaz, S., Vicent, T., Barceló, D., & Blánquez, P. (2017). Performance of a microalgal photobioreactor treating toilet wastewater: Pharmaceutically active compound removal and biomass harvesting. *Science of the Total Environment, 592*, 1–11.

Hu, Q., Sommerfeld, M., Jarvis, E., Ghirardi, M., Posewitz, M., Seibert, M., & Darzins, A. (2008). Microalgal triacylglycerols as feedstocks for biofuel production: Perspectives and advances. *The Plant Journal, 54*(4), 621–639.

Huang, J., Ying, J., Fan, F., Yang, Q., Wang, J., & Li, Y. (2016). Development of a novel multi-column airlift photobioreactor with easy scalability by means of computational fluid dynamics simulations and experiments. *Bioresource Technology, 222*, 399–407.

Huang, Q., Jiang, F., Wang, L., & Yang, C. (2017). Design of photobioreactors for mass cultivation of photosynthetic organisms. *Engineering, 3*(3), 318–329.

Janoska, A., Lamers, P. P., Hamhuis, A., van Eimeren, Y., Wijffels, R. H., & Janssen, M. (2017). A liquid foam-bed photobioreactor for microalgae production. *Chemical Engineering Journal, 313*, 1206–1214.

Jeong, D., & Jang, A. (2020). Exploration of microalgal species for simultaneous wastewater treatment and biofuel production. *Environmental Research, 188*, 109772.

Jiménez-González, A., Adam-Medina, M., Franco-Nava, M. A., & Guerrero-Ramírez, G. V. (2017). Grey-box model identification of temperature dynamics in a photobioreactor. *Chemical Engineering Research and Design, 121*, 125–133.

Jorquera, O., Kiperstok, A., Sales, E. A., Embiruçu, M., & Ghirardi, M. L. (2010). Comparative energy life-cycle analyses of microalgal biomass production in open ponds and photobioreactors. *Bioresource Technology, 101*(4), 1406–1413.

Khan, M. I., Shin, J. H., & Kim, J. D. (2018). The promising future of microalgae: Current status, challenges, and optimization of a sustainable and renewable industry for biofuels, feed, and other products. *Microbial Cell Factories, 17*(1), 1–21.

Kohlheb, N., van Afferden, M., Lara, E., Arbib, Z., Conthe, M., Poitzsch, C., Marquardt, T., & Becker, M. Y. (2020). Assessing the life-cycle sustainability of algae and bacteria-based wastewater treatment systems: High-rate algae pond and sequencing batch reactor. *Journal of Environmental Management, 264*, 110459.

Krichnavaruk, S., Powtongsook, S., & Pavasant, P. (2007). Enhanced productivity of *Chaetoceros calcitrans* in airlift photobioreactors. *Bioresource Technology, 98*(11), 2123–2130.

Kumar, K., Sirasale, A., & Das, D. (2013). Use of image analysis tool for the development of light distribution pattern inside the photobioreactor for the algal cultivation. *Bioresource Technology, 143*, 88–95.

Kumar, M., Sun, Y., Rathour, R., Pandey, A., Thakur, I. S., & Tsang, D. C. (2020). Algae as potential feedstock for the production of biofuels and value-added products: Opportunities and challenges. *Science of the Total Environment, 716*, 137116.

Liu, J., Pemberton, B., Lewis, J., Scales, P. J., & Martin, G. J. (2020). Wastewater treatment using filamentous algae–a review. *Bioresource Technology, 298*, 122556.

Mehrabadi, A., Craggs, R., & Farid, M. M. (2015). Wastewater treatment high rate algal ponds (WWT HRAP) for low-cost biofuel production. *Bioresource Technology, 184*, 202–214.

Mohan, S. V., Rohit, M. V., Subhash, G. V., Chandra, R., Devi, M. P., Butti, S. K., & Rajesh, K. (2019). Algal oils as biodiesel. In *Biofuels from Algae*, Pandey, A., Chang, J., Soccol, C. R., Lee, D., & Chisti, Y. (Eds.), 287–323. Elsevier, India.

Mohapatra, S., Dandapat, S. J. and Thatoi, H. (2017). Physicochemical characterization, modelling and optimization of ultrasono-assisted acid pretreatment of two *Pennisetum* sp. using Taguchi and artificial neural networking for enhanced delignification. *Journal of Environmental Management, 187*, 537–549.

Mohsenpour, S. F., Hennige, S., Willoughby, N., Adeloye, A., & Gutierrez, T. (2021). Integrating micro-algae into wastewater treatment: A review. *Science of the Total Environment, 752*, 142168.

Montemezzani, V., Duggan, I. C., Hogg, I. D., & Craggs, R. J. (2015). A review of potential methods for zooplankton control in wastewater treatment high rate algal ponds and algal production raceways. *Algal Research, 11*, 211–226.

Mtui, G., & Nakamura, Y. (2005). Bioconversion of lignocellulosic waste from selected dumping sites in Dar es Salaam, Tanzania. *Biodegradation*, *16*(6), 493–499.

Muradyan, E. A., Klyachko-Gurvich, G. L., Tsoglin, L. N., Sergeyenko, T. V., & Pronina, N. A. (2004). Changes in lipid metabolism during adaptation of the *Dunaliella salina* photosynthetic apparatus to high CO_2 concentration. *Russian Journal of Plant Physiology*, *51*(1), 53–62.

Nakanishi, A., Aikawa, S., Ho, S. H., Chen, C. Y., Chang, J. S., Hasunuma, T., & Kondo, A. (2014). Development of lipid productivities under different CO_2 conditions of marine microalgae *Chlamydomonas* sp. JSC4. *Bioresource Technology*, *152*, 247–252.

Ozkan, A., Kinney, K., Katz, L., & Berberoglu, H. (2012). Reduction of water and energy requirement of algae cultivation using an algae biofilm photobioreactor. *Bioresource Technology*, *114*, 542–548.

Park, J. B. K., Craggs, R. J., & Shilton, A. N. (2011). Wastewater treatment high rate algal ponds for biofuel production. *Bioresource Technology*, *102*(1), 35–42.

Perez-Garcia, O., Bashan, Y., & Esther Puente, M. (2011). Organic carbon supplementation of sterilized municipal wastewater is essential for heterotrophic growth and removing ammonium by the microalga *Chlorella vulgaris* 1. *Journal of Phycology*, *47*(1), 190–199.

Pham, H. M., Kwak, H. S., Hong, M. E., Lee, J., Chang, W. S., & Sim, S. J. (2017). Development of an X-Shape airlift photobioreactor for increasing algal biomass and biodiesel production. *Bioresource Technology*, *239*, 211–218.

Pittman, J. K., Dean, A. P., & Osundeko, O. (2011). The potential of sustainable algal biofuel production using wastewater resources. *Bioresource Technology*, *102*(1), 17–25.

Raftery, J. P. & Karim, M. N. (2017). Economic viability of consolidated bioprocessing utilizing multiple biomass substrates for commercial-scale cellulosic bioethanol production. *Biomass and Bioenergy*, *103*, 35–46.

Ramírez-Mérida, L., Zepka, L., & Jacob-Lopes, E. (2015). Current status, future developments and recent patents on photobioreactor technology. *Recent Patents on Engineering*, *9*(2), 80–90.

Ras, M., Steyer, J. P., & Bernard, O. (2013). Temperature effect on microalgae: A crucial factor for outdoor production. *Reviews in Environmental Science and Bio/Technology*, *12*(2), 153–164.

Sahle-Demessie, E., Bekele, S., & Pillai, U. R. (2003). Residence time distribution of fluids in stirred annular photoreactor. *Catalysis Today*, *88*(1–2), 61–72.

Sheng, A. L. K., Bilad, M. R., Osman, N. B., & Arahman, N. (2017). Sequencing batch membrane photobioreactor for real secondary effluent polishing using native microalgae: Process performance and full-scale projection. *Journal of Cleaner Production*, *168*, 708–715.

Sierra, E., Acién, F. G., Fernández, J. M., García, J. L., González, C., & Molina, E. (2008). Characterization of a flat plate photobioreactor for the production of microalgae. *Chemical Engineering Journal*, *138*(1–3), 136–147.

Singh, A., & Olsen, S. I. (2011). A critical review of biochemical conversion, sustainability and life cycle assessment of algal biofuels. *Applied Energy*, *88*(10), 3548–3555.

Singh, R. N., & Sharma, S. (2012). Development of suitable photobioreactor for algae production–A review. *Renewable and Sustainable Energy Reviews*, *16*(4), 2347–2353.

Stephenson, A. L., Kazamia, E., Dennis, J. S., Howe, C. J., Scott, S. A., & Smith, A. G. (2010). Life-cycle assessment of potential algal biodiesel production in the United Kingdom: A comparison of raceways and air-lift tubular bioreactors. *Energy & Fuels*, *24*(7), 4062–4077.

Sutherland, D. L., Howard-Williams, C., Turnbull, M. H., Broady, P. A., & Craggs, R. J. (2015). Enhancing microalgal photosynthesis and productivity in wastewater treatment high rate algal ponds for biofuel production. *Bioresource Technology*, *184*, 222–229.

Sutherland, D. L., Park, J., Heubeck, S., Ralph, P. J., & Craggs, R. J. (2020). Size matters–Microalgae production and nutrient removal in wastewater treatment high rate algal ponds of three different sizes. *Algal Research*, *45*, 101734.

Taherzadeh, M. J. & Karimi, K. (2007). Acid-based hydrolysis processes for ethanol from lignocellulosic materials: A review. *BioResources*, *2*, 472–499.

Vo, H. N. P., Ngo, H. H., Guo, W., Nguyen, T. M. H., Liu, Y., Liu, Y., Nguyen, D. D. & Chang, S. W. (2019). A critical review on designs and applications of microalgae-based photobioreactors for pollutants treatment. *Science of the Total Environment*, *651*, 1549–1568.

Xin, L., Hong-Ying, H., Ke, G., & Ying-Xue, S. (2010). Effects of different nitrogen and phosphorus concentrations on the growth, nutrient uptake, and lipid accumulation of a freshwater microalga *Scenedesmus* sp. *Bioresource Technology*, *101*(14), 5494–5500.

Yadala, S., & Cremaschi, S. (2014). Design and optimization of artificial cultivation units for algae production. *Energy*, *78*, 23–39.

Yen, H. W., & Brune, D. E. (2007). Anaerobic co-digestion of algal sludge and waste paper to produce methane. *Bioresource Technology*, *98*(1), 130–134.

Znad, H. (2020). Microalgae culture technology for carbon dioxide biomitigation. In *Handbook of Algal Science, Technology and Medicine*, Ozcan Konur (Eds.), Academic Press. United Kingdom, 303–316.

Part III

Micro Algae & Biofuel

9 LCA of an Algal Biomass Plant

Microalgae to Bio-Oil through Hydrothermal Liquefaction

Sudipta Sarkar
IIT

S. Baranidharan
NIT

Brajesh Kumar Dubey
IIT

CONTENTS

DOI: 10.1201/9781003197737-12

ABBREVIATIONS

ALOP Agricultural land occupation potential
ASTM American society for testing and materials
CC Climate change
EBP Ethanol-blended petrol
EROI Energy return on investment
FDP Fossil depletion
FETP Freshwater ecotoxicity
FEP Freshwater eutrophication
GHGs Greenhouse gases
GWP Global warming potential
HHV High heating value
HTP Human toxicity
HTL Hydrothermal liquefaction
IIRP-HE Ionizing radiation
ISO International standards organization
LCA Life cycle assessment
LCI Life cycle inventory
LCIA Life cycle impact assessment.
METP Marine ecotoxicity
MEP Marine eutrophication
MDP Metal depletion
NER Net energy ratio
NLTP Natural land transformation
ODP Ozone depletion
OPC Open-pond cultivation
FPP Flat-plate photobioreactor
PBRs Photobioreactors
PMFP Particulate matter formation
POFP Photochemical oxidant formation

PVC	Polyvinyl chloride
TAP	Terrestrial acidification
TETP	Terrestrial ecotoxicity
ULOP	Urban land occupation
WDP	Water depletion

9.1 INTRODUCTION

In today's world, fossil fuels are the major source of energy with a share of over 67.2%. In 1556, Georgius Agricola first introduced fossil fuel deposited in the Earth's crust. But it failed to catch people' attention until the industrial revolution in 18th century when Mikhail Lomonosov demonstrated the organic origin of fossil fuels. Starting from then, fossil fuel is the leading source of our energy needs. Fossil fuels, mainly coal, petroleum, and natural gas, are nonrenewable resources, as they are derived from million years' old plants and animal fossils. These fossil fuels generally contain high percentage of carbon and so on; burning of these fuels produce almost 22.8 billion tonnes carbon dioxide (CO_2) per annum along with a large amount of carbon monoxide (CO), nitrogen monoxide (NO), nitrogen di oxide (NO_2), sulfur dioxide (SO_2), etc. Being almost nonrenewable, this fuel source presently is in the verge of extinction. Several researchers have shown that crude oil and natural gas reservoirs will be exhausted around year 2051–2052 A.D., and coal sources may last a few longer, around 2100 A.D. (Shafiee and Topal 2010).

In comparison, these fuels contribute to environmental degradation by emitting harmful greenhouse gases (GHGs) and a variety of other noxious pollutants. As a result, research on biofuel is being conducted with the greatest urgency and significance as a potential replacement for these fossil fuels. Biofuels were already being used throughout our history's most significant periods, even before they were discovered. The voyage started a few thousand years ago with the discovery of fire, which is considered the first step toward human civilization. From there, the journey continued with the development of language. The very earliest kind of biofuel was the fire created by burning wood. Up until the middle of the nineteenth century, biofuels were used for both lighting lamps and as a cooking fuel. This was the first transportation fuel, and Henry Ford designed his first model "T" to run on ethanol in 1908. Though there were numerous fuels, biofuels were repeatedly reintroduced for various reasons: during World War I to address the oil shortage, in 1973 during the oil crisis caused by the OAPEC (Organization of Arab Petroleum Exporting Countries) oil export embargo, in 1979 during the oil crisis, in 1990 during the oil crisis, and now as the world's savior to mitigate climate change and restore fossil fuel sources.

Biofuels are energy produced by living organisms or their metabolic waste. Biofuel is a fuel made from renewable materials and derived from biological carbon fixation. Our primary energy concern today is the scarcity of fossil fuels and their rising cost as reserves dwindle. Biofuels are cheaper and quicker to prepare than fossil fuels. Biofuel manufacturing will minimize foreign energy dependence. First-generation biofuels are created by fermenting sugar, starch, or vegetable oil, which is unsustainable. These include ethanol, biodiesel, green diesel, biofuel gasoline, bioether, biogas, and syngas. Their major component is valuable crops, which might disrupt the food industry. First-generation biofuels are inefficient, harm rainforests, and need lots of fertilizer. Researchers sought a "greener" source. So reinvents biofuels. Second-generation lignocellulosic biofuels are "greener."

The production of a given quantity of fuel using second-generation biofuels requires less water and arable land, and the net energy ratio of these fuels is high (Schenk et al. 2008). Cellulosic ethanol, algae-based biofuels, biohydrogen, methanol, dimethylfuran, etc. are the examples of second-generation biofuel, and the effectiveness of the microalgae in producing bio-oil were so vast (Uffenorde and Much 1918) that it soon found a new, discrete new category named as "Third-Generation" biofuel. First-generation biofuels are environmentally and economically not friendly, and woody biomass feedstock of second-generation biofuels also requires costly technologies to convert them into fermentable sugars, whereas third-generation biofuel feedstock microalgae have many more advantages over them. Like, microalgae can produce 15–300 times more oil (Chisti 2007) than conventional crops like rapeseed and palm oil, harvesting cycle is also much more precise, 1–10 days, than other feedstocks with a higher growth rate, and it has a large amount of lipid content to convert into oil. Microalgae has also the ability to fix atmospheric CO_2, and it does not require herbicides or pesticides during its growth. For these enormous fruitful reasons, now-a-days microalgae as a feedstock for biofuels are widely used all over the globe. It is believed that biomass can contribute up to 20%–90% of global energy demand (Berndes, Hoogwijk, and Van Den Broek 2003).

The carbon footprint of biofuels is still not zero, though. Flue gas sequesters CO_2 throughout the whole manufacturing process of microalgae biofuel. Besides that, the environmental effect of land use, high temperature change, and carbon sink leaking must be taken into account. Microalgae's main drawback is that its ethanol is less stable than other feedstocks, necessitating the use of stabilizers. To determine the sustainability of the process of manufacturing bio-oil from microalgae utilizing flue gas from a power station, all possible consequences must be considered. As a result, in this article, researchers attempted to quantify the environmental effect and sustainability of microalgae biofuels. Environmental consequences may be assessed throughout the product's life cycle using LCA. The ISO-based LCA technique is the most accurate and transparent way for verifying possible consequences and determining the carbon footprint of a process. It shows producers, suppliers, and consumers where to make improvements to minimize GHG emissions.

9.2 LITERATURE REVIEW

9.2.1 Microalgae Cultivation

Microalgae or 'microphytes' are unicellular microscopic algae species typically found in aquatic systems; size ranges from a few micrometers (μm) to a few hundred micrometers. They produce biomass flocks using sunlight and CO_2 as energy and carbon source. Growth of microalgae is presented graphically in the research of Mata, Martins, and Caetano (2010) using Figure 9.1, wherein (1) represents lag phase, (2) exponential growth phase, (3) linear growth phase, (4) stationary growth phase, and (5) death phase. Microalgae use photosynthesis to get its food and carbon fixation in

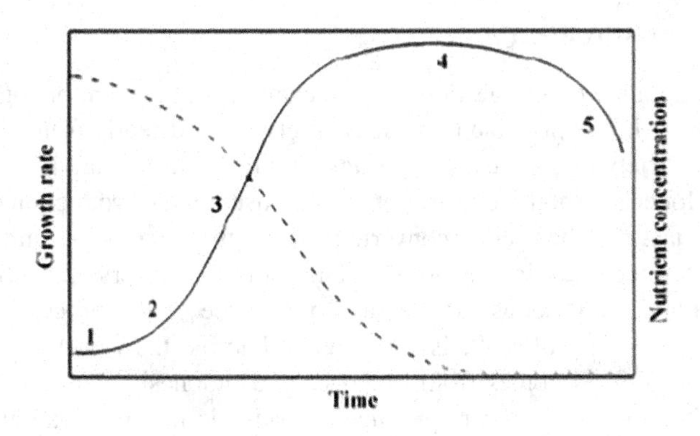

FIGURE 9.1 Microalgae growth curve.

its cells. In general, some nutrients (Nitrogen–Phosphorus–Potassium [NPK] based) for its growth. Microalgae's main constituents are protein, carbohydrate, lipid, and nucleic acid; the ratio of these can vary with different algae species. Solar energy intensity can vary depending upon the seasonal change and also time of day; another important parameter that affects solar radiation is geographic location. Additionally, it varies according to the type of cultivation process, which will be discussed later in this study. CO_2 is the most important nutrient for algal growth, which can be supplied externally via sequestration of power plant flue gas.

There are two ways to grow algae: open ponds and photobioreactors (PBRs). These systems are made to maximize solar radiation to microalgae. As a result, these ponds are kept shallow to allow for maximum sunlight penetration. The ponds are constantly fed CO_2 and nutrients, while algae-laden water is drained. Common open-pond systems include raceway ponds and shallow ponds. This system's main benefit is its ease of construction and use. The main drawbacks of open-pond systems are water evaporation, gas diffusion, large area requirements, and environmental contamination. A photobioreactor (PBR) is a closed system for growing microalgae in a controlled environment. Closed-system cultivation can reduce or eliminate open system disadvantages. PBR system has high surface-to-volume ratio and requires little space. More uniform gas transfer and temperature control reduces growth medium evaporation. A closed system reduces the risk of external contamination. Despite these advantages, the system's main flaw is the high initial cost, which has limited its widespread adoption. Tube, flat-plate, and vertical column PBRs are common. Tube PBRs are made of straight, coiled, or looped transparent glass or plastic tubes. However, tubular PBR has significant mass transfer issues (oxygen buildup). In 1986, Ramos de Ortega and Roux invented outdoor flat-plate PBR using thick, transparent PVC. Unlike tubular PBR, flat-plate PBR has low dissolved oxygen, allowing for high photosynthetic efficiency. Vertical column PBRs are ideal for large-scale cultivation. Algae feed on CO_2 and NO_2 from power plants. By feeding microalgae farms with waste heat from power plants, we can increase algae production and clean the air.

9.2.2 Biofuel and Bio-Oil

Agriculture and anaerobic digestion are two examples of current biological systems that create biofuel. It is possible to make biofuels either directly from plants or indirectly from a variety of sources, such as agricultural, commercial, home, and industrial waste. Modern carbon sequestration is often used in the production of renewable biofuels. Biomass may be used or converted into useful energy-containing molecules to manufacture other sustainable biofuels. This biomass conversion method may provide solid, liquid, or gaseous fuels as a consequence of the process (Hassan and Kalam 2013). A variety of biofuels are generated across the world. For instance, % of the world's palm oil comes from Malaysia and Indonesia (Mofijura et al. 2012). 105 billion liters of biofuels were produced worldwide in 2010 at an annual rate of 17% (28 billion gallons US). United States and Brazil produced over two-thirds of the global ethanol output in 2010 at 86 billion liters (or about 23 billion US gallons). With 53% of global biodiesel output, the European Union has overtaken the United States as the leading producer. First phase of Ethanol-Blended Petrol Program (EBP) was introduced by India's Ministry of Petroleum and Natural Gas in 2003, requiring 5% ethanol in gasoline in nine states and four union territories. (Hassan and Kalam 2013) presents the first- and second-generation biofuels goals, and plans are laid forth. With regard to biofuel production and usage, it also contrasts the current worldwide picture.

9.2.3 Generation of Biofuel

Two types of biofuels exist. The first generation of biofuels are produced through the fermentation of sugar, starch, or vegetable oil, which is not sustainable. Ethanol, biodiesel (bio alcohols), green diesel, biofuel gasoline, bio ether, biogas, syngas, and solid biofuels are all examples of this type of biofuel. They have the potential to wreak havoc on the food economy, as the primary ingredient in these fuels is derived from valuable crops. To address this issue, researchers began searching for a "greener" and more sustainable source. That is how the second generation of biofuels is innovated. The second generation of biofuels is composed of lignocellulosic biomass, which is "greener" and more sustainable for the environment. Second-generation biofuels include cellulosic ethanol, algae-based biofuels, biohydrogen, methanol, and dimethylfuran (Sims et al. 2010). Microalgae are increasingly being used as a biofuel source material in the modern era. Microalgae are unicellular photosynthetic microorganisms. They require light, sugars, CO_2, nitrogen, phosphorus, and potassium to grow. Microalgae are primarily composed of lipids, proteins, and carbohydrates, all of which must be produced in large quantities over a short period of time (Uffenorde and Much 1918). The primary advantages of microalgae for use as feedstock are their ability to be harvested regardless of seasonal variation, resulting in significantly higher productivity compared to conventional oil seed crops; they can be cultivated in arid land; they can be cultivated using salt or even waste water and thus can be used for waste water treatment (Schenk et al. 2008); they have a high tolerance for carbon content; they consume very little water; and they do not require herbicides (Uffenorde and Much 1918).

TABLE 9.1
Bio-oil Standard Characteristics

Tests	Results	Units
Water content	2.0	%wt
pH	2.2	
Density @ 15°C	1.207	kg/L
High heating value	17.57 (7,554 BTU/lb)	MJ/kg
Low heating value	15.83 (6,806 BTU/lb)	MJ/kg
Solids content	0.06	%wt
Ash content	0.0034	%wt
Pour point	−30	
Flash point	48.0	
Conradson carbon	16.6	%wt
Kinematic viscosity		
@ 20°C	47.18	mm^2/s
@ 50°C	9.726	mm^2/s
Carbon	42.64	%wt
Hydrogen	5.83	%wt
Nitrogen	0.10	%wt
Sulfur	0.01	%wt
Chlorine	0.012	%wt
Alkali metals	<0.003	%wt

9.2.4 Bio-Oil

With its high energy content and ability to partly replace petroleum crude, bio-oil or bio crude is a promising alternative fuel. Thermochemical conversion of microalgae may provide bio-oil. Most microalgae to bio-oil conversions are carried out through fast pyrolysis. Atmospheric pressure and moderate temperatures (usually 450°C–550°C) are required for fast pyrolysis to produce bio-oil (Stefanidis et al. 2015). The most promising method for converting wet algal biomass is hydrothermal liquefaction (HTL). Bio-oil typically has a high heating value (HHV) of 30–40 MJ/kg bio-oil or above (Bennion et al. 2015). Table 9.1 summarizes the standard characteristics of bio-oil as defined by the American Society for Testing and Materials (ASTM).

9.3 METHODOLOGY

9.3.1 Life Cycle Assessment (LCA)

LCA is tool for estimating the environmental impacts of a product's complete life cycle, beginning with the acquisition of raw materials from nature and ending with the product's usage, disposal, and recycling. With the "cradle to grave" method, energy and material intake and waste production are recorded throughout the product's life cycle. Identifying and quantifying the energy, raw materials used and the waste generated during the process aids in evaluating the environmental improvements opportunities;

Now, LCA is a part of everything from government policy and strategic planning to marketing and consumer education. Eco-labeling and environmental improvement potential may be discovered using LCA. ISO 14040, ISO 14044, ISO 14044, and ISO 14044 outline its goals and scope, inventory analysis, impact assessment, and interpretation, whereas ISO 14044 defines its requirements (ISO 14044).

9.3.1.1 Goal and Scope Definition

The purpose and scope of an LCA must be clearly stated and aligned with the intended application. Determine the goal and application of an LCA study, as well as who the study's findings are meant to be communicated to and whether the research's conclusions are intended for use in comparisons intended for public disclosure, to determine the purpose of the LCA. Bio-oil production from coal-fired power plants flue gas at an algal biomass facility in eastern India using a hydrothermal liquefaction method is the subject of this investigation. Gate-to-gate system boundary encompasses mainly the manufacturing process and co-products in this research. Input and output materials are examined as part of the research. Resources aren't included in this analysis, since they aren't used up and can't be sourced. The following items shall be considered and clearly described when defining the scope of an LCA.

9.3.1.2 Case Study Scenario

Using 15% carbon dioxide (CO_2) flue gas from their "Talcher" coal power plant, they develop microalgae in an open culture pond fitted with a heat recovery system and then dry microalgae slurry with 70% water content using the recovered heat in this thesis. In order to produce bio-oil, they collect and dry algal biomass, and then employ the HTL process. Microalgae cultivation requires a two-horsepower pump and a one-horsepower motor to run the system. This analysis is based on the Indian processing and electrical grid.

9.3.1.3 System Boundary

The system boundaries define which unit processes are to be included in the life cycle assessment. The system boundary denotes the unit processes that must be incorporated into the research system. It is highly dependent on the study's objective. The boundary Figure 9.2 in this study consists primarily of operational and maintenance phases. Beginning with the flue gas outflow from the stack, using it as a raw material for microalgal growth, recovering excess heat with a heat exchanger, and finally drying algal biomass, the open culture pond is the primary unit for this process consisting of a hydrothermal liquefaction unit, followed by another heat exchanger and depressurization unit, and finally an oil–water separator unit.

9.3.1.4 Functional Unit

In a life cycle assessment (LCA), a functional unit serves as a measurable unit of reference. Using this device, we may create a single reference flow for a variety of

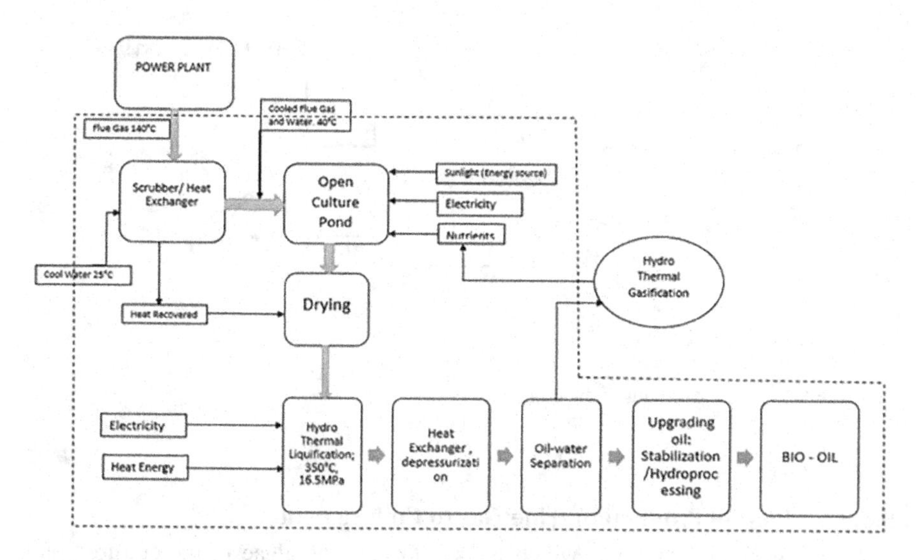

FIGURE 9.2 System boundary for algal biomass plant (HTL).

factors and scenarios. Because they must execute the same functions in order to be compared, systems must have identical functional units, such as reference flows, in order to do so. 1 kg of plant-produced bio-oil serves as a functional unit in this research. The output and input of the individual processes are listed on the system boundary.

Assumptions

- Phototrophic growth of microalgae occurs using sunlight as the primary source of energy.
- Algal biomass absorbs the sun's rays evenly.
- During a 20-day algae cultivation cycle, the climatic parameters stay unchanged.
- HTL plant is situated next to an algae production facility, which is not included in the LCIA research due to the low environmental effect of the operation of the facility.

9.3.2 Life Cycle Inventory Analysis (LCI)

LCI is used to collect quantified input and output data and to calculate them on a functional unit basis. LCI is divided into two phases: data collection (1) and data calculation (2). For each unit process contained within the system boundaries, qualitative and quantitative data for inclusion in the inventory shall be collected. Three stages of data collection are possible in our study: Flue gas to culture pond, algal biomass and microalgae cultivation to produce bio-oil. The majority of the data for this study came from the biomass plant authority, various technical literature reviews, and product manufacturers.

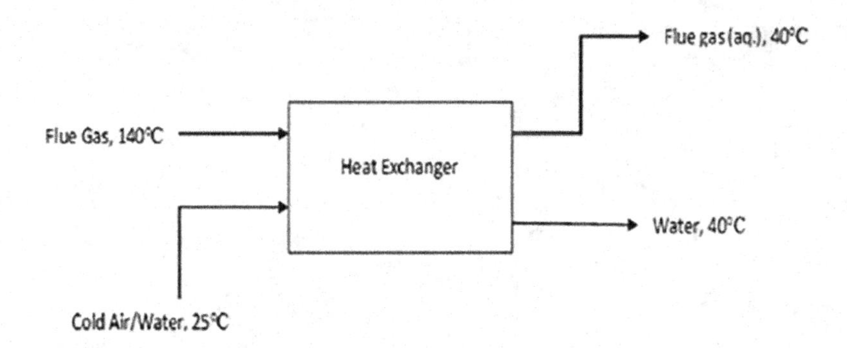

FIGURE 9.3 Heat exchange flow chart.

9.3.2.1 Bio-Oil Production (Flue Gas to Culture Pond)

The plant generates flue gas, which is used to feed the algae in the culture pond. Talcher coal is used to fuel this power plant. According to data provided by the Power Plant Authority, talcher coal is primarily a low-grade coal containing only 35% fixed carbon and nearly 25% ash. CO_2 is contained in 13% of flue gas. Flue gas from this power plant cannot be utilized directly in the culture pond because of its high temperature. Flue gas is collected through pipes and sent to a scrubber heat exchanger, where it is cleaned.

9.3.2.2 Heat Exchanger

In the Scrubber heat exchanger, 140°C hot flue gas is mixed with 25°C cold water to produce a 40°C flue gas composition flow (Figure 9.3). The result is a 40°C flue gas flow. Afterwards, the heat generated by this process is utilized to dry the biomass.

The heat exchanger uses the same amount of electricity as a 1 hp fan and a 0.5 hp pump. Cold air and cold water may be employed in a variety of ways to carry out this operation. Flue gas flow rate is 100 kg/h for the air system's cold air input flow rate of 2,170 kg/h. In the water system, 520 kg/h of cold water and 100 kg/h of flue gas are the input rates. Detailed information on the appendix may be found here. For open culture, this CO_2 13% water–air mixture must first be warmed up before injection.

9.3.2.3 Cultivation of Algal Biomass

Microalgae species were selectively enriched and cultivated in an open modified raceway pond system using local collections of blue-green algae at NALCO's flue gas-fed carbon sequestration plant in Angul, India. According to 16S/18S rRNA gene sequence analysis and subsequent biomass characterization, the algal biomass cultures were dominated by cyanobacterial species. Additional washes with water were performed on the algal biomass samples (distilled). Following that, samples are dried overnight in a hot air oven set to 80°C to determine loss on washing. Dry algae samples were crushed by hand and stored in an airtight bottle to be desiccated for all dry biomass analysis (Das et al. 2013).

9.3.2.4 Cultivation

In this stage, the process is typically carried out using a variety of microalgal cultivation techniques, including open-pond cultivation (OP), flat-plate photobioreactors (FPP), and combined open-pond and flat-plate cultivation systems (OP+FPP) (Tubular). A photo bioreactor (PBR) is a closed, controlled system that contains all of the raw materials required for the growth of microalgae. Because PBRs maximize light utilization, control gas transfer, and protect against external contamination, they are significantly more productive than open-pond systems. This system, however, has a high capital cost.

9.3.2.5 Open Culture Pond

In an open pond, microalgae (commercial strain) are cultured using the CO_2 content of flue gas as a feedstock and sunlight as an energy source. During this study cycle, algae biomass is generated via photoautotrophic growth at a retention time of approximately 20 days and a production rate of 22 tonnes of biomass per acre per year. The pond's nutrient supply includes the majority of micronutrients, as well as external nitrogen and phosphorous nutrients required for growth. After collecting the microalgae as a 70% water-content slurry, the recovered heat from the heat exchanger is used to dry the biomass, and the remaining water is recycled to the pond, resulting in a 5% operational loss in addition to pond evaporation. Similar conditions for *Chlorella vulgaris* cultivation in an open culture pond were reported in the literature, with a growth rate of 28 tons/year/acre and a solid concentration of 0.5 kg/m^3. Biomass with a crude oil yield of about 40%, the pond receives majority of its nutrients from nitrogen and phosphorus (Lardon et al. 2009; Chisti 2007).

9.3.2.6 Photo Bioreactors (PBRs)

The current study compared two distinct types of PBRs to open-pond cultivation—the most prevalent method at the moment. PBRs include FPP and tubular PBR (TP). Solar energy is the most frequently used source of energy in FPPs. According to Rilling et al. (2018) and colleagues, the growth rate of *Nannocloropsis* sp. in FPP is approximately 40 tons/year/acre. There is a 30% oil yield and a climate change impact of 6.71 kg CO_2-equivalent. In tubular PBR with a concentration of 0.5 kg/m^3, *Chlorella vulgaris* grows at a rate of 37 tons/year/acre. However, the oil yield is 40% with a negligible impact on climate change, around 1.2 kg CO_2-eq (Collet et al. 2010; Chisti 2007).

9.3.2.7 Drying

Microalgae are dried prior to being added to the conversion reactor. The HTL process does not require dry biomass; rather, it can be used directly with algal biomass slurry. Pyrolysis, on the other hand, requires only dried biomass. Although drying occurs in trays at the heat exchanger cum dryer, it is frequently assumed that it occurs via the rotary drum method, which requires up to 7.76 MJ of energy/kg of microalgae (Bennion et al. 2015).

9.3.2.8 Microalgae to Bio-Oil

9.3.2.8.1 Hydrothermal Liquefaction (HTL)

There are a few processes that can be followed to convert microalgae to bio-oil. The two most advantageous methods are slow pyrolysis of raw microalgae and direct lipid extraction -HTL. The study chose the HTL process to produce bio-oil, because the literature indicates that 30% of biomass slurry is hydrothermally treated in sub-critical water at temperatures of 300°C–350°C and pressures of 135–200 atm (Biddy et al. 2013). A catalyst, sodium bi carbonate (Na_2CO_3), is added in the amount of 39 g/kg of microalgae, and the process requires 5.919 MJ of energy/kg of microalgae (Bennion 2015). The yield of bio-oil produced by this method of conversion is approximately 0.37 kg/kg of microalgae.

9.3.2.8.2 Pyrolysis

Pyrolysis is carried out at a high temperature of around 400°C using a sodium bicarbonate catalyst that provides 27 g of sodium bicarbonate/kg of microalgae. This process requires a significant amount of energy to operate, 10.21 MJ, in comparison to the HTL process. The yield of bio-oil is low in comparison to the HTL method shown in Table 9.2, at approximately 0.293 kg/kg of microalgae. Although the HHV of oil from the pyrolysis process is approximately 38.7 MJ/kg of bio-oil, this is greater than the HHV of oil from the HTL method, which is approximately 34 MJ/kg of bio-oil.

9.3.2.8.3 Bio-oil Stabilization

Bio-oil stabilization is used to compensate for changes in the viscosity of the oil and to remove impurities. This is accomplished by adding supercritical propane to the oil in a 5:1 ratio. 2.15 MJ/kg of bio-oil is required to process the supercritical fluid. This procedure is carried out at a low temperature (23°C) and low pressure (3.5 MPa) (Bennion et al. 2015). The propane loss due to handling and processing is estimated to be 0.02 kg/kg of bio-oil, and the stabilized bio-oil yield is approximately 84.6%.

9.3.2.8.4 Hydroprocessing

The stabilized bio-oil obtained from supercritical fuel treatment is then treated further to obtain renewable bio-oil. This stage makes use of hydrogen. In the presence

TABLE 9.2
Inventory Comparison between HTL and Pyrolysis Subprocesses

STEP	Unit	Pyrolysis	HTL
Drying (Rotary drum method)	MJ/kg microalgae	7.76	NA
Temperature	°C	400	350
Catalyst (Na_2CO_3)	kg/kg of microalgae	0.027	0.039
Process energy intake	MJ/kg of microalgae	10.21	5.919
Heat recovery	%	85	85
Bio-oil yield	kg/kg of microalgae	0.293	0.37
Bio-oil HHV	MJ/kg of bio-oil	38.7	34

of zeolite catalyst (0.4 mg/kg stabilized bio-oil), hydrogen removes unwanted nitrogen and oxygen from the oil. The required amount of hydrogen varies according to the composition of the bio-oil. Hydrogen is required to refine and stabilize bio-oil at a rate of approximately 0.0317 kg/kg stabilized bio-oil (Fortier et al. 2014). Hydroprocessing requires approximately 0.84 MJ of energy/kg of stabilized bio-oil. However, the remaining portion of required energy is used to produce hydrogen, approximately 56.96 MJ/kg of hydrogen.

9.3.3 Microalgae to Biomethane

In biogas plants, biomethane is produced via anaerobic digestion. Wang et al. cultivated algal feedstock in a greenhouse using a glass flat-plate PBR system with a glass requirement of 0.27 m^2/GJ of biomethane energy. This value, 1 GJ of biomethane energy, is used as the study's functional unit. They used the algae *Spirulina* sp. for this purpose, with a growth rate of 1 kg/m^2/day. Around 53 W/m^3 electricity is used for the CO_2 injection, with a daily cycle period of 10 hours. Concentration of microalgal slurry via natural settling and centrifugation. Concentrating algae requires approximately 6.1 kWh of energy per functional unit. This microalga is digested anaerobically to produce biogas, which is then upgraded for market use. This section of the production requires approximately 101.95 kWh of electricity. The total GHG emissions from the process are approximately 49 kg CO_2-eq, with a net energy ratio of 1.54.

In the European Union and a few American countries, biodiesel derived from microalgae has already been used for transportation purposes. Generally, microalgae are converted to biodiesel via a process called transesterification. Lardon et al. conducted research on the *Chlorella vulgaris* microalgal species. The general composition of this type is as follows: protein (50%), carbohydrate (15%), lipids (25%), and others (10%). The algal slurry is dried to a solid content of 90% using 81.8 MJ of heat, and the process consumes approximately 8.52 MJ of electricity to produce 1 kg biodiesel. Oil from algae is extracted using a hexane solvent at a ratio of about 1:5, with a loss of about 2 g of oil/kg of biodiesel. The oil is transesterified in the presence of an alkaline catalyst in an alcoholic medium to produce biodiesel (Stephenson et al. 2010). This process consumes 0.6 MJ of energy and 114 g of methanol as an alcoholic medium/kg of biodiesel.

9.3.4 Life Cycle Impact Assessment (LCIA)

The Life Cycle Impact Assessment (LCIA) is a methodological tool for determining the significance and magnitude of potential environmental system impacts. It is a technique in which inventory data is associated with various impact categories in order to gain a comprehensive understanding of the features. The purpose and scope of the study dictate the impact assessment methods, and LCIA must be adjusted accordingly in some cases. There are numerous established methods for conducting an impact assessment study. The current study assessed the impact using the "ReCiPe" method and processed the data using the "SimaPro 8.0.3" software.

The primary objective of this method is to reduce a massive amount of inventory data to a manageable number of indicator category scores. The "ReCiPe" method differentiates between two types of indicators: (i) midpoint indicators and (ii) end-point indicators. The midpoint indicator includes 18 impact categories, whereas the end-point indicator includes three major impact categories: damage to human health, ecosystem damage, and resource availability damage. The ReCiPe midpoint indicator (Hierarchist version) assessment was used to determine the impact of the HTL and pyrolysis processes, and to compare them to two other published results: biodiesel conversion from algal biomass (Lardon et al. 2009) and the environmental impact of fossil fuel diesel (OBorn 2012).

9.4 RESULT AND DISCUSSION

9.4.1 Comparison between Microalgae Cultivation Processes

The results of the first objective indicate that the life cycle GHG emissions of various cultivation pathways range between 1.2 and 8.0 kg CO_2 equivalent/kg of biomass. Where the process under consideration, blue-green algae cultivation (red bars), produces approximately 1.37 kg CO_2-eq.—the second lowest amount produced by any other method. When compared to the other impact categories depicted in Figure 9.4, it is clear that this method has the lowest overall environmental impact. Carbon sequestration and the use of chemical-free fertilizer are the primary reasons for this impact analysis result, making this method the most preferable.

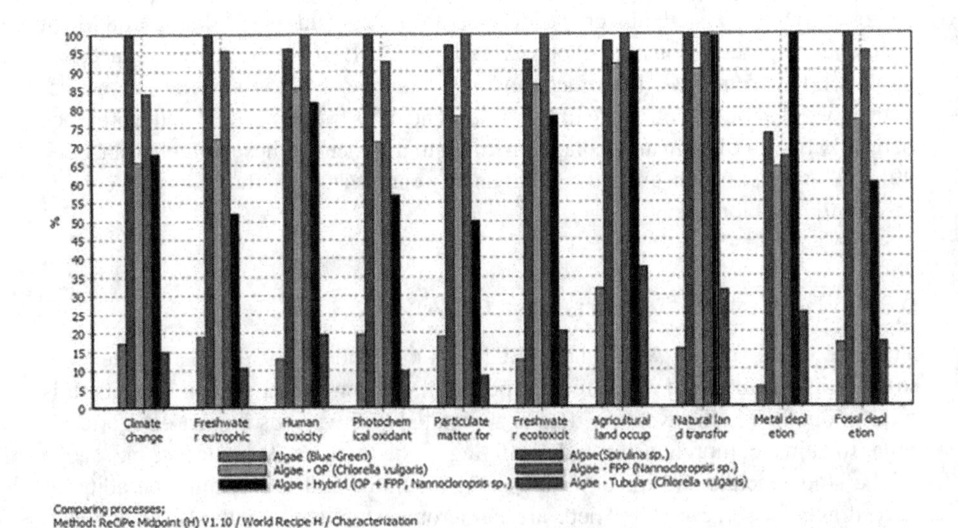

FIGURE 9.4 Comparative LCIA of different microalgae cultivation process.

9.4.2 Microalgae to Bio-Oil Production

This study has examined the LCA of two distinct thermochemical conversion methods for the production of bio-oil from microalgae: HTL and pyrolysis. Our analysis is based on data obtained from an established algal biomass plant in eastern India. The ReCiPe impact assessment method was used to calculate the total potential impact of the aforementioned conversion process, the IPCC GWP 100a was used to calculate the processes' carbon footprint over a 100-year period, and a NER study was used to calculate the processes' energy consumption; Figure 9.5 illustrates the outcome.

9.4.2.1 Microalgae Harvesting

The primary ingredient in algae harvesting is flue gas, which contains CO_2, NO_2, SO_2, and other greenhouse gases that contribute to climate change, acidification, particulate matter formation, and eutrophication. However, eutrophication is primarily a problem at this stage due to the use of fertilizer. Additionally, the gases emitted during this stage contribute significantly to ozone depletion by reacting with atmospheric humidity and other particles and gases to form ozone layer depleting particles. The coal used to generate energy and the flue gas have an effect on the fossil fuel and mineral depletion potential. In terms of human toxicity, the harvesting portion of the graph has a negative bar, indicating a positive effect on toxicity. Due to the fact that the process utilizes flue gas generated at the power plant, the carcinogenic effect of those gases is reduced to almost zero, which explains the negative bar on the impact chart.

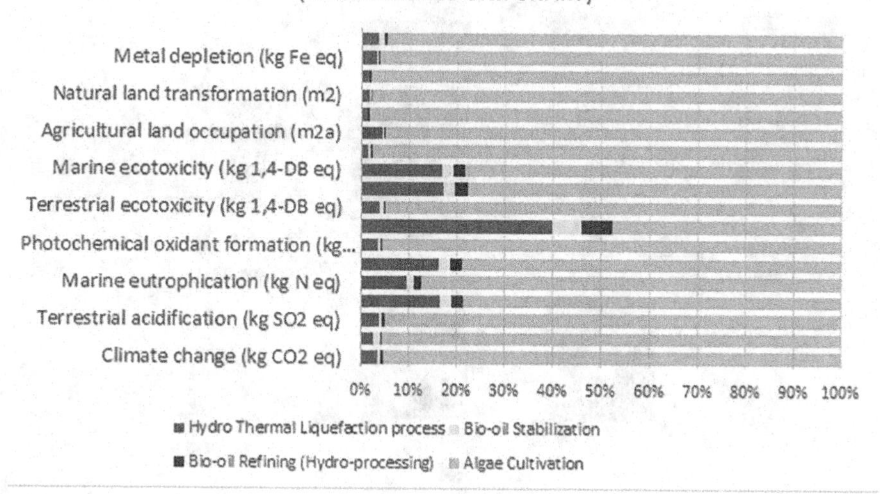

FIGURE 9.5 LCIA of Bio-oil production through HTL process.

9.4.2.2 Bio-Oil Stabilization and Hydroprocessing

Propane is used in the stabilization process, which results in the formation of bromo-tri-fluoro methane and other halons (CFC, HCFC, etc.), all of which have a significant depletion potential for the ozone layer. The primary component of propane, nitrogen oxides, contribute significantly to marine eutrophication. Carbon-14 and tritium (H-3) are the primary sources of ionizing radiation in the process. Metals such as barium, silver, and bromine are causing ecotoxicity in freshwater. However, because propane synthesis requires these metals as an ingredient or input, the picture depicts a negative ecotoxicity bar 12. During the hydroprocessing process, hydrogen is used to remove oxygen from the fuel. The interaction of oxygen with hydrogen and a trace of CO_2 results in the elimination of oxygen in the form of water. As a result, water is created rather than consumed, having a detrimental effect on water depletion.

9.4.3 Net Energy Ratio

Figure 9.6 illustrates the NER results for the two thermochemical pathways. The net energy is further subdivided into process subgroups to facilitate understanding of the process's energy consumption. It is used to demonstrate the energy-generating process's efficiency. NER is defined as the external energy required to convert the stored energy to usable energy. The overall NER values for the HTL and pyrolysis pathways are 0.914 and 2.383, respectively, indicating an unfavorable energy condition for these two pathways.

According to Bennion et al. (2015), the NER for biofuels should be less than 1.0, as the NER of petroleum diesel is 0.20, and, from other literature, the NER for sugar

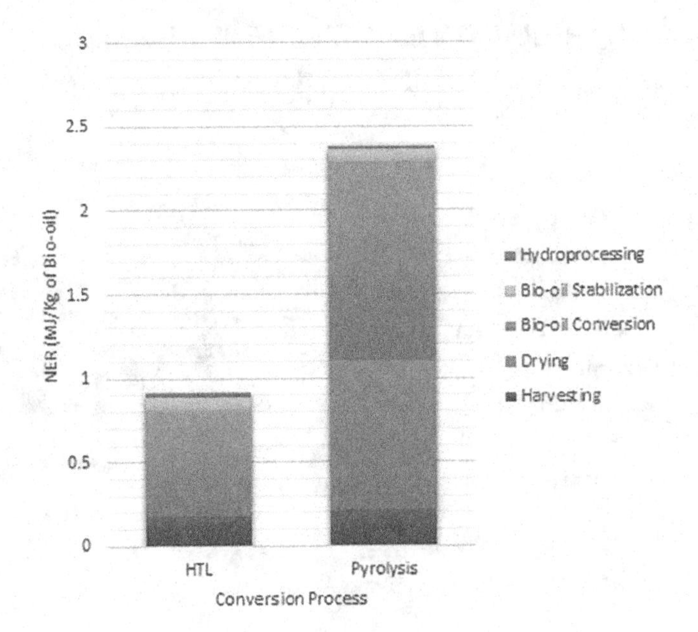

FIGURE 9.6 Net energy ratio.

cane ethanol ranges from 0.8 to 1.0, corn ethanol ranges from 0.8 to 1.6, and biodiesel NER is around 1.3. From this vantage point, the NER of the HTL process is approximately 0.9, and 1 kg of bio-oil contains approximately 34 MJ of energy. The conversion stage consumes the majority of the energy in the process, nearly 68% in the case of the HTL process and 49% in the case of the pyrolysis method. While pyrolysis appears to be less energy intensive than the HTL process, it was necessary to dry the microalgae feed, which consumed nearly another 38% of the energy required. While the HHV of Pyrolysis bio-oil is nearly 14% greater than that of HTL, the total energy requirement is 204% greater than that of HTL. When the HTL and pyrolysis subprocesses are compared in terms of efficiency, the pyrolysis subprocess is 26.4% efficient, while the HTL subprocess is only 17.4% efficient. Thus, it is abundantly clear that the HTL method is far more advantageous than pyrolysis for microalgal conversion.

9.4.4 Global Warming Potential (GWP)

The GWP is a relative measure of the amount of heat trapped in the atmosphere by a greenhouse gas. The amount of CO_2 gas required to trap an equal amount of heat in the atmosphere is denoted by this value. Because each step of a process can generates GHGs, it is the most critical parameter for environmental impact assessment and carbon footprint estimation. The GWP for the conversion of microalgae to bio-oil is depicted in Figure 9.7, broken down into subprocesses.

The GWP of the pyrolysis process is significantly greater than that of the HTL process (Figure 9.7). Pyrolysis contributes 38.33 kg CO_2-equivalent/kg bio-oil to the atmosphere, whereas HTL contributes only 10.225 kg CO_2-equivalent/kg bio-oil, nearly 73% less. Additionally, this process has an advantage in that it harvests

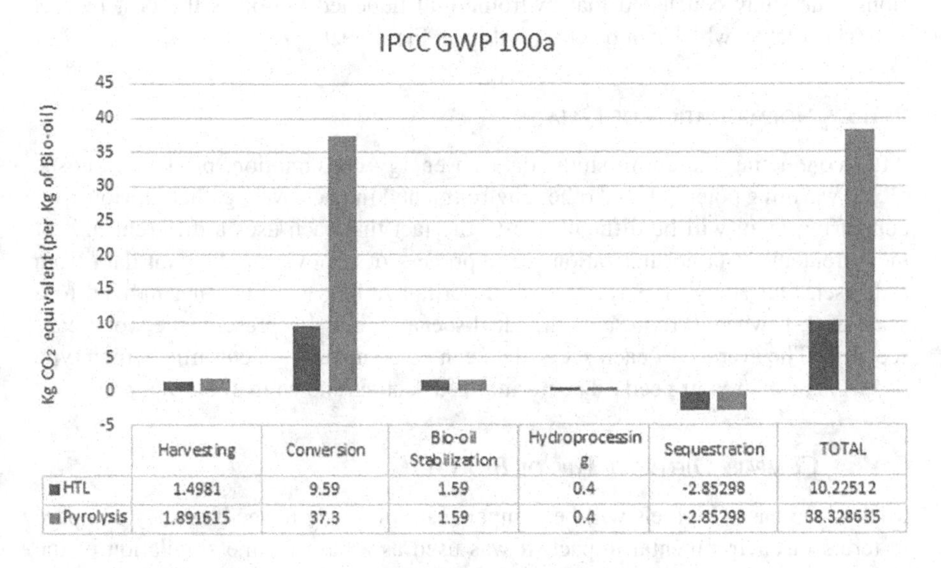

FIGURE 9.7 Global warming potential of conversion subprocesses.

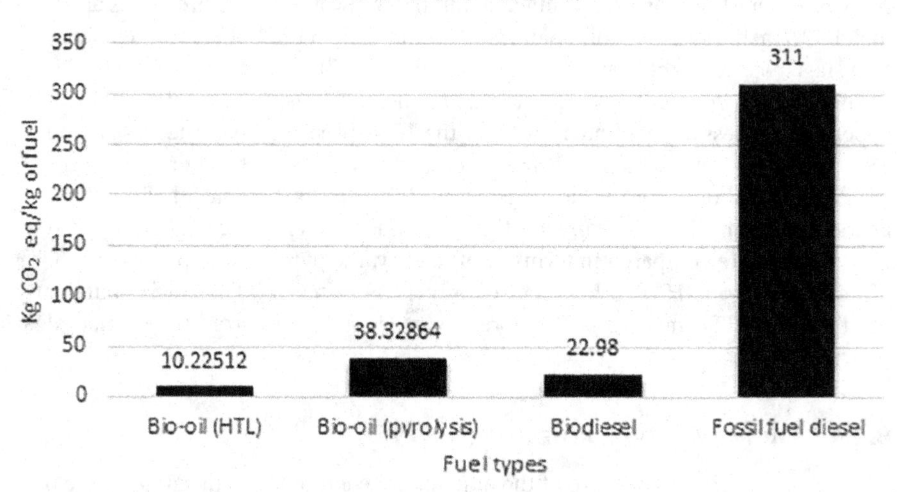

FIGURE 9.8 Global warming potential of four different fuel types.

microalgae using flue gas emitted by the power plant as its initial CO_2 demand. This process consumes 2.853 kgs of CO_2 to produce 1 kgs of bio-oil. Two additional processes, biodiesel from microalgae (Lardon et al. 2009) and fossil fuel diesel (OBorn 2012) have been studied in the literature and may help us better understand the situation. The final result is depicted in Figure 9.8.

As a result, it is clear that biofuels outperform fossil fuels in terms of GHG emissions. The study concluded that hydrothermal liquefied bio-oil is the best biofuel option available, which can be observed from Figure 9.9.

9.4.5 Normalization of Data

After conducting the entire study, data on energy consumption, production costs, global warming potential, and other environmental impacts were gathered. However, comparing them will be difficult due to the fact that each uses a different unit of measurement. Data normalization is the process of removing redundant data from a dataset. Our study employs max–min normalization, with data normalized to a scale of 0–1, where 0 represents the ideal scenario, and 1 represents the worst-case scenario. The average of each set of data can be compared to determine which type of fuel is best for your needs. The normalized score is shown in Table 9.3.

9.4.6 Compare Different Type of Biofuel

Microalgae-based biofuels will be compared in this study to see how they compare in terms of environmental impact. It was used as a base for the simulation of the conversion process according to the literature. The results show that bio-oil has the

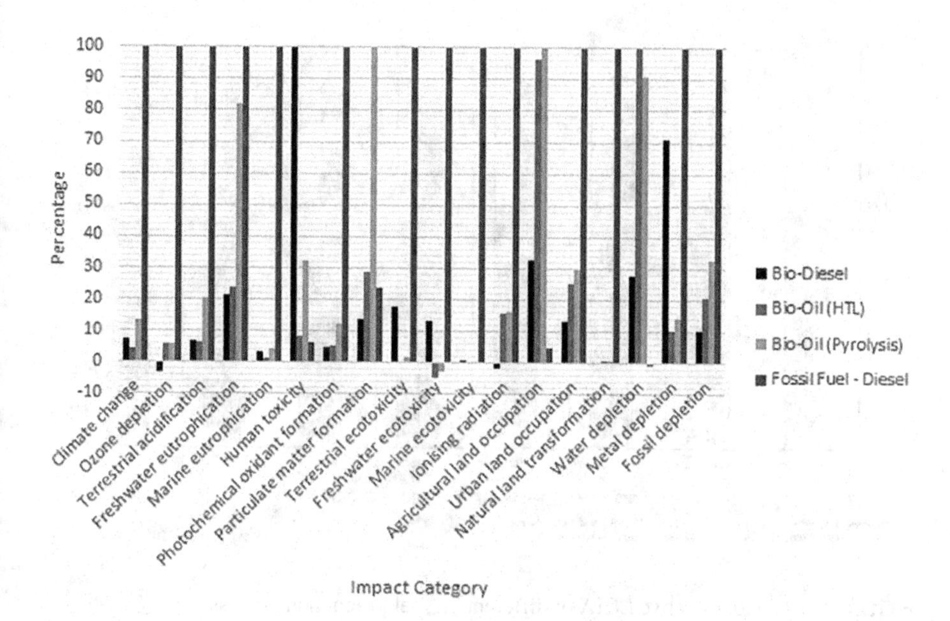

FIGURE 9.9 Life cycle impact assessment—ReCiPe midpoint indicator (H).

TABLE 9.3
Normalized Data Score of Four Fuel Types

Impact Category	Biodiesel	Bio-Oil – HTL	Bio-Oil – Pyrolysis	FossilFuel–Diesel
Climate change	0.033	0	0.093	1
Ozone depletion	0	0.087	0.089	1
Terrestrial acidification	0.007	0	0.147	1
Freshwater eutrophication	0	0.032	0.772	1
Marine eutrophication	0.022	0	0.031	1
Human toxicity	1	0.02	0.276	0
Photochemical oxidant formation	0	0.008	0.08	1
Particulate matter formation	0	0.175	1	0.115
Terrestrial ecotoxicity	0.175	0	0.013	1
Freshwater ecotoxicity	0.172	0	0.019	1
Marine ecotoxicity	0.005	0	0.002	1
Ionizing radiation	0	0.172	0.177	1
Agricultural land occupation	0.289	0.959	1	0
Urban land occupation	0	0.136	0.189	1
Natural land transformation	0	0.004	0.005	1
Water depletion	0.716	0	0.093	1
Metal depletion	0.676	0	0.044	1
Fossil depletion	0	0.119	0.252	1
Net Energy ratio (NER)	0.367	0.328	1	0
Average normalized score	0.182	0.107	0.278	0.796

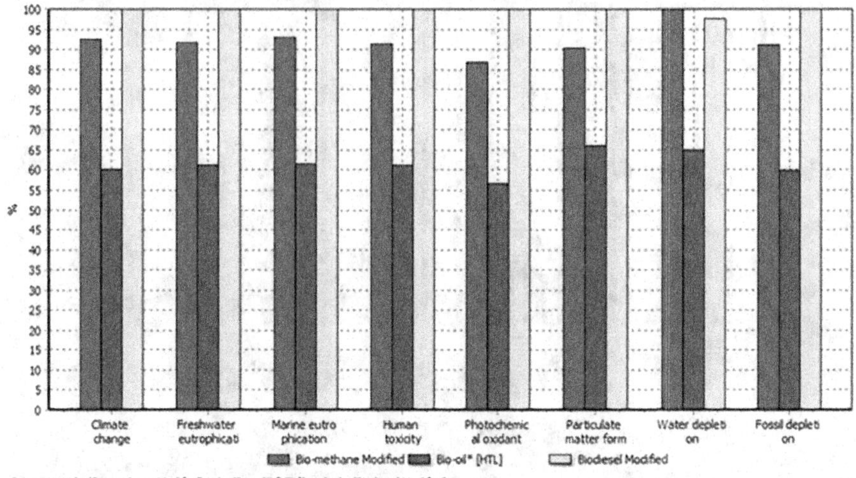

FIGURE 9.10 Comparative LCIA of different biofuel production process.

lowest environmental impact of all the alternatives. Figure 9.10 shows the results of an impact analysis study. Biomethane emits 393 kg CO_2-eq, biodiesel 425 kg CO_2-eq, and bio-oil 255 kg CO_2-eq of greenhouse gases, according to comparisons. Toxicities and resource depletion are also minimized during the production of bio-oil.

9.5 CONCLUSION

Hydrothermal liquefaction of microalgae to produce bio-oil is a viable solution to the global fuel crisis caused by fossil fuel depletion. Additionally, microalgae have the greatest potential for low-emission growth on nonarable land. The results of this study were compared to those of other existing fuel sources, such as biodiesel and fossil fuels, as well as a second conversion process called pyrolysis, in order to determine which was the best. Harvesting requires little energy or chemicals, and thus has a negligible impact. Additionally, this procedure made use of CO_2, which is emitted by a nearby thermal power plant and fed to microalgae. This process is referred to as carbon sequestration, and it is beneficial to the environment. All operations require energy from the power plant, which can be substituted with produced bio-oil. Pyrolysis produces more bio-oil than the HTL process per unit of energy consumed. Thus, pyrolysis is inferior to HTL. In comparison to microalgae biodiesel and fossil fuel diesel, the HTL technique offers the greatest benefits. Additionally, it compares the environmental impacts of microalgae to those of other fuels in order to determine the best alternative. The open-pond cultivation of the plant with carbon sequestration is more sustainable than conventional methods of cultivation. Biodiesel, bio-oil, and biomethane are all compared using the same microalgae feedstock. However, the investigation cannot be concluded without conducting a techno-economic analysis. Due to time and data constraints, this thesis did not address that section. It is strongly recommended, however, that a thorough economic analysis be conducted prior to presenting bio-oil as a viable transportation fuel option. Additionally, this research examines several alternative biofuels, including biohydrogen, syngas, and direct power generation from microalgae combustion. Comparative analysis of those will open up new opportunities for the sector.

REFERENCES

Bennion, E.P., D.M. Ginosar, J. Moses, F. Agblevor, and J.C. Quinn. 2015. "Lifecycle assessment of microalgae to biofuel: Comparison of thermochemical processing pathways." *Applied Energy* 154: 1062–1071. doi:10.1016/j.apenergy.2014.12.009.

Berndes, G., M. Hoogwijk, and R.V. Den Broek. 2003. "The contribution of biomass in the future global energy supply: A review of 17 studies." *Biomass and Bioenergy* 25 (1): 1–28. doi:10.1016/S0961-9534(02)00185-X.

Biddy, M., R. Davis, S. Jones, and Y. Zhu. 2013. *Whole Algae Hydrothermal Liquefaction Technology Pathway.* Technical Report Efficiency & Renewable Energy, Operated by the Alliance for Sustainable Energy, LLC. NREL/TP-5100-58051 PNNL-22314, no. March: 1–10. www.nrel.gov.

Chisti, Y. 2007. "Biodiesel from microalgae." *Biotechnology Advances* 25 (3): 294–306. doi:10.1016/j.biotechadv.2007.02.001.

Das, S., B. Das, B. Chhotaray, R.R. Pradhan, State Pollution, Control Board, National Aluminium, Company Limited, and Indocan Technology Solution. 2013. "Characterization of algal biomass cultivated in flue gas of coal fired power plant at," no. June: 2013. doi:10.13140/2.1.1897.0886.

Fortier, M.O.P., G.W. Roberts, S.M. Stagg-Williams, and B.S.M. Sturm. 2014. "Life cycle assessment of bio-jet fuel from hydrothermal liquefaction of microalgae." *Applied Energy* 122 (July 2011): 73–82. doi:10.1016/j.apenergy.2014.01.077.

Hassan, M.H., and M.A. Kalam. 2013. "An overview of biofuel as a renewable energy source: Development and challenges." *Procedia Engineering* 56: 39–53. doi:10.1016/j.proeng.2013.03.087.

Lardon, L., A. Hélias, B. Sialve, J.P. Steyer, and O. Bernard. 2009. "Life-cycle assessment of biodiesel production from microalgae." *Environmental Science and Technology* 43 (17): 6475–81. doi:10.1021/es900705j.

Mata, T.M., A.A. Martins, and N.S. Caetano. 2010. "Microalgae for biodiesel production and other applications: A review." *Renewable and Sustainable Energy Reviews* 14 (1): 217–32. doi:10.1016/j.rser.2009.07.020.

Mofijura, M., H.H. Masjuki, M.A. Kalam, M. Shahabuddin, M.A. Hazrat, and A.M. Liaquat. 2012. "Palm oil methyl ester and its emulsions effect on lubricant performance and engine components wear." *Energy Procedia* 14 (December): 1748–53. doi:10.1016/j.egypro.2011.12.1162.

OBorn, R. 2012. "From ground to gate: A lifecycle assessment of petroleum processing activities in the United Kingdom," no. June: 1–80.

Rilling, J.I., J.J. Acuña, M.J. Sadowsky, and M.A. Jorquera. 2018. "Putative nitrogen-fixing bacteria associated with the rhizosphere and root endosphere of wheat plants grown in an andisol from southern Chile." *Frontiers in Microbiology* 9 (Nov): 1–13. doi:10.3389/fmicb.2018.02710.

Schenk, P.M., S.R. Thomas-Hall, E. Stephens, U.C. Marx, J.H. Mussgnug, C. Posten, O. Kruse, and B. Hankamer. 2008. "Second generation biofuels: High-efficiency microalgae for biodiesel production." *BioEnergy Research* 1 (1): 20–43. doi:10.1007/s12155-008-9008-8.

Shafiee, S., and E. Topal. 2010. "A long-term view of worldwide fossil fuel prices." *Applied Energy* 87 (3): 988–1000. doi:10.1016/j.apenergy.2009.09.012.

Sims, R.E.H., W. Mabee, J.N. Saddler, and M. Taylor. 2010. "An overview of second generation biofuel technologies." *Bioresource Technology* 101 (6): 1570–80. doi:10.1016/j.biortech.2009.11.046.

Stefanidis, S.D., E. Heracleous, D.T. Patiaka, K.G. Kalogiannis, C.M. Michailof, and A.A. Lappas. 2015. "Optimization of bio-oil yields by demineralization of low quality biomass." *Biomass and Bioenergy* 83: 105–15. doi:10.1016/j.biombioe.2015.09.004.

Stephenson, A.L., E. Kazamia, J.S. Dennis, C.J. Howe, S.A. Scott, and A.G. Smith. 2010. "Life-cycle assessment of potential algal biodiesel production in the United Kingdom: A comparison of raceways and air-lift tubular bioreactors." *Energy and Fuels* 24 (7): 4062–77. doi:10.1021/ef1003123.

Uffenorde, and Much. 1918. "Eine Kriegsepidemiologische Beobachtung. I. Klinischer Teil." *Deutsche Medizinische Wochenschrift* 44 (3): 57–59. doi:10.1055/s-0028-1134195.

APPENDIX

Inventory Table: Inventory Data Sheet for HTL method

Step	Value	Unit
CO_2 from Flue Gas (water)		
Energy intake	1.5	hp
	1.119	kW
Water		
Flow in	520	kg/h
Temperature	25	°C
	298	K
Enthalpy	−7.816187	MMBtu/h
Flue Gas		
Flow In	100	kg/h
Temperature	140	°C
	413	K
Enthalpy	−0.2630599	MMBtu/h
CO_2	13	kg/h
N	68.99579	kg/h
O_2	4.999693	kg/h
CO	0.0003999756	kg/h
NO_2	0.004499725	kg/h
SO_2	0.001199927	kg/h
Water	12.9992	kg/h
CO_2 from Flue Gas (AIR)		
Energy intake	1.5	hp
	1.119	kW
Air		
Flow in	2170	kg/h
Temperature	25	°C
	298	K
Enthalpy	−3.976559	MMBtu/h
Flue Gas		
Flow in	100	kg/h
Temperature	140	°C
	413	K
Enthalpy	−0.26306	MMBtu/h
CO_2	12.99921	kg/h
N	68.99577	kg/h

(Continued)

Step	Value	Unit
O_2	4.999695	kg/h
CO	0.00039999757	kg/h
NO_2	0.004499726	kg/h
SO_2	0.001199927	kg/h
Moisture Content	0.13	
Open-Pond Cultivation		
Energy intake	3 (2 hp fan + 1 hp motor/pump)	hp
	2.238	kW
CO_2 uptake efficiency	70–90	%
Evaporative water loss	150[b]	cm/year
Operative water loss	0.05	
Biomass productivity	22	Tons/acre/year
	18.894	$g/m^2/day$
Biomass lipid content	0.1–0.15	
Cycle period	20	days
Fertilizer (BG-11)[a]	50	g/kg of dry biomass
1. $NaNO_3$		
2. K_2HPO_4		
3. $MgSO_4,7H_2O$		
4. $CaCl_2.2H_2O$		
5. Na_2CO_3		
6. Na_2Mg-EDTA		
7. Citric acid		
8. Ferric ammonium citrate		
1.	46.1	g
	0.965	g
	1.106	g
	1.127	g
	0.615	g
	0.031	g
	0.172	g
	0.184	g

[a] http://www.himedialabs.com/TD/M1541.pdf, HI MEDIA LAB.
[b] Water Resource Systems Division, National Institute of Hydrology, Roorke – 247767.

Hydrothermal Liquefaction			**Source**
Catalyst ($NaCO_3$)	0.039	kg/kg of Algae	Bennion et al. (2015)
Temperature	350	°C	
Energy intake	5.919	MJ/kg of Algae	Bennion et al. (2015)
Heat recovery	70 (60–80)	%	Fortier et al. (2014)
Bio-Oil yield	0.37	kg/kg of Algae	Bennion et al. (2015)
Bio-Oil stabilization			
Super critical fluid processing	2.15	MJ/kg of Bio-oil	Bennion et al. (2015)
Propane losses	0.02	kg/kg of Bio-oil	Bennion et al. (2015)
Stabilized oil yield	84.6	%	Bennion et al. (2015)
Hydroprocessing			
Hydrogen	0.0317	kg/kg of Stabilized Bio-oil	Fortier et al. (2014)
Hydrogen production	56.96	MJ/kg of Hydrogen	Bennion et al. (2015)
Hydroprocessing	0.8381	MJ/kg of Stabilized Bio-oil	Bennion et al. (2015)
Catalyst (Zeolite)	0.0004	kg/kg of Stabilized Bio-oil	Bennion et al. (2015)
Refined oil yield	90	%	

SimaPro 8.0.3.14 Impact assessment

Project: Bio-oil

Results: Impact assessment

Product: 1 kg Bio-oil by HTL process

Method: ReCiPe Midpoint (H) V1.10/World Recipe H

Impact Category	Unit	Total	Hydrothermal Liquefaction Process	Bio-oil Stabilization	Bio-oil Refining (Hydroprocessing)	Algae Cultivation
Climate change	kg CO_2 eq	253.678	8.811781	1.656579	1.365052	241.8446
Ozone depletion	kg CFC-11 eq	5.14E-07	1.31E-08	8.48E-09	1.9E-09	4.9E-07
Terrestrial acidification	kg SO_2 eq	1.598406	0.060396	0.009837	0.009248	1.518925
Freshwater eutrophication	kg P eq	0.119671	0.019685	0.002938	0.003052	0.093996
Marine eutrophication	kg N eq	0.052381	0.005061	0.000777	0.000775	0.045768
Human toxicity	kg 1,4-DB eq	78.89783	12.75044	1.897563	1.969505	62.28032
Photochemical oxidant formation	kg NMVOC	0.7998	0.02731	0.004936	0.004206	0.763348
Particulate matter formation	kg PM10 eq	0.951705	0.380578	0.057071	0.059089	0.454966
Terrestrial ecotoxicity	kg 1,4-DB eq	0.002803	0.000111	1.89E-05	1.57E-05	0.002658
Freshwater ecotoxicity	kg 1,4-DB eq	1.896036	0.322537	0.047988	0.049856	1.475654
Marine ecotoxicity	kg 1,4-DB eq	1.846219	0.308478	0.046061	0.047665	1.444015
Ionizing radiation	kBq U235 eq	2.456648	0.038115	0.013733	0.005931	2.398869
Agricultural land occupation	m2a	8.03834	0.357803	0.01723	0.017923	7.645384
Urban land occupation	m2a	2.205142	0.028851	0.003031	0.003153	2.170107
Natural land transformation	m2	0.010286	0.00018	1.81E-05	1.88E-05	0.010069
Water depletion	m3	49.76047	0.807553	0.095223	0.098921	48.75877
Metal depletion	kg Fe eq	1.821548	0.059338	0.006112	0.005885	1.750212
Fossil depletion	kg oil eq	55.16977	1.974358	0.670766	0.305909	52.21874

Basic Composition of Different Microalgae Species

Microalgae Species	Protein (% wt)	Carbohydrates (% wt)	Lipids (% wt)	Nucleic Acid (% wt)	Others (% wt)	Source
Chlorella vulgaris	51–58	12–17	12–14	4–5	5–20	Chisti (2007)
Spirulina sp.	46–71	8–16	4–9	2–5	0–20	Lardon et al. (2009)
Nannochloropsis sp.	27–38	25–38	12–19	–	12–20	Rilling et al. (2018)
Local blue-green microalga	30.2	48.4	13.3	–	–	NALCO India

Inventory of Different Cultivation Pathways

Item	Open Ponds	Tubular	Flat-Plate PBR
Electricity, kWh	0.42	0.23	1.94
CO_2, kg	1.97	1.17	–
N, kg	0.07	0.01	0.07
P, kg	0.01	0.0002	0.01
Water, m^3	7.1	0.25	0.37
Growth, $g/m^2.d$	19	25	27
Concentration, kg/m^3	0.5	0.5	2.7
Oil content, % wt	40	40	30
Source	Lardon et al. (2009), Chisti (2007)	Collet et al. (2010), Chisti (2007)	Rilling et al. (2018)

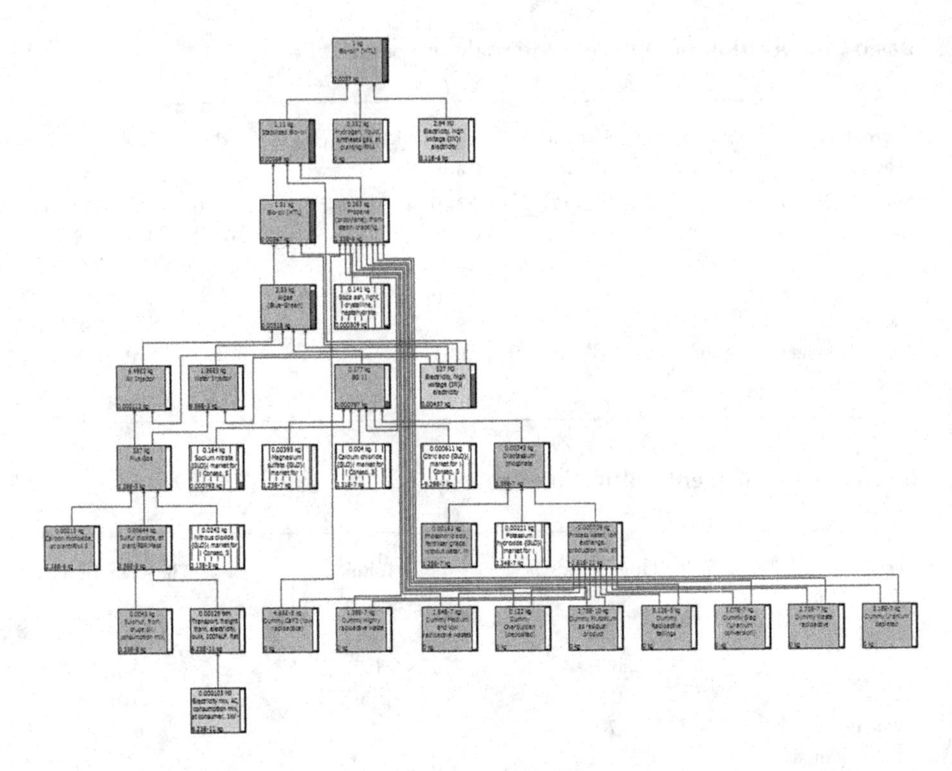

System tree generate by SimaPro 8.0.3 for Bio-oil production through Hydrothermal Liquefaction Method.

10 Constraints for Biofuel Production from Microalgae

C.K. Madhubalaji
Ozone Research Applications India Private Limited

CONTENTS

DOI: 10.1201/9781003197737-13

10.1 INTRODUCTION

The dependence on fuels is enormously increasing along with the rise in the global population. Various modes of transport facilities have increased dependence on fuels. Fossil fuel reserves are continuously decreasing. Fossil fuels are unsustainable energy sources that impact air, water, land degradation, and climate change; reducing fossil fuel reserves increases the competitiveness among the countries. The amount of fossil fuel consumption through transport challenges the world to look for alternatives. As fuel is one of the essential requirements for transport, there are no other thoughts, all the countries are ready to pay and import the fuels, and most of the countries in the world are importing the fuels. A drastic increase in fuel prices strongly influences biofuel production and supports alternative energies. The US was mandated to reduce its dependence on foreign oil imports during these price increases. Instead of fossil fuels, marine algae-based production of biofuels can provide environmental benefits, economic opportunities, and employment; therefore, interest in marine resources for biofuel production has increased. Hence, sustainable resources are effective alternatives to achieving future biofuel needs. However, microalgae are proposed as a sustainable biofuel resource (Biodiesel, bioethanol, biogas, bio-oil, biohydrogen, bioelectricity, etc.) to overcome future fossil fuel scarcities. Aspects that encouraged the researchers and entrepreneurs on biofuel production using marine resources such as algae are: (1) It can be grown throughout the year, (2) grow with natural sunlight and atmospheric CO_2, (3) higher productivity per acre of land, (4) in third-generation biofuel, nonfood-based bio stocks are considered, (5) usage of nonarable land, (6) grow in any water (marine, fresh, brackish, and wastewater), (7) in addition to biofuel, it will give other valuable co-products (metabolites), (8) biofuel is environment friendly.

Microalgal biofuels have become one of the world's leading research areas with enormous benefits for humans and the environment. Over the past four decades, significant research has been conducted on microalgae and their role in biofuel production. Researchers have explored the various aspects such as from identification to cultivation, harvesting to purification to use algae as renewable and alternative biomass feedstock. There are a series of steps involved in microalgae-based biofuel production; each stage has certain constraints. However, critical scrutiny of the issues facing during the microalgal biofuel production has not been addressed, resulting in an unclear focus and direction for developing microalgal biofuels. In theory, microalgae can convert an average of 12.8%–14.4% of solar radiation into biomass. The yield is 77 $g/m^2/day^1$, providing about 280,000 kg/ha/year of biomass. However, algal biomass yields achieved to date are much lower. Further, the major problem involved is that no sustainable, scalable, and commercially viable systems have been established to use marine algae for biofuel production in the global market, indicating no benchmark for commercial biofuel production.

10.2 SIGNIFICANT CONSTRAINT STEPS IN THE MICROALGAL BIOFUEL PRODUCTION PROCESS

Microalgae based biofuel production facing different types of challenges are presented in Figure 10.1.

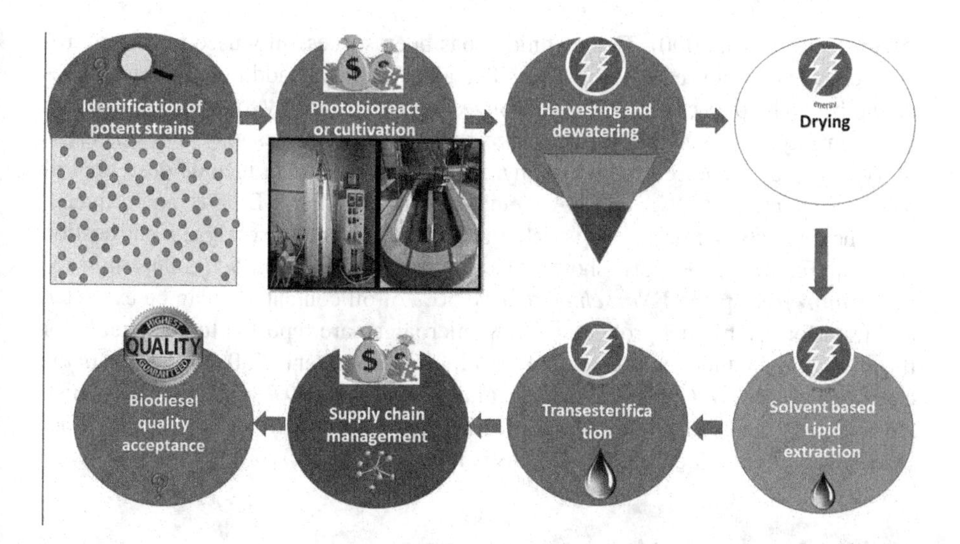

FIGURE 10.1 Challenges at different stages of microalgal-based biofuel production process.

10.2.1 Potential Microalgal Biomass Identification for Biofuel Production

Microalgae grow majorly in natural environments, *viz.*, water, soil, and rocks. In addition, they also can be found and collected in general aquatic ecosystems, such as lakes, rivers, oceans, and extreme environments. The strain selection for higher biofuel production is one of the challenging tasks. Even though many species of microalgae have been isolated and identified for the production of lipid, there is no accord on which species is the most productive. Different species are anticipated to perform greatest in different aquatic and geographic conditions. Further, value-added products are currently used for commercial production, which requires the choice of the best performing algae appropriate for use in multiproduct algal biorefineries. Combining traditional and modern methods may be the most competent, from isolation to pilot-scale cultivation. Selecting microalgal strains with biodiesel production potential, and isolation is necessary for acquiring pure culture and is the first step towards pure culture. Traditional isolation techniques include separation under a microscope using a micropipette or dilution of cells and culturing in liquid agar plates or nutrient broth . Conventional methods of isolating single cells based on raw samples are time consuming. They require sterilized media and equipment, and this laborious process results in a pure culture that is often easily identifiable. Another approach used in the lab scale is to enrich some microalgal strains by adding nutrients to increase algal growth. The essential nutrients for algal growth are phosphate and nitrogen, and certain types of algae may require trace elements to grow. Although automated separation techniques have advantages over traditional methods, single-cell separation using micropipettes is a very efficient method for various samples. An automated single-cell isolation method was developed, i.e., Flow cytometry, and widely popularized and used for cell counting and sorting (Davey and Kell

1996; Reckermann 2000). This technique has been successfully used to sort microalgal cells from water-containing diversified algal strains. In addition, bioinformatics methods can help discover new algal isolates having the ability to produce biodiesel, and their phylogenetic analysis can suggest which other species may have this capability. The steps required to obtain phylogenetic analysis data include designing of primer, extraction of DNA or RNA, amplification using PCR, DGGE, and sequencing. The diversity of microalgae is rich; therefore, lipid-rich microalgae identification is a major challenge. Reports showed that *Botyococcus braunii*, *Schizochytrium* sp., *Nanochloropsis* sp., and *Nitzschia* sp. have 50% of oil content that can be extracted and used for use biofuel production. Few microalgae are reported to be suitable for the production of biodiesel because of their high lipid content of 50%–70% and in the case of microalga *B. braunii*, which accumulates oil up to 80% of its biomass (Chisti 2007; Powell and Hill 2009; Mata, Martins and Caetano 2010). The list of other microalgae and producing lipid contents is mentioned in Table 10.1.

10.2.2 Nutritional Modes of Cultivation

Microalgae are majorly cultivated by using three nutritional modes: (1) autotrophic, (2) heterotrophic, and (3) mixotrophic conditions. Autotrophs use an energy source as light and an inorganic carbon source as CO_2. At the same time, heterotrophs do not rely on light as an energy source and use organic substrates such as acetate, glucose, glycerol as energy as well as carbon sources. In mixed nutrient cultures, microalgae can be grown by autotrophic and/or heterotrophic pathways based on the light intensity and concentration of organic carbon sources (Mata, Martins, and Caetano 2010). To date, photoautotrophy alone is economically and technically feasible for the cultivation of microalgae on a pilot scale, usually in sunny and free outdoor environments. In addition, photoautotrophic microalgae can sequester carbon dioxide (CO_2) from flue gas and act as a premium carbon source to the culture system. But CO_2 supplementation has its limitations. Further, especially in moderate countries, the right intensity of sunlight is not continuously available throughout the year.

Researchers have practiced various induction strategies for higher biomass and lipid production in microalgal cultivation, and they are briefly described.

10.2.3 Strategies for Induction of Lipid Content in Microalgal Biomass

The strategies for maximized biomass and lipid production are critical to improve the economics of microalgae-based biofuels production. Traditional methods to increase microalgal biomass and lipid productivity include controlling nutrients (N and P) and environmental factors (i.e., temperature, light, and salinity).

10.2.3.1 Nutrient Stress

Nitrogen starvation is the most successful lipid-inducing method in microalgae. But the lipid accumulation affected by nitrogen nutrition starvation manifests only after 2–5 days and is often accompanied by slow biomass growth rates. The use of nitrogen assimilation chemical inhibitors is another route to induce environments similar to

TABLE 10.1
Potential Microalgal Species Reported to Use as a Source of Biofuel

S.No	Microalgae	Specific Growth Rate (day^{-1})	Lipid (%)/ Lipid Productivity (mg/L/day)	Biomass Productivity (g/L/day)	Reference
1	*Chlorella vulgaris*	0.45	16.01%	0.15	Madhubalaji et al. (2020)
2	*Parachlorella kessleri*	–	66%	0.82	Takeshita et al. (2018)
3	*Scenedesmus quadricauda*	–	38.61%	0.36	Song and Pei (2018)
4	*Botryococcus braunii*	0.14	17.85%	0.346	Órpez et al. (2009)
5	*Botryococcus braunii*	0.11	36.14%	0.034	Sydney et al. (2011)
6	*Chlamydomonas reinhardtii*	0.564	25.25%	2	Kong et al. (2010)
7	*Chlorella protothecoides*	–	55.2%	2.02	Xu, Miao, and Wu (2006)
8	*Chlorella sp,*	0.86	66.1%	-	Hsieh and Wu (2009)
9	*Isochrysisgalbana*	–	22.34%	0.17	Su et al. (2007)
10	*Neochloris Oleabundans*	–	52%	0.03	Gouveia et al. (2009)
11	*Amphora sp.* MACC4	0.45	105[a]	0.105	Sabu, Bright Singh, and Joseph (2017)
12	*Biddulphia sp.* MACC6	0.35	60.82[a]	0.06	Sabu, Bright Singh, and Joseph (2017)
13	*N. phyllepta* MACC8	0.58	114.38[a]	0.431	Sabu, Bright Singh, and Joseph (2017)
14	*Nitzschia sp.* MACC11	–	49[a]	0.224	Sabu, Bright Singh, and Joseph (2017)
15	*Picochlorum sp.* MACC13	0.31	36.4[a]	0.151	Sabu, Bright Singh, and Joseph (2017)
16	*Prymnesium sp.* MACC15	0.38	40.82[a]	0.194	Sabu, Bright Singh, and Joseph (2017)

[a] Indicates lipid productivity.

nitrogen limitation/starvation, as shown in *Chlamydomonas reinhardtii*. For example, treatment with methionine sulfoximine inhibitor doubled the neutral lipid content after one day of treatment in *Chlamydomonas reinhardtii,* while treatment with nitrogen starvation showed an effect after three days. Further, a study found that only triacyl glycerides accumulated in *T. obliquus* after the starch synthesis rate are inhibited after nitrogen limitation (León-Saiki et al. 2017).

10.2.3.2 Salt Stress

Salt stress is one of the strategies used for microalgae to increase lipid productivity. In large-scale two-phase cultures, adding NaCl rather than changing the medium with a nitrogen-free medium induces stress conditions more efficiently. High salt stress imposed by late NaCl supplementation was shown to increase *Monorapidium dybowskii* LB50 lipid productivity during two-stage cultivation (Yang et al. 2014). Likewise, the application of salt stress at a concentration of 400 mM for three days can increase the lipid and carbohydrate content of *Scenedesmus* CCNM 1077 with the most negligible reduction in biomass based on two-step cultivation (Pancha et al. 2015). Further, it has been reported that microalgae, viz., *Chlorella sorokiana,* as well as *Desmodesmus* sp., can increase lipid production by increasing the concentration of $CaCl_2$ (Srivastava and Goud 2017). A mechanistic explanation is given, whereby Ca triggers a signaling pathway for accumulating neutral lipids, which act as a defense mechanism against osmotic stress-induced cellular damage. Salt treatment affects the microalgae fatty acid profile, especially increased oleic acid levels. In another study, control of carbonate, nitrate, and iron levels as a feasible strategy for increasing the biomass and lipid levels of three microalgae (i.e., *Scenedesmus, Chlorella,* and *Chlamydomonas*) isolated from quarry pond water (Sivaramakrishnan and Incharoensakdi 2017). And concluded that *Scenedesmus* sp. is the best suited for the production of biodiesel among the three strains, as it has the desired fatty acid profile and tolerates higher concentrations of sodium carbonate.

10.2.3.3 Temperature Stress

Temperature is another critical factor for microalgal growth, directly affecting biochemical processes in algal cell factories, including photosynthesis. Every individual strain has its optimal temperature for its growth. Raising the temperature to the ideal range can increase exponential algal growth; however, beyond the optimal point raising/lowering the temperature delays or even stops algal growth (Béchet et al. 2017). A temperature range of 20°C–30°C is reported as optimal for most algae (Singh and Singh 2015). However, thermophilic algae, i.e., *Chaetoceros* and *Anacystis nidulans,* can withstand temperatures as high as 40°C, and algae grow in hot springs at temperatures close to 80°C (Covarrubias et al. 2016). Growing microalgal cultures at suboptimal temperatures results in substantial biomass loss, especially in open-air culture systems (Lee et al. 2015; Hu, Zhang, and Sommerfeld 2006; Alabi, Tampier, and Bibeau 2009). Temperature is an essential factor in pilot-scale cultivation of microalgae; majorly open pond cultivation requires careful monitoring as algae experience significant changes in temperature over time (Bechet et al. 2010). As a stress treatment, the temperature can also be used to induce the production of lipids and conversion of valuable metabolites. *Chlorella vulgaris* cultures grown at 25°C produced more carbohydrates and lipids than those grown at 30°C. Temperatures between 27°C and 31°C were optimal for several microalgae species (Converti et al. 2009). Low temperatures reduce carbon assimilation activity and impair photosynthesis, while excessively high temperatures disrupt the energy balance in cells and inactivate photosynthetic proteins, which reduce photosynthesis. Higher temperatures also decrease respiration and cell size. The decreased photosynthesis results in a reduced growth rate (Atkinson, Ciotti, and Montagnes 2003). The main temperature effect on photosynthesis is because of the decreased activity of the bifunctional enzyme rubisco (ribulose 1,5-diphosphate). Further, based on the presence of relative O_2 and CO_2 quantities in the chloroplast, rubisco can act as a carboxylase/oxygenase. An increase in temperature can increase Rubisco enzymes' CO_2 fixation activity up to a level; after that, it decreases (Salvucci and Crafts-Brandner 2004). Therefore, the temperature is considered a limiting factor for algal growth rate by affecting the affinity of ribulose towards CO_2.

10.2.3.4 pH

The algal cultivation medium pH is another essential factor that affects microalgal growth. Microalgal strains have different requirements for pH; however, most of the algal optimum pH range is 6–9 (Lam and Lee 2012). Different growth media of microalgae have various pH based on their composition. Most algal strains are pH-sensitive, and few tolerate a wide pH range. *C. vulgaris* can grow over a wide pH range, but maximum growth rates and biomass productivity have been reported at pH 9–10 (Daliry et al. 2017). pH increases the salinity of the medium and is very detrimental to algal cells (Juneja, Ceballos, and Murthy 2013).

10.2.3.5 Light Stress

Controlling light conditions according to light intensity and wavelength is one of the critical strategies to increase microalgal biomass and lipid production. Light-emitting diodes (LEDs), color covers, and dyes are existing practices that provide light with the specific intensity and wavelength that maximize microalgae growth (Ramanna, Rawat, and Bux 2017), for example, exposure to red wavelength LEDs increased *Chlorella* biomass and lipid production in experiments (Severes et al. 2017). Exposing *T. oblique* to different LED flashlight (red and blue light) frequencies have increased lipid accumulation (Choi, Moon, and Kang 2015). Further, incubation of *Cyclotella cryptica* at yellow wavelengths, i.e., 580 nm, and alternating exposure to blue wavelengths, i.e., 450 nm, resulted in an increase of lipid content twofold and fourfold, respectively (Shih et al. 2014). Organic dyes used to change the spectrum are another strategy to improve microalgal growth and lipid accumulation (Seo et al. 2015). Besides wavelength, light intensity is another factor that affects the growth and lipid production of microalgae. High light-induced in *Chlorella* sp. and *Monoraphidium* sp. showed that carbon distribution for lipid synthesis (primarily neutral lipid) was higher than in carbohydrates (He et al. 2015).

10.2.3.6 CO_2 Supplementation

More than 50% of algae biomass is made of carbon, and it is essential to provide adequate carbon for microalgae cultivation (Show et al. 2017). In phototrophic microalgae cultivation, carbon is majorly provided as carbon dioxide. It is assessed that 1.7–1.8 g of CO_2 is required for 1 g of algal biomass production, carbon dioxide is necessary to produce lipid-rich microalgae, and 3 g of Carbon dioxide may be needed per gram of biomass (Morweiser et al. 2010). Low carbon dioxide concentrations in the air are generally insufficient to ensure the high productivity of algal biomass. While 1%–5% of carbon dioxide concentrations can usually support maximum growth of microalgae and 5%–15% CO_2 is provided to the laboratory experimental cultures. Several research studies have shown that high levels of CO_2 can increase microalgal biomass and lipid production. Higher CO_2 concentrations, i.e., 30%–50%, can promote lipids, especially polyunsaturated fatty acids accumulation in *Chlorella pyrenoidosa* SJTU-2 and *Tomopterus obliquus* SJTU-3 (Tang et al. 2011).

Latest strategies to increase microalgal biomass and lipid productivity include various approaches, i.e., genetic and metabolic manipulation, the addition of phytohormones, and co-cultivation of microalgae with yeast and bacteria.

10.2.3.7 Synthetic Biology for Improving Microalgal Properties

For biofuel production, unmodified model microalgae identification, which is suitable for already established harvesting, extraction, and purification infrastructure, is economically viable. But the expected scenario is to identify a different species that have emerged, each exhibiting one or more of these necessary characteristics. When all these characteristics are incorporated into a single microalgal strain, it may be an economically viable option for biofuel production. Besides improving strain for fuel production, genes identified from other algal strains can improve the expression of heterologous proteins, resulting in higher biomass and lipid production. Transgenic microalgae can be produced that can grow at high CO_2 concentrations, are resistant to pollutants present in flue gas/sewage, and can produce cells with high lipid content. Progress in this area of research is urgently needed to achieve significant breakthroughs in producing greener, more sustainable microalgal biofuels.

10.2.4 Microalgal Cultivation in Photobioreactors

In addition to nutritional supplements, systems used for microalgae cultivation also have an essential role in industry success or failure. An efficient and effective cultivation system should meet the following conditions: (1) Ease of operation, (2) effective lighting area, (3) minimal land requirements, (4) low pollution levels, (5) optimal gas–liquid transfer, and (6) low capital and production cost (Xu et al. 2009; Madhubalaji et al. 2019).

Microalgae are cultivated in raceway ponds as well as photobioreactors. These photobioreactors are costly, and this can be overcome by cultivating the microalgae in raceway ponds. Only the raceway pond will be commercialized for biofuel production from microalgae. The open raceway pond system contains a paddle wheel that avoids the precipitation of microalgal biomass and injects carbon dioxide from the bottom of the airstrip as a carbon source. Even though the raceway pond system is a comparatively easy operation and requires less energy, the system is more prone to contamination by harmful microorganisms that ultimately threaten microalgae survival. In addition to this, an open raceway pond system results in more water loss because of evaporation. The use of closed and open cultivation system facilities for commercial algal production has advantages and disadvantages, but both necessitate high capital investments. Comparatively closed photobioreactors are more expensive but not as mature as reactors used in commercial practice, so chances for substantial cost reductions may arise. Further, neither closed photobioreactors nor open ponds are developed technologies as a result; many uncertainties remain, until large industries are up and running for several years. Running at pilot scale can addresses open and closed system culture issues, such as reactor material of construction, optimal culture scale, mixing, evaporation, heating/cooling, oxygen enrichment, and carbon dioxide management. Further, another practice to enhance algal biomass and lipid production is using hybrid systems like combined system containing open ponds and a closed photobioreactor.

10.2.4.1 Contamination during Microalgal Cultivation

Contamination is the most typical problem faced during microalgae cultivation in open pond systems. Contamination can come in pathogens (viruses and bacteria), herbivores (copepods, cladocerans, rotifers, etc.), and competing with microalgal growth. To reduce the effects of herbivores and pathogens, chemical and ecological methods are in practice. Contamination from competing microbes for algal species is an inevitable sign of operating a microalgal culture system. Contamination with bacteria oxidizes organic matter and actively competes for nutrients, leading to culture spoilage. Heterotrophic bacteria can be controlled by increasing pH. The optimum pH for aerobic bacteria commonly found in the microalgal pond system is 8.3. Raising pH above this level results in preventing competition by affecting nitrogen efficiency and effective inhibition. Open pond systems are also vulnerable to grazing animals in the form of zooplankton and protozoa. Although culture practices combining both biomass and lipid production are being studied, ideal microalgal culture strategies include a high nutrient phase for biomass production followed by a nutrient-poor phase for lipid production. The first phase created an environment where rapid growth paid off; in the second stage, lipids are produced.

After microalgal growth in ponds and bioreactors, when the microalgae reach their stationary phase, it needs to be processed through the following steps: harvesting, dehydration, and extraction of lipids and carbohydrates. However, these steps are considered critical and energy-intensive processes. Further, algal biodiesel is impossible because of the high cost associated with harvesting, conversion, operating, and maintenance.

10.2.5 Biomass Harvesting

Harvesting of microalgae is considered the main challenging and complex technical task in producing biodiesel using microalgae, because these are small microorganisms (usually 1–20 μm). Microalgae are usually single-cellular, low-density microbes and are in suspension form that makes separating cells from the medium as challenging. Several methods are used for microalgal harvesting: (1) Bulk Harvesting—the separation of microalgae through gravity settling, flocculation, and flotation; (2) thickening—the algal slurry concentration using flotation and centrifugation (Chen et al. 2011). Under known processing strategies, cultures containing as low as 0.02%–0.07% algae (~1 g algae/5,000 g water) must be concentrated in a slurry containing at least 1% algae. The final slurry concentration depends on the extraction method and affects the required energy input. As the proportion of dry biomass required increases, energy costs rise sharply. The final slurry concentration is also influenced by transportation, water quality, and recycling issues. Therefore, viable algal fuel strategies must consider energy costs and site issues related to harvesting and drainage. In algal biomass production downstream unit processes, an essential economic factor is associated with cell harvesting. It is assessed that the harvesting process contributes to 20%–30% of the overall cost of biomass production. In phototrophic cultures, the algal biomass concentration typically reaches up to 0.5–1.0 g/L in open ponds, while in closed bioreactor systems the algal biomass concentration typically reaches up to 5–10 g/L, whereas in case of yeast or bacterial fermentation can reach cell concentration beyond 100 g/L. Therefore, microalgal biomass production is still low compared to bacterial or yeast fermentation. To produce 1 g/L of algal biomass, 1,000 kg of water must be used to sequester 1 kg of microalgal biomass. In harvesting biomass, the major challenge is low cell densities and concentrating cells so that lipids can be extracted (up to 1,000 times) with a low-cost unit process. Hence, centrifugation, like an energy-intensive process, is practicable for high-value products. However, it is reported that the most promising low-cost method for producing biofuels from algae is to use gravity settling, further enhanced by flocculation, using flocculants. Using chemical flocculants at lower amounts can be cost-effective to aid this process, depending on the amount used.

10.2.6 Drying

In contrast to terrestrial energy crops, the production of biofuels requires extensive drying of microalgal biomass due to the water presence, which inhibits downstream processes, such as extraction of lipid and conversion (via transesterification) processes. However, since drying through solar energy is considered an excellent method for drying wet microalgal paste after harvesting, no significant concerns have been raised. However, because of limited solar radiation at different seasons, drying through solar energy is not possible in temperate countries and should be taken seriously. Considering this scenario, fossil fuel-generated heat is needed to continuously dry the microalgal biomass to ensure optimal biomass retention in each culture cycle. Further, LCA studies have emphasized that natural gas as a fuel for dry microalgal biomass consumes approximately 69% of the total input energy, resulting in a negative energy balance in microalgal biofuel production (Sander and Murthy 2010).

10.2.7 Solvent-Based Extraction of the Biofuels

The dried microalgal biomass lipid extraction process contributes energy expenditure relatively small fraction (approx.5%–10%), but the process is still essential. Low lipid-containing microalgae especially require efficient extraction of lipids, as the loss of lipids during extraction substantially impacts microalgal biofuels production cost (Ranjan, Patil, and Moholkar 2010). Unlike terrestrial crops, microalgal biomass lipid extraction is comparatively difficult because of their thicker cell walls, which avoids intralipid release. Therefore, generally used mechanical presses to extract oil from terrestrial crops efficiently are not suitable to use with microalgal biomass. The two lipid extraction methods are primarily in practice: (1) for dry microalgal biomass solvent-based extraction and (2) for wet pasty microalgal biomass supercritical liquid-based extraction.

In microalgal biodiesel production, through the transesterification process, algal oil using acids or bases as catalysts convert it into fatty acid methyl esters (the chemical constituents of biodiesel). Mechanical pulverization of algal biomass for extraction of oil is also challenging to implement with existing pulverization equipment. These limitations can be overcome by producing biodiesel by directly transesterifying oily feedstocks. Compared with the extractive transesterification process, oily biomass direct transesterification results in a higher biodiesel yield and higher FAME content. FAME composition and biodiesel yield were significantly affected by transesterification conditions: catalyst loading, temperature, reaction time, and the ratio of biomass to methanol. In addition, reactants' complete mixing is another important factor to consider affecting fuel quality. It has been reported that direct transesterification through produced biodiesel meets most specifications of ASTM such as free glycerol, total glycerol, acid number, soap content, corrosiveness to copper, flash point, viscosity (Johnson and Wen 2009).

10.3 LCA STUDIES

Currently, LCA is extensively recognized as a powerful tool that provides researchers and policy makers with a clear idea of how to uncover the true potential of a given process/product. It can also indicate whether particular product manufacturing processes have any negative environmental phenomena, such as global warming, eutrophication, ocean toxicity, ozone depletion, terrestrial competition, photochemical oxidation, etc.; therefore, preventive measures can be taken. It is recommended to mitigate negative impacts to reduce impact. Besides, energy balances can be calculated to identify and demonstrate energy hotspots for all phases within the LCA process system boundaries. Some reported LCA studies on microalgal biofuels have limited data with parameters relevant to microalgal biofuel production, *viz.*, biomass and lipid content, and their productivity and energy efficiency in downstream processes such as harvesting, drying, and transesterification, which were obtained only from small-scale laboratory experimental data (Sander and Murthy 2010). While the data used in these LCA assessments may not be relevant when applied to pilot-scale production, most LCA studies have determined that biofuel production from microalgae is very energy intensive. Currently, the energy demands of microalgal biomass are high compared to current terrestrial crops and thus are not yet a viable option for commercial biofuel production. LCA studies have highlighted that microalgal biomass harvesting and drying contribute significant energy consumption in microalgal biodiesel production (Sander and Murthy 2010). The results suggest that large-scale harvesting techniques can be essential in reducing energy consumption during microalgal sludge thickening. Multiple LCA studies found that compared to jatropha, oil palm, and rapeseed, the energy conversion efficiency of microalgae is relatively lower, signifying that microalgae-based biofuel production is not sustainable. Further, LCA studies of microalgae biofuel production showed four critical energy-intensive hotspots: (1) source of nutrients (Yang et al. 2011), (2) design of photobioreactor (Jorquera et al., 2010; Stephenson et al., 2010), (3) biomass dewatering and drying (Sander and Murthy 2010), and (4) extraction of lipid (Sander and Murthy 2010).

Energy efficiency ratio (EER) is used to express a sustainable energy index for making a particular product through LCA. EER is defined as the ratio of output energy to input energy. EER ratio greater than 1 represents net positive energy produced and vice versa. EER of biofuels obtained from different feedstocks, including microalgae, was considered in a comparative study. But, it should be noted that rough guide and LCA studies are based on various assumptions and system boundaries. Unexpectedly, they reported that biodiesel made from oleaginous plants was more energy efficient than microalgae showing EER values for oil crop biodiesel greater than 1, while the EER for microalgae-based biodiesel was only 0.07. These quantitative results suggest that microalgae culturing for biofuel production will not necessarily yield positive results and may pose significant risks to unsustainable biofuel production. Furthermore, several aspects such as the water reusability for microalgae recultivation, contaminated wastewater used as a nutrient source, and the conversion efficiency of extraction and transesterification were not clearly defined in these LCA studies (Sander and Murthy 2010). Taking these factors into account, EER values should continue to decline significantly. Active and long-term research efforts are required for commercial viability demonstration, hence it is clear that a comprehensive evaluation of algal fuel research and development and commercialization is incomplete without examining the potential cost of algal fuel technology. However, the costs of algal biofuels production are highly dependent on feedstock prices, TAG content, yield, and several other factors that will help determine algal lipids' performance in a current market.

10.4 CHALLENGES IN LARGE-SCALE SUPPLYING ALGAL BIOMASS

A significant aspect to consider in the supply chain is raw material production. Sustainable biomass is required for large-scale productions, therefore industries need to be developed to produce biomass raw materials in sufficient quantities. Bioengineering is necessary to develop improved algal strains (higher biomass and lipid productivity); year-round biofuel production needs a reliable and constant supply of raw materials (Yue, You, and Snyder 2014). Efficient separation of waste feedstocks remains a problem.

Further, one of the typical problems in the operating supply chain of biofuel is the seasonal and annual variation in supplying biomass. However, fuel demand for transportation is present throughout the year. Hence, it is challenging to operate and manage biomass storage for continuous biomass supply. Algal biomass resource storage can be done in different ways. In the biomass supply chain, ambient storage is the cheapest option and can significantly reduce storage costs. However, it may have side effects like degradation of biomass, reduction in calorific value, and potential health risks, mainly due to higher water content. In the bioenergy supply chain, biomass logistics have an essential role. Biomass logistics integrates timely management of operations like raw material collection, storage, and delivery efficiency throughout the year and a supply system that consistently provides biomass with high quality. Biomass collection and delivery need an extensive selection of equipment and methods for inexpensive harvesting, storage, and pretreatment. Shipping costs are one of the current hurdles. In biofuel distribution, infrastructure components' (storage, blending, transport, and delivery) compatibility is critical.

One of the main issues associated with large-scale biomass deployment is energy density. In addition, the form of biomass feedstock significantly affects bulk density biomass transportation economics. Due to this reason, for efficient biomass supply, densification and compaction are considered essential.

10.5 BIODIESEL CHALLENGES AFTER ENTERING THE MARKET

After all the above challenges, when microalgal biodiesel enters the market, biodiesel faces quality challenges associated with its performance.

10.5.1 Engine Performance Challenges Associated with Biodiesel

The biodiesel quality has a significant role in the engine life and performance. The biodiesel quality may not be offered an immediate effect on the engine. But it can be noticeable in the long run, like corrosion, deposits, and damages can accumulate, resulting in the engine's catastrophic failure. However, it is challenging to discover the exact variance between good-quality and poor-quality biodiesel, and reports showed it could have a tremendous effect; **Engine Power:** When using biodiesel, torque and engine power tend to be reduced by 3%–5%. This is because of less energy per unit volume for biodiesel fuel than conventional diesel. **Fuel Efficiency:** Due to the lower energy content of the fuel, fuel efficiency inclines to be lower with biodiesel. Typically, the reduction is the same as the motor peak power reduction (approximately 3%–5%); **Engine Wear:** Biodiesel usage causes short-term engine wear, less than mineral oil diesel. While long-term testing has yet to be announced, biodiesel is expected to reduce engine wear in the long run. **Sediments and blockages:** Biodiesel deposits and blockages have been widely reported but are usually due to poor quality or oxidized biodiesel. Usually, deposit formation in the engine is not a problem due to good fuel quality. **Engine exhaust pollution:** Biodiesel causes far less air pollution because of its higher O_2 content as well as lack of sulfur and "aromatic compounds." Further higher nitrogen oxide emissions result with biodiesel usage. However, by properly regulating the engine, this problem can be minimized. **Performance in Cold Weather:** Cold weather-tested engines often show severe operational challenges, primarily caused by injector choking and/or clogged filters like petrodiesel. Additives used for flow improving and a "winter blend" of biodiesel with kerosene have proven effective in extending the operating temperature range of biodiesel fuels. Pure biodiesel usually works well down to temperatures as low as 5°C (varies widely based on the used oil type). These additives decrease this range by about 5°–8°, and winter mixes have been shown to be operative in temperatures below −20°C. Further, other challenges like residue trace chemicals (methanol) presence in the biodiesel, incomplete processing of the biodiesel, washing of the biodiesel may remain the water residues in the biodiesel, presence of by-products in the production of biodiesel (glycerin), oxidization or polymerization of the biodiesel because of exposing biodiesel to modest to high temperatures or long-term storage. Several pollutants can be reduced when biofuels from different (micro)algae are used, but, in general, an increased NO_x has been reported with increasing combustion chamber temperature. There is a possibility that the usage of biofuels in engines will reduce the engine's performance, critical component damage, and some engine failures. However, experiments conducted with blending in the range of 20%–30% showed similar characteristics of the diesel on engine performance. Further, the evaluation of biodiesel produced by algae or microalgae is an under-researched area, and, to date, some published papers have shown conflicting results. Further, large-scale supply of biomass is hindered by various bottlenecks, including initial raw material costs, involvement of environmental regulations, biomass producers, and sustainability. Therefore, the discovery of solutions to all these problems results in finding solutions to create the raw material biomass of the future around the world.

10.6 POSSIBLE SOLUTIONS FOR BIOFUEL PRODUCTION ASSOCIATED CHALLENGES

Major constraints in algal-based biofuel production are presented in Figure 10.2. At every step understanding each challenge in biofuel production from microalgae gives us the most required solutions.

Unavailability of raw materials (Feedstock): Incompetent resource administration (Government) and management of nonintervention are essential factors for obstructing the expansion of the biomass industry, creating reliable algae biomass feedstock supply chain, and establishing pilot-scale algae production units at various places. **Seasonal and Regional availability and storage issues for biomass:** Seasonal fluctuations lead to fuel prices; due to the biomass with low density, it is difficult to obtain land for harvesting and storage. **The burden on the transport segment:** The transport of moist biomass from the harvesting to the production site becomes disadvantageous and expensive with increasing distance due to biomass moisture. **Conversion plant inefficiencies, shortages of core technologies, and equipment:** Technical barriers are due to a lack of efficient bioenergy systems. Proper pretreatment is needed to prevent loss of calorific value and biodegradation, which raises production costs and equipment investment. **Undeveloped industrial chain:** Long-term raw material supply contracts at reasonable prices is almost impossible. Low profitability is also why many upstream companies lack the impetus for technological reform.

Economic challenges: Raw material procurement costs: To reduce the transportation costs, biomass projects strive to occupy land near the source, resulting in the concentration of biomass projects. **Limited funding, high investment costs:** Due to scattered funds, regular variations in crude oil prices in the international market, poor profitability, and increased market risks, investors rarely invest in a power-producing industry using biomass because these are associated with a higher investment as well as operating costs. Technologies for pretreatment of biomass have additional charges that may not be affordable by farmers and small fuel companies.

Social challenges: Conflicting decisions: Decisions when choosing locations, routes, suppliers, and technologies are critical and require proper communication. Enhanced management and accountability make stakeholders aware of resource use's economic, ecological, and social benefits. **Environmental impacts:** Biomass cultivation can deplete soil nutrients, endorse aesthetic degradation, and increase

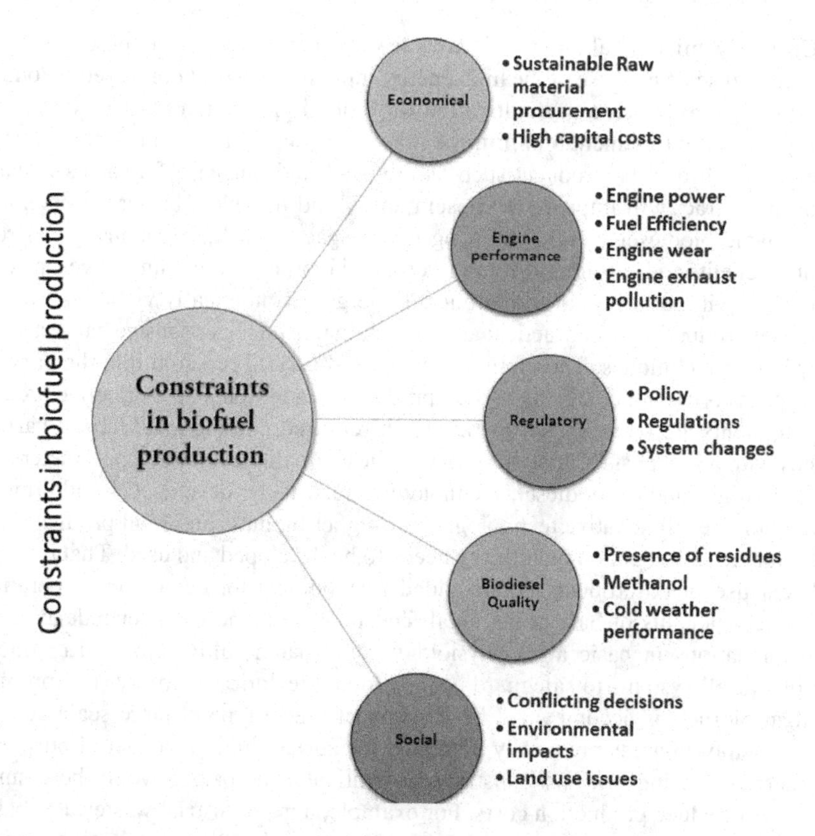

FIGURE 10.2 Overall constraints in biofuel production from microalgae.

biodiversity loss. Installing energy farms in rural areas has other social effects, such as increased service demand, traffic, etc. The possible negative social impact seems adequate to ignore the benefits of creating new jobs (Raychaudhuri and Ghosh 2016).

Regulatory and Policy Challenges: Policy: Currently, the government subsidizes domestic fuel prices, generating electricity. The cost of generating electricity from conventional energy sources is lower than from renewable fuels. **System:** No specific regulations on the utilization of biomass resources and detailed penalties for behaviors that should not be widely used. **Regulations:** There is no unique management mechanism for developing the industry of biomass resources, and there is no special division to implement and manage biodiesel relevant guidelines and national standards.

Microalgal biodiesel production can be an energy-intensive process. Huge quantities of glycerol are formed as a by-product, potentially flooding the market and driving down prices. Crude oil is the source for producing the methanol used in the transesterification process. These problems can be overcome by implementing dealings via state-of-the-art designs and biodiesel storage tanks. The biorefinery concept can overcome the high energy input and convert biomass into energy, enabling a waste-free process.

Currently, microalgal biomass is not a feasible option for biofuel production at a commercial scale because of the high energy input requirement compared to today's terrestrial crops. Microalgae cultivation for biofuel production must be integrated with wastewater treatment to minimize substantial depending on inorganic nutrient sources. High-quality products such as omega-3 and omega-6 fattyacids isolated from the extract will improve biodiesel quality and provide a buffer for economically viable biodiesel production using microalgae. To achieve enduring environmental benefits and sustainability, all microalgal biofuels' processing stages must be simplified without using large amounts of energy. Economically viable microalgal biodiesel production is impractical and unsustainable due to expensive harvesting or dehydration techniques. Therefore, extensive and careful research into the harvesting process is essential. During biofuel production, additional cost-effective process developments are required for operations. It required making local labs at various places with a low-cost estimation of the biodiesel quality evaluating parameters. To produce high-quality biodiesel, technologies need to be developed in addition to developing new innovative technologies for higher biomass and lipid productivities at a pilot scale. Integrated biorefinery needs to be developed and used. That requires efficient use of co-products (value-added metabolites) for economical operation. The safety benefits of harnessing algal feedstocks and energy independence need key innovations in basic algal physiology, algal batch cultivation, and engineering of overall systems to safeguard economic and technical viability. The promise of algal biofuels is accompanied by a vision of a new type of large-scale cultivation, possibly in areas previously untapped for agricultural or industrial purposes. Furthermore, using low-cost substrates to cultivate microalgae would be a smart strategy to reduce production costs. For example, agro-industrial wastewater-based production systems can generate high biomass as well as lipid productivity, and it is also helping wastewater bioremediation and waste CO_2 removal. Therefore, when this technology is in the primary development stages, it is important to consider the environmental impact of work and public acceptance, societal implications, and regulatory issues. Across the globe, researchers, policymakers, industry experts, financiers, nation regulators, scientists, and engineers should come together and discuss the possible solutions for the major constraints facing the commercialization of biofuel productions. Further, in the context of end uses of biofuels, the availability of biofuel-compatible vehicles needs to expand (Hoekman 2009). Biofuel-compatible vehicles should show higher performance than conventional vehicles to increase consumer satisfaction.

10.7 CONCLUSION

Algae-based biodiesel had high expectations over the years, but the amount of biodiesel currently made from algae is negligible compared to the amount of conventional diesel. In addition, there is no comprehensive analysis of all factors' impact on biodiesel production. The quality and biodiesel supply chain need to be developed in every country. Integrating basic, applied, and engineering aspects for biofuel production could lead to the best possible solutions to achieve scalable and sustainable algae-based biofuels. Scientists, industrialists, governments, and policymakers must work together to develop sustainable biofuel solutions. The number of articles published on algae cultivation, oil extraction, and biodiesel production varied widely from the number of contradictory articles reporting biodiesel engine performance. For biofuels to be commercially successful at scale, careful examination and favorable economic conditions are needed at every point in the chain.

REFERENCES

Alabi, A.O., M. Tampier, and E. Bibeau. 2009. "Microalgae technologies and processes for biofuels/bioenergy production in British Columbia: Current technology, suitability and barriers to implementation: Executive summary" Final report submitted to The British Columbia innovation council. Victoria, British Columbia Cambridge: Seed Science Press Ltd.

Atkinson, D., B.J. Ciotti, and D.J.S. Montagnes. 2003. "Protists decrease in size linearly with temperature: Ca. 2.5% C-1." *Proceedings of the Royal Society of London. Series B: Biological Sciences* 270 (1533): 2605–11.

Béchet, Q., M. Laviale, N. Arsapin, H. Bonnefond, and O. Bernard. 2017. "Modeling the impact of high temperatures on microalgal viability and photosynthetic activity." *Biotechnology for Biofuels* 10 (1): 1–11.

Bechet, Q., A. Shilton, O.B. Fringer, R. Munoz, and B. Guieysse. 2010. "Mechanistic modeling of broth temperature in outdoor photobioreactors." *Environmental Science & Technology* 44 (6): 2197–2203.

Chisti, Y. 2007. "Biodiesel from microalgae." *Biotechnology Advances* 25 (3): 294–306.

Choi, H.G., B.Y. Moon, and N.J. Kang. 2015. "Effects of LED light on the production of strawberry during cultivation in a plastic greenhouse and in a growth chamber." *Scientia Horticulturae* 189: 22–31.

Chen, C. Y., K. L.Yeh, R. Aisyah, D. J. Lee, and J. S. Chang. 2011. "Cultivation, photobioreactor design and harvesting of microalgae for biodiesel production: a critical review". *Bioresource Technology* 102(1), 71–81.

Converti, A., A.A. Casazza, E.Y. Ortiz, P. Perego, and M. Del Borghi. 2009. "Effect of temperature and nitrogen concentration on the growth and lipid content of *Nannochloropsis oculata* and *Chlorella vulgaris* for biodiesel production." *Chemical Engineering and Processing: Process Intensification* 48 (6): 1146–51.

Covarrubias, Y., E.A. Cantoral-Uriza, J.S. Casas-Flores, and J.V. García-Meza. 2016. "Thermophile mats of microalgae growing on the woody structure of a cooling tower of a thermoelectric power plant in central Mexico." *Revista Mexicana de Biodiversidad* 87 (2): 277–87.

Daliry, S., A. Hallajisani, R.J. Mohammadi, H. Nouri, and A. Golzary. 2017. "Investigation of optimal condition for *Chlorella vulgaris* microalgae growth." *Global Journal of Environmental Science and Management* 3 (2): 217–30.

Davey, H.M., and D.B. Kell. 1996. "Flow cytometry and cell sorting of heterogeneous microbial populations: The importance of single-cell analyses." *Microbiological Reviews* 60 (4): 641–96. doi:10.1128/mr.60.4.641-696.1996.

Gouveia, L., A.E. Marques, T.L. Da Silva, and A. Reis. 2009. "*Neochloris oleabundans* UTEX# 1185: A suitable renewable lipid source for biofuel production." *Journal of Industrial Microbiology and Biotechnology* 36 (6): 821–26.

He, Q., H. Yang, L. Xu, L. Xia, and C. Hu. 2015. "Sufficient utilization of natural fluctuating light intensity is an effective approach of promoting lipid productivity in oleaginous microalgal cultivation outdoors." *Bioresource Technology* 180: 79–87.

Hoekman, S.K. 2009. "Biofuels in the US–challenges and opportunities." *Renewable Energy* 34 (1): 14–22.

Hsieh, C.-H., and W.-T. Wu. 2009. "Cultivation of microalgae for oil production with a cultivation strategy of urea limitation." *Bioresource Technology* 100 (17): 3921–26.

Hu, Q., C. Zhang, and M. Sommerfeld. 2006. "Biodiesel from algae: Lessons learned over the past 60 years and future perspectives." *Journal of Phycology*, 42: 12–12.

Johnson, M.B., and Z. Wen. 2009. "Production of biodiesel fuel from the microalga *Schizochytrium limacinum* by direct transesterification of algal biomass." *Energy & Fuels* 23 (10): 5179–83.

Jorquera, O., A. Kiperstok, E. A. Sales, M. Embiruçu, and M. L. Ghirardi, M. L. 2010. "Comparative energy life-cycle analyses of microalgal biomass production in open ponds and photobioreactors". *Bioresource Technology* 101(4), 1406–1413.

Juneja, A., R.M. Ceballos, and G.S. Murthy. 2013. "Effects of environmental factors and nutrient availability on the biochemical composition of algae for biofuels production: A review." *Energies* 6 (9): 4607–38. doi:10.3390/en6094607.

Kong, Q.-x., L. Li, B. Martinez, P. Chen, and R. Ruan. 2010. "Culture of microalgae *Chlamydomonas reinhardtii* in wastewater for biomass feedstock production." *Applied Biochemistry and Biotechnology* 160 (1): 9–18.

Lam, M.K., and K.T. Lee. 2012. "Potential of using organic fertilizer to cultivate *Chlorella vulgaris* for biodiesel production." *Applied Energy* 94 (June): 303–8. doi:10.1016/j.apenergy.2012.01.075.

Lee, C.G., D.H. Seong, S.M. Yim, and J.H. Bae. 2015. "A novel *Tetraselmis* Sp. and method for preparing biodiesel with this strain." *Korean Patent* 10: 1509562.

León-Saiki, G.M., I.M. Remmers, D.E. Martens, P.P. Lamers, R.H. Wijffels, and D. van der Veen. 2017. "The role of starch as transient energy buffer in synchronized microalgal growth in *Acutodesmus obliquus*." *Algal Research* 25: 160–67.

Madhubalaji, C.K., T.S. Chandra, V.S. Chauhan, R. Sarada, and S.N. Mudliar. 2020. "*Chlorella vulgaris* cultivation in airlift photobioreactor with transparent draft tube: Effect of hydrodynamics, light and carbon dioxide on biochemical profile particularly omega-6/omega-3 fatty acid ratio." *Journal of Food Science and Technology* 57 (3): 866–76.

Madhubalaji, C. K., A. Shekh, P. V. Sijil, S. Mudliar, V.S. Chauhan, R. Sarada, A.R. Rao, and G.A. Ravishankar. 2019. "Open cultivation systems and closed photobioreactors for microalgal cultivation and biomass production." In *Handbook of Algal Technologies and Phytochemicals*, 178–202. CRC Press, Boca Raton, FL.

Mata, T.M., A.A. Martins, and N.S. Caetano. 2010. "Microalgae for biodiesel production and other applications: A review." *Renewable and Sustainable Energy Reviews* 14 (1): 217–32.

Morweiser, M., O. Kruse, B. Hankamer, and C. Posten. 2010. "Developments and perspectives of photobioreactors for biofuel production." *Applied Microbiology and Biotechnology* 87 (4): 1291–1301.

Órpez, R., M.E. Martínez, G. Hodaifa, F.E. Yousfi, N. Jbari, and S. Sánchez. 2009. "Growth of the microalga *Botryococcus braunii* in secondarily treated sewage." *Desalination* 246 (1–3): 625–30.

Pancha, I., K. Chokshi, R. Maurya, K. Trivedi, S.K. Patidar, A. Ghosh, and S. Mishra. 2015. "Salinity induced oxidative stress enhanced biofuel production potential of microalgae *Scenedesmus* Sp. CCNM 1077." *Bioresource Technology* 189: 341–48.

Powell, E.E., and G.A. Hill. 2009. "Economic assessment of an integrated bioethanol–biodiesel–microbial fuel cell facility utilizing yeast and photosynthetic algae." *Chemical Engineering Research and Design* 87 (9): 1340–48.

Ramanna, L., I. Rawat, and F. Bux. 2017. "Light enhancement strategies improve microalgal biomass productivity." *Renewable and Sustainable Energy Reviews* 80: 765–73.

Ranjan, A., C. Patil, and V.S. Moholkar. 2010. "Mechanistic assessment of microalgal lipid extraction." *Industrial & Engineering Chemistry Research* 49 (6): 2979–85.

Raychaudhuri, A., and S.K. Ghosh. 2016. "Biomass supply chain in Asian and European countries." *Procedia Environmental Sciences* 35: 914–24.

Reckermann, M. 2000. "Flow sorting in aquatic ecology." *Scientia Marina* 64 (June): 235–46. doi:10.3989/scimar.2000.64n2235.

Sabu, S., I. S. Bright Singh, and V. Joseph. 2017. "Molecular identification and comparative evaluation of tropical marine microalgae for biodiesel production." *Marine Biotechnology* 19 (4): 328–44.

Salvucci, M.E., and S.J. Crafts-Brandner. 2004. "Relationship between the heat tolerance of photosynthesis and the thermal stability of rubisco activase in plants from contrasting thermal environments." *Plant Physiology* 134 (4): 1460–70.

Sander, K., and G.S. Murthy. 2010. "Life cycle analysis of algae biodiesel." *The International Journal of Life Cycle Assessment* 15 (7): 704–14.

Seo, Y.H., Y. Lee, D.Y. Jeon, and J.-I. Han. 2015. "Enhancing the light utilization efficiency of microalgae using organic dyes." *Bioresource Technology* 181: 355–59.

Severes, A., S. Hegde, L. D'Souza, and S. Hegde. 2017. "Use of light emitting diodes (LEDs) for enhanced lipid production in micro-algae based biofuels." *Journal of Photochemistry and Photobiology B: Biology* 170: 235–40.

Shih, S.C.C., N.S. Mufti, M.D. Chamberlain, J. Kim, and A.R. Wheeler. 2014. "A droplet-based screen for wavelength-dependent lipid production in algae." *Energy & Environmental Science* 7 (7): 2366–75.

Show, P.L., M.S.Y. Tang, D. Nagarajan, T.C. Ling, C.-W. Ooi, and J.-S. Chang. 2017. "A holistic approach to managing microalgae for biofuel applications." *International Journal of Molecular Sciences* 18 (1): 215.

Singh, S. P., and P. Singh. 2015. "Effect of temperature and light on the growth of algae species: A review." *Renewable and Sustainable Energy Reviews* 50: 431–44.

Sivaramakrishnan, R., and A. Incharoensakdi. 2017. "Enhancement of total lipid yield by nitrogen, carbon, and iron supplementation in isolated microalgae." *Journal of Phycology* 53 (4): 855–68.

Song, M., and H. Pei. 2018. "The growth and lipid accumulation of *Scenedesmus quadricauda* during batch mixotrophic/heterotrophic cultivation using xylose as a carbon source." *Bioresource Technology* 263 (September): 525–31. doi:10.1016/j.biortech.2018.05.020.

Srivastava, G., and V.V. Goud. 2017. "Salinity induced lipid production in microalgae and cluster analysis (ICCB 16-BR_047)." *Bioresource Technology* 242: 244–52.

Stephenson, A. L., E. Kazamia, J. S. Dennis, C. J. Howe, S. A. Scott, and A. G. Smith. 2010. "Life-cycle assessment of potential algal biodiesel production in the United Kingdom: A comparison of raceways and air-lift tubular bioreactors". *Energy & Fuels* 24(7): 4062–4077.

Su, C.-H., R. Giridhar, C.-W. Chen, and W.-T. Wu. 2007. "A novel approach for medium formulation for growth of a microalga using motile intensity." *Bioresource Technology* 98 (16): 3012–16.

Sydney, E.B., T.E. Da Silva, A. Tokarski, A.D. Novak, J.C. De Carvalho, A.L. Woiciecohwski, C. Larroche, and C.R. Soccol. 2011. "Screening of microalgae with potential for biodiesel production and nutrient removal from treated domestic sewage." *Applied Energy* 88 (10): 3291–94.

Takeshita, T., I.N. Ivanov, K. Oshima, K. Ishii, H. Kawamoto, S. Ota, T. Yamazaki, et al. 2018. "Comparison of lipid productivity of *Parachlorella kessleri* heavy-ion beam irradiation mutant PK4 in laboratory and 150-L mass bioreactor, identification and characterization of its genetic variation." *Algal Research* 35 (November): 416–26. doi:10.1016/j.algal.2018.09.005.

Tang, D., W. Han, P. Li, X. Miao, and J. Zhong. 2011. "CO_2 biofixation and fatty acid composition of *Scenedesmus obliquus* and *Chlorella pyrenoidosa* in response to different CO_2 levels." *Bioresource Technology* 102 (3): 3071–76.

Xu, H., X. Miao, and Q. Wu. 2006. "High quality biodiesel production from a microalga *Chlorella protothecoides* by heterotrophic growth in fermenters." *Journal of Biotechnology* 126 (4): 499–507.

Xu, L., P.J. Weathers, X.-R. Xiong, and C.-Z. Liu. 2009. "Microalgal bioreactors: Challenges and opportunities." *Engineering in Life Sciences* 9 (3): 178–89.

Yang, H., Q. He, J. Rong, L. Xia, and C. Hu. 2014. "Rapid neutral lipid accumulation of the alkali-resistant Oleaginous *Monoraphidium dybowskii* LB50 by NaCl induction." *Bioresource Technology* 172: 131–37.

Yue, D., F. You, and S.W. Snyder. 2014. "Biomass-to-bioenergy and biofuel supply chain optimization: Overview, key issues and challenges." *Computers & Chemical Engineering* 66: 36–56.

Part IV

Global Policies and Storage System for Biofuels

11 Renewable Energy Directives and Global Policies for Different Generation of Biofuel

Dayanand Sharma
Sharda University

Aneesh Mathew
National Institute of Technology

Dhamodharan Kondusamy
Thapar Institute of Engineering and Technology

Anudeep Nema
Eklavya University

Rajnikant Prasad
G. H. Raisoni Institute of Engineering and Business Management

Mritunjay
National Institute of Technology

Loganath Radhakrishnan
Institute for Globally Distributed Open Research and Education
Indian Green Service

Pranav Prashant Dagwar
Bharathidasan University

DOI: 10.1201/9781003197737-15

CONTENTS

11.1 INTRODUCTION

The increase in the energy demand has resulted in energy crisis in many countries (Malik, Rena, and Kumar 2021). Renewable energy technologies are the best solutions for climate change and global warming-related issues. The sustainable development of the country completely depends on how the country implements the renewable energy technologies in their infrastructure. Several renewable energy options are available to the country based on their geographical location, economical status, and renewable source availability. Many countries have prepared the policies to incorporate the renewable energy technologies in their energy sector. Solar, wind, and geothermal energy are in higher consideration of the renewable energy technologies due to its clean source and abundant availability. According to ministry of new and renewable energy (MNRE), Government of India, apart from large hydropower plants, India has installed and attained a cumulative renewable energy capacity of 92.54 GW. And from the period April 2020 till January, 2021 out India installed for 5.47 GW power. MNRE, 2021 reported that the renewable energy installation capacity has increased by 2.5 folds in a period of just 6.5 years from April 2014 to January 2021, in the same time the solar energy capacity has increased 15 folds. It also declares that India stances fourth in renewable energy capacity, fourth position in Wind power, and fifth position in solar energy installed capacity. In 17 goals of the United Nations sustainable development program, goal number 7 is to achieve affordable and clean energy. The United Nations Regional Information Center (UNRIC) for western Europe stated that every country in the globe is improving on the way to goal 7, with a motivational approach that energy is available abundant everywhere, and this motivation made the poorer countries access to electricity in higher percentage, and renewable energy is creating some notable improvements in the electricity sector.

With the current rate of fossil fuel consumption, it may run out between the year 2069–2088, making it one of the priority to switch for alternative fuel for the future (Kumar et al. 2019). Renewable energy is considered or conceptualized as the most important and vital solution for the global climate change. Every set of people from politics, business, and academia are framing the same, but the concept of renewable energy is problematic and is arguable, and it should be abandoned in favor of more explicit conceptualization (Harjanne and Korhonen 2019). The transition of concept from conventional energy sources to the renewable energy sources is facing a tremendous challenge globally. Still, the countries that depend on fossil reserves for their economy face many challenges compared to others (Alizadeh et al. 2020). Alizadeh et al. (2020) also provide a framework that comprises the identification and confiscation of renewables for energy purposes . The framework is based on two models' combination, i.e., Benefit, Opportunity, Cost, Risk (BOCR) and Analytic Network Process (ANP) models. From the experts it's assured that the evolution of renewable energy is basically a political struggle, and they assure that only by opposing and destabilizing of major energy power systems the shift to renewable energy sources is possible (Burke and Stephens 2018). This chapter discuss about the renewable energy technologies available, policies around the world along with barriers to implement, and bottlenecks with solution.

11.1.1 Renewable Energy Technologies

- **Solar Energy**: It is considered as vital and ecofriendly energy resource for the future world (Alanne and Saari 2006). Solar outcompetes the other renewable resources due to the fact that it is an abundantly available resource. Many studies showed that the global energy demand can be met by the solar energy, since it is available abundantly with no cost. It is a promising resource, since it is not exhaustible. Rather the solar radiation and intensity are the two factors determining the efficiency of the solar PV industry. The solar system can be effectively used in the villages, for industrial operations, and homes due to its affordability and easy applicability. The world energy demand is increasing at a rapid rate. Many sources are used to meet the energy demand. Around 77.9% of the world's energy demand are met by fossil fuels and nuclear power plants, and the contribution of solar PV is only 0.7% (REN21 2014). The thermal energy produced by solar collectors was 326 GW in the year 2013. Many countries around the world rely on the solar power for power generation, since it saves money used for power generation. The solar energy definitely offers a long-term solution to meet the energy demands.

- **Wind Energy**: It is an indirect form of solar energy that is continuously replenished by sun (Joselin Herbert et al. 2007). The estimated energy from the wind is approximately 10 million MW of energy, which is always available. As per greenspace, approximately 10% of the electricity can be supplied by wind energy, and this will contribute to 5% of world energy market with improved technology. The global wind capacity in the year 2004 was 6,614 MW, which has grown to a cumulative power of 46,048 MW. Denmark ranks world's largest manufacturer and exporter of wind turbines, which expects to meet 50% of the domestic energy demand by 2030. However, the cost of wind turbines is more, incurring heavy initial investments. The cost effectiveness of these turbines depends on the dynamics, design, and size.
- **Hydropower Energy**: It has generated energy for the civilization for more than a century, after the development of wheels. As compared to other renewable technologies, it has advanced to a great level. Hydropower can be used to generate energy as well as power machinery at the same time. Power generated form the hydropower is one of the cheapest form of electricity in terms of cost (Kaunda, Kimambo, and Nielsen 2012) and has capacity to cope-up with the fluctuations in the energy demand. The hydropower is generated from either run-of-river, reservoir, or pumped storage, depending upon water impoundment. Run-of-river hydropower project generates electricity from the flow of river without impoundment. For example, in Canada, 64% of the electric power production was sourced from run-of-river project generating 27% of the total electricity generation (Douglas 2007). A storage hydropower uses stored water for power generation. This system is used to meet the variable flow in the middle of a river system (Kaunda, Kimambo, and Nielsen 2012). A pumped storage plant is hydraulic energy storage device. In this water is pumped to give hydraulic energy from the lower reservoir to the upper reservoir. The advantages of hydropower include reduction in air pollutants and increased economic opportunities leading to sustainable development (Mohan, Paranjothi, and Prince Israel 1997).

- **Bioenergy**: Wide variety of feedstock are available in abundance that can be converted into biofuels. The feedstock mainly include waste from agriculture, industrial, and municipal solid waste. Biofuels are used in the transportation sector, electricity, and in the production of biogas. Biofuels contribute a significant fraction of the transportation sector worldwide. The advantages of using biofuels include easy extraction, sustainability, combustion based on the carbon dioxide cycle, and is environment friendly (Gaurav et al. 2017). The national average target of India is to blend 20% ethanol with gasoline by 2025. India also targets to blend 5% biodiesel with conventional biodiesel by 2030. (Chandra 2021). Biomass is the fourth largest available energy resource globally, which is natural, inexpensive, and available in abundance in a short duration of time. A wide range of biomass can be used for biofuel production. Rena et al. (2020) used various agricultural residues for the biohydrogen and biomethane production. The annual global production of biomass is about 220 billion tonnes on a dry weight basis, equivalent to 4,500 EJ of solar energy captured each year corresponding to 270 EJ of annual bioenergy market (Kumari and Singh 2018). Similar to fossil fuels, biofuels exist in three phases of solid, liquid and gas. Solid fuels include fuels like wood chips, wood pallets, and animal waste. Liquid biofuels include ethanol, biodiesel, green diesel, methanol, and green gasoline (Chandra 2021). Gaseous biofuels include biomethane and biohydrogen. The primary biomass includes food crops rich in sugar and starch, such as sugarcane and corn; nonfood biomass includes crop residue and lignocellulose material obtained from the municipal and industrial solid waste, and algal biomass (Ho, Ngo, and Guo 2014). The energy from the biofuels is stabilized during biological carbon fixation, where carbon dioxide is converted into sugar that is present in the plant biomass and living organism (Alaswad et al. 2015). Rena et al. (2020) showed capability of agricultural residue for production of biofuels through anaerobic digestion.
- **Wave Energy**: Ocean wave offers a huge potential for extracting renewable energy compared to other methods. It offers the highest energy density compared to different renewable resources (Clément et al. 2002). It offers the least environmental damages with benefits like it can be used to meet the energy demand in temperate climate areas (Clément et al. 2002). The estimated potential worldwide of wave power is around 2 TW (Drew, Plummer, and Sahinkaya 2009).

11.2 BIOFUEL TECHNOLOGIES DEVELOPMENTS AND ADVANCEMENTS

To handle the climate change issues and exhaustion of fossil fuels, the alternate renewable and sustainable sources are found to be the better solution. Alternate Liquid fuel for the transportation section seems to be bigger task, and it's found to be possible by the conversion of lignocellulose biomass to liquid fuels. It shows the benefits from the first-generation biofuels and fossil fuels (Ibarra-Gonzalez and Rong 2019). The environmental benefits and sustainable production option methods made the world to produce biofuel to replace the fossil fuels. One drawback toward the biodiesel production is the availability of feedstock around the globe that leads the global powers to identify the alternative and abundant feedstock. In this way, the biomass from the industries and domestic considered as waste or by product will be more beneficial (Gumisiriza et al. 2017). Some of technologies are discussed as follows:

11.2.1 Thermochemical

There are four thermochemical biofuel production technologies: Combustion, pyrolysis, gasification, and hydrothermal gasification. In this four technologies, the first three are dry technologies and the last one is wet method, and it's suitable for high moisture content wastes. Pyrolysis process is considered to be the base method for thermochemical method of biofuel production. Combustion is the process where the biomass reacts with the oxidants and produces heat and other chemical products as a final product. The heat produced during the combustion process is due to the exothermic reaction between the organic matter consists of both carbon and hydrogen molecules reacted with air or oxygen (Bhaskar et al. 2011). If heat is the main requirement in the industry or process, then combustion of biomass will be the better option than pyrolysis and gasification process, because combustion process produces more heat energy than liquid fuel. Combustion is a low cost and highly reliable technology and also the most available in market. For the conversion of biomass into a liquid fuel in a faster and efficient way, pyrolysis is the stand-alone option without any conflicts. Multiple biofuel options have been produced from algal biomass, i.e., biohydrogen, biodiesel, bioethanol, and biogas through different pathways: biochemical and thermochemical pathways. In this case, thermochemical conversion is found as the most efficient way to avoid long production time and improve recovery efficiency than other biochemical methods (Raheem et al. 2015). Dote et al. (1994) studied the thermochemical liquefaction of *Botryococcus braunii* and achieved 64% yield of bio-oil at 300°C with a higher heating value (HHV) of 45.8% MJ/kg. Similar study was conducted by Minowa et al. (1995), with *Dunaliella tertiolecta* recovering a bio-oil yield of 42% on dry basis and HHV of 34.9 MJ/kg. The major issue faced by the biodiesel industries is the low supply of biomass feedstock for production; this can be solved by contribution of agroforestry sector. The supply chain of various residues and byproducts such as rice husks, almond husks, kiwi pruning, vine pruning, olive pomace, and pine woodchips would be the best alternative source. Nunes et al. (2020) characterized these feedstocks and showed the improved calorific value & fixed carbon content by Torre faction at 300°C. But they also found a loss in total mass for pyrolysis process (Nunes et al. 2020).

11.2.2 Chemical and Biological

The usual method for biodiesel production is transesterification of fresh and used cooking oil as feedstock. The drawbacks in the biodiesel is low reaction efficiency, and that can be enhanced by integrating the green chemistry principles and process intensification effects. The green chemistry principles can reduce or exclude the use and formation of hazardous byproducts (Gude and Martinez-Guerra 2018). The material that enhances the conversion efficiency and increases the fuel burning value in transesterification process is mentioned as catalysts. It performs a substantial role in transesterification of vegetable oils. The waste oils contain large amount of free fatty acids (FFA); it reduces the process efficiency of transesterification. Some of the catalysts that are more used in transesterification studies for improving the biodiesel recovery efficiency are sodium hydroxide (NaOH), sodium methoxide (CH_3ONa) and potassium hydroxide (KOH), beryllium oxide (BeO), magnesium oxide (MgO), calcium oxide (CaO), strontium oxide (SrO), barium oxide (BaO), and radium oxide (RaO) (Thangaraj et al. 2019). From the study using two different microalgal species identified from the sewage treatment plant for converting carbon source from CO_2 sequestration process into biodiesel under optimized culture conditions, total lipid content and lipid yield was found higher in Arthronema sp. (180 mg/L; 32.14%) than Chlorella sp. (98 mg/L; 29.6%) in 50 mm $NaHCO_3$ (Maheshwari et al. 2020). Chi et al. (2018) studied the biodiesel extraction and transesterification from the microbial lipids; the study proves that benefits in biodiesel production by direct addition of *n*-hexane and methanol to the dehydrated excess sludge and also the parameters favorable at 75°C for 7 hours of in situ transesterification. They also confirmed that the ultrasonic sludge pretreatment was not effective for biodiesel production by in situ acid-catalyzed transesterification. Production of biogas and biohydrogen from the biomass is the another option for the fuel, but there are lots of constraints in usage such as storage, less calorific value, longer fuel production period, and more sensitivity toward operational parameters. Biogas and biohydrogen production is suitable for the low calorific value biomass feedstock with higher moisture content and the place where there is no need of higher colorific fuel, either in room heating purpose or cooking.

11.3 ADVANCES IN THE PRODUCTION OF BIOFUELS

Biofuels are the alternative of nonrenewable fuels. These are basically derived from the organic compounds (biomass) by the decomposition of plants and living things. These sources of biofuels can be grown or cultivated again and again. Biofuels have been accepted widely because of their nontoxic, sulfur-free, biodegradable nature, originating from the renewable sources. Biofuels are developed year by year with several changes in processing method as well as in the base materials (Ullah et al. 2018). Table 11.1 shows the advancement in production of biofuels with its advantages and disadvantages

Biofuels are developed year by year with several changes in processing methods as well as in the base materials (Ullah et al. 2018). The current shift of trend and constant rise and fall in the costs of fossil fuel with the shift toward green energies and technologies have significantly contributed to the augmentation in the application of conversion techniques concerned with organic compounds (Malode et al. 2020). It has been studied that the developing countries face challenges interims biofuels due to their high cost of production and due to lack of infrastructure, inefficient conversion technologies, and funding options; it turns difficult for developing nations to support the high-end production of biofuels and paradigm the shift from fossil fuels.

TABLE 11.1
The Advancement in Production of Biofuels with Its Advantages and Disadvantages

	First Generation	Second Generation	Third Generation	Fourth Generation
Materials used	Sugar, starch, vegetable oil	Lignocellulosic crop, agricultural waste, woodland residue	Algae	Genetically modified algae, engineered cyanobacterial
Techniques	Fermentation of materials by microorganisms	Fermentation of cellulose		
Commercially acceptance	Yes	Yes	No	No
Advantages	Easy conversion; easy harvesting; the capital cost is low	No food-energy conflict; easy harvesting; the capital cost is low	No food-energy conflict; use of nonarable land; wastewater can also be used; CO2 fixation; application in wastewater treatment	No food-energy conflict; Use of nonarable land; wastewater can also be used; CO2 fixation; application in wastewater treatment
Disadvantages	Causing food security problem; requirement of arable land; requirement of potable water; Not a sustainable approach; use of pesticide and fertilizer	Requirement of arable land; not a sustainable approach	Harvesting of microalgae is expensive; availability of sunlight only during day time; High capital cost	Harvesting of microalgae is expensive; availability of sunlight only during day time; Very high capital cost; requirement of skilled person to operate

Economic globalization has turned out to be the dominant feature of economic performance, as it plays an essential responsibility for environmental quality. So this has carved out a pathway for the developing countries to reduce their dependency on the traditional sources of fuel, i.e., fossil fuels and increase in the use of biological energy, i.e., biofuels. (Avinash et al., 2018).

It has been suggested that the global necessity for biofuels has set to in the near future to 41 billion liters, i.e., 28%, alongside the policies formed by governments is considered a principal driving factor for the expansion of the fuel demand and influence the growth. The demand for renewable energy has nearly increased three times between 2020 and 2026 due to the various new policies that have been implemented by the US and Europe. Asia accounts for the largest production that even surpasses Brazil and makes Asia the second-largest biofuel producer globally. Europe and USA are considered the flag bearers of renewable energy and play an important role in the world. In the past recent years, China has turned out to be a leading developer in the field of renewable heat and power worldwide, whereas India is one of the emerging competitors in the field of biofuels, and it has also reached the third spot in ethanol blending after US and Brazil. The US and Brazil are considered as the largest markets for biofuels, as they account for possible expansion of demand in the upcoming years (IEA December 2021; IRENA, IEA 21).

In Asia, the region accounts for increased demand and supply, which has surpassed Europe in its demand and production. The countries in the Asia Pacific have formulated new policies to decrease the dependency on their transport fuel from fossil fuels to biofuels. The nations like China, India, Indonesia, and Singapore have formulated their plans for producing and supplying biofuel by setting up targets for 2030 so as to fulfill the rising demands from many developed nations (IEA December 2021).

With the increase in demand and prices for feedstocks and fuels, there has been a major trend of policy interplay across countries and various regions. The new policies for biofuels are very positively impactful in environmental concerns, as it has better GHG profiles and lower pollution levels; alongside the shift has been noticed regarding the use of various fuels in regions. In biofuels the fuels like ethanol, biodiesel, biojet have a multiplying potential for growth, which has a region-wise influence. Figure 11.1 shows the demand of biofuel in the order of Asia to Europe. Asia is on the top of chart in terms of demand of biofuels. The different flexible policy and directives are now promoting the use of biofuels over the fossil fuels.

Following the expansion of renewable biological energy, i.e., biofuel, the production increases the competition for the compliant feedstock, which boosts the cost. As the production of biofuel is dependent on the agricultural system, this increases the intensity of the agriculture system, so many countries have their own policies in terms to regulate the feedstock for the production of fuels. The improvement of technology has also addressed the scientific approach to maximize productivity in

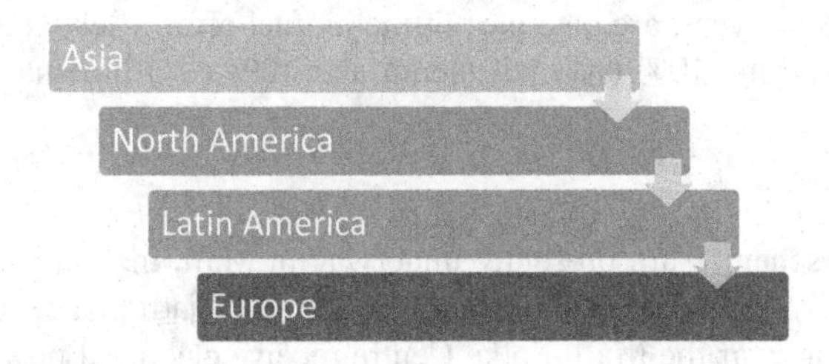

FIGURE 11.1 The figure represents the growth of demand of biofuel throughout the world; which starts with Asia, followed by North America, Latin America then Europe.

simultaneously reducing the use of land waste and chemicals (Biofuels: Policies, Standards and Technologies World Energy Council 2010; International Energy Agency December 2021).

With the technological advancements and increased concern of nations toward the conservation of the environment and promotion of sustainability, there has been a gradual deployment of renewable energy sources as options that have led to the decline in the use of fossil fuels.

11.4 DIRECTIVES AND POLICIES AROUND THE GLOBE ON BIOFUELS

The main goal of the section is to offer a quick outline of different strategies that support the development of the global biofuel industry. The world's prime producers are discussed individually. The past decade's fundamental laws, different action strategies, and incentive schemes are examined.

11.4.1 African Countries

11.4.1.1 Ghana

Ghana adopted a bioenergy policy in 2010, intending to replace 10% of its petroleum fuel products with biofuels by the year 2020 and 20% by the year 2030 (Kemausuor et al. 2011). This policy utilized the vast majority of Ghana's biomass resources to generate transportation fuels and electrical power. In Ghana, various biomass resources, such as, maize, sugarcane, cassava, and jatropha oil seeds have been recognized as potential bioethanol and biodiesel feedstock (Kemausuor et al. 2011).

11.4.1.2 Malawi

A USD 8 million biodiesel manufacture plant unit has been inaugurated in Lilongwe, Malawi's head, in 2006 (Kemausuor et al. 2011). Dutch investors spearheaded the project, and it now utilizes up to 250.0 tons of oil seeds of jatropha to generate 5,000.00 L of biodiesel per day. About 1,000 of local agriculturalists have profited from the initiative, which has planted 100 lakhs jatropha plants during 5 years period. Those plants are used as raw resources in manufacturing plants (Kemausuor et al. 2011). The Dwangwa estate bioethanol plants, which produce 15–20 million liters of sugar cane molasses bioethanol per year, and the Nchalo facility, which produces 12 million liters per year, are also part of the biofuel effort (Sekoai and Yoro 2014). These plants combine 10% (v/v) petroleum and 10% (v/v) bioethanol (Sekoai and Yoro 2014).

11.4.1.3 Mali

Several biofuel schemes are presently underway in Mali, including the Mali-Folke Center, Nyetaa—a local NGO supporting local farmers' jatropha oilseed production. Societies existing near the Mali–Folke Centre receive electrical power produced by power stations that use oil seeds from jatropha. A 15-year electrification program was started to provide about 300 kW of electrical power to 10,000 rural people. In addition, over 100 ha of jatropha were planted as a source of feedstock for the factory (Cotula et al., 2008). It improves the living conditions of these communities.

11.4.1.4 Mozambique

In recent years, Mozambique has implemented a number of biofuel production initiatives. For example, the Ndzilo ethanol production plant can handle 2 million liters of cassava-processed ethanol (Amigun, Musango, and Stafford 2011). Jatropha seeds have become increasingly important in biodiesel production in Mozambique due to their advantages, including the ability to withstand extreme dry conditions. In the recent years, companies such as Petromoc and Sun Biofuels have established biodiesel production factories to help the nation's energy division (Amigun, Musango, and Stafford 2011). In addition, the government of Mozambique has blended 5%–10% (v/v) bioethanol with gasoline (American Petroleum Institute 2016).

11.4.1.5 Burkina Faso

Burkina Faso is one of the nations that are strongly influenced by the energy- or power-related problems (Jumbe, Msiska, and Madjera 2009). Therefore, biofuel development has exploded in the recent years in an attempt to address this issue. In 2009, for instance, 70,000 jatropha oil seeds were planted. Since 2003, the governments of Burkina Faso and France have worked together to develop this sector in their respective countries with the European Union Biofuel Directive. As a result, biofuel development schemes like the Fondation Faso Biocarburant (FFB) have been founded and funded (Tatsidjodoung, Dabat, and Blin 2012).

11.4.1.6 Senegal

Over the last several years, Senegal's administration has established several initiatives to develop biofuels to progress the nation's power and energy sector. Collaboration with biofuel professionals from India and Brazil made this possible. Senegal has implemented several biofuel projects, including planting jatropha oil seeds on 4,000 acres in Tuda. For example, in Kolda and Tambacounda, more than 50,000 hectares of land are assessed and cultivated for castor oil and sun flowers.

11.4.1.7 Tanzania

Biofuel programs are being implemented in Tanzania by several firms, NGOs, and minor-scale agriculturalists. Tanzania's Traditional Energy and Environmental Organization—a Tanzanian nongovernmental organization (NGO) that supports small-scale rural farmers, collaborates on projects in Dar-es Salaam and Monduli District. It promotes the cultivation of jatropha oil seeds.

Diligent Tanzania Ltd. specializes in the manufacture of jatropha oil and biodiesel, and provides advice to jatropha agriculturalists.

Kakute Ltd, a company based in Japan, focuses on jatropha seed planting, oil processing, and the ARI-Monduli project, which enlists the help of local women farmers to grow jatropha oil seeds.

MVIWATA is a Tanzanian firm of farmers made up of about 2000 small-scale farmers who wanted to plant jatropha oil seeds.

Peter Burland owns and operates the farm Kikuletwa, but it discovered the significant potential in jatropha oil seeds in 2002 and began planting large hectares of it on its land area. Since then, he's been hired by several firms to produce oil from jatropha seeds.

Jatropha Products Tanzania Limited: This organization, which has been around for a long time, promotes jatropha production in Tanzania (Sekoai and Yoro 2016).

11.4.1.8 South Africa

Biofuels are one of the most important renewable energy sources in South Africa, accounting for 9%–14% of total renewable energy. As the country continues to strengthen the biofuel industry and thus reduce its reliance on imported fuel, the South African Energy Department announced in 2013 that it intends to make the compulsory combination of diesel and petrol with biofuels like bioethanol and biodiesel as of October 1, 2015. Furthermore, the nation has planned a 5-year preliminary project to produce 2% biofuel. As a result, five companies in South Africa have been approved permits to produce bioethanol and biodiesel. Also, there is fuel excise exclusion for ethanol. Biodiesel producers obtain a refund of 50% on the over-all fuel levy (Ebadian et al. 2020). According to a review of possible feedstock, Sorghum can be utilized to make bioethanol, and Sorghum can be used to make bioethanol. However, with an assessed yearly generation of about 8 million tons, maize has not been used as one of the most stable foods in the country, posing a threat to food security. Furthermore, massive biomass inflows are affecting South Africa's agriculture, civic, and manufacturing divisions. As a result, other biofuel alternatives like, bioelectricity, biohydrogen, and biomethane production would dominate the country's power mix (Sekoai and Yoro 2016).

11.4.1.9 Nigeria

In Nigeria, a variety of biofuel and biodiesel production start-ups have been implemented, ranging from possibility research to refinery plant construction (Boynton 2016; Ohimain 2010). Nigeria presently has five primary commercial ethanol distilleries that harvest up to 0.134 billion liters of ethanol per year. All the three companies have begun production: In Lagos, Biodiesel Nigeria Limited, Aura Biocorporation (Located at Cross River State), and Shashwat Jatropha (located at Kebbi State) are among the companies (Ohimain 2010).

11.4.2 North America Countries

11.4.2.1 Canada

Bill C-33, Canada's Environmental Protection Act by 2010, a 5% renewable portion in gasoline is required, and a 2% renewable portion in diesel and heating oil is required by the year 2012. Given the current gasoline sales trends, to come across the projected goals, at least 1.90 billion liters of ethanol must be generated. By 2012, 520 million liters of biodiesel per year production capacity is expected to be required to meet the federal mandate of 2% biodiesel.

Almost all ethanol is made from cereal grains. Corn contributed 69% of ethanol production in 2009, while wheat contributed 30%. Animal fats are used to make biodiesel. Tallow grease was estimated to be the most common biodiesel feedstock in 2009 (49%); after that comes yellow grease (37%). Canola has become increasingly important in biodiesel production, accounting for 14% of entire production during 2009 (Dessureault 2009). The production of renewable fuels has an annual economic impact of CAN$2 billion on the Canadian economy. Biofuel usage has produced a considerable savings or funds comparative to a situation where no biofuel has been utilized, about $4.3 billion (2017 CAD) from 2011 to 2018 (Wolinetz, Hein, and Moawad 2019).

Since April 2008, there have been direct incentive payments for production, in addition to the federal share mandate. From 2008 to the end of 2010, ethanol producers were eligible for the maximum incentive rate of CAN$0.10 per liter through the Eco ENERGY for Biofuels scheme. Every year, the payment decreases by CAN$0.01, until it reaches CAN$0.04 in 2015 and 2016. From 2008 to 2010, the maximum biodiesel incentive rate was CAN$0.20 per liter (Dahman et al., 2019).

11.4.2.2 The United States

The Renewable Fuel Standards (RFS1) in the Energy Policy Act of 2005 were the primary to mandate specific biofuel volumes. The primary goal was to use 4,000 million gallons of renewables in transportation fuels during the year 2006 to increase their share over time. The novel regulations (RFS2) raise the biofuel goals, encourage second-generation production, and introduce novel standards to guarantee additional ecologically sustainable manufacturing.

The updated RFS2 was implemented in July 2010, bringing the Energy Independence and Security Act of 2007s proposals to a close. By 2022, 36,000 million gallons of renewables should be utilized in transportation fuels. The capacity of conventional biofuels (e.g., corn-ethanol) must be about 15 million gallons, starting in 2015. Cellulosic biofuels have been projected to grow from 0.10 billion gallons during the year 2010 to 16.0 billion gallons during the year 2022, with advanced biofuels contributing fewer than 21.0 billion gallons by that period.

The reduction of the country's reliance on oil has been the guiding principle of biofuel policies. One of the goals outlined in the 2007 Energy Independence and Security Act is to cut gasoline usage by 20% over the subsequent 10 years. The 2008 Biomass Program had two principal goals. To begin, by 2030, reduce gasoline usage by 30% than 2004 levels. Also, corn-derived ethanol is used to make cellulosic ethanol.

Historically, the ethanol industry has been the focus of US biofuel policies. The Energy Tax Act of 1978 created tax credits for ethanol blenders. Various levels of credit have been guaranteed over the last decade. Since then, however, some form of subsidy has been in place. Currently, corn is the primary source of ethanol. Twenty percent of the USA corn supply has been used as an ethanol feedstock during 2006, which demonstrates its significance to the corn production (IEA, 2021). Figure 11.2 represents the biofuel generation and consumption in the US from 2007 to 2016.

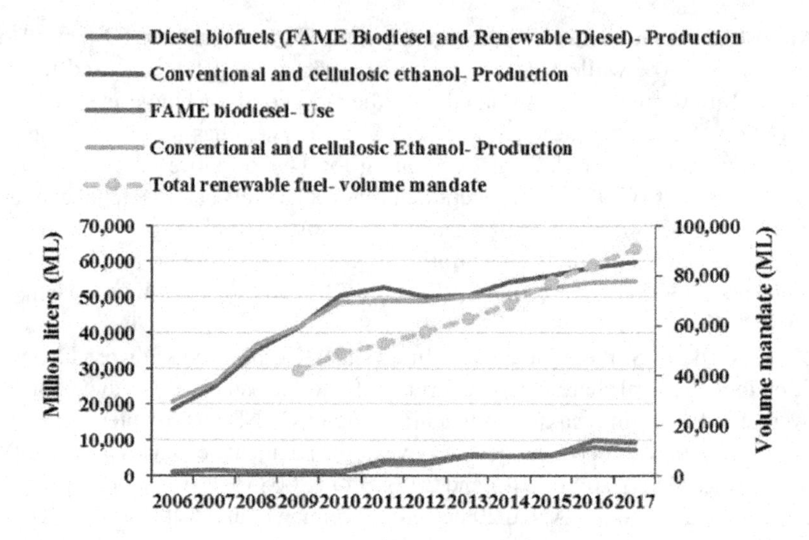

FIGURE 11.2 A line chart showing million liters on Y-axis increasing as time in year (from 2006 to 2017) increases on the X-axis, continuously increasing with year for different biofuel generation and consumption in United States.

11.4.3 South America

11.4.3.1 Argentina

Due to Argentine Biofuel Law 26.093 executed in February 2007, by January, 2010, gasoline and diesel contained a 5% biofuel content. In 2016, this proportion of blend was increased to 12% and 10% for ethanol and biodiesel, respectively.

With Resolution 1295/2008, In November 2008, the technical and quality requirements for ethanol as fuel were established. However, the quality specifications for biodiesel were only formalized in February 2010 with Resolution 6/2010.

Biodiesel is made from soybeans and is centered in Rosario's existing soybean processing cluster. Biodiesel production had increased by 433% between 2008 and 2007. With a capacity of 2.4 billion liters, production reached 880 million liters in 2009 (Joseph, 2009). However, ethanol production is still in its infancy (1.4 billion liters in 2019), and it is heavily reliant on the sugar and corn production industry.

Biofuel producers who export the product items are not eligible for tax breaks. On the other hand, biodiesel and ethanol vended on the domestic marketplace are eligible for monetary assistance. Producers have the option of receiving a refund of their value-added tax or benefiting from augmented depreciation over capital fund investments. Besides, the government guarantees that the biofuel yield will be procured for the 15-year period stipulated in the Biofuel Law. On the other hand, annually, financial incentives were reviewed (As a result, they cannot be certain.). Costs were set by the government.

11.4.3.2 Brazil

Brazil's biofuels programme is the most advanced and comprehensive in the world. Its origins could be traced back to the 1970s oil crisis. Brazil launched the National

Alcohol Program Proalcool in 1975, aiming at cane sugar ethanol. The goal was to reduce energy supply limitations, create a steady inner demand for sugar cane surplus production, and compensate for global sugar price fluctuations. The commercialization of biofuels proved successful over the next decade, with ethanol powering 96% of vehicles vended in Brazil during 1985 (Colares 2008).

The commercialization of biofuels proved successful over the next decade, with ethanol powering 96% of vehicles vended in Brazil during 1985. By the late 1990s, ethanol-powered vehicle sales had dropped to 1%, and in the 1990s, energy and fuel markets were deregularized. However, the Brazilian currency increased ethanol production costs from 1994 to 1999. The government attempted to counteract these disadvantages by enacting legislation during 1993 that needed gasoline to contain 22% ethanol. This percent was increased to 25.0% in 2003.

More deregulation declarations in the energy and fuel markets were executed. During 1998, the Brazil liberalized the cost of hydrated alcohol utilized in fuels, and, in 1999, it mandated that hydrated ethanol fuel deals take place at public auctions.

The victory of the Proalcool programme has been replicated in the significance of sugar and ethanol manufacture in the Brazilian economy. The two production industries employ 3.60 million people and account for 3.50% of GDP, with ethanol manufacture only consuming half of the sugar cane source (de Almeida et al., 2008).

Brazil is currently leading in biodiesel production as a result of the success of the bioethanol plant. In 2005, the National Program on Biodiesel Production and Use (PNPB) was established. From 2008 to 2012, to replace biodiesel, the PNPB needed 2.0% of petrol-based diesel, which was increased to 5% from 2013 onwards (Colares 2008). The Brazilian administration has already authorized a 4.0% biodiesel combination portion since July 1, 2009. According to the current legislation, the 5% biodiesel target will be met before 2013. In 2009, 1.5 billion liters of biodiesel were expected to be produced (Barros, 2009). To meet the essential utilized shares, a significant increase in capacity is being implemented. There are currently 65 industrial plants in operation, with another 12 awaiting official approval. The total capacity of the plant is 4 billion liters. Figure 11.3 displays the biofuel generation and consumption in Brazil from 2008 to 2017.

11.4.3.3 Colombia

In 2001, for cities with population above 500,000, the government introduced a 10% bioethanol blend (Law 963). The law was implemented during 2005 (Huete et al. 2008). By 2009, 10% ethanol content was present in almost 75% of the fuel being used in Columbia. The administration hoped to raise this to 100% by 2010 and has plans to raise the ethanol portion to 25%. The government sets the price of ethanol made from sugarcane, and it is determined by international sugar prices. The government also exempts VAT for ethanol used as fuel.

Colombian biofuel policies have made, since 2008, a re-configuration to the sugar and palm monetary divisions. Ethanol manufacture from sugar cane produces a substitutive manufacture association with sugar in Colombia (Palacio-Ciro and Vasco-Correa 2020).

The Law 939 of 2004 and Resolution 1289 of 2005 were legislations that encouraged the production and consumption of Biodiesel in Columbia. The latter mandated

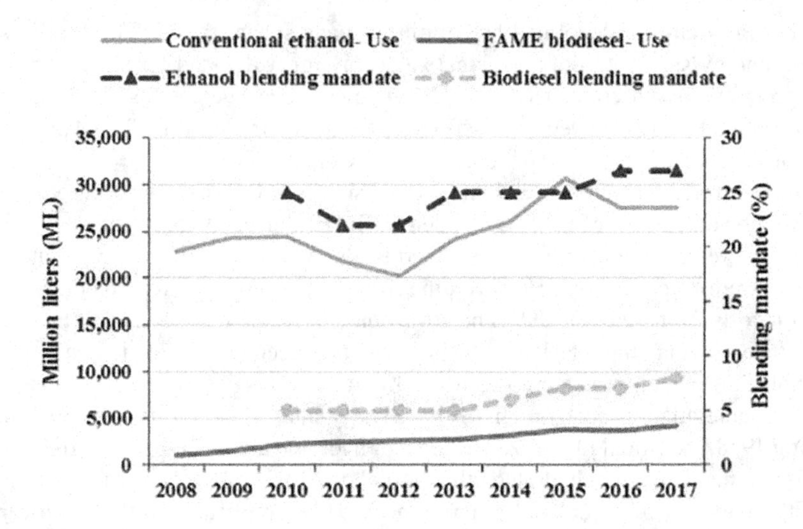

FIGURE 11.3 A line chart showing million liters and blending mandate in percentage on Y-axis increasing as time in year (from 2008 to 2017) increases on the X-axis, continuously increasing with year for different biofuel generation and consumption in Brazil.

a blend of 5.0% biodiesel by the year 2008 (Huete et al. 2008). The promotion package for biodiesel intends to increase the blending up to 10.0% during 2010 and 20.0% during 2012. Oil from palm trees, which is used to produce biodiesel, has fuel tax and sales tax exemptions and the advantage of tax exemption for income-obtained crops that are used for making biodiesel.

11.4.4 Asian Countries

11.4.4.1 China

China's biofuel regulations are mostly geared toward ethanol production. Because China is one of the net importers of vegetable oils, the government has only moderately pushed biodiesel. Biofuels contribute a small portion in the entire fuel utilization, and the consumption of biofuels is to upsurge progressively in the upcoming years. One of the significant drivers for the biofuel manufacture development industries and concurrent attenuation of biofuels in the energy market is the nation's policies (Saravanan, Pugazhendhi, and Mathimani 2020). The charges of fossil fuels, the significant gap between energy supply and demand, energy safety, ecological problems, and the fall of GHG emissions usually encourage the progress of biomass energy production in China (Yang et al. 2021). The government approved denatured fuel ethanol and bioethanol gasoline regulations for vehicles during 2001. In 2002, the Ethanol Promotion scheme was established to make use of surplus maize stockpiles. Following closely behind was the state programme of pilot missions on Bioethanol Gasoline for vehicles. The National Development and Reform Commission (NDRC) expanded the pilot project's scope to the national level in 2004, launching the State Scheme of Extensive Pilot Projects on Bioethanol Gasoline for Automobiles (The state programme of pilot missions on Bioethanol Gasoline for vehicles) (SSEPP). The government has strictly regulated ethanol manufacture and supply, appointing CNCP or Sinopec as the preferred distributors. Extensive pilot operations in five provinces and 27 cities have attained the 10.0% mixing target by the beginning of 2006 (Dong 2007). There is no official biofuel nationwide mandate for ethanol and biodiesel usage in the transport division (Ebadian et al. 2020). Due to technology and market obstructions, the real commercialization measure is far inferior than earlier estimated. In case of biodiesel, the unlawful consumption of waste oil must be sternly banned, which can be helpful to upsurge the feedstock delivery, and reduce the cost (Hao et al. 2018).

The NDRC established a Medium-Term and Long-Term Development Plan for Renewable Energy in August 2007. By 2010, renewable energy consumption as a percentage of whole prime energy usage must have risen to 10.0%, and it should have risen to 15% by 2020. Biofuels are projected to play a critical role in meeting these objectives. The yield of ethanol was expected to increase to 2.0 million tons by the year 2010 and 10.0 million tons by the year 2020. By 2010, biodiesel usage should be 200,000 tons, and it should be 2.0 million tons by the year 2020. The authentic administration keeps the price of fuel ethanol low enough that ethanol manufacturing would be unsustainable without outside funding. Producers were given a 200 USD subsidy per tonne of ethanol in 2007 (equal to US$0.158 per litre). Since 2008, subsidies based on a yearly appraisal of individual plant performance have replaced the fixed subsidy (IEA 2021). The ethanol manufacturers who have been granted a license are exempt from the 5.0% consumption tax and the 17.0% VAT. Intermediate inputs such as cereals and fertilizers are also given financial support. Furthermore, the Ministry of Finance directly supports second-generation cultivations by providing US$438 per hectare for jatropha cultivations and US$394 per hectare for cassava plantations. Biodiesel does not currently receive any direct subsidies. Biodiesel cannot be blended and distributed across the country, because there are no national biodiesel requirements. In the year January 2020, China has deferred a national wide rollout of a gasoline blend comprising of 10.0% ethanol scheduled resulting a severe drop in the nation's corn stocks and restricted biofuels manufacture volume (Biofuels Digest 2020).

11.4.4.2 India

In September 2008, Government of India (GoI) sanctioned the National Policy on biofuels, which proposed blending 20% bioethanol and biodiesel with gasoline and mineral diesel, respectively, by 2017. A key element of biodiesel production is the use of nonedible oil seeds grown on waste and marginal land plots. To assist cultivators, the government would guarantee a regularly updated seeds for biodiesel oil that have a minimum support price (MSP). For bioethanol and biodiesel, a Minimum Purchase Price (MPP) is also being imposed (Altenburg et al. 2009).

In four of the seven union territories and nine states, mandated ethanol blends of 5% were needed (out of 28). From November 2006, the 5% target was expanded to include 20 states. The decision of Government of India to increase the ethanol contribution supplies to 10.0% during 2008 has been postponed due to changes in the supply of sugar molasses related with sugar cane output—the major ethanol feedstock.

Furthermore, imported ethanol is not permitted to meet the EBP program's standards for promoting the domestic biofuel economy.

The National Mission on Biodiesel was also launched in 2009. To meet a 10% blending target, the GoI planned to cultivate jatropha on 11.20 million hectares of waste land by 2012. However, the cost of producing biodiesel outweighed the cost of purchasing it (which is set by nationwide regulators on a 6-month basis) (Singh 2009). Biofuel policy of India also adopted mandate in ethanol blending and biodiesel scheme. Though, there is not much achievement that was gained out of the biodiesel scheme primarily due to the fuel vs. food conflict and price of the biomass agriculture (Saravanan, Pugazhendhi, and Mathimani 2020).

The absence of the central excise tax (4%) is one of the fiscal incentives for biodiesel production; however, most state governments have kept the state excise fee. However, there are no direct financial tax benefits for ethanol manufacture. Nonetheless, the administration provides subsidized loans to sugar mills to build an ethanol production plant.

11.4.4.3 Indonesia

Indonesia implemented obligatory biofuel usage limits in October 2008. Biodiesel's market share should reach 2.5% by 2010 and 20.0% by 2025. In 2010, the ethanol component of gasoline must be 3%, with an upsurge to 4% in 2011. By 2025, the percentage might be as high as 15%. The previous government projections that expected 10% biofuels share by 2010 were rewritten in light of the new restrictions. Biofuel blending has been impeded by price fluctuations in feedstocks. Pertamina—the state-owned oil and fuel distribution firm and the only biofuel supplier—began marketing a blend with 5.0% biodiesel in 2006. The cost of palm oil has risen over time, and biodiesel's blending share has declined to roughly 1% in the first half of 2009 (Kifayat et al., 2018).

Fuel prices are subsidized by the government. It was calculated that gasoline subsidies had totalled additional than US$14.5 billion by October 2008. End users pay the same price for ethanol and biodiesel blends as they do for petroleum-based diesel and gasoline. Because Pertamina is a government-owned firm, any losses it experiences to equal the obligatory biofuel shares can be deemed biofuel subsidies. The losses suffered by Petromina as a result of biofuel mixing totalled US$40 million between 2006 and June 2008. In addition, importing countries like as the United States and the European Union provide indirect financial assistance through their support policies (Kifayat et al 2018).

11.4.4.4 Malaysia

The Malaysian administration intended to make advantage of rising biodiesel demand and the country's dominant focus in palm oil manufacture. The National Biofuel Policy (BNP) was established in 2005, planning to impose a 5% biodiesel (B5) mandate. The April 2007 Biofuels Industries Act established Biofuel legislation to govern and assist the industry. The 5% biodiesel mandate, on the other hand, has yet to be enforced. There are two forms of biodiesel generated. Envodiesel is made by mixing petroleum diesel with raw palm oil in a direct process. Although automotive manufacturers discourage its use, the latter is used for residential consumption.

Biodiesel made from palm methyl esters (PME) is made by transesterifying palm oil. PME is produced for overseas markets and accounts for the majority of biofuel output. Malaysia exported 95 thousand tons of PME in 2007. Approximately 75% of all biodiesel is produced in this country (Lopez and Laan 2008). Although automotive manufacturers discourage its use, the latter is used for residential consumption. Palm methyl esters (PME) Biodiesel, on the other hand, is made by transesterifying palm oil. PME is produced for overseas markets and accounts for the majority of biofuel output. Malaysia exported 95,000 tons of PME in 2007, 75% of entire biodiesel manufacture (Lopez and Laan 2008).

11.4.5 Australian Countries

In 2001, the Australian administration fixed a nonbinding goal of producing 350 million liters of biofuel per year by 2010. Although the federal government's targets are not mandatory, New South Wales fixed a blending 10.0% ethanol content in gasoline by 2011 in 2006, while Queensland fixed a 5.0% ethanol proportion in gasoline by 2011.

Biofuel production in Australia is comparatively inadequate. In 2007, ethanol production capacity was about 140.0 million liters, and biodiesel processing capacity was about 323.0 million liters, with future biodiesel and ethanol processing capacities expected to exceed 1 billion litres [AGRIRDC (Australian Government Rural Industries Research and Development Corporation), 2007]. In the fiscal years 2006–2007, it produced 83.0 million liters of ethanol and 77.0 million liters of biodiesel.

11.4.6 European

11.4.6.1 The European Union

In April 2009, the European Union's parliament endorsed a least binding goal of 10.0% biofuels in transportation sector by the year 2020, as part of the EU Directive 2009/28/EC on renewable energy. The rule also required biofuels to achieve at least 35.0% drop in GHG releases throughout their life cycle, with a goal of at least 50% by 2017. Indirect land-use change sustainability criteria are also offered. Principal forests, extremely biodiverse grass land, threatened regions, and carbon-rich regions cannot be used as biofeedstock.

Though the allocation was not mandatory, and different nations have been free to circulate their nationwide action strategies, Directive 2003/30/EC established a 5.75% market penetration goal for biofuels by 2010. Each country has been asked to aim for a 2% market share by 2005. Biofuels, on the other hand, computed for barely 1.0% of transportation fuels in 2005. Likewise, with a projected share of 4.2%, the 2010 objective is likely to be missed. At EU level, the huge mainstream of the present inadequate supply of liquid biofuels still contains in first-generation biofuels, which are to be phased out. Innovative and food waste centered biofuels are unlikely to fulfill the gap rapidly (Cadillo-Benalcazar et al. 2021).

Several governments have used tax reductions or exemptions to encourage production or consumption. The Energy Taxation Directive 2003/96/EC established the incentives that might be used to promote biofuels and meet the common agenda's goals. After the EU Commission approves tax exemptions, single nations can implement them. To avoid overcompensation, exemptions should be proportional to combination levels and must take raw material charges into account. Furthermore, they are limited to a 6-year term but may be renewed.

11.4.6.2 France

France's biofuel regulations include mandatory bioethanol and biodiesel blends, and financial incentives. The total percentage of biofuels in the transportation division was 1.2% in 2005, and it is predicted to rise to 7% by 2010. In terms of net calorific values, the aggregate biofuel percentages were computed. In 2010, biodiesel usage is predicted to reach 2.70 million tons (equal to 7.60% of entire diesel capacity), while bioethanol demand is expected to meet 0.810 million tons (equivalent to 10.70% of over-all gasoline capacity). The preceding administration's initial efforts to achieve a 10% goal for biofuels integration were shelved, and there are currently no concrete biofuel goals beyond 2010.

Despite recent reductions in economic incentives, biofuel output has risen dramatically. Bioethanol is made from weed, maize, and sugar beet. Biodiesel is made from vegetable oils (especially rapeseed). With 2.7 million tons of biodiesel consumed in 2010, France is expected to overtake Germany for the first time. With a projected output of 2.66 million tons, Germany continues to lead the European biodiesel industry, whilst French businesses are projected to generate 2.30 million tons in 2010. France remains the Euro region's top ethanol producer, with 700,000 tons of ethanol expected in 2010. In the medium future, Germany and France should continue to be the primary European biofuel markets, both ethanol and biodiesel have similar consumption and production patterns. Biodiesel production in France is about 2206.0 millions of liters per year.

11.4.6.3 Germany

The German government adopted the vision of its mandated biofuel targets in July 2009. For 2009, the whole mandated biofuels contribution in the transportation division was increased to 5.3%, and it shall remain at that level until 2014 (all the proportions are computed in net calorific standards). Furthermore, the least biodiesel percentage in diesel transportation has been retroactively fixed at 4.40% for 2009, and it shall remain at that level until 2014. Likewise, the percentage of ethanol in gasoline has been set at 2.80% from 2009 to 2014. Biofuel quotas will be established based on GHG emission reductions starting in 2015 (Sorda, Banse, and Kemfert 2010). The German administration is presently debating on how to reduce CO_2 emissions while also monitoring their implementation. Germany is the primary EU nation to recommend biofuel quotas based on reductions in greenhouse gas emissions. Biodiesel production in Germany is about 2859.0 millions of liters per year. From 2010 to 2014, there was a 6.3% biofuels production counting both ethanol and biodiesel. There is a significant reduction in GHG emissions of 3.50%, 4.0%, 6.0%, respectively, in the fuel mixture for the overall fuel division from 2015, 2017, 2020, onwards. No tax relaxation for FAME biodiesel, HVO/ HEFA fuels, vegetable oils, and ethanol. The fuel tax for CNG and biomethane is € 0.0139/ kWh till 2023. A carbon tax was indirectly applicable via CO_2 tax for passenger cars (Ebadian et al. 2020). Figure 11.4 represents the biofuel generation and consumption in Germany from 2007 to 2016.

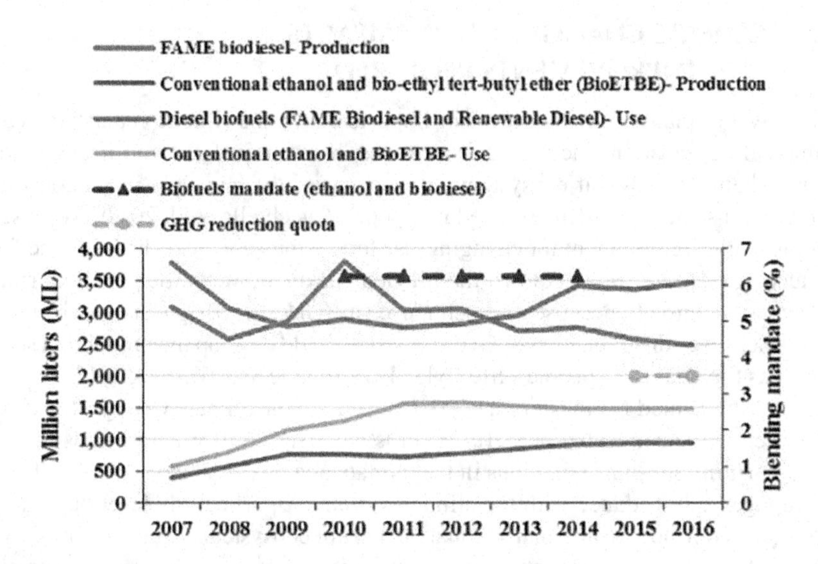

FIGURE 11.4 The figure represents the biofuel generation and consumption in Germany from 2007 to 2016.

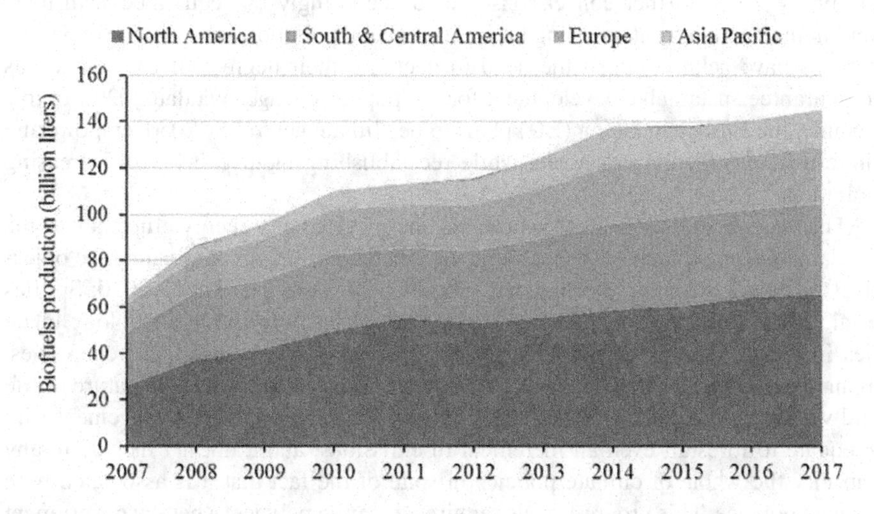

FIGURE 11.5 Area chart showing biofuels production in billion liters for North America, South and Central America, Europe and Asia Pacific on Y-axis increasing as time in year (from 2007 to 2017) increases on the X-axis, continuously increasing with year.

Biofuels manufacture has augmented at a yearly increase rate of 11.40%, from over 64.0 billion liters' production occurred in the year 2007 to over 145.0 billion liters in the year 2017 (BP, 2018). Figure 11.5 shows the global biofuels production from 2007 to 2017.

11.5 CLIMATE CHANGE ACTION IMPACTS AND IMPROVEMENTS REQUIRED

Currently, our plant is having the major impacts due to the climate change and environmental degradation. Increasing heat waves, heavy rain falls, and heavy storms are not changing only our ecosystem but also causes the huge loss in economy and increase the scale of challenges faced by people (Cucchiella et al. 2018). At present, there is a huge requirement for changing our food habits and also changing the food production methods to breakdown the previous nexus between the global warming and economic growth because of greenhouse gas (GHG) emission. As per the recent studies, it proves that reduction of GHG emission could develop the production sector as well as the quality of people's life style also (Bojanic and Warnick 2012).

Under the United Nations Framework Convention on Climate Change, the essential target of the connected peaceful accords endorsed to date is the adjustment of ozone-harming substance fixations in the climate at a level that would forestall risky anthropogenic impedance with the climate system. A particularly level ought to be accomplished inside a time span adequate to permit ecosystems to adjust normally to climate change, to guarantee that food creation isn't compromised, and to empower monetary improvement to continue in a reasonable way (Kyoto Protocol, Copenhagen Accord, Paris Agreement). As per the United Nations, while there are more than 7.7 billion individuals living on Earth in 2020, this figure is relied upon to arrive at 9.7 billion by 2050; further conventions should accordingly be acquainted with forestall human action front speeding up environmental pollution and resource repletion. Nations have acknowledged the need to decrease their degrees of CO_2 emissions to guarantee sustainable development for the populace (Jäger-Waldau 2020). In this manner, the European Union (EU) plans to be climate neutral by 2050, empowering the effective utilization of assets while reestablishing biodiversity and decreasing pollution.

The various studies focus to renewable energy (RE) and energy efficiency as the fundamental apparatuses for checking climate change, and accomplishing the objectives set at the different international submits held (Østergaard et al. 2020; Spillias et al. 2020; Xu, Chou, and Zhang 2019). The intricacy of the methodology taken lies in the need to change current energy systems through a shift to renewables, along these lines guaranteeing energy security just as enhancements in nature of life and wellbeing (Buonocore et al. 2016). Nonetheless, renewables' arrangement isn't adequate to forestall even an increment in emissions, and it doesn't include in any capacity the whole of climate policies in spite of the fact that it is associated with climate policies. This progress cycle requires dynamic policies, whereby government officials distribute funds to venture projects pointed toward moderating the reasons for global warming, while at the same time supporting and giving motivators to the organization of renewable technologies (Bhardwaj et al. 2019). The course of progress requires the private sector and public institutions to cooperate to pinpoint issues, put out objectives, and distinguish potential collaborations between sectors (van de Ven, González-Eguino, and Arto 2018).

The outcomes lead more noteworthy straightforwardness to global policy issues, working with the comparison of endeavors to control climate change. Also, they give data on the accomplishment of the objectives set in the Paris Agreement, in light of the examination of the below indicators (Puertas and Marti 2021).

GHG Emissions (40%): Quantitatively evaluates the measures taken by nations to minimize the GHG emissions—an objective set for all nations to make preparations for unsafe climate change. It is allocated a higher load than different parts, in light of the fact that subject matter authorities agree, it bears the best liability regarding global warming. The CCPI utilizes the PRIMAP information base to evaluate all GHG emissions.

Renewable Energy (20%): Measures activities focused on expanding the utilization of REs in each of the nations evaluated. The CCPI utilizes measurable data given by the International Energy Agency.

Energy Use (20%): Evaluates upgrades in energy efficiency, and control depends on domestic energy utilization. The CCPI utilizes factual data given by the International Energy Agency.

Climate Policy (20%): Evaluates the viability of climate policies executed in the various nations. Assessments of nations' exhibition in climate strategy depend on a yearly refreshed study of national climate and energy specialist's front civil society.

The overall index shows nations score in the range of 0–100, where higher values show more "climate-friendly" nation. The last CCPI ranking is determined from the weighted average scores accomplished in each indicator, by the following equation (Puertas and Marti 2021).

$$\text{CCPI} = \sum_{i=1}^{n} Wi * Xi$$

Where,

Xi = Normalized indicator

Wi = Weighing of Xi

The GHG emission values in the range of 0 and 40, and the remainder indicators in the range of 0 and 20. The studies shows that no nation has had the option to accomplish the most extreme value allocated to every part of the indicator, with more prominent opportunity to get better seen in Renewable Energy and GHG emissions (Puertas and Marti 2021).

11.6 CONCLUSION

The overview on renewable directives and global policies from different research-based studies empathize that the implementation of renewable energy is important for the pollution-free economic growth of a country. Energy is the highest sector where the highest amount of GHGs are producing so that the International conventions and sustainable goals from the united nation organization compels or forces the nations for the transition of energy from conventional to the renewable source by encouraging the carbon trading system—clean development mechanism to prevent from drastic climate change. Transition from conventional to renewable energy is more beneficial for the countries where there are less fossil fuel reserves compared to its usage, but it's bad for the fossil fuel exporting countries. Fossil fuel exporting countries have to shift their economy generation to different sectors, such as product manufacturing, service oriented, or tourism based.

REFERENCES

Alanne, K., and A. Saari. 2006. "Distributed energy generation and sustainable development." *Renewable and Sustainable Energy Reviews* 10 (6): 539–58. Doi: 10.1016/j.rser.2004.11.004.

Alaswad, A., M. Dassisti, T. Prescott, and A. G. Olabi. 2015. "Technologies and developments of third generation biofuel production." *Renewable and Sustainable Energy Reviews.* Doi: 10.1016/j.rser.2015.07.058.

Alizadeh, R., L. Soltanisehat, P.D. Lund, and H. Zamanisabzi. 2020. "Improving renewable energy policy planning and decision-making through a hybrid MCDM method." *Energy Policy* 137 (February): 111174. Doi: 10.1016/J.ENPOL.2019.111174.

Altenburg, T., H. Dietz, M. Hahl, N. Nikolidakis, C. Rosendahl, and K. Seelige. 2009. *Biodiesel in India*. Se2Isnch. Vol. 49.

Amigun, B., J.K. Musango, and W. Stafford. 2011. "Biofuels and sustainability in Africa." *Renewable and Sustainable Energy Reviews* 15 (2): 1360–72. Doi: 10.1016/J.RSER.2010.10.015.

Areal, F.J., L. Riesgo, and E. Rodríguez-Cerezo. 2013. "Economic and agronomic impact of commercialized GM crops: A meta-analysis." *Journal of Agricultural Science* 151 (1): 7–33. Doi: 10.1017/S0021859612000111.

Avinash, A, P. Sasikumar, and A. Murugesan. 2018 "Understanding the interaction among the barriers of biodiesel production from waste cooking oil in India- an interpretive structural modeling approach." *Renewable Energy* 127: 678–84.

Bhardwaj, A., M. Joshi, R. Khosla, and N.K. Dubash. 2019. "More priorities, more problems? Decision-making with multiple energy, Development and climate objectives." *Energy Research & Social Science* 49 (March): 143–57. Doi: 10.1016/J.ERSS.2018.11.003.

Bhaskar, T., B. Bhavya, R. Singh, D. Viswanath Naik, A. Kumar, and H.B. Goyal. 2011. "Thermochemical conversion of biomass to biofuels." *Biofuels*, January, 51–77. Doi: 10.1016/B978-0-12-385099-7.00003-6.

Bojanic, D.C., and R.B. Warnick. 2012. "The role of purchase decision involvement in a special event." *Journal of Travel Research* 51 (3): 357–66. Doi: 10.1177/0047287511418364.

Buitrón, G., J. Carrillo-Reyes, M. Morales, C. Faraloni, and G. Torzillo. 2017. "Biohydrogen production from microalgae." *Microalgae-Based Biofuels and Bioproducts: From Feedstock Cultivation to End-Products*, January, 209–34. Doi: 10.1016/B978-0-08-101023-5.00009-1.

Buonocore, J.J., P. Luckow, G. Norris, J.D. Spengler, B. Biewald, J. Fisher, and J.I. Levy. 2016. "Health and climate benefits of different energy-efficiency and renewable energy choices." *Nature Climate Change* 6 (1): 100–106. Doi: 10.1038/nclimate2771.

Burke, M.J., and J.C. Stephens. 2018. "Political power and renewable energy futures: A critical review." *Energy Research & Social Science* 35 (January): 78–93. Doi: 10.1016/J.ERSS.2017.10.018.

Cadillo-Benalcazar, J.J., S.G.F. Bukkens, M. Ripa, and M. Giampietro. 2021. "Why does the european union produce biofuels? Examining consistency and plausibility in prevailing narratives with quantitative storytelling." *Energy Research & Social Science* 71 (January): 101810. Doi: 10.1016/J.ERSS.2020.101810.

Chandra, A. 2021. "Report Name: Biofuels Annual."

Chen, C.Y., K.L. Yeh, R. Aisyah, D.J. Lee, and J.S. Chang. 2011. "Cultivation, photobioreactor design and harvesting of microalgae for biodiesel production: A critical review." *Bioresource Technology* 102 (1): 71–81. Doi: 10.1016/J.BIORTECH.2010.06.159.

Chi, X., A. Li, M. Li, L. Ma, Y. Tang, B. Hu, and J. Yang. 2018. "Influent characteristics affect biodiesel production from waste sludge in biological wastewater treatment systems." *International Biodeterioration and Biodegradation* 132 (19): 226–35. Doi: 10.1016/j.ibiod.2018.04.010.

Claxton, L.D. 2015. "The history, genotoxicity, and carcinogenicity of carbon-based fuels and their emissions. Part 3: Diesel and gasoline." *Mutation Research/Reviews in Mutation Research* 763 (January): 30–85. Doi: 10.1016/J.MRREV.2014.09.002.

Clément, A., P. McCullen, A. Falcão, A. Fiorentino, F. Gardner, K. Hammarlund, G. Lemonis, et al. 2002. "Wave energy in Europe: Current status and perspectives." *Renewable and Sustainable Energy Reviews* 6 (5): 405–31. Doi: 10.1016/S1364-0321(02)00009-6.

Colares, J.F. 2008. "Case western reserve University school of law scholarly commons a brief history of Brazilian biofuels legislation."

Cotula, L., Dyer, N., and Vermeulen, S. 2008. Fuelling exclusion? The biofuels boom and poor people's access to land, IIED, London. ISBN: 978-1-84369-702-2.

Cucchiella, F., I. D'Adamo, M. Gastaldi, and M. Miliacca. 2018. "Efficiency and allocation of emission allowances and energy consumption over more sustainable European economies." *Journal of Cleaner Production* 182 (May): 805–17. Doi: 10.1016/J.JCLEPRO.2018.02.079.

Dahman, Y., K. Syed, S. Begum, P. Roy, and B. Mohtasebi. 2019. "Biofuels: Their characteristics and analysis." *Biomass, Biopolymer-Based Materials, and Bioenergy: Construction, Biomedical, and Other Industrial Applications*, January, 277–325. Doi: 10.1016/B978-0-08-102426-3.00014-X.

Daroch, M., S. Geng, and G. Wang. 2013. "Recent advances in liquid biofuel production from algal feedstocks." *Applied Energy* 102 (February): 1371–81. Doi: 10.1016/J.APENERGY.2012.07.031.

Dong, F. 2007. "Food security and biofuels development: The case of China." *Director*, no. October.

Dote, Y., S. Sawayama, S. Inoue, T. Minowa, and S.Y. Yokoyama. 1994. "Recovery of liquid fuel from hydrocarbon-rich microalgae by thermochemical liquefaction." *Fuel* 73 (12): 1855–57. Doi: 10.1016/0016-2361(94)90211-9.

Douglas, T. 2007. "'Green' hydro power - understanding impacts, approvals, and sustainability of run-of-river independent power projects in British Columbia." *Watershed Watch Salmon Society*, no. August.

Drew, B., A.R. Plummer, and M.N. Sahinkaya. 2009. "A review of wave energy converter technology." *Proceedings of the Institution of Mechanical Engineers, Part A: Journal of Power and Energy* 223 (8): 887–902. Doi: 10.1243/09576509JPE782.

Ebadian, M., S. van Dyk, J.D. McMillan, and J. Saddler. 2020. "Biofuels policies that have encouraged their production and use: An international perspective." *Energy Policy* 147 (December). Doi: 10.1016/j.enpol.2020.111906.

Gaurav, N., S. Sivasankari, G.S. Kiran, A. Ninawe, and J. Selvin. 2017. "Utilization of bioresources for sustainable biofuels: A review." *Renewable and Sustainable Energy Reviews* 73 (November 2016): 205–14. Doi: 10.1016/j.rser.2017.01.070.

Greenwell, H.C., L.M.L. Laurens, R.J. Shields, R.W. Lovitt, and K.J. Flynn. 2010. "Placing microalgae on the biofuels priority list: A review of the technological challenges." *Journal of the Royal Society Interface* 7 (46): 703–26. Doi: 10.1098/rsif.2009.0322.

Gude, V.G., and E. Martinez-Guerra. 2018. "Green chemistry with process intensification for sustainable biodiesel production." *Environmental Chemistry Letters* 16 (2): 327–41. Doi: 10.1007/s10311-017-0680-9.

Gumisiriza, R., J.F. Hawumba, M. Okure, and O. Hensel. 2017. "Biomass waste-to-energy valorisation technologies: A review case for banana processing in Uganda." *Biotechnology for Biofuels* 10 (1): 1–29. Doi: 10.1186/s13068-016-0689-5.

Hannon, M., J. Gimpel, M. Tran, B. Rasala, and S. Mayfield. 2010. "Biofuels from algae: Challenges and potential." *Biofuels* 1 (5): 763–84. Doi: 10.4155/bfs.10.44.

Hao, H., Z. Liu, F. Zhao, J. Ren, S. Chang, K. Rong, and J. Du. 2018. "Biofuel for vehicle use in China: Current status, future potential and policy implications." *Renewable and Sustainable Energy Reviews* 82 (February): 645–53. Doi: 10.1016/J.RSER.2017.09.045.

Harjanne, A., and J.M. Korhonen. 2019. "Abandoning the concept of renewable energy." *Energy Policy* 127 (April): 330–40. Doi: 10.1016/J.ENPOL.2018.12.029.

Ho, D.P., H.H. Ngo, and W. Guo. 2014. "A mini review on renewable sources for biofuel." *Bioresource Technology* 169: 742–49. Doi: 10.1016/j.biortech.2014.07.022.

Huete, S., B. Flach, K. Bendz, B. Dahlbacka, B. Flach, M. Hanley, P. Rucinski, and J. Wilson. 2008. "GAIN Report," 1–6.

Ibarra-Gonzalez, P., and B.G. Rong. 2019. "A review of the current state of biofuels production from lignocellulosic biomass using thermochemical conversion routes." *Chinese Journal of Chemical Engineering* 27 (7): 1523–35. Doi: 10.1016/J.CJCHE.2018.09.018.

Inderwildi, O.R., and D.A. King. 2009. "Quo vadis biofuels?" *Energy and Environmental Science* 2 (4): 343–46. Doi: 10.1039/b822951c.

International Energy Agency. 2021, 89–107. www.iea.org.

IRENA, IEA and REN21. 2018. *Renewable Energy Policies in a Time of Transition.* IRENA, OECD/ IEA and REN21, 17–20, ISBN 978-92-9260-061-7.

Jäger-Waldau, A. 2020. "The untapped area potential for photovoltaic power in the European Union." *Clean Technologies* 2 (4): 440–46. Doi: 10.3390/cleantechnol2040027.

Joselin Herbert, G. M., S. Iniyan, E. Sreevalsan, and S. Rajapandian. 2007. "A review of wind energy technologies." *Renewable and Sustainable Energy Reviews* 11 (6): 1117–45. Doi: 10.1016/j.rser.2005.08.004.

Jumbe, C.B.L., F.B.M. Msiska, and M. Madjera. 2009. "Biofuels development in sub-saharan Africa: Are the policies conducive?" *Energy Policy* 37 (11): 4980–86. Doi: 10.1016/J.ENPOL.2009.06.064.

Kaunda, C.S., C.Z. Kimambo, and T.K. Nielsen. 2012. "Hydropower in the context of sustainable energy supply: A review of technologies and challenges." *ISRN Renewable Energy* 2012: 1–15. Doi: 10.5402/2012/730631.

Kemausuor, F., G.Y. Obeng, A. Brew-Hammond, and A. Duker. 2011. "A review of trends, policies and plans for increasing energy access in Ghana." *Renewable and Sustainable Energy Reviews* 15 (9): 5143–54. Doi: 10.1016/J.RSER.2011.07.041.

Kifayat, U., V.K. Sharma, M. Ahmad, P. Lv, J. Krahl, Z.W. Sofia. 2018. "The insight views of advanced technologies and its application in bio-origin fuel synthesis from lignocellulose biomasses waste, a review." *Renewable and Sustainable Energy Reviews*, 82 (3): 3992–4008, ISSN 1364-0321, Doi: 10.1016/j.rser.2017.10.074. (https://www.sciencedirect.com/science/article/pii/S1364032117314417).

Kumar, P., C., Rena, A. Meenakshi, A.S. Khapre, S. Kumar, A. Anshul, L. Singh, S. Hyoun Kim, B.D. Lee, and R. Kumar. 2019. "Bio-hythane production from organic fraction of municipal solid waste in single and two stage anaerobic digestion processes." *Bioresource Technology* 294 (August): 122220. Doi: 10.1016/j.biortech.2019.122220.

Kumari, D., and R. Singh. 2018. "Pretreatment of lignocellulosic wastes for biofuel production: A critical review." *Renewable and Sustainable Energy Reviews* Doi: 10.1016/j.rser.2018.03.111.

Li, M., N. Luo, and Y. Lu. 2017. "Biomass energy technological paradigm (BETP): Trends in this sector." *Sustainability (Switzerland)* 9 (4): 1–28. Doi: 10.3390/su9040567.

Lopez, G.P., and T. Laan. 2008. "Government support for biodiesel in Malaysia." *Global Subsidies Initiative.*

Maheshwari, N., P.K. Krishna, I.S. Thakur, and S. Srivastava. 2020. "Biological fixation of carbon dioxide and biodiesel production using microalgae isolated from sewage waste water." *Environmental Science and Pollution Research* 27 (22): 27319–29. Doi: 10.1007/s11356-019-05928-y.

Malik, S.N., Rena, and S. Kumar. 2021. "Enhancement effect of zero-valent iron nanoparticle and iron oxide nanoparticles on dark fermentative hydrogen production from molasses-based distillery wastewater." *International Journal of Hydrogen Energy* 46 (58): 29812–21. Doi: 10.1016/j.ijhydene.2021.06.125.

Malode, S.J., K.K. Prabhu, R.J. Mascarenhas, N.P. Shetti, and T.M. Aminabhavi. 2021. "Recent advances and viability in biofuel production." *Energy Conversion and Management: X* 10 (September 2020): 100070. Doi: 10.1016/j.ecmx.2020.100070.

Minowa, T., S.y. Yokoyama, M. Kishimoto, and T. Okakura. 1995. "Oil production from algal cells of *Dunaliella tertiolecta* by direct thermochemical liquefaction." *Fuel* 74 (12): 1735–38. Doi: 10.1016/0016-2361(95)80001-X.

Mohan, M.R., S.R. Paranjothi, and S.P. Israel. 1997. "Use of pumped-hydro as peak-load management plant in optimal scheduling of power systems." *Electric Machines and Power Systems* 25 (10): 1047–61. Doi: 10.1080/07313569708955796.

Nunes, L.J.R., L.M.E.F. Loureiro, L.C.R. Sá, and H.F.C. Silva. 2020. "Waste recovery through thermochemical conversion technologies: A case study with several portuguese agroforestry by-products." *Clean Technologies* 2 (3): 377–91. Doi: 10.3390/cleantechnol2030023.

Ohimain, E.I. 2010. "Emerging Bio-ethanol projects in Nigeria: their opportunities and challenges." *Energy Policy* 38 (11): 7161–68. Doi: 10.1016/J.ENPOL.2010.07.038.

Østergaard, P.A., N. Duic, Y. Noorollahi, H. Mikulcic, and S. Kalogirou. 2020. "Sustainable development using renewable energy technology." *Renewable Energy* 146 (February): 2430–37. Doi: 10.1016/J.RENENE.2019.08.094.

Palacio-Ciro, S., and C.A. Vasco-Correa. 2020. "Biofuels policy in Colombia: A reconfiguration to the sugar and palm sectors?" *Renewable and Sustainable Energy Reviews* 134 (December): 110316. Doi: 10.1016/J.RSER.2020.110316.

Puertas, R., and L. Marti. 2021. "International ranking of climate change action: An analysis using the indicators from the climate change performance index." *Renewable and Sustainable Energy Reviews* 148 (September): 111316. Doi: 10.1016/J.RSER.2021.111316.

Raheem, A.W. A.K.G. Wan Azlina, Y.H. Taufiq Yap, M.K. Danquah, and R. Harun. 2015. "Thermochemical conversion of microalgal biomass for biofuel production." *Renewable and Sustainable Energy Reviews* 49 (September): 990–99. Doi: 10.1016/J.RSER.2015.04.186.

REN21, P.S. 2014. "Renewables 2014: Global Status Report. 2014: Secretariat Renewable Energy Policy Network for the 21st Century(REN21) Paris." Doi: 10.1016/j.ces.2019.03.057.

Rena, K.M.B. Zacharia, S. Yadav, N.P. Machhirake, S.H. Kim, B.D. Lee, H. Jeong, L. Singh, S. Kumar, and R. Kumar. 2020. "Bio-hydrogen and bio-methane potential analysis for production of bio-hythane using various agricultural residues." *Bioresource Technology* 309 (February): 123297. Doi: 10.1016/j.biortech.2020.123297.

Saravanan, A.P., A. Pugazhendhi, and T. Mathimani. 2020. "A comprehensive assessment of biofuel policies in the BRICS nations: Implementation, blending target and gaps." *Fuel* 272 (July): 117635. Doi: 10.1016/J.FUEL.2020.117635.

Saravanan, A., P.S. Kumar, S. Jeevanantham, S. Karishma, D.-V.N. Vo 2022 "Recent advances and sustainable development of biofuels production from lignocellulosic biomass." *Bioresource Technology* 344: 1–3. ISSN 0960-8524, Doi: 10.1016/j.biortech.2021.126203. (https://www.sciencedirect.com/science/article/pii/S0960852421015455).

Sekoai, P.T., and K.O. Yoro. 2016. "Biofuel development initiatives in Sub-Saharan Africa: Opportunities and challenges." *Climate* 4 (2). Doi: 10.3390/cli4020033.

Sorda, G., M. Banse, and C. Kemfert. 2010. "An overview of biofuel policies across the world." *Energy Policy* 38 (11): 6977–88. Doi: 10.1016/J.ENPOL.2010.06.066.

Spillias, S., P. Kareiva, M. Ruckelshaus, and E. McDonald-Madden. 2020. "Renewable energy targets may undermine their sustainability." *Nature Climate Change* 10 (11): 974–76. Doi: 10.1038/s41558-020-00939-x.

Tandon, P., and Q. Jin. 2017. "Microalgae culture enhancement through key microbial approaches." *Renewable and Sustainable Energy Reviews* 80 (December): 1089–99. Doi: 10.1016/J.RSER.2017.05.260.

Tatsidjodoung, P., M.H. Dabat, and J. Blin. 2012. "Insights into biofuel development in burkina Faso: Potential and strategies for sustainable energy policies." *Renewable and Sustainable Energy Reviews* 16 (7): 5319–30. Doi: 10.1016/J.RSER.2012.05.028.

Thangaraj, B., P.R. Solomon, B. Muniyandi, S. Ranganathan, and L. Lin. 2019. "Catalysis in biodiesel production - A review." *Clean Energy* 3 (1): 2–23. Doi: 10.1093/ce/zky020.

Ullah, K., V.K. Sharma, M. Ahmad, P. Lv, J. Krahl, and Z. Wang. 2018. "The insight views of advanced technologies and its application in bio-origin fuel synthesis from lignocellulose biomasses waste, a review." *Renewable and Sustainable Energy Reviews* 82 (October): 3992–4008. Doi: 10.1016/j.rser.2017.10.074.

Vangrysperre, W., M. Callens, H. Kersters-Hilderson, and C.K. De Bruyne. 1988. "Evidence for an essential histidine residue in D-Xylose isomerases." *The Biochemical Journal* 250 (1): 153–60. Doi: 10.1042/bj2500153.

van de Ven, D.J.v.d., M. González-Eguino, and I. Arto. 2018. "The potential of behavioural change for climate change mitigation: A case study for the European union." *Mitigation and Adaptation Strategies for Global Change* 23 (6): 853–86. Doi: 10.1007/s11027-017-9763-y.

Wolinetz, M., M. Hein, and B. Moawad. 2019. "Biofuels in Canada 2019." *Navius Research.*

Xu, R., L.C. Chou, and W.H. Zhang. 2019. "The effect of CO_2 emissions and economic performance on hydrogen-based renewable production in 35 European countries." *International Journal of Hydrogen Energy* 44 (56): 29418–25. Doi: 10.1016/J.IJHYDENE.2019.02.167.

Yang, Y., Z. Tian, Y. Lan, S. Wang, and H. Chen. 2021. "An overview of biofuel power generation on policies and finance environment, applied biofuels, device and performance." *Journal of Traffic and Transportation Engineering (English Edition)* 8 (4): 534–53. Doi: 10.1016/J.JTTE.2021.07.002.

Yogeeswari, S., T.A. Masron. 2021. "The impact of economic globalization on biofuel in developing countries." *Energy Conversion and Management*, 10: 1–2, 100064, ISSN 2590-1745, Doi: 10.1016/j.ecmx.2020.100064. (https://www.sciencedirect.com/science/article/pii/S2590174520300362).

Zeraatkar, A.K., H. Ahmadzadeh, A.F. Talebi, N.R. Moheimani, and M.P. McHenry. 2016. "Potential use of algae for heavy metal bioremediation, a critical review." *Journal of Environmental Management* 181 (October): 817–31. Doi: 10.1016/J.JENVMAN.2016.06.059.

Zhu, B., G. Chen, X. Cao, and D. Wei. 2017. "Molecular characterization of CO_2 sequestration and assimilation in microalgae and its biotechnological applications." *Bioresource Technology* 244 (November): 1207–15. Doi: 10.1016/J.BIORTECH.2017.05.199.

12 Biofuels Storage and Transport Systems
Challenges and Role of Type-IV Composite Overwrapped Vessels

Mukesh Kumar
Indian Institute of Technology Ropar

CONTENTS

Biofuel storage could be the central component in successfully implementing a clean and efficient biofuel-based economy. The current storage devices are degraded with biofuels because of the self-degradation properties that lead to leakage, corrosion, sludge formation, and oxidation with time. Compared to other vessels, Type-IV composite overwrapped vessels (COVs) have favorable characteristics, such as high fatigue resistance, corrosion resistance, and a large strength-to-weight ratio. It consists of a polymeric liner, metallic boss to hold the filling valve, and high-strength composite over the liner boss assembly to provide adequate strength. The polymeric liner plays a vital role in designing Type-IV COVs and acts as a barrier to biofuels to resist explosive decompression and permeation-based system failures.

So, in this chapter, the effect of operational and environmental parameters (temperature, pressure, humidity, exposure duration, and storage time), and biofuel properties (molecular size, density, viscosity, and diffusivity) during storage on metal, polymer,

DOI: 10.1201/9781003197737-16

and composite vessels materials are investigated. In addition, potential challenges, safety standards, scope, and cost of hydrogen storage and transportation are revealed.

12.1 INTRODUCTION

The primary energy sources on the earth are conventional fuels such as coal, petroleum, natural gas, etc. The significant concerns with traditional fuels are the severe issues related to climatic fluctuations, global warming, ozone layer depletion, greenhouse gas (GHGs) emissions, acid rain, etc. Further, these traditional resources have limited availability due to excessive utilization. In this context, the focus of the world has shifted towards an urgent requirement for clean, sustainable, and efficient alternative (renewable) energy sources (biofuels, hydrogen, wind, solar, geothermal, compressed, and liquefied natural gas, etc.) to conserve the resources. These nonconventional energy sources have the potential to eliminate some of the dismissive limitations, such as eradicating the surplus usage of fossil fuel, depletion of natural resources, and heavy dependence on manufacturers. On the other side with the applicability of nonconventional energy resources, Biofuels are considered to be essential energy resources for achieving sustainable development goals on a worldwide basis, as they are generally produced from organic matter (food crops, non-food crops, wastes using microbes, and so on) and renewable resources (Axelsson et al. 2012). With the reference to the above-mentioned concrete merits for the utilization of nonconventional energy resources, which further resulted in improving the farmers' financial situation by enhancing agricultural production that could indeed assist in driving the financial economy of the nation for progressive growth and development. Moreover, as referred from the existing literature, advanced (third-generation) biofuels have several advantages over first- and second-generation biofuels (Gaudreau 2009). As far as the latest scientific advancements are concerned, biofuels such as biohydrogen have been found to be more efficient than conventional fossil fuels and are the ones that have gained more attraction among advanced biofuels due to their excellent properties. As hydrogen has produced through the action of living organisms and other biological processes, named biohydrogen (Kumar et al. 2021). Moreover, the hydrogen produced through renewable energy sources, biomass, and so on methods can be used to reduce emissions or contaminants (unburnt hydrocarbons) in numerous applications in transport, agriculture, and so on sectors. As far as the application prospects for the same are concerned, hydrogen has emerged as a true zero-emission solution for powering light to heavy-duty vehicles and stationary equipment. In addition, it finds a wide range of applications from transportation to electricity generation. Nowadays, there is widespread interest in this recent technology, as biohydrogen is a clean, sustainable, and carbon-free fuel gas and can be readily produced from biomass of certain kinds. From the referred sources in context with the production methods of advanced biofuel technologies—the antibacterial mechanism (Ghimire et al. 2017), microbes are grown in a bioreactor (Zaidi et al. 2019), metallic nanoparticles (gold, silver, and so on), KTCC1737 (Khan et al. 2021), and fermentation-based nanoparticles from black liquor (Tawfik et al. 2021) are also able to produce biohydrogen. Numerous hydrogen production and storage methods are reliable, economical, safe, and efficient (Kayfeci and Keçeba 2019).

Hydrogen stored using compressed, cryo-compressed, and materials-based storage is generally used worldwide in various industrial applications. The other new type of energy has also emerged, and the applications of hydrogen energy and usage have grown up with an efficient way to avoid footprint constraints. However, the storage and transportation of hydrogen in large volumes and mass is still a topic of ongoing research. In commercially available light and heavy-duty fuel cell vehicles (FCVs), Type-IV composite overwrapped vessels (COVs) are preferred for storing gaseous hydrogen in compressed form at high pressures (more than 70 MPa) following due safety standards and protocols. The first 70 MPa hydrogen storage Type-IV COV has been demonstrated in 2001. The first FCV has been developed in Japan by Toyota, as described by Toyota Motor Corporation (Aso et al. 2007; Yumiya et al. 2015). Such high-pressure Type-IV COVs consist of a polymeric liner that provides the hydrogen permeation barrier, a metallic boss that holds the high-pressure filling valve, and the carbon fiber-reinforced polymer composite windings around the liner-boss assembly to ensure adequate strength to hold highly pressurized hydrogen gas with proper safety, as shown in Figure 12.1.

Still, efforts are being made day to night for sustainable mobility to transform classic vehicles into hydrogen-powered electric bikes/cars/buses/trucks/boats/trains, etc. High-performance COVs are generally used in aerospace, astronautics, automotive industries, spacecraft, and launch vehicle applications. It provides safer, cost-effective, and lightweight storage systems for the fluids such as compressed natural gas, nitrogen, oxygen, hydrogen, etc. The Type-IV COV has the advantages of being lightweight, having high resistance to fatigue, excellent corrosion resistance, high strength/stiffness-to-weight ratio, and service life.

This book chapter includes the most renewable and efficient type of biofuel (biohydrogen) energy storage and transportation challenges. The role of the storage and transportation system is summarized in terms of critical operating conditions, specific fuel characteristic issues, and challenges related to biohydrogen. The main problem is to design a failure-resistant storage device against high pressure, temperature, corrosion, permeation, leakage, etc. Understanding the cause and prevention strategy to minimize these critical challenges for biofuel storage is essential. Failure can be controlled by suitable engineering design specifications, material compatibility,

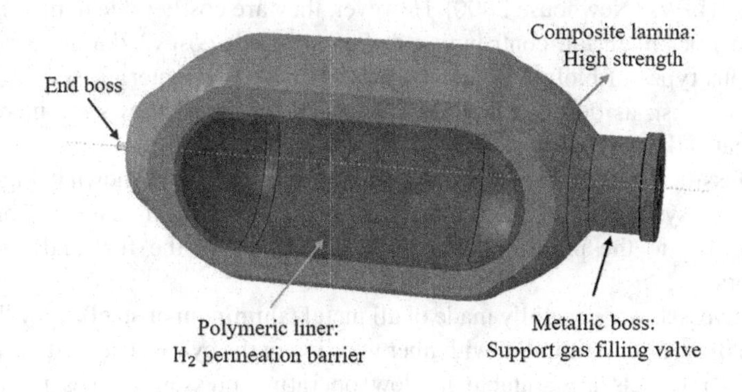

FIGURE 12.1 Schematic of Type-IV COV with integrated parts.

coatings, cathodic protection, consideration of the change in environmental conditions, online-real time monitoring, and so on (Groysman 2017). Here, suitable design standards, material specifications, design, material-related failure, and prevention strategies are described based on the given operating conditions for biohydrogen storage. Type-IV COVs technologies provide an efficient opportunity for the storage of biohydrogen. In order to meet the standards of the energy requirement, it has a substantial advantage of transporting and storing biohydrogen. This chapter also includes the potential challenges with biohydrogen and other biofuel storage.

12.2 CHALLENGES OF HYDROGEN STORAGE AND RELEVANT FAILURES OF TYPE-IV COVs MATERIALS

As acquired from the previously published sources, the high-pressure hydrogen storage with Type-IV COV is currently the only commercially available solution to provide a competitive driving range with superior results for FCV applications. Such high-pressure Type-IV COVs at 70 MPa working pressure (WP) are the best alternatives compared to other pressure vessels (PV) types. Type-IV COVs storage systems are the preferred mode of storing alternative fuels, rather than metallic or metal-based (such as stainless steel, carbon steel, Inconel, and more) storage vessels. The options for biofuel storage in materials such as glass-reinforced plastic, thermoplastic, and polyethylene COVs increase daily, rather than in steel and concrete materials. A good energy storage and transportation system should have lower manufacturing costs and resistance to chemical corrosion, the barrier to permeation, and so on. Most metallic PVs are prone to corrosion, hydrogen embrittlement (HE), catastrophic failure, and lesser strength-to-weight ratio capacity at critical operational parameters (such as pressure, temperature, and so on). The polymeric liner and carbon fiber reinforced polymer material used in Type-IV COVs are generally the best suited for FCV applications and also the best possible solution applicable to other biofuel storage and transportation as per safety standards. While designing the Type-IV COVs, the polymeric liner is the key component that works as a barrier for biofuels through the wall of COVs to avoid leakage, corrosion, and other related failures. Type-IV COVs have high hydrogen storage capacity and weight ratio compared to Type-I, Type-II, and Type-III PVs (Newhouse 2000). However, they are costlier due to the high usage of carbon fiber material, contributing to 70% of their cost (Azkarate et al. 2016). The various types of biofuel storage vessels are based on material, WP, limitations, weight, and cost, as described in Table 12.1. The sign "+" shows the increase and "–" indicates the decrease.

The classification of compressed biohydrogen storage PVs is shown in Figure 12.2. The storage system for a particular fuel, such as hydrogen, is commonly chosen according to the physicochemical characteristics of the fuel and operational parameters.

Type-I vessels are generally made of all metal (aluminum or steel). Type-II vessels are made of steel or aluminum, with fiber winding in the cylindrical part only. Type-1 and Type-II vessels are suitable for low operating pressure ranges in stationary applications. Type-III vessels are generally aluminum liners with a full-composite

TABLE 12.1
Types of Hydrogen Storage Vessels

	Type of PV, Liner	Limitations, Pressure	Weight, Cost
Type-I	Noncomposite, all metal	HE, not portable, 10–20 MPa	++, –
Type-II	Composite only on the cylindrical part, all metal	HE, only for stationary use, 25–30 MPa	+, 0
Type-III	Composite on cylindrical and dome parts, all metal	HE, portable, used in transportation, 35–70 MPa	–, +
Type-IV	Composite on cylindrical and dome parts, polymer	Hydrogen permeation, used in vehicles, 70 MPa	–, ++
Type-V	Composite toroidal tank, disposable mandrel	Liner-less, 15%–20% less weight than Type-IV COVs	––, ++

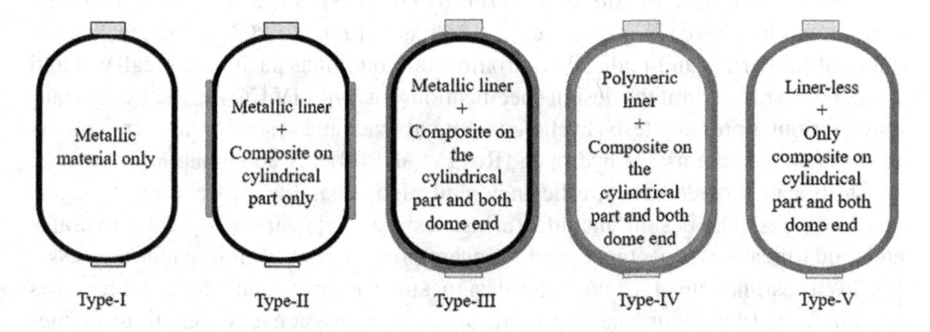

FIGURE 12.2 Schematic of types of hydrogen storage vessels.

overwrap. These vessels show catastrophic failure. Hydrogen interaction with most metals leads to the degradation of their mechanical properties caused by the phenomenon known as HE under the environment of high-pressure gaseous hydrogen (Das et al. 2020; Gobbi et al. 2019; Lin et al. 2022; Charles et al. 2019). Type-IV vessels are all polymeric liners with full composite wrapping. Type-III and Type-IV vessels are the most efficient and light-in-weight. Type-V vessels are linerless and have composite winding all over the cylindrical and dome part. Type-V is still under development for high-pressure sustainability. It is used in cryogenic and space launch vehicles.

The space shuttle challenger (Jones 1991), O-ring failure (Dalal et al. 1989), and Hindenburg disaster (Potter 2007) are well-known hydrogen-based failures. The high-pressure hydrogen storage pressure leads to an explosion with a minor cause and can have crucial consequences. The release of stored energy for pressurized systems depends on the critical operational parameters, such as quantity, pressure, temperature, and release of hazardous fluids failing a plastic liner and liner-less COVs. Therefore, there is a requirement for stringent testing protocols and safety standards to choose storage for hydrogen fuel (Standard et al. 2000). Before transitioning to a hydrogen-based economy, a number of technological and financial obstacles must be reduced, including cost-competitive manufacturing, secure transportation, safe storage, effective systems, and so on.

The selection of material and design standards for hydrogen storage is critical based on compatibility criteria (Ronevich and Marchi 2021). The failure of Type-IV COVs can be attributed to the critical operational parameters for hydrogen storage, such as pressure, temperature, concentration, and characteristics such as density, size of hydrogen, and so on, respectively. Hydrogen storage is challenging due to its chemical and physical properties under critical operations (Williams 1980). It can be stored in Type-IV COVs in gaseous or liquid form in a wide range of operational parameters. The high-pressure compressed hydrogen at more than 70 MPa WP holds various challenges, some of which are described in this section.

12.2.1 Design Limitations to Store Gaseous Hydrogen at High Pressure

The low volumetric capacity of hydrogen makes the requirement to achieve an optimal value of hydrogen density of 42 kg/m^3 by compressed hydrogen storage at high pressure up to 70 MPa WP to drive the acceptable range to FCVs. The mechanical material properties and realistic environmental conditions such as critical WP and temperature range limit the design specifications for Type-IV COVs. The safety standards for burst pressure tests, cyclic operational load, and system weight are considered to enhance the usable hydrogen (Read et al. 2007). A combination of analysis and tests can provide safety, efficiency, and high reliability by adhering to rigorous processes. The design, manufacturing, testing, maintenance, operation parameters, qualification, acceptance, and stringent process control throughout the vessel life cycle estimation of COVs. The design and various manufacturing processes are influenced by endurance requirements and vessel geometry (length-to-diameter ratio, cylindrical composite lamina orientation, hemispherical to the geodesic dome). The testing methods, operational and design parameters need safety factors and design criteria such as stress ratio for dome safety, strain, failure indices (Souza and Tarpaniet 2021), etc. Testing is performed to reduce failure against rupture, safe–unsafe mode of failure, and leak-before-burst test for COVs and demonstration. In addition, finite element analysis (FEA) can estimate approximated safe life concerning the possibility of liner fatigue failure due to crack initiation and growth (Saulsberry et al. 2007; Building et al. 2021). Various destructive and nondestructive tests (NDT) are employed to detect damage (flaws, notches) in the liner, and overwrap for safer design and manufacturing of these Type-IV COVs. NDT methods for detecting damage in liners and the overwrap include: visual (Blanc-vannet 2017), dye penetrant, X-ray, laser ultrasonic (Gupta et al. 2008), eddy current, borescope inspection, acoustic emission (AE) (Achdjian et al. 2016; Dahmene et al. 2018; Building et al. 2021; Munzke et al. 2020), flash/infrared thermography, fiber-optic and resistance strain gauge (Gupta et al. 2008), optical backscatter reflectometer (Souza and Tarpani 2021), laser shearography, digital image correlation (DIC) of overwrap strains (Munzke et al. 2020), optical strain sensing (Munzke et al. 2020), and Raman spectroscopy to measure residual fiber stress (Munzke et al. 2020; Saulsberry et al. 2007).

To make hydrogen-based technology operational, several tests (approximately more than 80) are used to perform before a Type-IV COVs can be put into commercial use (Heitsch et al. 2011; Saulsberry et al. 2007). The safety of the vehicles for the given operational parameters is the primary concern for which the various tests are performed. These tests and technology make the substantial initial investment costs stagnant to reach the market (Azkarate et al. 2016). Materials' selection based on an international organization for standardization (ISO) 11119-3 such as high-strength carbon fibers and advanced polymers is the heart of Type-IV COVs. ISO standards (11119-1, 11119-2, 11119-3, 11515, and 17519) address the re-qualification of COVs' design and establishment. COVs (composite cylinders/tubes) tests are listed in the department of transport special permits (e.g., DOT-SP 14951, DOT-SP 20391) (Standard 2000). FE simulations can also be a valuable tool in solving several critical and minor concerns for the storage and transportation of hydrogen. Various tests and FE simulations are performed to store the hydrogen energy in Type-IV COVs using the most efficient and safe manner. Extensive studies related to failure analysis in Type-IV COV have been reported in the literature. It is essential to select optimal Type-IV COV materials based on the biofuel chemo-physical properties. The designer should consider the size, weight, points of failure, and the way required to minimize them (Azeem et al. 2022). The various studies design and oversight of Type-IV COVs with different polymeric liner materials such as high-density or low-density polyethylene HDPE/LDPE blend (Neto et al. 2011; Hocine et al. 2013), HDPE (Hua et al. 2017; Alam et al. 2020; Jois et al. 2021; Landi et al. 2020; Alam et al. 2020; Yeh and Liu 2017; Antunes et al. 2008), polyamide (PA) (Nebe et al. 2020; Jebeli and Rarani 2022; Alcántar et al. 2017; Nebe et al. 2021), polyethylene (PE) (Zhang et al. 2020), polyurethane (PU) (Gentilleau et al. 2014) are tested for burst pressure. The selection of polymeric liner materials depends on the application, the process of manufacturing (rotational molding or blow molding), tool cost, quantity required, geometry complexity, life cycle, size, material, and so on.

The high-pressure and corresponding higher-density of the requirement of hydrogen storage can be achieved by the robust design of Type-IV COV. The robust design of Type-IV COVs can be achieved by the selection of design and modeling, efficient utilization of storage system requirements, and considering all constraints of manufacturing parameters.

12.2.2 Limitations of High Temperature Developed During Cyclic Filling of Type-IV COVs

To compete with conventional gasoline vehicles, hydrogen FCVs require high storage capacity (>1.5 kg of H_2), sufficient drive range (>100 km), and minimum refueling time (<5 minutes). The processing of H_2 as fuel for FCVs applications at high storage and dispensing pressures causes rapid thermal and pressure changes. The critical temperature for a given flow rate and pressurization rate predicts the amount of precooling needed for hydrogen gas during the fueling–refueling cycle of the COVs at 70 MPa WP. A hydrogen-filling station system with a Type IV COV and filling station tank with corresponding operating pressure and temperature is shown in Figure 12.3.

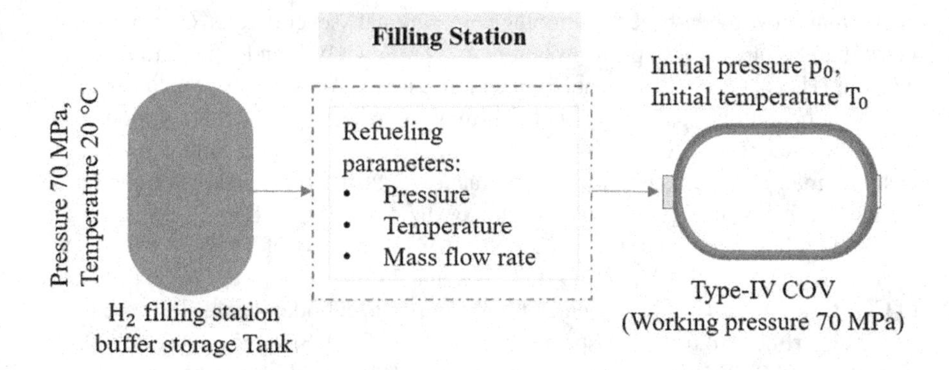

FIGURE 12.3 Schematic of the filling system of Type-IV COV with integrated parts.

The filling system contains a filling station tank at a pressure of 70 MPa, at a temperature of 20°C. A Type IV COV could be filled up to 70 MPa WP, considering refilling parameters such as initial temperature, initial pressure, and mass flow rate. It is necessary to predict the Type-IV COVs materials' (polymeric liner, composite fiber, metallic boss, resin matrix) characteristics under extreme temperature conditions. The fast-filling parameters and thermomechanical analysis are performed to estimate the temperature, and pressure on the liner and composite materials of Type-IV COVs with the polymeric liner material PE (Xiao et al. 2019), HDPE (De Miguel et al. 2015; Acosta et al. 2014; Sapre et al. 2020; Liu et al. 2020), and PA (Heitsch et al. 2009). The filling cycles temperature range is −40°C and 85°C during pressurized to 43 or 86 MPa for 1,500 cycles.

The cyclic filling-de-filling of hydrogen for Type-IV COVs reaches the critical temperature. A selection of higher conductivity material for Type-IV COVs can release the critical temperatures. The parameters such as precooling, flow rate, and so on should be considered to fulfill the requirements to enhance the life and capacity of the storage system as per department of energy (DOE), United States (US) targets.

12.2.3 Permeation-Based Failure in Materials and Multi-material Interface of Type-IV COVs

Hydrogen is the smallest molecule in the periodic table, and it can easily pass through most materials. It needs a large volume of space to store at atmospheric pressure and has a lesser density. The highest hydrogen density possible at high pressure is needed to meet the energy demands for FCV applications. Type-IV COVs polymeric liner material is in direct contact with hydrogen. The small molecular size of hydrogen causes permeability to most materials. The accumulation of hydrogen at interfaces and inside the material of Type-IV COVs at critical operations causes various types of failure. So, robust and leak-proof Type-IV COVs per US DOE, US, 2025 permeation limits can be achieved by considering all the causes of permeation-based failures.

As discussed earlier, the purpose of polymeric liner material is to be used as a permeation barrier and mandrel for composite wrapping for Type-IV COVs. Polymers are also used in high-pressure hydrogen environments as O-ring, polymer electrolyte membrane (PEM) FCVs as membrane material, etc. Generally, in PEM FCVs, the operational parameters such as pressure, temperature, voltage, swelling, and permeation cause failure at the interface and inside the material with respect to time (Husar et al. 2007). Hydrogen storage in the polymer is challenging and risky in terms of permeation and mechanical properties, which cause degradation-based failure at critical pressure and temperature (Bhardwaj et al. 2014). These failures can be described as (Azkarate et al. 2016):

1. The gas adsorption causes polymer swelling and mass shrinkage (Standard 2000).
2. Excess hydrogen dissolution and rapid cycling to exposure cause material degradation such as blistering, crack formation, extrusion, etc.
3. Hydrogen impurities can deteriorate the polymer's behavior in a hydrogen environment.
4. The excess amount of hydrogen in gas and polymer in the presence of a tiny spark causes fire.
5. The tiny behavior of hydrogen causes permeation through most polymeric and metallic materials.
6. The properties of mechanical transport, such as permeability, diffusivity, and solubility, are responsible for hydrogen permeation.
7. Temperature-dependent material properties of the polymer could cause a permeation-related failure (Mahl et al. 2019).
8. Gas permeation through polymeric liner material of Type-IV COVs is caused by gas dissolution, diffusion, and interactions with polymers and gases (Standard 2000).

High mechanical strength (especially flexural mode), good toughness/impact resistance, and processability are the three main characteristics of a PE grade for vessel production. HDPE is an ideal option that can meet the requirements of vessel production, provided it has the appropriate molecular weight, molecular weight distribution, and engineered and well-distributed short and long branches. Recently, double-ring vessels (containing rings at the top and bottom of the vessel) are produced with very high mechanical strength and require unique grades of HDPE. Decisions made in the early stages of field development can significantly impact asset economics and longevity. The sample scale testing of the polymeric liner is done for the maximum flow rate or residual pressure to avoid buckling or collapse for permeation and decompression tests (Papin et al. 2019). Nonsteady state measurement of thermal desorption analysis (TDA) and steady-state high-pressure hydrogen gas permeation test (HPHP) is used up to 90 MPa to estimate the permeability coefficient for LDPE LLDPE, HDPE, UHMWPE, MDPE, PE100 (Fujiwara et al. 2021). The PA materials have higher yield stress than that HDPE due to better resistance to blistering (Yersak et al. 2017). It is found that loss of useable hydrogen remains below 0.05 g/h/kg H_2 (Smith 2014). Explosive decompression failure (XDF) with effects of cavity size, cavity location, and pressure inside the cavity on damage initiation and evolution inside the ethylene propylene diene monomer (EPDM) polymer material with hyperelastic modeling is performed (Kulkarni et al. 2021). The nondimensional parameters are evaluated to estimate cavitation due to decompression failure in HDPE, PA, and EPDM polymers for the maximum hydrogen pressure of 87.5 MPa (Melnichuk et al. 2020). By depressurizing hydrogen gas, the diffusivity coefficient for Type-IV COVs of HDPE liner material was lowered and its solubility was raised (Prachumchon 2012). Vinyl triethoxysilane has been exposed to pressurized hydrogen from 0.1 to 27 MPa, saturated, and submitted to gas decompression over a wide range of controlled pressure rates (between 0.2 and 90 MPa/min) (Jaravel et al. 2011). The HDPE and LDPE permeation properties are evaluated for the given crystallinity and pressure, respectively (Kane 2008). PTFE and HDPE show an increase in Young's modulus and little change in the stiffness with hydrogen exposure (Menon and Brooks 2017). Hydrogen gas permeability characteristics up to 100 MPa were measured for permeation test setup of permeation area, permeation distance, avoiding the effect of filters, and keeping the stability of pressure without gas leakage (Fujiwara et al. 2020). Unfilled, silica-filled, and carbon black-filled EPDM composites are exposed to hydrogen gas at up to 10 MPa. The effects of fillers on the critical hydrogen pressure at crack initiation are investigated (Yamabe and Nishimura 2010). The leakage of gaseous hydrogen and damage through composite lamina has been performed to estimate the empirical relation (Grenoble and Gates 2005). The parameters for hydrogen gas sorption in polymeric material of shapes like a cylinder with reference to diameter and thickness are able to predict the value for sorption and desorption during the experiment (Exp.). The sorption and desorption parameters are evaluated, such as diffusion coefficient, time of equilibrium, and concentration for the given gas pressure (Jung et al. 2022). Permeation tests for PE100 material at various pressure, hydrogen content, and temperatures are evaluated (Klopffer et al. 2015). The stress and concentration gradient techniques are compared for HDPE for 4 hours. It is observed that the diffusion coefficient is independent of pressure and concentration for HDPE and highly dependent on gas behavior (Gay et al. 2020). The FE modeling has been performed for the various rate of decompression, residual pressure, and temperature to prevent collapse for rapid unloading of 2.4 L PV that is preliminary loaded up to 70 MPa (Pépin et al. 2018). It was discovered that permeability is also pressure-dependent at high pressure (Naito et al. 1996). The polymeric liner material of Type-IV COVs is generally made up of LLDPE, HDPE, PA6, and PU. The polymeric liner material-based permeation and failure analysis is summarized in Table 12.2.

The extensive literature study related to the failure of polymeric liner materials in presence of pressurized hydrogen gas provided insight into material selection for biofuel storage and transportation. The possible types of failures due to hydrogen permeation and XDF are described based on the operating pressure, temperature limit, and time of exposure. The smallest size of hydrogen after entering into polymer material causes design and material failure of Type-IV COVs. The experimental and FE simulation framework is available from the literature to help with the estimation of design and materials parameters required for the storage and transportation of hydrogen in extreme operating conditions.

TABLE 12.2
Hydrogen Permeation Studies Used in Type-IV COVs Material

Pressure, Temperature, Liner, Time	Test, Failure Parameters	Method, Software
70 MPa, 27°C, 55°C, 72 hours	Collapse, buckling, Liner-composite interface	Exp. and FE (Papin et al. 2019)
90 MPa, 30°C, PE grades, 24 hours	Permeation wrt pressure and crystallinity	HPHP and TDA (Fujiwara et al. 2021)
87.5 MPa, 25°C, HDPE, PA, 13 hours	Depressurization-induced blistering	COMSOL (Yersak et al. 2017)
86 MPa, −30°C to 85°C, HDPE, PET, PA6, 16.38 hours	Permeability vs. pressure for various Young's modulus	Exp. HPT (Smith et al. 2014)
27 MPa, 25°C, EPDM, 13 hours	XDF with volume fraction of cavity	FE Abaqus (Kulkarni et al. 2021)
87.5–0.1 MPa, 50°C–25°C, HDPE, PA, EPDM, 780 13 hours	XDF causes cavitation, depressurization	Numerical (Melnichuk et al. 2020)
0.71 MPa, 25°C, HDPE, 5.83 hours	Permeation and fracture	COMSOL, Exp. (Prachumchon 2012)
9 MPa, 25°C, elastomer, 13 minutes	Decompressed at different rates, pressure	Exp. (Jaravel et al. 2011)
10.3 MPa, 25°C, HDPE, LDPE, 1 second	Permeability and diffusivity wrt pressure	Numerical (Kane 2008)
100 MPa, −40°C–85°C, HDPE, 7 days	Permeability and strength wrt pressure	Exp. (Menon and Brooks 2017)
90 MPa, 30°C, HDPE, 24 hours	HTM properties	Exp. (Fujiwara et al. 2020)
10 MPa, 30°C, EPDM, 24 hours	Failure in polymer	Exp. (Yamabe and Nishimura 2010)
−196 MPa, 23°C, Orthotropic	Permeability, damage, orientation	Exp. (Grenoble and Gates 2005)
5.75 MPa, 25°C, EPDM, 13.9 hours	Concentration	Exp. (Jung et al. 2022)
2–0.5 MPa, 20°C–50°C, PE100, PA11, 450 days	Permeability vs time, temperature	Exp. (Klopffer et al. 2015)
2 MPa, 20°C–50°C, HDPE, 4 hours	Permeability, diffusivity formulation	Exp. (Gay et al. 2020)
35 MPa, 65°C, PA6, PA12, 7 days	CTE difference, delamination	Abaqus (Pépin et al. 2018)

12.3 EFFECT OF OTHER BIOFUELS ON STORAGE MATERIAL SYSTEM

Biofuel storage could be the central aspect behind the successful implementation of the biofuel-based economy. There are several factors that are responsible for the efficient and sustainable storage of biofuels. The storage and transportation of first-generation biofuels are challenging due to specific properties that cause the failure of storage devices with time (Komariah et al. 2017). Presently, storage devices have degraded the quality of biofuels due to their self-degradation properties, and further cause changes in chemical and physical properties because of the larger time gap between storage and utilization. This leads to corrosion failure (Groysman 2017), sludge formation, and contamination of the storage devices due to leakage, lack of maintenance, and exposure to air resulting in critical disasters (Komariah et al. 2021). Therefore, the storage devices should have corrosion-resistant, chemical-resistant, degradation-resistant, and antioxidation properties.

The biofuels get exposed to various materials (metals and nonmetals) during storage and transportation. Therefore, it is necessary to account for biofuels' environmental parameters and operating conditions in the storage systems. The common nonmetal materials in the existing diesel engine vehicles are elastomers, rubber, and plastics (Bhardwaj et al. 2014). Depending on the applications, the materials as discussed in the former statement are usually present as hoses, tubes, fuel lines, sealants, and fuel tanks where these components come into contact with fuels during the idling process. From the failures perspective, biofuel can cause chemical degradation, swelling, cracking, and extraction of additives of elastomers or rubber due to long-time storage and exposure to sunshine. The absorption of fluid into the polymer network of the elastomer or rubber can cause swelling. The effect of biodiesel on various nonmetallic materials such as HDPE, EPDM, Teflon, natural rubber, and so on are well studied in multiple works of literature available (Saat et al. 2017). The main issue of the effect of biofuel storage on metal is corrosion, which results in a mass loss (Papavinasam et al. 2011). Biodiesel has specific properties such as density, viscosity, cetane number, flash point, pour point, saponification number, acid number, etc., which get affected by the inclusions such as water, sediment, carbon residue, sulfated ash, sulfur, copper strip corrosion, free glycerin, total glycerin, ester content, etc. These inclusions present in biodiesel after burning may cause health and environmental impact (Howell et al. 1997). The performance of metallic components (Bhardwaj et al. 2014; Alves et al. 2017; Fathallah and Pinto 2022) gets degraded, and the engine life of automotive materials has been reduced by corrosion failures.

As a result of the discussion mentioned in the subsection, Biofuels can be used in engines and power trains without significant changes in the existing infrastructures. These are available for customers and industries at affordable prices. Biofuels can provide energy balance and security to the government and society worldwide. These can also increase the country's agriculture production with the usage of biofuels. It has already been proved that biofuels are safer than conventional fuels due to their low volatility.

12.4 COST OF STORAGE AND TRANSPORTATION OF HYDROGEN

The onboard and offboard performance has included the cost and design parameters for the storage and transport system of Type-IV COVs to meet the US DOE 2010, 2015, and 2020 standards, and to fulfill the ultimate targets for FCV applications (Johnson et al. 2017). The onboard performance has included the cost of manufacturing, storage system, the volume of production, and performance in terms of weight and volume. The offboard performance has included the cost of infrastructure required to refuel the onboard storage system, well-to-tank (WTT) efficiency, and GHG emissions (Hua et al. 2011). The onboard performance has comprised of the balance-of-plant (BOP) and bill-of-material (BOM) components for Type-IV COVs are NWP, the number of tanks, liner material and thickness, dome shape, the maximum filling pressure, empty pressure, usable hydrogen, internal volume, safety factor, L:D ratio, carbon fiber type, composite tensile strength, life cycle, and so on factors. The seasonal variation of hydrogen storage and transport is a challenge for FCV applications, which may hinder the offboard performance. In terms of the storage & transport of hydrogen fuel concerned, the liquid organic hydrogen carriers (LOHC) model is considered to be the most efficient supply chain as compared to storage in salt caverns (geological method of storage), liquid hydrogen (LH2), and compressed gaseous hydrogen (GH2) based on GHG emission (Reuß et al. 2017). In addition, the investment cost of LH2 and GH2 tanks is based on the 2015 US DOE targets. As LOHC storage is assumed to be 50 €/kg of hydrogen, which includes LOHC-carrier material and two tanks. The transport module contains pipe (high investment cost and low operational cost) and truck (limited batch capacity doesn't benefit from an increase in the demand for hydrogen) related costs. The cost comparison is followed by different pathways as represented as model-1 (storage1, transport1: without conversion model), model-2 (storage2, transport2: without seasonal GH2 tanks), and model-3 (storage3, transport3: using different heat sources for LOHC station) for hydrogen storage and transportation model for 250 km distance and 50 tons/day hydrogen demand. The cost of hydrogen as per well-to-wheel (WTW) analysis for different pathways is shown in Figure 12.4 (Reuß et al. 2017).

The number of parameters is responsible to make the hydrogen economy, which plays an important role to meet the objectives of sustainable development of hydrogen-based infrastructure for storage and transportation. To make hydrogen more sustainable, there is an optimized supply chain has been required for storage and transportation for efficient utilization.

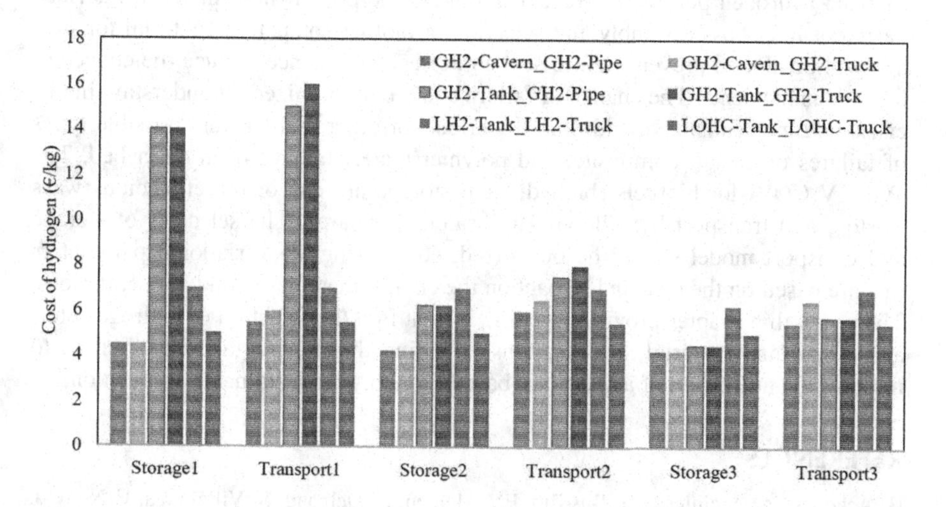

FIGURE 12.4 Cost of hydrogen storage and transportation for various supply chain models. Reproduced with permission and adapted under a creative commons license from the reference (Reuß et al. 2017). Copyright 1969, Elsevier.

12.5 CONCLUSION AND FUTURE SCOPE

In this chapter, the insight of challenges with biofuel storage and transportation have been discussed. Biohydrogen is a third-generation zero-emission technology for meeting a wide range of energy needs. The storage and transportation of hydrogen and other biofuels are vital due to critical operating parameters (pressure, temperature, concentration), environmental conditions (moisture, chemical environment, and so on), molecular size, time of exposure, and their effect on the output responsive parameters such as corrosion, permeation-based failure, and change in thermophysical properties.

As far as the novel findings are concerned, this chapter has marked a valuable impact by considering the extensive literature studies where Type-IV COVs for hydrogen storage has been reported to be safer and more efficient storage system as per the safety standards and protocols for FCV applications. From these results, it has also been observed that specific types of materials for the polymeric liner, metallic boss, and composite material can be used for manufacturing Type-IV COV. The robust and failure-resistant design with proper testing should be strictly followed by considering the realistic environmental and operational parameters as per standards such as ISO 91111-3. The challenges of high-pressure hydrogen storage due to a critical temperature of 85°C developed during cycling filling of Type-IV COVs material. Pre-cooling at approximately −40°C and highly thermally conductive material usage is one of the best ways to release the maximum temperature in a shorter duration. The methods of failure prevention strategies and protocols should be critically revised based on real-time situations. The hydrogen interaction with materials has caused permeability due to various failure types at the interfaces of polymer–liner–boss materials that further have been examined by multiple experimental and FE simulations studies. In context with the geometrical design, and related constructional features, the selection of an HDPE or PE100-like specific grade of materials for the polymeric liner is important to avoid hydrogen permeation-related failures. The proper multi-material interface design of liner-boss assembly and selection of biofuel-compatible material for storage will be able to prevent failure at the interface and, hence, reduce the chance of hazardous incidents. The chance of failure could be minimized by understanding its cause and mechanism. In addition, the effects of other biofuels and possible types of failures on metal, composite, and polymer materials have been described. The Type-IV COVs for biofuels (biohydrogen) storage are one of the efficacious ways to store and transport by following the safety standards. The selection of storage and transport model should be optimized, considering the variational parameters that are based on the cost and impact on the environment based on GHG emissions. Therefore, this chapter provides specific insight into the selection of hydrogen storage design and material specifications including the transportation challenges to improve the utilization of a hydrogen-based economy for sustainable applications.

REFERENCES

H. Achdjian, A. Arciniegas, J. Bustillo, F.V. Meulen, L. Delnaud, S. Villalonga, F. Nony, J. Fortineau, and G. Umr. Type IV composite pressure vessel characterization by measurement of acoustic reverberation. *22nd International Congress on Acoustic, Buenos Aires*, Argentina, 1–8, 2016.

B. Acosta, P. Moretto, N. De Miguel, R. Ortiz, F. Harskamp, and C. Bonato. JRC reference data from experiments of onboard hydrogen tanks fast filling. *International Journal of Hydrogen Energy*, 39(35):20531–20537, 2014.

S. Alam, G.R. Yandek, R.C. Lee, and J.M. Mabry. Design and development of a filament wound composite overwrapped pressure vessel. *Composites Part C: Open Access*, 2: 100045, 2020.

V. Alcántar, S.M. Aceves, E. Ledesma, S. Ledesma, and E. Aguilera. Optimization of Type 4 composite pressure vessels using genetic algorithms and simulated annealing. *International Journal of Hydrogen Energy*, 42(24): 15770–15781, 2017.

S.M. Alves, V.S. Mello, and F.K. Dutra-Pereira. Biodiesel compatibility with elastomers and steel. *Frontiers in Bioenergy and Biofuels Chapter 16. Intech Open,* Rijeka, 2017. Doi: 10.5772/65551.

P.J. Antunes, G.R. Dias, J.P. Nunes, F.W.J. Van Hattum, and T. Oliveira. Finite element modeling of thermoplastic matrix composite gas cylinders. *Journal of Thermoplastic Composite Materials*, 21(5): 411–441, 2008.

S. Aso, K. Mikio, and N. Yasuhiro. Development of Fuel Cell Hybrid Vehicles in TOYOTA, *TOYOTA MOTOR CORPORATION, 1, Toyota-Cho,* Toyota, Aichi, Japan, IEEE, 471-8571, 2007.

L. Axelsson, M. Franzén, M. Ostwald, G. Berndes, G. Lakshmi, and N. H. Ravindranath. Perspective: Jatropha cultivation in southern India: Assessing farmers' experiences. *Biofuels, Bioproducts and Biorefining*, 6(3):246–256, 2012.

I. Azkarate, H. Barthelemy, F. Dolci, P. Hooker, T. Jordan, J. Keller, F. Markert, P. Moretto, M. Steen, and A. Tchouvelev. Research Priority Workshop on Hydrogen Safety. *European Commission, Joint Research Centre*, Petten, Netherlands, 2016. Doi: 10.2760/77730.

M. Azeem, H. Haji, M. Kumar, L. Gemi, R. Khan, T. Ahmed, Q. Ma, R. Sadique, A. Akmar, and M. Mustapha. Application of filament winding technology in composite pressure vessels and challenges: A review. *Journal of Energy Storage*, 49: 103468, 2022.

M. Bhardwaj, P. Gupta, and N. Kumar. Compatibility of metals and elastomers in biodiesel: A review. *International Journal of Research (IJR)*, 1(7): 376–391, 2014.

P. Blanc-vannet. Residual performance and non-destructive testing of composite pressure vessels with plastic liner after mechanical impact. *20th International Conference on Composite Materials*, Paris, France, 2017.

P. Blanc-Vannet, P. Papin, M. Weber, P. Renault, J. Pepin, E. Lainé, G. Tantchou, S. Castagnet, and J.-C. Grandidier. Sample scale testing method to prevent collapse of plastic liners in composite pressure vessels. *International Journal of Hydrogen Energy*, 8682–8691, 2019. Doi: 10.1016/j.ijhydene.2018.10.031.

Y. Charles, M. Gaspérini, N. Fagnon, K. Ardon, A. Duhamel. Finite element simulation of hydrogen transport during plastic bulging of iron submitted to gaseous hydrogen pressure. *Engineering Fracture Mechanics,* 218:106580, 2019. Doi: 10.1016/j.engfracmech.2019.106580.

S.R. Dalal, E.B. Fowlkes, B. Hoadley. Risk analysis of the space shuttle: Pre-challenger prediction of failure. *Journal of American Statistical Association*, 84:945–57, 1989. Doi: 10.1080/01621459.1989.10478858.

F. Dahmene, S. Yaacoubi, S. Bittendiebel, O. Bardoux, P. Blanc-Vannet, A. Maldachowska, M. Barcikowski, M. Panek, N. Alexandre, F. Nony, K. Lasn, A. Echtermeyer, P.S. Heggem. Use of acoustic emission for inspection of composite pressure vessels subjected to mechanical impact. *33nd European Conference on Acoustic Emission Testing*, Senlis, France, 1–11, 2018.

T. Das, E. Legrand, S.V. Brahimi, J. Song, S. Yue. Evaluation of material susceptibility to hydrogen embrittlement (HE): An approach based on experimental and finite element (FE) analyses. *Engineering Fracture Mechanics*, 224:106714, 2020. Doi: 10.1016/j.engfracmech.2019.106714.

N. De Miguel, R. Ortiz Cebolla, B. Acosta, P. Moretto, F. Harskamp, and C. Bonato. Compressed hydrogen tanks for on-board application: Thermal behavior during cycling. *International Journal of Hydrogen Energy*, 40(19):6449–6458, 2015.

E. Building. Modal acoustic emission (MAE) Examination Specification for Requalification of Composite Overwrapped Pressure Vessels (Cylinders and tubes), *U.S. Department of Transportation*, New Jersey Avenue S.E. Washington, 2021.

A.Z.M. Fathallah and F. Pinto. The Influence of NaCl dissolved on biodiesel of used cooking oil on performance and its degradation of main components of diesel engine. *IOP Conference Series: Earth and Environmental Science*, 972(1): 012030, 2022.

H. Fujiwara, H. Ono, K. Ohyama, M. Kasai, F. Kaneko, and S. Nishimura. Hydrogen permeation under high pressure conditions and the destruction of exposed polyethylene-property of polymeric materials for high-pressure hydrogen devices (2)-. *International Journal of Hydrogen Energy*, 46(21): 11832–11848, 2021.

H. Fujiwara, H. Ono, K. Onoue, and S. Nishimura. High-pressure gaseous hydrogen permeation test method property of polymeric materials for high-pressure hydrogen devices (1). *International Journal of Hydrogen Energy*, 45(53): 29082–29094, 2020.

K. Gaudreau. Biofuel basics. *Alternatives Journal*, 35(2): 14–19, 2009.

N. Gay, T. Lamouchi, F. Agostini, C.A. Davy, and F. Skoczylas. Hydrogen diffusion through polymer membranes. *MATEC Web of Conference*, Batumi, Georgia, 322: 01044, 2020.

B. Gentilleau, F. Touchard, and J. C. Grandidier. Numerical study of the influence of temperature and matrix cracking on type IV hydrogen high pressure storage vessel behavior. *Journal Composite Structures*, 111(1): 98–110, 2014.

A. Ghimire, G. Kumar, P. Sivagurunathan, S. Shobana, G.D. Saratale, H.W. Kim, V. Luongo, G. Esposito, and R. Munoz. Bio-hythane production from microalgae biomass: Key challenges and potential opportunities for algal bio-refineries. *Bioresource Technology*, 241:525–536, 2017.

R.W. Grenoble and T.S. Gates. Hydrogen permeability of polymer matrix composites at cryogenic temperatures. *Collection of Technical Papers -AIAA/ASME/ASCE/AHS/ASC Structures, Structural Dynamics and Materials Conference*, American Institute of Aeronautics and Astronautics, U.S.A, 5: 3507–3514, 2005.

G. Gobbi, C. Colombo, S. Miccoli, L. Vergani. A fully coupled implementation of hydrogen embrittlement in FE analysis. *Advances in Engineering Software*. 135:102673, 2019. Doi: 10.1016/j.advengsoft.2019.04.004.

A. Groysman. (2017). 1651770175271.pdf, 61(3): 100–117, 2017. Doi: 10.1515/kom-2017-0013.

K. Gupta, A. McClanahan, K. Erikson, and R. Zoughi. Show Me the Road to Hydrogen Non-Destructive Evaluation (NDE). *Rolla Missouri University of Science and Technology*, Rolla, Phelps, 47 p, 2008.

M. Heitsch, D. Baraldi, and P. Moretto. Numerical investigations on the fast filling of hydrogen tanks. *International Journal of Hydrogen Energy*, 36(3): 2606–2612, 2011.

M. Heitsch, D. Baraldi, and P. Moretto. Simulation of the fast filling of hydrogen tanks. *Proceedings of the 3rd International Conference on Hydrogen Safety*, Ajaccio, France, 1–12, 2009.

A. Hocine, A. Ghouaoula, F.K. Achira, S.M. Medjdoub, F.K. Achira, and S.M. Medjdoub. Analysis of failure pressures of composite cylinders with a polymer liner of type IV CNG vessels. *International Journal of Mechanical, Industrial Science and Engineering*, 7(1):148–152, 2013.

S. Howell. Biodiesel use in underground metal and non-metal mines. *Dieselnet*, (May): 1–3, 1997.

T.Q. Hua, H.S. Roh, and R.K. Ahluwalia. Performance assessment of 700-bar compressed hydrogen storage for light duty fuel cell vehicles. *International Journal of Hydrogen Energy*, 42(40): 25121–25129, 2017.

T.Q. Hua, R.K. Ahluwalia, J.K. Peng, M. Kromer, S. Lasher, K. McKenney, K. Law, and J. Sinha. Technical assessment of compressed hydrogen storage tank systems for automotive applications. *International Journal of Hydrogen Energy*, 36(4): 3037–3049, 2011. Doi: 0.1016/j.ijhydene.2010.11.090.

A. Husar, M. Serra, and C. Kunusch. Description of gasket failure in a cell PEMFC stack. *Journal of Power Sources*, 169(1):85–91, 2007.

J. Jaravel, S. Castagnet, J.C. Grandidier, and G. Benoît. On key parameters influencing cavitation damage upon fast decompression in a hydrogen saturated elastomer. *Polymer Testing*, 30(8): 811–818, 2011.

M.A. Jebeli and M. Heidari-Rarani. Development of Abaqus WCM plugin for progressive failure analysis of type IV composite pressure vessels based on Puck failure criterion. *Engineering Failure Analysis*, 131(2021): 105851, 2022.

K. Johnson, M.J. Veenstra, D. Gotthold, K. Simmons, K. Alvine, B. Hobein, D. Houston, N. Newhouse, B. Yeggy, A. Vaipan, and T. Steinhausler. Advancements and opportunities for on-board 700 bar compressed hydrogen tanks in the progression towards the commercialization of fuel cell vehicles. *SAE International Journal of Alternative Powertrains*, 6(2), July 2017.

K.C. Jois, M. Welsh, T. Gries, and J. Sackmann. Numerical analysis of filament wound cylindrical composite pressure vessels accounting for variable dome contour. *Journal of Composites Science*, 5(2): 56, 2021.

L. Jones, M. Fisher, A. Mccool, J. Mccarty. Propulsion at the Marshall Space Flight Centre-A brief history 1991, Alabama, U.S. *AIP Conference Proceedings,* 61, 1991. Doi: 10.2514/6.1991-2553.

J.K. Jung, K.T. Kim, U.B. Baek, and S.H. Nahm. Volume dependence of hydrogen diffusion for sorption and desorption processes in cylindrical-shaped polymers. *Polymers*, 14(4): 756, 2022.

M.C. Kane. Permeability, solubility, and interaction of hydrogen in polymers- an assessment of materials for hydrogen transport. pages WSRC-STI-2008-00009, Rev. 0, Savannah River Site, Aiken, SC (United States), 2008.

M. Kayfeci and A. Keçebas. Hydrogen Storage: Processes, Systems, and Technologies. *Hydrogen and Fuel Cell Technologies Office,* San Diego, London, 85–110, 2019. Doi: 10.1016/B978-0-12-814853-2.00004-7.

I. Khan, P. Anburajan, G. Kumar, J.J. Yoon, A. Bahuguna, A.G.L. de Moura, A. Pugazhendhi, S.H. Kim, and S.C. Kang. Comparative effect of silver nanoparticles (AgNPs) derived from actinomycetes and henna on biohydrogen production by *Clostridium beijerinckii* (KTCC1737). *International Journal of Energy Research*, 45(12): 17269–17278, 2021.

M.H. Klopffer, P. Berne, and É. Espuche. Development of innovating materials for distributing mixtures of hydrogen and natural gas. Study of the barrier properties and durability of polymer pipes. *Oil and Gas Science and Technology*, 70(2): 305–315, 2015.

J.A. Ronevich, C.S. Marchi. Materials compatibility concerns for hydrogen blended into natural gas. *American Society of Mechanical Engineers, Pressure Vessel and Piping Conference.* 4: 6, 2021. Doi: 10.1115/PVP2021-62045.

L.N. Komariah, S. Aprisah, and Y.S.L. Rosa. Storage tank materials for biodiesel blends; The analysis of fuel property changes. *MATEC Web of Conference*, Bangka Island, Indonesia, 101, 2017.

L.N. Komariah, S. Arita, B.E. Prianda, and T.K. Dewi. Technical assessment of biodiesel storage tank; A corrosion case study. *Journal of King Saud University - Engineering Sciences*, 1018–3639, 2021. Doi: 10.1016/j.jksues.2021.03.016.

S.S. Kulkarni, K.S. Choi, W. Kuang, N. Menon, B Mills, A. Soulami, and K. Simmons. Damage evolution in polymer due to exposure to high-pressure hydrogen gas. *International Journal of Hydrogen Energy*, 46(36): 19001–19022, 2021.

Y. Kumar, P. Yogeshwar, S. Bajpai, P. Jaiswal, S. Yadav, D.P. Pathak, M. Sonker, and S.K. Tiwary. Nanomaterials: Stimulants for biofuels and renewables, yield and energy optimization. *Materials Advances*, 2(16): 5318–5343, 2021.

D. Landi, A. Vita, S. Borriello, M. Scafa, and M. Germani. A methodological approach for the design of composite tanks produced by filament winding. *Computer Aided Design and Application*, 17:1229–1240, 2020. Doi: 10.14733/cadaps.2020.1229-1240.

M. Lin, H. Yu, X. Wang, R. Wang, Y. Ding, A. Alvaro. A microstructure informed and mixed-mode cohesive zone approach to simulating hydrogen embrittlement. *International Journal of Hydrogen Energy*. 47(39):17479–17493, 2022. Doi: 10.1016/j.ijhydene.2022.03.226.

J. Liu, S. Zheng, Z. Zhang, J. Zheng, and Y. Zhao. Numerical study on the fast filling of on-bus gaseous hydrogen storage cylinder. *International Journal of Hydrogen Energy*, 45(15): 9241–9251, 2020.

M. Mahl, C. Jelich, and H. Baier. On the temperature-dependent non-isosensitive mechanical behavior of polyethylene in a hydrogen pressure vessel. *Procedia Manufacturing*, 30: 475–482, 2019.

M. Melnichuk, F. Thiébaud, and D. Perreux. Non-dimensional assessments to estimate decompression failure in polymers for hydrogen systems. *International Journal of Hydrogen Energy*, 45(11), 6738–6744, 2020. Doi: 10.1016/j.ijhydene.2019.12.107.

N.C. Menon and K. Brooks. Behavior of Polymers in High-Pressure Environments as Applicable to the Hydrogen Infrastructure. *Proceedings of ASME 2016 Pressure Vessel and Piping Conference*, Vancouver, Br Columbia, Canada, 646–56, 2017. Doi: 10.1115/1.861387_ch74.

D. Munzke, E. Duffner, R. Eisermann, M. Schukar, A. Schoppa, M. Szczepaniak, J. Strohhäcker, and G. Mair. Monitoring of type IV composite pressure vessels with multilayer fully integrated optical fiber based distributed strain sensing. *Materials Today: Proceedings*, 34(1): 217–223, 2020.

M. Nebe, A. Soriano, C. Braun, P. Middendorf, and F. Walther. Analysis on the mechanical response of composite pressure vessels during internal pressure loading: FE modeling and experimental correlation. *Composites Part B: Engineering*, 212(2020): 108550, 2021.

M. Nebe, T.J. Asijee, C. Braun, J.M.J.F. Van Campen, and F. Walther. Experimental and analytical analysis on the stacking sequence of composite pressure vessels. *Composite Structures*, 112429, 2020.

E.S.B. Neto, M. Chludzinski, P. B. Roese, J. S.O. Fonseca, S. C. Amico, and C. A. Ferreira. Experimental and numerical analysis of a LLDPE/HDPE liner for a composite pressure vessel. *Polymer Testing*, 30(6): 693–700, 2011.

N.L. Newhouse. Application of composites in natural gas vehicle fuel containers. *Comprehensive Composite Materials*: 6, 581–597, 2000, NE, USA. Doi: 10.1016/b0-08-042993-9/00195-9.

S. Papavinasam, A. Anand, M. Paramesh, J. Krausher, J. Li, P. Liu, S. Mani, and S. Krishnamurthy. Corrosion of metals in biofuels. *ECS Transactions*, 33(14): 1–19, 2011.

J. Pépin, E. Lainé, J.C. Grandidier, G. Benoit, D. Mellier, M. Weber, and C. Langlois. Replication of liner collapse phenomenon observed in hyperbaric type IV hydrogen storage vessel by explosive decompression experiments. *International Journal of Hydrogen Energy*, 43(9): 4671–4680, 2018.

S. Prachumchon. *A Study of HDPE in High Pressure of Hydrogen Gas — Measurement of Permeation Parameters and Fracture Criteria. The University of Nebraska-Lincoln*, Lincoln, Nebraska, 2012.

Y. Naito, Y. Kamiya, K. Terada, K. Mizoguchi, J.-S. Wang. Pressure dependence of gas permeability in a rubbery polymer. *Journal of Applied Polymer Science*, 61(6): 945–50, 1996. Doi: 10.1002/(sici)1097-4628(19960808)61:6<945::aid-app8>3.0.co;2-h.

M. Reuß, T. Grube, M. Robinius, P. Preuster, P. Wasserscheid, and D. Stolten. Seasonal storage and alternative carriers: A flexible hydrogen supply chain model. *Applied Energy*, 200: 290–302, 2017. Doi: 10.1016/j.apenergy.2017.05.050.

N. Saat, A. Samsuri, K. Hamzah, and H. Sulaiman. Review on the compatibility of non-metal materials in automotive components of diesel engine vehicles with blended biodiesel fuel. *Advanced Science Letters*, 23(5):4728–4732, 2017.

S. Sapre, K. Pareek, and M. Vyas. Investigation of structural stability of type IV compressed hydrogen storage tank during refueling of fuel cell vehicle. *Energy Storage*, 2(4): 1–11, 2020.

R. Saulsberry, N. Greene, K. Cameron, E. Madaras, L. Grimes-Ledesma, J. Thesken. Nondestructive methods and special test instrumentation supporting NASA composite overwrapped pressure vessel assessments. Collection of Technical Papers - AIAA/ASME/ASCE/AHS/ASC *Structure, Structural Dynamics and Materials Conference*, United States, 8:7543–7560, 2007. Doi: 10.2514/6.2007-2324.

D.B. Smith. Lifecycle verification of tank liner polymers. ORNL/TM-2014/48. Oak Ridge National Lab (ORNL), Oak Ridge, TN (United States), 27, 2014.

G. Souza and J.R. Tarpani. Using OBR for pressure monitoring and BVID detection in type IV composite overwrapped pressure vessels. *Journal of Composite Materials*, 55(3): 423–436, 2021.

A. Tawfik, M. Nasr, A. Galal, M. El-Qelish, Z. Yu, M.A. Hassan, H.A. Salah, M.S. Hasanin, F. Meng, A. Bokhari, M.A. Qyyum, and M. Lee. Fermentation-based nanoparticle systems for the sustainable conversion of black liquor into biohydrogen. *Journal of Cleaner Production*, 309: 127349, 2021.

C. Read, G. Thomas, G. Ordaz, and S. Satyapal. U.S. Department of Energy's System targets for on-board vehicular hydrogen storage. *Material Matters*, 2:3–4, 2007.

S. Potter. Retrospect: May 6, 1937: The Hindenburg Disaster. *Weatherwise, New York*, 60:16–17, 2007. Doi: 10.3200/wewi.60.3.16-17.

ISO Standard. ISO 11114-2, 2000(E) Transportable gas cylinders-Compatibility of cylinder and valve materials with gas contents-Part 2: Non-metallic materials, 2000.

L.O. Williams. Chemical and physical properties of hydrogen. *Hydrogen Power*, I: 15–36, 1980.

J. Xiao, S. Ma, X. Wang, S. Deng, T. Yang, and P. Bénard. Effect of hydrogen refueling parameters on the final state of charge. *Energies*, 12(4):1–10, 2019.

J. Yamabe and S. Nishimura. Estimation of critical pressure of decompression failure of EPDM composites for sealing under high-pressure hydrogen gas. *18th European Conference on Fracture: Fracture of Materials and Structures from Micro to Macro Scale*, Dresden, Germany, 1–8, 2010.

M.-K. Yeh and T.-H. Liu. Finite element analysis of graphite/epoxy composite pressure vessel. *Journal of Materials Science and Chemical Engineering*, 05(07): 19–28, 2017.

T.A. Yersak, D.R. Baker, Y. Yanagisawa, S. Slavik, R. Immel, A. Mack-Gardner, M. Herrmann, and M. Cai. Predictive model for depressurization-induced blistering of type IV tank liners for hydrogen storage. *International Journal of Hydrogen Energy*, 42(48):28910–28917, 2017.

H. Yumiya, M. Kizaki, and H. Asai. Toyota fuel cell system (TFCS). *28th International Electric Vehicle Symposium and Exhibition*, Goyang, South Korea, 7: 85–92, 2015.

A.A. Zaidi, R. Feng, A. Malik, S.Z. Khan, Y. Shi, A.J. Bhutta, and A.H. Shah. Combining microwave pre-treatment with iron oxide nanoparticles enhanced biogas and hydrogen yield from green algae. *Processes*, 7(1):24, 2019.

Q. Zhang, H. Xu, X. Jia, L. Zu, S. Cheng, and H. Wang. Design of a 70 MPa type IV hydrogen storage vessel using accurate modeling techniques for dome thickness prediction. *Composite Structures*, 236: 111915, 2020.

Part V

Anaerobic Biotechnology

13 Anaerobic Biotechnology
A Useful Technique for Biofuel Production

Saurabh Singh
Banaras Hindu University

Amit Kumar Mishra
Dr. Ram Manohar Lohia Avadh University

Jay Prakash Verma
Banaras Hindu University

Shashi Arya
Imperial College London

Vijay Nimkande
CSIR—National Environmental Engineering Research Institute

CONTENTS

DOI: 10.1201/9781003197737-18

13.1 INTRODUCTION

Rapid growth in population, industrial development, suburbanization, and economic growth leads to a noteworthy increase in residual municipal solid waste (MSW). Industrial development and major migration of people from rural to urban places and consequently urbanization is increasing rapidly. The per capita generation of municipal solid waste (MSW) has also increased exponentially with the recent scientific developments. More importantly, these solid wastes need more land to get disposed of properly, due to which issues related to disposal have become highly challenging. Alternative energy strategies have gained attraction in the recent times mainly for future world stability. Nowadays, with this characteristic, renewable energy sources (mainly waste materials to energy) are the frontrunner to solve the world energy problem and contribute majorly to sustainable waste management and circular economy. There are several options of wastes from different sources like agricultural (plant and animal wastes), industrial (sugar refinery, dairy wastes, tanneries and slaughterhouses), and household (kitchen waste and garden waste) sectors that are the potential renewable energy sources to attain sustainability and can make the paradigm switch to waste-to-energy routes (WTERs).

There are several methods to manage residual waste, and composting is one of the oldest. Although, it has its limitations as composting is a challenging process, generally, the waste material arrives in a mixed form and contains inorganic material as well. It is reported that when mixed waste is composted without segregating, the end product is mostly of poor value. In the Indian context, to manage the residual waste incineration is a meagre option, as most of the time waste consists of high organic material (40%–60%), along with high inert content (30%–50%) and low calorific value content (800–1,100 kcal/kg, whereas moisture content is between 40% and 60% in MSW, and the large capital costs are required to setting up and running the plants. Gasification is mainly residual solid waste incineration that is performed under oxygen-deficient conditions; the end product results into the fuel gas. Gasifiers are mostly used for the burning of biomass like agro-residues, sawmill dust, and forest wastes. Gasification is mainly performed by some specific steps like drying, removing the inert, and shredding for size reduction of municipal solid waste (Management & Handling Rules, 2004). A landfill is a commonly practised area of land onto or into which residual waste is deposited. The goal is to avoid every type of contact between the waste and the surrounding environment, mainly the groundwater. Landfills continue to be the cost effective and widely adopted practice in India, at the same time certain improvements need to be made to ensure safe sanitary landfilling. Recently, development of landfill technology is upgraded to the bioreactor landfill. Bioreactor landfills are specifically designed, constructed, and operated to enhance moisture content that increases the reaction rate of anaerobic biodegradation. Most importantly, bioreactor landfills are different from conventional landfills due to leachate recirculation. This approach leads to diminishing the landfill maintenance time and reducing the period of monitoring. There are several other potential substitutes for waste to energy routes, for instance, biofuel (BF) is an advanced version of the conventional treatment process, as it only produces the renewable fuel, so BF has an added advantage, and it can be considered as a better option due to its cleaner operation and better product range (i.e., gas works as an energy source, whereas leftover is used as solid waste manure). Methane is reported to be the main constituent of biogas, as a majority of energy is retained in methane (Surendra et al., 2014).

Different waste sources like urban, agronomy, industrial sectors, vegetable markets, etc., are the source for the large quantities of solid waste that mainly contains biodegradable organic matter along with varied MSW having maximum quantity. It is reported that, if processed anaerobically, it can generate a significant quantity of biogas, i.e., about 250–350 m^3/tonne of waste and manure, and will significantly help in reducing the load on excessively burdened landfills and will, in turn, prevent the degradation of soil and water. The anaerobic digester can be used specifically for this purpose, as it can supplement the fuel & energy requirements. Furthermore, in a work carried out by Srivastava et al. (2020), agro-waste that is left after the crop harvest can be used as a potential source for bioenergy production of different fuels in form of Hydrogen, bioelectricity, ethanol, and algal diesel or other types of fuel sources by microbial action that can provide a sustainable source of bioenergy. The bioreactor option is a new way forward towards environmentally sound landfills among which biofuel production is receiving major attention.

The increased climate change concerns together with increasing fuel demands have triggered an unprecedented rate of production. Biofuel before the 19th century was restricted to the use of wood in the form of firewood, woodchips, and charcoal (Guo et al., 2015). The modern-day biofuel production slowly started in early 1900s with the use of peanut oil in internal combustion engines by Rudolf Diesel. Later, in early 1940s, Brazil, which is among the leading producers of second-generation biofuel, attempted for oils and fats in internal combustion engines in 1940s. Further developments in the biofuel production industry led to its market being driven mainly by biodiesel and bioethanol in the late 1900s, with other forms of biofuel making a less dominant impact. Today, with the advent of many new technological developments along with increased knowledge on the processing of biomass, new and more refined forms of biofuels are in existence.

13.2 SECOND-GENERATION BIOFUEL PRODUCTION

Second-generation biofuel technologies utilize more flexible and non-food lignocellulosic feedstock. Agricultural biomass is composed of lignocellulosic biomass, mainly cellulose, hemicellulose, and lignin (Singh et al., 2018, 2021). The attempts to produce biofuel from different oilseeds (first-generation biofuel) led to agricultural land competition for food crops (Singh et al., 2020b). The second-generation biofuel was the advanced type using agricultural waste for the production of biofuel, mainly bioethanol (Singh and Verma, 2019). Lignocellulosic biomass production potential is around 220 billion tonnes worldwide with a capacity of satisfying 13% of the world energy requirements. Biomass from agricultural and forest residues as well as feedstock such as trees, jatropha, straw, bagasse, and purpose energy crops are grown on marginal lands (Buyx and Tait, 2011; Senauer, 2008). These feedstocks are converted into ethanol and methanol (Buyx and Tait, 2011; Gupta and Verma, 2015). The main feedstocks included in this type of category are the waste parts of the plants, such as leaves, barks, seeds, and even the wood residues. These types of materials basically serve as the main feedstock for this kind of biofuel production (Naik et al., 2010). In a study done by Naik et al., it is reported that, except for municipal and industrial wastes, all other types can be used as a feedstock for biofuel production as lignocellulosic biomass (2010).

Panicum virgatum, perennial rhizomatous grasses, miscanthus, reed canary grass, and giant weed can be used as a feedstock for second-generation biofuel production. These are crops grown solely for the production of energy from it, termed as bioenergy crops. Their intensity of production can compete with food crops production. These crops also have a very low amount of lignin and can grow even on the most stressful soils. Thus, they become good feedstock for second-generation biofuel production. On an average, the production efficiencies of different crop residues are found to be 0.41 litre per kilogram of dry biomass. Different crop residues such as barley straw, wheat straw, oat straw, and sorghum straw have comparatively higher bioethanol production efficiencies with respect to other food crops' residues (Kim and Dale, 2004).

The biggest advantage presented of second-generation biofuel is that they can give a greater supply of biofuel when compared with that of the first-generation biofuel mainly because of their feedstock. It has been assumed that they can do so in a more sustainable and reasonably economical manner. Lignocellulosic biomass is also used directly as a fuel in the rural areas by burning it directly as firewood for the production of heat. But the use of lignocellulosic biomass in such a way is not considered as second-generation biofuel. The possibility to use cellulosic and heterogeneous biomass in an efficient and sustainable manner suggests lower costs and a better environmental performance (Granda et al., 2007; Hill, 2009). C4 plants may be used as a feedstock for second-generation biofuel more effectively (Byrt et al., 2011). Feedstock for second generation can be a by- or co-product or even waste (Cantrell et al., 2008), or be supplied by dedicated plantations. In general, sugarcane (*Saccharum officinarum*) and maize (*Zea mays*) are used in the production of bioethanol, i.e., second-generation biofuel. This is because the efficiency of conversion of solar energy into biomass is 3.7% and 2.4% in a C4 and C3 plant, respectively (Granda et al., 2007). Considering the difference in the efficiencies of C4 and C3 plants, it becomes very much advantageous for C4 plants to be served as a feedstock, which is a significant 60% increase. This is the very reason why C4 plants have been referred to as the most productive plants on the planet. The length of the growing season also affects the C4 plants' yield and thus makes it a preferable choice over C3 plants.

Among the major biofuels produced, bioethanol, biobutanol, biomethane, biohydrogen, and biodiesel have made significant gains in terms of production. Global bioethanol market in 2019 was USD 43.2 bn and is projected to increase at a whopping 14% to USD 64.8 bn by 2025 (Globenewswire, 2020). Bioethanol production rose to almost 41,000 M gal (Niphadkar et al., 2018), while another agency estimated it to 29,026 M gal in 2019 from 13,123 M gal in 2007 (Center, 2020). The overwhelming increase in bioethanol production can be credited to various technological developments in the past decade (Lugani et al., 2020; Rastogi and Shrivastava, 2017).

13.3 ANAEROBIC FERMENTATION AND RESPIRATION

In oxygen-deficient conditions, microbes employ two mechanisms for survival with respect to energy conservation—anaerobic fermentation and anaerobic respiration. Anaerobic fermentation refers to a redox reaction that occurs in the absence of exogenous electron acceptor, and oxidation is coupled with reduction of the compound derived from an electron donor. During this substrate-level phosphorylation occurs, catabolizing an organic compound to produce ATP required for growth. Anaerobic respiration is a process in which the electron acceptor is other than oxygen, i.e., it can be NO_3^-, SO_4^{2-}, etc. It is enabled by the ETS containing cytochromes, quinones, iron–sulphur proteins, and other ETS proteins analogous to aerobic microbes. The amount of energy produced for the growth during Embden–Meyerhof–Parnas (EMP) is 2 molecules of ATP per molecule of glucose, while in Entner–Duodoroff (ED) pathway only 1 molecule of ATP is produced per molecule of glucose (Jurtshuk, 1996).

Anaerobic bacteria have been widely used as tools for the production of bioproducts at the industrial level. These are mainly used in the production of fuels and chemicals from renewable sources. Further, research is being carried out in order to modify the microbes to prevent or modify the production of by products to increase their specificity on the substrate. Anaerobic bacteria and archaebacteria are known to have potentially unique properties within themselves, referring to the type of environment they inhabit. They are known to inhabit extreme environments and thus would produce enzymes capable of surviving extreme situations for example thermostability, salinity stress resistivity, and many other important aspects. Thermostability of the enzymes from thermophilic microbes is a key aspect with respect to the production of biofuels from lignocellulosic biomass (Singh et al., 2019a). The current interest in thermophilic bacteria is in the exploration and analysis of cellulosomes that are known to be the key system involved in the degradation of the lignocellulosic biomass. Cellulosomes are complex multidomain enzyme systems or free enzyme systems associated with anaerobic bacterial cell wall. Mostly the complex multidomain enzyme systems are found in the anaerobic bacteria used for the degradation of lignocellulosic biomass. There are many positives associated with the anaerobic microbes for which they are preferred in the biofuel industry.

Anaerobic bacteria with regards to the lignocellulolytic activity have also been explored in the endophytes of ruminants' gut, where the possibility of finding obligate anaerobes capable of producing cellulolytic enzymes is very high. This is because the ruminants feed on high cellulosic diet, which increases the possibility of finding culturable cellulolytic microbes. Rumen microbes are known to produce various enzymes capable of degrading cellulosic biomass (Table 13.1).

13.4 ANAEROBIC BIOTECHNOLOGY FOR BIOFUEL PRODUCTION

Biofuel production from lignocellulosic biomass involves three major steps—removal of lignin from the lignocellulosic matrix, degradation of complex polysaccharides

TABLE 13.1
List of Some Anaerobic Bacteria from the Rumen of Animals (Ruminants)

Name of Microbe/s	Rumen Source	Enzyme Secreted	Reference
Pseudomonas aeruginosa, Bacillus, Micrococcus and Streptococcus, Fusarium, Penicillium, Aspergillus, and Mucor	Goat, Sheep, and Cow	Cellulose degrading enzyme. Specific enzyme activity not mentioned.	Oyeleke and Okusanmi (2008)
Firmicutes, U-Lanchospiraceae, Butyrivibrio, Streptococcus, Enterococcus, Anaerovibrio, Selenomonas, U-Clostridiales, Saccharofermentans, Ruminococcus, Clostridium cluster IV, Proteobacteria, Spirochaetes, Bacteroidetes, Prevotella, Fibrobacteres, Actinobacteria	Cow	FPase, CMCase, and xylanase	Nyonyo et al. (2014)
Bacillus, Enterobacter and *Bacteriodes*	Sheep	FPase and CMCase	Guder and Krishna (2019)
Escherichia coli	Inner Mongolia bovine	Cellulase, endoglucanase, exoglucanase, and beta-glucosidase	Pang et al. (2017)
Enterobacter	Aceh cattle	Cellulase	Sari et al. (2017)
Fibrobacter succinogenes S85	Herbivores (specific not mentioned)	Cellulases, endoglucanase, exoglucanase, xylanase, beta-galactosidase	Raut et al. (2019)
Bacillus subtilis CDB9, CDB 10	Not mentioned	CMCase, FPase	Rawway et al. (2018)

into monomer sugars, and then into fuel like products such as bioethanol, biomethanol through fermentation and distillation, respectively. Biofuel production from lignocellulosic biomass involves the use of complex set of enzymes that are produced by microbes. Microbes have been harnessed since ages for the production of enzymes and conversion of raw materials into products (Table 13.2). In the anaerobic bacteria, enzymes involved have an advantage over aerobic ones that in the former they can be used for consolidated bioprocessing, as the hydrolysis step can also be performed anaerobically. Microbes including bacteria and fungi employ a complex set of enzymes that degrade plant cell walls and bring about the degradation of lignocellulosic matrix are called carbohydrate active enzymes (CAZymes). These carbohydrate-active enzymes involve a lot of groups of enzymes such as cellulases, hemicellulases and related glycoside hydrolases (133), polysaccharide lyases (23), glycosyl transferases (97), carbohydrate esterases (16), other auxillary enzymes (13), and carbohydrate-binding molecules (71) (Park et al., 2018).

TABLE 13.2
List of Anaerobic Microbes Producing Enzymes for Different Biofuel Production

Name of Anaerobic Microbe/s of Group of Microbes/Consortia/Community	Process in which Used	Enzymes Estimated during the Study	Reference
Anaerobic fungal isolate strain C1A	Lignocellulose biomass degradation	Cellulase, xylanase	Ranganathan et al. (2017)
H. saccincola and three novel anaerobic thermophilic bacteria	lignocellulose degradation	Cellulase, endoglucanase, and xylanase	Shikata et al. (2018)
Alginate anaerobic hydrolytic bacteria	alginate conversion	oligo-alginate lyase	Zhang et al. (2019)
Clostridium sp. DBT-IOC-C19	Consolidated bioethanol production	NA	Singh et al. (2017)
Anaerobic microbial consortium: *Clostridiaceae* and *Peptostreptococcaceae*	1,3-propanediol production	NA	Zhou et al. (2017)
Bacillus sp. BMP01 and *Ochrobactrum oryzae* BMP03	Lignocellulosic biomass conversion	Cellulase and xylanase	Tsegaye et al. (2019)
Acetogenic bacteria NJUST19	Acetate production	NA	Huang et al. (2020)
Spathaspora yeasts	Bioethanol production	Xylanase	Cadete and Rosa (2018)

Glycoside hydrolases hydrolyse the glycosidic bonds in the chemical structure of carbohydrates within or attached to some other moiety. Extracellular glycoside hydrolases of the anaerobic bacteria, in general, have a catalytic module apart from the complex set of arrangement of the many modules. Thus, enzymes can have one or more than one catalytic module/s that may include carbohydrate-binding molecule, immunoglobulin-like dockerin, fibronectin type III, and surface layer homology. These modules are linked by amino acids through covalent bonding. The catalytic modules in combination with the non-catalytic module bring about the hydrolysis of the glycosidic bonds. The mechanism of action of glycosyl hydrolases involves general acid catalysis with a proton donor and a nucleophile base (Davies and Henrissat, 1995). Further this can be categorized into types, the retaining mechanism and the inverting mechanism. In retaining mechanism, the acid catalyst protonates glycosidic oxygen followed by its hydrolysis by water molecule. Two nucleophilic substitutions take place in this process. In the inverting mechanism, simultaneous protonation of glycosidic oxygen and aglycon departure is followed by based residue activated water molecule. This follows a single nucleophilic substitution reaction with the production of a product opposite to the substrate stereochemistry (Davies and Henrissat, 1995). Carbohydrate binding modules bring about the degradation of insoluble polysaccharides. CBMs are divided into families based on their substrate specificity such as starch-, chitin-, xylan-, and mannose-binding domains.

13.5 APPLICATIONS OF ANAEROBIC FERMENTATION

The anaerobic fermentation has widespread applications and expanded very rapidly in 21st century. It is being used for production of several industrial chemicals, such as ethanol, lactic acid, butyl alcohol acetic acid, hydrogen gas, and different nutraceuticals. Anaerobic biotechnology is a sustainable method that combines renewable biofuels production along with treatment of waste with recovery of useful by products. A broad range of application of anaerobic technology could comfort increasing energy insecurity and substantially reducing the emission of toxic air pollutants (Khanal, 2008). Anaerobic biotechnology fulfils the three basic needs of underdeveloped and developing nations, which are as follows:

a. To provide digested materials as a source of biofertilizer for agriculture sector;
b. To generate renewable energy for several activities, such as small-scale business and household activities;
c. Sanitation and overall improvement in health through pollution control.

Anaerobic fermentation plays a significantly greater role not only in controlling pollution but also in supplementing valuable resources. Anaerobic fermentation is becoming immensely popular due to its potential to produce renewable biofuels and value-added products from low-value feedstock such as waste streams. Young and McCarty published first report in 1969, elaborating the application of an anaerobic filter that is biofilm for the treatment of wastewater. This technique plays a significant role in the production of biodiesel, biomethane, butanol, biohydrogen, and electricity generation, which have been discussed in detail in the next section, i.e., anaerobic fermentation in bioenergy production.

Furthermore, anaerobic fermentation also helps in the recovery of various value-added products, such as acetic acid, nisin, and lactic acid. In a properly managed condition, anaerobic fermentation process produces acetic acid from H_2 and CO_2. The control strategies include redox potential, pH, periodic reduction of hydrogen, and addition of other substances, such as bile salts, protein (Vandevivere, 1999). Miller and Wolin, 1995 reported that coculture of cellulolytic bacterium and a reductive acetogen yields high concentration of acetate from cellulose substrate from H_2 and CO_2.

Bacteriocin, particularly, Nisin is commercially produced by fermentation using lactic acid bacteria, mainly *Lactococcus lactis*. This bacteriocin is used as a food preservative and has a strong ability against gram-positive pathogens. Due to its significant role towards a wide range of pathogens, it is getting substantial interest. Several food-processing industries such as corn-processing, cheese-processing, and waste streams from soy-processing industries serve as a perfect feedstock for nisin production. These waste streams are nutritionally rich and fit for the growth of *Lactococcus lactis*, as it contains carbohydrate, protein, phosphorus, and numerous trace elements.

Apart from this, anaerobic fermentation also provides efficient support in waste treatment process, such as it lowers the requirement of nutrients and significantly reduces the quantity of sludge generation. The anaerobic process requiring just 20% of the nutrients required for the aerobic process, because of the lower biomass synthesis rate considerably lower nutrients are required. The wastewater treatment by aerobic process, more specifically activated sludge process, produces huge amounts of sludge. On the other hand, anaerobic treatment processes utilize almost 90% of the organic matter for the production of methane; remaining 10% or less is converted to biomass. In addition, the anaerobic sludge does not undergo further treatment other than dewatering, as it is already well stabilized.

Along with this, anaerobic fermentation requires less energy, as it needs no air/O_2 as compared to aerobic treatments that are energy-intensive, demanding energy for aeration for the removal of organic matter (Van Haandel and Lettinga, 1994). Moreover, anaerobic treatment delivers a net economic gain by energy generation from methane gas.

13.6 ANAEROBIC FERMENTATION IN BIOENERGY PRODUCTION

Burning of fossil fuel leads to increase in atmospheric CO_2 concentration that poses great risk worldwide. With the beginning of industrial revolution and increasing economic growth, this trend is going to hike in near future. Anaerobic fermentation is a process of conversion of biomass into different bioenergy form (Sialve et al., 2009). The production of biogas and anaerobic fermentation are probable means of producing an energy carrier from renewable resources and also accomplish multiple benefits in terms of environment (Adams, 1990).

Bioenergy can include energy from biomass in the form of biofuels, biopower, or other bioproducts. Biofuels are basically transportation fuels like bioethanol, biodiesel, biohydrogen, and biomethane. Bio power consists of either direct biomass burning or converting it into oil or gaseous fuel to generate electricity. All aspects of the biomass-to-bioenergy conversion process are presented in Figure 13.1.

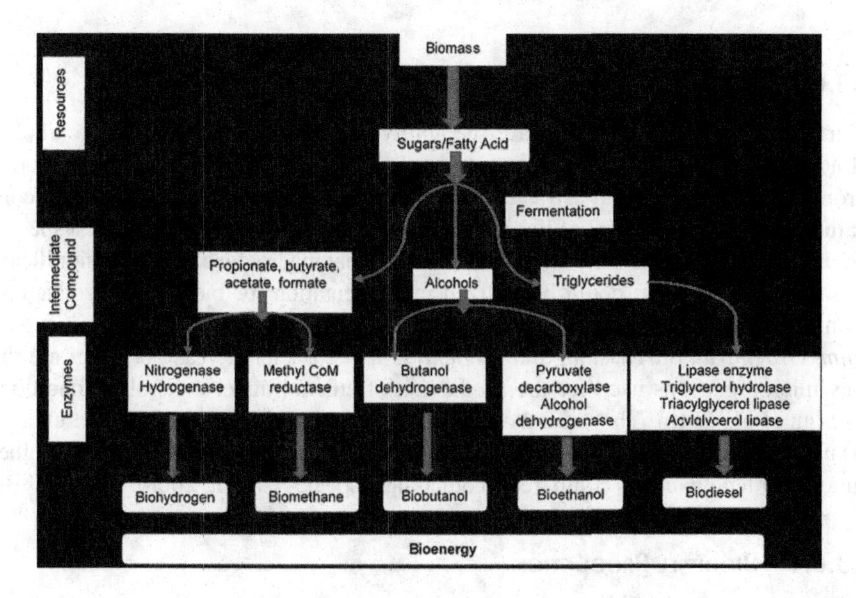

FIGURE 13.1 Biomass to Bioenergy.

13.6.1 Biomethane Production

Biomass consists of a mixture of organic and inorganic matter. The organic part consists of polymeric macromolecules, such as proteins, lipids, polysaccharides, and nucleic acids. Methane gas is a major by-product of anaerobic digestion of organic solid and liquid wastes. Mixtures of bacteria are used to hydrolyze and break down the organic biopolymers (i.e., carbohydrates, lipids, and proteins) into monomers, which are then converted into a methane-rich gas via fermentation (CH_4= 50%–75%). CO_2 is the second main component of biogas (CO_2= 25%–50%).

13.6.2 Biohydrogen Production

Biological hydrogen production is considered as one of the most auspicious green alternatives for sustainable energy production from biomass. Hydrogen is a clean fuel that does not produce carbon dioxide as a by-product when it is burnt for electricity generation in fuel cells (Park et al., 2009). Biomass can be converted in bio-H_2 by photo fermentation as well as dark fermentation using anaerobic organisms. Cyanobacteria and algae split the water molecule into oxygen and hydrogen with the help of hydrogenase enzyme—carries out photo evolution of hydrogen (Adams, 1990).

From a comprehensive environmental viewpoint, production of hydrogen from renewable organic wastes symbolises a significant area of bioenergy production. Miyake et al. (1984) illustrate the production of hydrogen using glucose as a model substrate by the following equations:

$$C_6H_{12}O_6 + 2H_2O \rightarrow 2CH_3COOH + 2CO_2 + 4H_2\_G^\circ = -184 \text{ kJ}$$

$$C_6H_{12}O_6 \rightarrow CH_3CH_2CH_2COOH + 2CO_2 + 2H_2\_G^\circ = -255 \text{ kJ}$$

13.6.3 Bioethanol Production

Certain species of microalgae have the ability of producing high levels of carbohydrates as compared to lipids as storage polymer. Hence, ethanol could be produced from the fermentable sugar extracted from these storage carbohydrates. Seaweeds contain no or very less amount of lignin and are, therefore, considered as the best species for ethanol production. Wi et al. (2009) reported that macroalgae contain a significant amount of sugars that could be utilized in fermentation for bioethanol production. Some of the promising species of algae for ethanol production are *Prymnesium parvum, Gracilaria, Sargassum, and Euglena gracilis.* Bacteria, yeast, or fungi are the key microorganisms used for the production of ethanol under anaerobic conditions (Harun et al., 2010). They metabolize carbohydrates and produce CO_2 and ethanol as metabolic end products in an anaerobic condition. Brown seaweed produce higher amount of bioethanol as compared to other algal species (Moen, 2008).

13.6.4 Biodiesel Production

The biogas generated during anaerobic fermentation of organic waste can be converted into liquid fuel i.e., biodiesel. The biogas is first converted into liquid methanol using a thermal catalytic process. Biodiesel is produced by the process of transesterification of fats or oil with methanol in the presence of a catalyst (sodium or potassium hydroxide).

13.6.5 Butanol Production

Butanol is also a potential substitute for fossil fuel and is considered a superior fuel to ethanol for different reasons:

- More favourable physical properties,
- better economics, and
- safety.

In addition, the butanol eliminates the need for engine modification that has been running on gasoline. Butanol is produced by fermentative bacteria, including *Clostridium acetobutylicum* (Qureshi et al., 2006) and *Clostridium beijerinkii*. The ratio of acetone, butanol, and ethanol is 3:6:1, with butanol being the major fermentation by-product. Low butanol yield through fermentation coupled with cheap petroleum feedstock is the major impediment to the widespread development of butanol fuel. Carbohydrate-rich waste stream could serve as a perfect feedstock for butanol production (Rathour et al., 2018). Hydrogen gas is another by-product of butanol fermentation and can also be recovered as a renewable energy.

13.7 ELECTRICITY GENERATION

Biochemical energy stored in the carbohydrate and other organic matter in wastewater can be directly converted into electricity by microbial fuel cells (MFC) (Dalvi et al., 2011). An MFC contains two chambers consisting of an anode and cathode, similar to hydrogen fuel cell, separated by a cation exchange membrane. The organic matter is oxidized by anaerobic microbes in the anode chamber and electrons are released (Liu et al., 2005). These electrons are then transferred to the anode and flowed to the cathode through a conductive material, such as a resistor or to an external load. The electrons in the cathode combine with protons that diffuse through the cation exchange membrane and oxygen. The oxygen is reduced to water. In the MFC, the driving force is the redox reaction of substrates mediated by anaerobic microorganisms (Geelhoed et al., 2010). Thus, MFC research has a potential to treat the wastewater and produce electricity.

13.8 CONCLUSIONS AND FUTURE PERSPECTIVES

Great strides have been made in the field of anaerobic fermentation with employment of microbes. Though at the microbial level there are certain ways through which biofuel production can be increased from lignocellulosic biomass, such as thermophilic microbes. Genetic modifications in the microbe can also be seen as one of the promising aspects through which biofuel production from lignocellulosic biomass can be enhanced (Singh et al., 2020a). Another promising aspect in biofuel production industry is third-generation biofuels from algae (Singh et al., 2019b). Lignocellulosic biomass can hugely serve as the resource for the production of biofuels. Anaerobic microbes due to their extremophilic adaptations can be harnessed greatly to serve for the production of biofuels. The enzymes produced by the anaerobic microbes due to their extremophilic properties are stable at varying conditions and would benefit at the industrial level.

ACKNOWLEDGEMENTS

SS is highly grateful to the University Grants Commission (UGC Ref No.3819)/(NET-JULY-2018)) for providing the Junior Research Fellowship and Senior Research Fellowship to carry out the research work.

REFERENCES

Adams, M.W. (1990) The structure and mechanism of iron-hydrogenases. *Biochimica et Biophysica Acta (BBA)-Bioenergetics* 1020, 115–145.

Buyx, A.M., Tait, J. (2011) Biofuels: Ethics and policy-making. *Biofuels, Bioproducts and Biorefining* 5, 631–639.

Byrt, C.S., Grof, C.P., Furbank, R.T. (2011) C4 Plants as biofuel feedstocks: Optimising biomass production and feedstock quality from a lignocellulosic perspective free access. *Journal of Integrative Plant Biology* 53, 120–135.

Cadete, R.M., Rosa, C.A. (2018) The yeasts of the genus Spathaspora: Potential candidates for second-generation biofuel production. *Yeast* 35, 191–199.

Cantrell, K.B., Ducey, T., Ro, K.S., Hunt, P.G. (2008) Livestock waste-to-bioenergy generation opportunities. *Bioresource Technology* 99, 7941–7953.

Center, A.F.D., (2020) *Global Ethanol Production by Country or Region*. U.S. Department of Energy, Energy Efficiency and Renewable Energy.

Dalvi, A.D., Mohandas, N., Shinde, O.A., Kininge, P.T. (2011) Microbial fuel cell for production of bioelectricity from whey and biological waste treatment. *International Journal of Advanced Biotechnology and Research* 2, 263–268.

Davies, G., Henrissat, B. (1995) Structures and mechanisms of glycosyl hydrolases. *Structure* 3, 853–859.

Geelhoed, J.S., Hamelers, H.V., Stams, A.J. (2010) Electricity-mediated biological hydrogen production. *Current Opinion in Microbiology* 13, 307–315.

Globenewswire, (2020) *Global Bioethanol Market (2020 to 2025) - Rising Demand from the Pharmaceutical Industry Presents Opportunities*. Research and Markets, Dublin.

Granda, C.B., Zhu, L., Holtzapple, M.T. (2007) Sustainable liquid biofuels and their environmental impact. *Environmental Progress* 26, 233–250.

Guder, D.G., Krishna, M. (2019) Isolation and characterization of potential cellulose degrading bacteria from sheep rumen. *Journal of Pure and Applied Microbiology* 13, 1831–1839.

Guo, M., Song, W., Buhain, J. (2015) Bioenergy and biofuels: History, status, and perspective. *Renewable and Sustainable Energy Reviews* 42, 712–725.

Gupta, A., Verma, J.P. (2015) Sustainable bio-ethanol production from agro-residues: A review. *Renewable and Sustainable Energy Reviews* 41, 550–567.

Harun, R., Singh, M., Forde, G.M., Danquah, M.K. (2010) Bioprocess engineering of microalgae to produce a variety of consumer products. *Renewable and Sustainable Energy Reviews* 14, 1037–1047.

Hill, J. (2009) Environmental costs and benefits of transportation biofuel production from food- and lignocellulose-based energy crops: A review. *Sustainable Agriculture*, 125–139.

Huang, C., Wang, W., Sun, X., Shen, J., Wang, L. (2020) A novel acetogenic bacteria isolated from waste activated sludge and its potential application for enhancing anaerobic digestion performance. *Journal of Environmental Management* 255, 109842.

Jurtshuk, P. (1996) Bacterial metabolism. *Medical Microbiology,* 4.

Khanal, S.K. (2008) Overview of anaerobic biotechnology. *Anaerobic Biotechnology for Bioenergy Production*, 1–27.

Kim, S., Dale, B.E. (2004) Global potential bioethanol production from wasted crops and crop residues. *Biomass and Bioenergy* 26, 361–375.

Liu, H., Grot, S., Logan, B.E. (2005) Electrochemically assisted microbial production of hydrogen from acetate. *Environmental Science & Technology* 39, 4317–4320.

Lugani, Y., Rai, R., Prabhu, A.A., Maan, P., Hans, M., Kumar, V., Kumar, S., Chandel, A.K., Sengar, R. (2020) Recent advances in bioethanol production from lignocelluloses: A comprehensive review with a focus on enzyme engineering and designer biocatalysts. *Biofuel Research Journal* 7(4), 1267–1295.

Miyake, J., Mao, X., Kawamura, S. (1984) Photoproduction of hydrogen from glucose by a coculture of a photosynthetic bacterium and *Clostridium butyricum. Hakko Kogaku Zasshi;(Japan)* 62 (6).

Moen, E. (2008) Biological degradation of brown seaweeds. The potential of marine biomass for anaerobic biogas production. *Argyll, Scotland: Scottish Association for Marine Science Oban* 9, 157–166.

Naik, S.N., Goud, V.V., Rout, P.K., Dalai, A.K. (2010) Production of first and second generation biofuels: A comprehensive review. *Renewable and Sustainable Energy Reviews* 14, 578–597.

Niphadkar, S., Bagade, P., Ahmed, S. (2018) Bioethanol production: Insight into past, present and future perspectives. *Biofuels* 9, 229–238.

Nyonyo, T., Shinkai, T., Mitsumori, M. (2014) Improved culturability of cellulolytic rumen bacteria and phylogenetic diversity of culturable cellulolytic and xylanolytic bacteria newly isolated from the bovine rumen. *FEMS Microbiology Ecology* 88, 528–537.

Oyeleke, S., Okusanmi, T. (2008) Isolation and characterization of cellulose hydrolysing microorganism from the rumen of ruminants. *African Journal of Biotechnology* 7 (10).

Pang, J., Liu, Z.-Y., Hao, M., Zhang, Y.-F., Qi, Q.-S. (2017) An isolated cellulolytic Escherichia coli from bovine rumen produces ethanol and hydrogen from corn straw. *Biotechnology for Biofuels* 10, 1–10.

Park, J.-I., Lee, J., Sim, S.J., Lee, J.-H. (2009) Production of hydrogen from marine macroalgae biomass using anaerobic sewage sludge microflora. *Biotechnology and Bioprocess Engineering* 14, 307–315.

Park, Y.-J., Jeong, Y.-U., Kong, W.-S. (2018) Genome sequencing and carbohydrate-active enzyme (CAZyme) repertoire of the white rot fungus *Flammulina elastica. International Journal of Molecular Sciences* 19, 2379.

Qureshi, N., Li, X.L., Hughes, S., Saha, B.C., Cotta, M.A. (2006) Butanol production from corn fiber xylan using *Clostridium acetobutylicum. Biotechnology Progress* 22, 673–680.

Ranganathan, A., Smith, O.P., Youssef, N.H., Struchtemeyer, C.G., Atiyeh, H.K., Elshahed, M.S. (2017) Utilizing anaerobic fungi for two-stage sugar extraction and biofuel production from lignocellulosic biomass. *Frontiers in Microbiology* 8, 635.

Rastogi, M., Shrivastava, S. (2017) Recent advances in second generation bioethanol production: An insight to pretreatment, saccharification and fermentation processes. *Renewable and Sustainable Energy Reviews* 80, 330–340.

Rathour, R.K., Ahuja, V., Bhatia, R.K., Bhatt, A.K. (2018) Biobutanol: New era of biofuels. *International Journal of Energy Research* 42, 4532–4545.

Raut, M.P., Couto, N., Karunakaran, E., Biggs, C.A., Wright, P.C. (2019) Deciphering the unique cellulose degradation mechanism of the ruminal bacterium *Fibrobacter succinogenes* S85. *Scientific reports* 9, 1–15.

Rawway, M., Ali, S.G., Badawy, A.S. (2018) Isolation and identification of cellulose degrading bacteria from different sources at assiut governorate (Upper Egypt). *Journal of Environment and Health* 6, 15–24.

Sari, W.N., Safika, D., Fahrimal, Y. (2017) Isolation and identification of a cellulolytic Enterobacter from rumen of Aceh cattle. *Veterinary World* 10, 1515.

Senauer, B. (2008) Food market effects of a global resource shift toward bioenergy. *American Journal of Agricultural Economics* 90, 1226–1232.

Shikata, A., Sermsathanaswadi, J., Thianheng, P., Baramee, S., Tachaapaikoon, C., Waeonukul, R., Pason, P., Ratanakhanokchai, K., Kosugi, A. (2018) Characterization of an anaerobic, thermophilic, alkaliphilic, high lignocellulosic biomass-degrading bacterial community, ISHI-3, isolated from biocompost. *Enzyme and Microbial Technology* 118, 66–75.

Sialve, B., Bernet, N., Bernard, O. (2009) Anaerobic digestion of microalgae as a necessary step to make microalgal biodiesel sustainable. *Biotechnology Advances* 27, 409–416.

Singh, N., Mathur, A.S., Tuli, D.K., Gupta, R.P., Barrow, C.J., Puri, M. (2017) Cellulosic ethanol production via consolidated bioprocessing by a novel thermophilic anaerobic bacterium isolated from a Himalayan hot spring. *Biotechnology for biofuels* 10, 73.

Singh, S., Gaurav, A.K., Verma, J.P. (2020a) Genetically modified microbes for second-generation bioethanol production. *Fungal Biotechnology and Bioengineering*, 187–198.

Singh, S., Jaiswal, D.K., Krishna, R., Mukherjee, A., Verma, J.P. (2020b) Restoration of degraded lands through bioenergy plantations. *Restoration Ecology* 28, 263–266.

Singh, S., Jaiswal, D.K., Sivakumar, N., Verma, J.P. (2019a) Developing efficient thermophilic cellulose degrading consortium for glucose production from different agro-residues. *Frontiers in Energy Research* 7, 61.

Singh, S., Kumar, A., Verma, J.P., (2021) Techno-economic analysis of second-generation biofuel technologies. *Bioprocessing for Biofuel Production*, 157–181.

Singh, S., Pereira, A.P., Verma, J.P. (2019b) Research and Production of Third-Generation Biofuels. *Bioprocessing for Biomolecules Production*, 401–416.

Singh, S., Verma, J.P., (2019) *Bioethanol Production from Different Lignocellulosic Biomass, Sustainable Biofuel and Biomass*. Apple Academic Press, New York, 281–300.

Singh, S., Verma, J.P., de Araujo Pereira, A.P., Sivakumar, N. (2018) Production of biofuels and chemicals from lignin. *Journal of Cleaner Production* 203, 966–967.

Srivastava, R.K., Shetti, N.P., Reddy, K.R., Kwon, E.E., Nadagouda, M.N. and Aminabhavi, T.M. (2021). Biomass utilization and production of biofuels from carbon neutral materials. *Environmental Pollution*, 276, 116731.

Surendra, K.C., Takara, D., Hashimoto, A.G. and Khanal, S.K. (2014). Biogas as a sustainable energy source for developing countries: Opportunities and challenges. *Renewable and Sustainable Energy Reviews*, 31, 846–859.

Tsegaye, B., Balomajumder, C., Roy, P. (2019) Isolation and characterization of novel lignolytic, cellulolytic, and hemicellulolytic bacteria from wood-feeding termite *Cryptotermes brevis. International Microbiology* 22, 29–39.

Van Haandel, A.C., Lettinga, G. (1994) Anaerobic sewage treatment. John Wiley & Sons, London, England.

Vandevivere, P. (1999) New and broader applications of anaerobic digestion. *Critical Reviews in Environmental Science and Technology* 29, 151–173.

Wi, S.G., Kim, H.J., Mahadevan, S.A., Yang, D.-J., Bae, H.-J. (2009) The potential value of the seaweed Ceylon moss (*Gelidium amansii*) as an alternative bioenergy resource. *Bioresource Technology* 100, 6658–6660.

Zhang, F., Zhang, W., Qian, D.-K., Dai, K., van Loosdrecht, M.C.M., Zeng, R.J. (2019) Synergetic alginate conversion by a microbial consortium of hydrolytic bacteria and methanogens. *Water Research* 163, 114892.

Zhou, J.-J., Shen, J.-T., Jiang, L.-L., Sun, Y.-Q., Mu, Y., Xiu, Z.-L. (2017) Selection and characterization of an anaerobic microbial consortium with high adaptation to crude glycerol for 1,3-propanediol production. *Applied Microbiology and Biotechnology* 101, 5985–5996.

Part VI

Advance Green Fuel

14 Biohythane Production
Future of Biofuel

Ali Asger and Abhishek Mote
Institute of Chemical Technology

M. Shahbaz Khan
CSIR—National Environmental Engineering
Research Institute (CSIR—NEERI)

Sameena N. Malik
Institute of Chemical Technology

CONTENTS

DOI: 10.1201/9781003197737-20

14.1 INTRODUCTION

The use of fossil fuels as the prime source for the world's energy results in severe threat to the environment due to the rising greenhouse gas emissions (Rena et al. 2020; Kongjan et al. 2018). Modernization and advancement in the extraction techniques has caused the depletion of fossil fuels at a remarkable pace. This rapid depletion creates a warning towards global energy security (Rena et al. 2020; Tiwari and Pandey 2012). Huge growth in population along with paced urbanization across the globe has resulted in an increased energy consumption of 37% from 8.5 billion tones oil equivalent in the year 1992 to 13.5 billion tones oil equivalent in 2017. The environmental impacts and rising concerns for energy security has compelled the researchers to innovate and develop new processes/methods that can harness energy from renewable energy sources (Martinez-Jimenez et al. 2017; Mishra et al. 2017; Ghimire et al. 2015; Ellabban et al. 2014). These new processes to harness energy from renewable sources should be developed such that they preserve the environment and also cater to the future energy needs (Maaroff et al. 2019). One such source can be organic waste, which can future generate renewable energy due to the high presence of carbohydrates and ample availability (Vo et al. 2019; Bolzonella et al. 2018). The conventional technique of using microorganisms to convert organic waste into bioenergy helps in producing energy as well as in waste management (Singh et al. 2012). India is the second largest producer of vegetable and fruits across the globe (National Horticulture Board 2018). Almost close to one-third of the total food production of the globe in the year 2011 was wasted or was not consumed according to food and agricultural organization (FAO) (Rena et al. 2020; Galanakis 2018). Lignocellulosic waste of 200 billion tones is generated every year globally (Ji et al. 2017; Ren et al. 2009). Incorrect disposal of this organic waste leads to emission of carbon, which further contributes to global warming. Textile industry is another large-scale generator of harmful wastewater (Gadow and Li 2020a; Holkar et al. 2016; Cao et al. 2019; Yaseen and Scholz 2019). This industrial wastewater from pharmaceutical, cosmetics, and textile industry largely consist of azo dyes. The colored mixes in these azo wastewaters are considered to be the main culprit for environmental contamination (Gadow and Li 2020). Animal manure, organic waste, agricultural residues, algae, industrial wastewater, etc., can be a good source of bioenergy (Kongjan et al. 2018). Renewable energy generation can prove to be an important step in waste processing for both biomass and wastewater. Waste-generating resources are not efficiently used. Due to this it is difficult to achieve the Sustainable Development Goals, i.e., SDG, which are adopted by the United Nations (UN) organization in the year 2015 (Kumar et al. 2019).

The term Hythane® was trademarked by Hydrogen Component Inc. (HCI) in one of their patented study, where they used a blend of hydrogen and methane to fuel an internal combustion engine. They found that combustion of this new fuel blend liberates approximately same amount of energy as that of compressed natural gas (CNG). Also, this new fuel blend resulted in reduced NO_x emissions than CNG (Bolzonella et al. 2018; Hans and Kumar 2019). This is due to the fact that H_2 has high heat content (122 kJ/g) than methane (55 kJ/g) (Maaroff et al. 2019; Mishra et al. 2017; Zhang et al. 2020; Roy and Das 2016). Hydrogen exhibits good energy density and has high efficiency in its conversion to usable power. This new fuel blend "Hythane" is also referred as H_2 enriched CNG (Liu et al. 2018). Storage of hythane does not need any significant change in the existing storage equipment of CNG (Zhang et al. 2020). Use of physical and chemical processes for hythane production is not a good approach because of the requirement of input energy. Hythane that is produced using organic wastes as substrate are referred as "Biohythane." For this biohythane, hydrogen and methane both are produced using organic/renewable sources. This blend of bio-H_2 and bio-CH_4 can mask each other's drawbacks. The idea of using fermentation to get hydrogen-enriched CNG has appealed many scientists. Biohythane generally consists of hydrogen, methane, and CO_2 of about 5%–15%, 48%–62%, 30%–10%, respectively (Hans and Kumar 2019; Kumar et al. 2019). Biohythane reduces the ignition time and temperature, and also enhances the engine performance. Methane has low range of flammability, which can be easily enlarged by the addition of hydrogen. This is because of the two-and-half times more heating value (mass specific) of 119,930 kJ/kg for hydrogen, compared to that of 50,020 kJ/kg for methane. Also, hydrogen with a flame speed between 2.6 and 3.2 m/s (which is approx. seven times more than CH_4) can significantly reduce the combustion time when compared to methane (Kumar et al. 2018). Robust reducing capability shown by hydrogen drives methane combustion and improves combustion at less temperatures. In the year 1995, the "Montreal hythane bus project" was the pioneer to use hythane as a fuel in an internal combustion engine. This reduced the nitrous oxide gas by about 45% (Kumar et al. 2018).

Anaerobic digestion method can be used to produce biohythane (Hans and Kumar 2019). In this method, dark fermentation step is used to produce H_2, whereas another step of anaerobic digestion is used for CH_4 production (Krishnan et al. 2016). Both biohydrogen and biomethane are produced in two separate digesters and then further mixed together (Ta et al. 2020). Organic matter like wheat straw, potato wastes, rice straw, paper waste, corn stalk, olive pulp, etc., along with industrial waste water such as palm oil, mill effluent, cheese whey stalk, molasses, etc., can be used for biohythane production (Maaroff et al. 2019; Seengenyoung et al. 2019; Mamimin et al. 2015, 2019). Biohythane was produced at a yield of 396, 225, 287, 523, 669, and 360 mL/g from wheat straw, potato waste, rice straw, corn stalk, food waste, and starch waste-water, respectively (Mamimin et al. 2019). Biomass from algae can also be used as a substrate for biohythane production due to its good photosynthetic efficiency and high content of organic matter (Jehlee et al. 2017). Solid wastes (both organic and industrial wastewater) also prove to be very good substrate for biohythane production. One of the major obstructions in biohythane production is the shortage of green hydrogen. Therefore, special efforts need to be given towards the development of technologies that can produce biohydrogen in an environment-friendly manner.

14.2 PROCESS MECHANISM

Use of anaerobic digestion for the production of biomethane (b-CH_4) as well as biohydrogen (b-H_2), and further blending them with each other can prove to be sustainable path for biohythane production (Vo et al. 2019). In the second step of anaerobic fermentation (AD), hydrogen is made through the dark fermentation (DF) process, which takes place without presence of light. High manufacturing rate and a broad scale of raw materials (biomass) make anaerobic digestion process a practical solution. Anaerobic digestion also converts organic substrates into liquid fertilizers (Meena et al. 2020). Obtaining the theoretical output of hydrogen in dark fermentation is not practical because of the increased operating cost, generation of volatile fatty acids VFAs, i.e., inhibitory metabolites, and the presence of methanogens and hydrogen producers. Hydrogen is solely not adequate to reduce the biomass, as not more than 20% of COD was transformed along with less recovery of energy during the individual hydrogen production process. A significant chunk of energy of biomass is unreacted at the end as VFAs. These VFAs are then used to recover the remaining energy. This recovery can be done by transforming these VFSs into methane using anaerobic digestion (Hans and Kumar 2019).

A single-stage anaerobic digestion process gives only methane in large quantities of biogas along with few traces of hydrogen. A two-stage anaerobic digestion process produces methane and hydrogen at the same time. This way we can improve the b-CNG efficiency, easy engine ignition, and reduce the requirement of energy than dark fermentation (DF) process (Hans and Kumar 2019). Reduction in waste generated and increase in rate of output are some advantages of two-stage AD over DF (Vo et al. 2019). However, the two-stage anaerobic digestion process is costly as it demands a separate reactor, additional reactor for blending methane and hydrogen. Less recovery of energy, unstable production of gas, and incomplete digestion/fermentation is the downside of the single-stage anaerobic digestion process (Vo et al. 2019). Due to all the above reasons, two-stage anaerobic digestion process is the most feasible and practical process for the production of biohythane (Hans and Kumar 2019). DF process can be used for H_2 production by using substrates rich in carbohydrate. But there is a kinetic limitation for H_2 production of theoretical yield that should not be more than 498 mL H_2/g-sugars, which is less than 33.33% of the total content of energy in the substrate (rich in carbohydrates). The un-transformed 66.67% of energy is still left in the form of VFAs. Due to this DF process is not practically cost-effective. The acetogenesis stage followed by methanogenesis of the two-stage AD process can solve this issue of unrecovered VFAs and be feasible for biohythane production from biomass substrates. Around 11%–16% of H_2 (vol. %) is present in biohythane, which comes from the DF process (Kongjan et al. 2018). Butyrate, acetate along with ethanol produce accompany biohythane production (Vo et al. 2019).

14.2.1 Anaerobic Digestion

It is a process that is used to degrade the organic biomass; this is done by using biochemicals reactions mediated by microorganisms and in the absence of O_2 to transform energy in the biomass into ATP (Martinez-Jimenez et al. 2017). Anaerobic digestion involves four steps: (1) hydrolysis, (2) acidogenesis, (3) acetogenesis, and (4) methanogenesis. First step involves conversion of complex organic strata like lipids, proteins, cellulose, carbohydrates, etc., into glucose, fatty acids, amino acids, etc., like simple monomers (Hans and Kumar 2019). (Equations 14.1–14.3).

$$(C_6\ H_{10}O_4)_n + 2H_2O \rightarrow C_6\ H_{12}O_6 + O_2 \tag{14.1}$$

$$(-\ RCH(NH_2)COO-)_n + (n-1)H_2O \rightarrow nRCH(NH_2)COOH \tag{14.2}$$

$$\left(H_2COOC(CH_2)_n\ CH_3\right)_n \rightarrow n\left(H_2COOC(CH_2)_n\ CH_3\right) \tag{14.3}$$

In the second step of acidogenesis, the simple monomers from the hydrolysis step are converted into CO_2 and VFAs. This is done using acidogens (Hans and Kumar 2019). (Equations 14.4–14.7).

$$C_6\ H_{12}O_6 \rightarrow 2CH_3CH_2OH + 2CO_2 \tag{14.4}$$

$$C_6\ H_{12}O_6 + 2H_2 \rightarrow 2CH_3CH_2COOH + 2H_2O \tag{14.5}$$

$$C_6\ H_{12}O_6 \rightarrow CH_3CH_2CH_2COOH + 2CO_2 + 2H_2 \tag{14.6}$$

$$C_6\ H_{12}O_6 \rightarrow 3CH_3COOH \tag{14.7}$$

In the third step of acetogenesis, the volatile fatty acids from the previous stage are digested to hydrogen and carbon dioxide using the acetogenic bacteria (Hans and Kumar 2019). (Equations 14.8–14.10).

$$CH_3CH_2COO^- + 3H_2O \leftrightarrow CH_3COO^- + H^+ + HCO_{3-} + 3H_2 \tag{14.8}$$

$$C_6\ H_{12}O_6 + 2H_2O \leftrightarrow 2CH_3COOH + 2CO_2 + 4H_2 \tag{14.9}$$

$$CH_3CH_2OH + 2H_2O \leftrightarrow CH_3COO^- + 2H_2 + H^+ \tag{14.10}$$

In the last step of methanogenesis, hydrogen and CO_2 from the earlier stage are digested into methane and CO_2; this can only be possible due to the activity of the methanogen bacteria (Hans and Kumar 2019).

(Equations 14.11–14.13).

$$2CH_3CH_2OH \rightarrow CH_4 + 2CO_2 \tag{14.11}$$

$$CO_2 + 4H_2 \rightarrow CH_4 + 2H_2O \tag{14.12}$$

$$2CH_3CH_2OH + CO_2 \rightarrow CH_4 + 2CH_3COOH \tag{14.13}$$

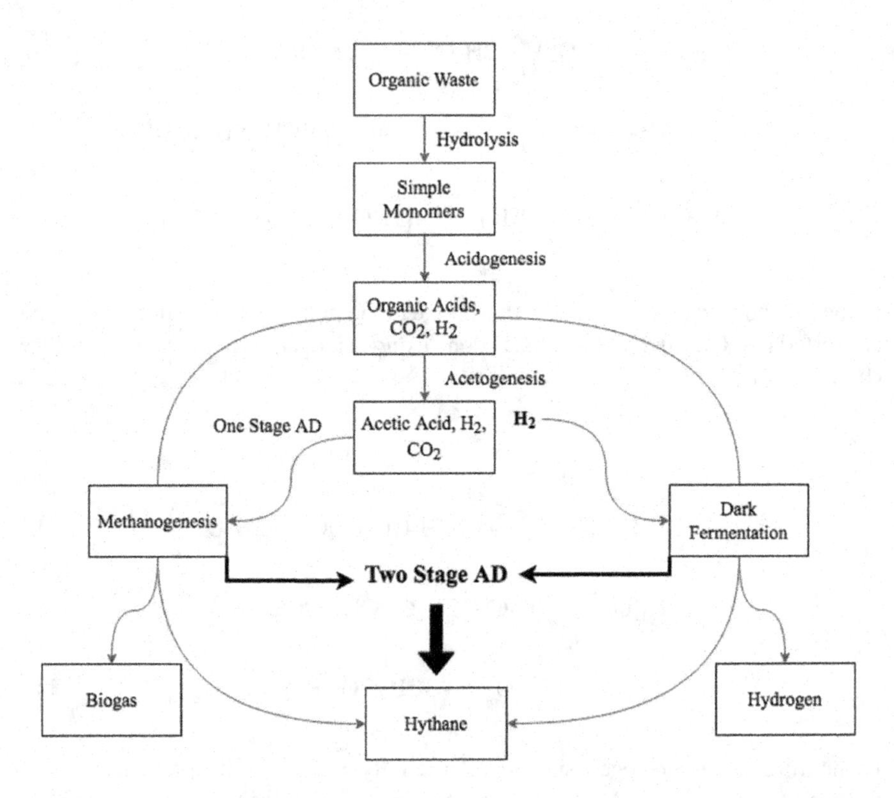

FIGURE 14.1 Anaerobic digestion process for the production of hythane (Hans and Kumar 2019).

The glucose-like simple monomers obtained from these steps are further metabolized to form pyruvate through glycolysis process. The formed ATP in this glycolytic path is consumed by the microorganisms/bacteria. The pyruvate formed is digested to give butyrate and acetate along with hydrogen, since the electrons are neutralized by the protons, using Hydrogenase enzyme (Hydrogenase) (Figure 14.1). Pathways like facultative and obligate anaerobes produce hydrogen. In obligate anaerobes pyruvate is oxidized to acetyl co-enzyme A, i.e., CoA. This is done by using PFOR or pyruvate–ferredoxin–oxidoreductase. This is followed by reducing Fd, i.e., ferredoxin. This is later oxidized, as hydrogen formation takes place by hydrogenase enzyme (Hans and Kumar 2019). (Equations 14.14 and 14.15).

$$C_3\ H_3O_{3-} + CoA + 2Fd(ox) \rightarrow acetyl - CoA + 2Fd(red) + CO_2 \quad (14.14)$$

$$2H^+ + Fd(red) \rightarrow H_2 + Fd(ox) \quad (14.15)$$

In facultative anaerobes, pyruvate is oxidized in formate and acetyl-CoA along. This is due to the PFL, i.e., pyruvate–formate–lyase. The formate formed here is later split into hydrogen and carbon dioxide. This is done using the FHL i.e., formate–hydrogen–lyase (Hans and Kumar 2019). (Equations 14.16 and 14.17).

$$C_3\,H_3O_{3-} + CoA \rightarrow acetyl - CoA + CHOO^- \tag{14.16}$$

$$HCOOH \rightarrow CO_2 + H_2 \tag{14.17}$$

If only acetate is produced as the end product, then one mole of glucose gives four moles of hydrogen. On the other hand, if only butyrate is produced as the end product, then one mole of glucose gives two moles of hydrogen (Equations 14.6 and 14.9). In methanogenesis, MCR, i.e. methyl-coenzyme-M-reductase is the prime enzyme, which initiates methane formation followed by its anaerobic oxidation. Here, methyl-thioether-methyl-coenzyme-M along with thiol-coenzyme-B acts as substrate. These substrates are then digested reversibly to methane and equivalent heterosulfide. The catalytic mechanism is not known. Anaerobic digestion process can be classified into one-stage AD and two-stage AD based on their final output products. The stage one AD has methane in dominance with some trace products. On the other hand, the two-stage AD produces hydrogen and methane at the same time.

Figure 14.2 shows the difference between both these stages of AD (Hans and Kumar 2019). Bacteria are the main agents that carry digestion of both first and

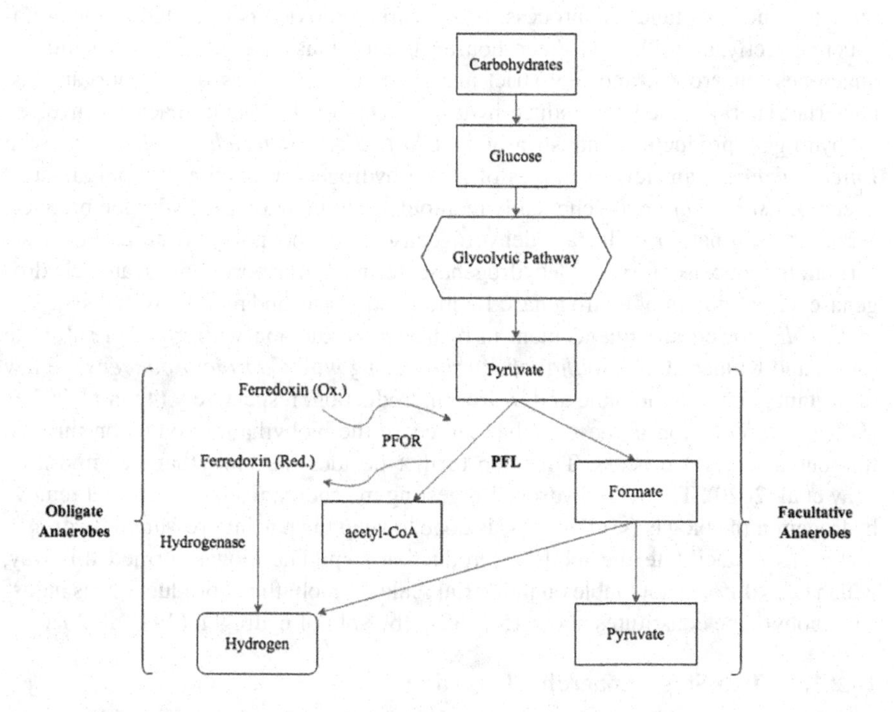

FIGURE 14.2 Two methods of anaerobic digestion processes: single-stage AD and two-stage AD (Hans and Kumar 2019).

second stages. Production of hydrogen in a stable manner is a crucial parameter for the biohythane production through AD process (Qin et al. 2019). High biogas output, dissolution of the toxic components into substrate, improved diversity in microbial selection, enhancement in the nutrient balance, and high load of the biomass are few advantages that AD process carries along (Jehlee et al. 2019). The two methods of AD process are the prime methods used for the production of biohythane.

14.2.1.1 One-Stage Anaerobic Digestion

In this method, all the four steps namely (1) hydrolysis, (2) acidogenesis, (3) acetogenesis, and (4) methanogenesis are carried out in a single bioreactor/digester vessel. This method yields 61%–66% of methane in dominance along with hydrogen, carbon dioxide, and hydrogen sulfide in small quantities. This method is efficient and occurs in nature; even after this it is not widely accepted due to one setback. This drawback is the formation of VFAs during acetogenesis and acidogenesis, which further acidifies the reactor, i.e., decreases the pH, which then kills the bacterium present in the bioreactor. This reduces the methane output yield. Hydrogen producers cannot undergo this unnecessary VFA. This ceases the digestion at acidogenesis and acetogenesis steps. Buytric acid, propionic acid, and acetic acid are few examples of VFAs formed here. Upper limits for the VFAs is 1,800, 900, and 2,400 mg/L for buytric acid, propionic acid, acetic acid, respectively. However, build-up of unnecessary hydrogen also leads to cease the methanogenesis step (Hans and Kumar 2019).

Methane and carbon dioxide (about 30%–40%) form a large chunk of the total output of the one-stage AD process. If this carbon dioxide is liberated in the atmosphere directly, it will lead to green-house gas emissions on a large scale. Facultative anaerobes, anaerobes, and few strict anaerobes are the types of microorganisms/bacteria generally used to produce hydrogen. One of the crucial bacteria involved in hydrogen production industrially is *Clostridia*. *Escherichia coli* along with *Enterobacter* sp. are few examples of prime hydrogen-producing microorganisms/bacteria. *Escherichia coli* genetically manipulates to increase the hydrogen produce. Strategic elimination of lactate dehydrogenase, phosphoenolpyruvate carboxylase, furmarate reductase, formate dehydrogenase, formate transporter, pyruvate dehydrogenase, large subunit of hydrogenase 1, nitrate reductase, and methylglyoxal synthase in *E. coli* showed an enhancement in hydrogen production with glycerol substrate (Hans and Kumar 2019). *Methanobacterium* along with *Caproiciproducens* are few important genera in methane and hydrogen production, respectively (Ta et al. 2020). Hydrogen production is the step that can cease the biohythane production through this one-stage AD process. This step further decides the biohythane composition (Lay et al. 2020). There are hydrogen-digesting microorganisms as well that remove hydrogen in the biogas (<0.1%). This is done to keep the partial pressure of hydrogen low so as to facilitate the methane-production step. The biogas formed this way, which has <0.1%, is not viable on industrial scale for biohythane production, as industrial biohythane constitutes about 15%–30% (by vol.) of hydrogen (Ta et al. 2020).

14.2.1.2 Two-Stage Anaerobic Digestion

In this method of AD, the initial three steps are (1) hydrolysis, (2) acidogenesis, and (3) acetogenesis. Here, these steps are executed/completed in the first-stage in a single bioreactor. Whereas, the last step of methanogenesis takes place in the second stage in another separate vessel of bioreactor. Growth kinetics—physiological differences along with reactivity towards environmental conditions in methanogens and acidogens—forms the base for CH_4 and H_2 production via two-stage AD process (Mamimin et al. 2015). This method, i.e., two-stage AD process involves two different bioreactors (one for acid-producing and second for methane-producing phases). This leads to high stability in operation, increased rate of volumetric load, and methane and hydrogen production at the same time (Chen et al. 2021).

In the first stage of this process, the complex biomass is digested into simple monomers by using the hydrolytic bacteria. These simple monomers are transformed in carbon dioxide, hydrogen, and other organic acids. This is done by the acidogens. Further, the carbon dioxide, hydrogen, and other organic acids are digested to carbon dioxide, hydrogen, and acetic acid, using acetogens (Hans and Kumar 2019) (Equations 14.1–14.10). In the second stage of this AD process, effluents and carbon dioxide, hydrogen and acetic acid are used to produce methane along with carbon dioxide. This is accomplished with the help of methanogens. Volatile fatty acids from the first-stage method are fed to the methanogens in second bioreactor of this two-stage method. This increases the methane yield of second stage (Hans and Kumar 2019).

An increase of 26% for methane along with 23% for hydrogen was observed in two-stage AD process over one-stage AD process through biomass containing solid waste from municipalities, where Kvesitadze et al. (2012) and Malave (Luongo et al. 2015) performed two experiments. The first experiment was between a one-stage AD process and two-stage AD process using glucose. In the experiment, the two-stage process with glucose gave an increase of over 48% in efficiency, when compared to the one-stage AD process. This experiment ensures that the two-stage AD process gives higher output energy over conventional single-stage AD process. The study done by Massanet and team (2013) in their experiment between one-stage AD and two-stage AD from pellets of wheat feed and inoculum as effluent from sewage treatment plant (STP), compared these two types of ADs with the help of real-time data of gas production. An increase in methane yield of about 359 L/kg-VS from two-stage AD over 261 L/kg-VS from the single-stage AD in a period of 20 days was observed. The total time period required to complete one cycle of one-stage AD is of 28–30 days. However, the hydraulic retention time for the acidogenic phase only requires 1–3 days of time to hydrogen as well as to generate acids. Due to this, in two-stage anaerobic digestion process, the hydrogen produced can be separated immediately once produced. Also, the effluents generated are fed to the second bioreactor vessel to convert it into methane. This takes 12–15 days. The complete process takes a total time period of 13–18 days (Hans and Kumar 2019). Schievano et al. (2014) analyzed the bio-CH_4 as well as bio-H_2 generated through organic biomass using a two-stage AD process. He concluded that an increase of about 8%–43% in recovered energy can be achieved. The two-stage AD process exhibits a trustworthy process; this is due to its good self-pH adjusting ability. This process also eliminates the breakouts because of excessive accumulation of volatile fatty acids (Hans and Kumar 2019).

14.3 FACTORS AFFECTING BIOHYTHANE PRODUCTION

The whole two-stage anaerobic digestion process depends on microbes. The activity of these microbes is greatly dependent on the conditions during the process. Process conditions like pretreatment, inoculum characteristics, rate of organic loading, retention time, pH, temperature, reducing equivalents, feedstock/substrate, bioreactor configuration, etc., need to be very carefully optimized for the efficient and stable operation of the bioreactors. Few factors impact the process individually, whereas others have a collective impact, which makes it even more necessary to have a hold over these conditions so as to improve the process (Hans and Kumar 2019; Lay et al. 2020).

14.3.1 Hydraulic Retention Time (HRT)

HRT is one of the most important process conditions that can affect the population of microbes, efficiency of uptake of substrate, and the biological pathways in the bioreactors. To design a small reactor that has high rate of substrate loading, the hydraulic retention time needs to be less. This directly affects the reactor costs (Vo et al. 2019). The yield of bio-H_2 should be elevated to create a sustainable biohythane economy. This can be done by optimizing HRT (Maaroff et al. 2019); compact retention time can efficiently improve the bio-H_2 yield. Although there is limitation to this retention time, because if the retention time is shortened below a certain limit, then it becomes difficult to keep up with the solid retention time, as the biomass content at cellular levels is vulnerable to get washed at such compact hydraulic retention time (Maaroff et al. 2019).

14.3.2 Lowering the Equivalent Content

At the time of hydrogen making, reducing equivalents such as NADH and NADPH are put away using reduction of protons (Hb) as well as carriers of electrons. This further leads to generation of hydrogen (Hans and Kumar 2019). The hydrogen yield is depended upon the reducing equivalents, so to increase the hydrogen yield, it becomes necessary to increase or recover the reducing equivalents by obstructing the pathway, which is used for the formation of metabolites. Recent studies have reported the enhancement in the hydrogen generation. This was done by elevating the reducing equivalents by means of mutagenesis method in thermophilic hydrogen producers by obstructing the production of alcohols (Hans and Kumar 2019). Roy and Das (2016) synthesized a pathway obstructing mutant for alcohol generation. It is named as *Thermoanaerobacterium thermosaccharolyticum* IIT BTST1. It was synthesized by barely exposing it to ethyl methane sulfonate. This mutant increased the yield for hydrogen by approximately 13.5%. Similar pathway-obstructing mutant for production of alcohol known as *Enterobacter aerogens* was introduced (Ito et al. 2004). This mutant that was developed by barely exposing it to *N*-methyl-*N*-nitro-*N*-nitrosoguanidine showed a 15% improvement in the yields of hydrogen. Such studies show that lowering the alcohol generation by improving the number of reducing equivalents also increases hydrogen production.

14.3.3 Carbon-to-Nitrogen Ratio

The significance of carbon-to-nitrogen ratio (C/N) is very nicely explained by Farghaly and Tawfik (2017). They used the effluent from a paper mill for biohythane production using an anaerobic digestion process. They reported that this C/N ratio had a significant relevance to biohythane production, as it affects the species of microorganisms that manages the reaction kinetics of biomethane as well as of biohydrogen (Lay et al. 2020).

14.3.4 Pretreatment Technique

It is one of the major steps involved in anaerobic digestion. It is responsible to get high biohythane yields. Cellulose, which is the prime substrate for digestion, is enclosed within a layer of lignin and hemicellulose. The removal of layer of lignin and opening the packing of hemicellulose is important for the efficient digestion of cellulose. Inoculum along with feedstock are generally treated for biohythane production (Hans and Kumar 2019). Their pretreatment is discussed below.

14.3.4.1 Pretreatment of Inoculum

Pretreatment of Inoculum is a crucial process that is being utilized in the reactor at Stage I (acidogenic) because of the existence of hydrogen producers along with hydrogen consumers. Hydrogen producers like species of Bacillus and Clostridium that produce endospores. These endospores again germinate, once the favorable conditions are set up. These conditions include elevated temperatures, toxicity of chemicals, and desiccation. On the other hand, the consumers of hydrogen are very much reactive to temperature, pH, and concentration of feedstock. So, to increase the hydrogen yield, these consumers in the reaction mixture should be separated by pretreatment of the inoculums. Heat treatment of the inoculum at 100°C–110°C usually deactivates the consumers. However, a study showed that no significant variation was observed between two sludge samples—one heat treated and another untreated sample. When the effluent is recirculated from methanogenic to the acidogenic reactor, heating the fraction of the sludge that is large is recommended. The frequency of this heating should be less. This was done to increase the germination of the spores of the hydrogen producers, which takes 65–95 minutes of time for germination (Hans and Kumar 2019).

14.3.4.2 Pretreatment of Feedstock

For the efficient digestion of cellulose, it is necessary to separate the rigid layers of hemicelluloses and lignin. Pretreatment of feedstock can be done by using biological, physio-chemical, as well as by physical methods. This pretreatment significantly improves the biohythane generation. The feedstock Lignocellulosic biomass is generally pretreated by treatment with hot water, microwave, acid, ionizing radiation, explosion of stream, alkali, etc. Lignin layer is separated by breaking bonds among hemicelluloses and lignin. This leads to the release of cellulose for further digestion with hydrolytic enzymes (Hans and Kumar 2019). To produce hydrogen from sugarcane bagasse, it needs to be first hydrolyzed at around 121°C using approx. 2%

H_2SO_4. This H_2 generation stops when the sulfuric acid concentration exceeds 2%. This is because of the constant rise of furfural along with acetic acid, which leads in to hindrance to the process (Rai et al. 2014). To enhance hydrogen yield, pretreatment of lignocellulosic biomass using physiochemical routes can result in an increase in cellulose hydrolysis rate and efficient changes in the structure of the biomass. Although, the use of high amount of energy, specific instruments along with related inhibitors make this physio-chemical treatment less sustainable, both from economics point of view and also for the ecology. However, using biological treatment with microbes or enzymes to produce lignocellulosic biomass-based fuels, has proved to be an environment-friendly technique, since use of chemicals is not mandatory. In anaerobic digestion, to enhance the production of biofuel, formation of VFAs, recovery through NH_3, and to enhance digestibility of the lignocellulosic biomass—a process known as bioaugmentation is used (Hans and Kumar 2019). In hydrothermal pretreatment technique, the biomass is heated at elevated temperatures by using H_2O as a medium. Irradiation with ultrasound is a type of pretreatment where waves of ultrasound are made to penetrate through the biomass that is submerged in an aqueous slurry or solution. This ultrasound treatment produces regions of expansion along with compression within the biomass, which results in enlarged surface area. This further results into easy depolymerization of lignin (Hans and Kumar 2019).

14.3.5 Variety of Feedstock

Using pure sugar for the production of hydrogen is not economically practical and feasible. Earlier, only simple substrates like sucrose, fructose, starch, and glucose were used as substrates. However, the use of complex feedstock such as fats, carbohydrates, proteins, and lipids has resulted to be more efficient in the generation of biohythane because of their wide availability and economical cost. Using complex organic wastes is ecologically, financially and structurally suitable, since they are widely available, and they also create difficulty in their disposal. The biomass that has high content of carbohydrate has more probability for hydrogen generation when compared to lipids and proteins. Feedstock-containing lipids have chance for methane production. So, fats along with starch are favored feedstock followed with proteins for biofuel generation. Higher content of starch in feedstock leads to decreased hydrogen generation. Feedstock having low starch content generates more hydrogen because of incomplete hydrolysis at high starch content (Hans and Kumar 2019). Organic waste from the kitchen is an encouraging substrate for hydrogen generation because of the high content of carbohydrates. This food waste also contains some volatile fatty acids like lactate. Effluent from palm oil mill (POME) combined with bunches of fruits and decanter cake were used in the feedstock to produce biohythane. The yield of hydrogen was recorded to be 16.52 and 16.26 L/kg-VS, respectively. Anaerobic digestion of effluent from palm oil mill along with decanter cakes gave the highest yield for methane of 391.62 L/kg-VS (Hans and Kumar 2019). Weeds from marine sources like water hyacinth is a rapid growing aquatic weed that has wide, broad, and shiny leaves having spongy, long, and bulbous stalk. This weed is nontoxic and is easily accessible. Such aquatic weeds can be a valuable feedstock for biohythane production. Algae like *Scendesmus, Chlorella and Anabaena* are commercially grown to generate biofuel. The less content of lignin in algae results in higher extraction of bioenergy and also requires less energy to degrade (Hans and Kumar 2019).

14.3.6 pH

In any chemical process, pH is one most crucial process conditions, which can have an impact on the growth of microbes. It also plays a major role in stabilizing the process by optimizing the efficiency of the enzymes. pH value in both the bioreactors of the two stages of anaerobic digestion is different so that the microbes and enzyme can grow. In the first reactor of AD for hydrogen production, there occurs buildup of volatile fatty acids. These VFAs damage the membrane of microbes and further interferes in the balance of pH. Hydrogen production ceases in the first reactor, if pH is in the range of 3.8–4.2. Because of the existence of bicarbonate and acetate ions in the effluent in the second reactor, i.e., methanogenic reactor, the pH is kept in the range of 7.29–7.55. Low value of pH can lead to the generation of solvents other than hydrogen like lactic acids and alcohols. Low value of pH is poisonous towards the methanogens. However, low value of pH is useful for the producers of hydrogen, because the ideal value of pH for the activity of hydrogenase is 5.5 (Micolucci et al. 2014; Hans and Kumar 2019).

14.3.7 Temperature

The impact of temperature on anaerobic digestion to produce methane and hydrogen is very significant. Features of the microbes and nutrient demand are affected by temperature variation in the digestion process. Enzymes and microbes have a specific threshold temperature. They display the highest microbial activity at this temperature. A deflection in the threshold temperature range generally results in the inactivation or denaturization of the microorganisms and enzymes. This inactivation can lead to the obstruction in the process. A rise of one degree in temperature doubles the activity of enzymes only up to the threshold range (Hans and Kumar 2019). A drop of 2°C–3°C can result in the buildup of VFAs inside the bioreactor, which will further lead to a drop in pH value. Under thermophilic conditions, the hydrolysis rate for cellulose is more than that was in the mesophilic condition. The thermophilic conditions require small volume of the bioreactor (Hans et al. 2018). The best range of temperature thermophilic and mesophiliuc methanogens is 50°C–65°C and 31°C–34°C, respectively.

Abreu et al. (2016) studied the impact of co-cultures of *C. bescii* and *C. saccharolyticus*. An increase in hydrogen generation of approx. 98 L/kg-VS was reported. The methane yield was found to be 322 L/kg from the total COD content. The principal enzyme that produces hydrogen is hydrogenase. It is highly affected by temperature variation. This enzyme hydrogenase gives the highest microbial activity at approximately 55°C. In acidogenic reactor, thermophilic conditions increase the value of pH. This will impact the activity of microbes. Thermophilic conditions improve the degradation of lipids because of the improved liquid solubility at elevated temperatures (Hans and Kumar 2019).

14.4 BIOHYTHANE FROM SOLID WASTES

In the recent years, there has been great interest in biohythane, and its performance has been gaining lot of attention. Majorly, two categories of solid waste commonly chosen for experiments include sewage sludge and food wastes. A simple reason for this is their widespread availability. It is estimated that 72% of single-state anaerobic digestion plants are treating food wastes, for a complete of 80 lakh tons digested amongst the EU nations, and rest 28% of the anaerobic digestion plants intake feed as sewage sludge (Bolzonella et al. 2018). Yeshanew et al. (2016) performed a two-stage anaerobic digestion by using a CSTR for dark fermentation and using an AFBR for methanogenic phase. The reactors operated for a total of 200 days and were fed with household food waste (HFW). The composition of HFWs was prepared in accordance to "European Average Household Food Waste," which consisted of 79% of fruits and vegetables, 6% of bakery and bread, 2% dairy products, 8% of fish and meat, and 5% rice and cooked pasta. The production of hydrogen resulted in a better performance in CSTR, which recorded the maximum of 115 L H_2/kg VS fed (Bolzonella et al. 2018). The key observation from this test was the understanding of the crucial role of pH. The accumulation of volatile fatty acids' because of very less HRT or OLR during accelerated synthesis results in pH decrease. Further, this section throws light on various works in literature of biohythane production from solid-waste feeds, including co-digested regimes.

14.4.1 HOUSEHOLD FOOD WASTE CO-DIGESTION WITH SEWAGE SLUDGE

Household food wastes are frequently co-digested with sewage sludge for enhancing the production of biohythane. Sewage sludge has proven to be more attractive in biohythane fermentation due to it being a stable source, low cost, huge amount, and high organic content of greater than 60% of dry matter (Yang et al. 2015). From the sludge the hydrogen yield is usually in the range of 10–90 mL/gVS_{added}, which is comparatively very less than rest of the feedstocks such as crude glycerol (29.2–219.1 mL/gVS_{added}) and household food wastes (100–250 mL/gVS_{added}) (Yang et al. 2015; Kim et al. 2013). Moreover, the C/N factor of sewage sludge is usually in the range of 4–9, which turns out to be much lesser than the optimum value for hydrogen fermentation (Yang et al. 2015). It was reported that the concentration of ammonia may increase in the AD process due to its small C/N ratio, which causes microorganisms inhibition (Bolzonella et al. 2018). Instead household food wastes have greater C/N ratio (normally superior than 20) and subsequently may be considered for co-digestion with sewage sludge. Riviere et al. (2009) reported that the differences in sewage sludge ratio and HFWs muddle the microorganisms' interaction. Xu et al. (2019) elucidated further that individual parameters like, pH or TVFAs/TA ratio, are usually not able to indicate the start of the process' inhibition.

Cavinato et al. (2012) reported a long-term method that takes place at pilot scale and was carried out for a duration of 90 days; the start-up phase was recognized from the day 0 to 49, when the steady state condition runs for around 40 days, which means 13 hydraulic retention times for the DF and 3 HRTs for AD process. The first phase is inoculated with tap water and food waste, which allows the initial solubilization of the organic material that has a pH drop till about 4, resulting in inhibition of bacteria that produce hydrogen. The AD phase attained steady state condition after around 50 days, which showed total alkalinity (average) of 9,806 mg$CaCO_3$/L, 8.4 pH being the average and total volatile fatty acid content of 1,107 mg COD/L (Cavinato et al. 2012). After being under operation for 70 days, the level of ammonia concentration reached 916 mg/L. Cavinato et al. (2012) reported that not just methane but also the production rate of biohydrogen was majorly dependent on the pH value and the concentration of ammonia above 2 gN/I. Another study was carried out that lasted for about 310 days with the goal of determining the best recirculation ratio, which allows for system control and also best yield of biohythane (Micolucci et al. 2014). For the first period of day 1–day 90, the highest gas production yield, in terms of methane production and specific hydrogen was achieved: 0.52 and 0.085 $m^3H_2/kgTVS_{fed}$, respectively. A clear indication of the process upset is easily observed in the latter part of the period, around days 81–90 in which the trends of hydrogen-specific productions and ammonia concentration are reversed. When the ammonia passes a threshold of 2 gN/I, the hydrogen-specific productions decreases from 0.075 to 0.03 $m^3H_2/kgVS$. Also, a similar trend was depicted by the methane-specific production (Micolucci et al. 2014). In the third period, which is from day 171 to day 310, following an adjustment period, the entire system reaches a stable production of biogas, which meets the biohythane characteristics; in detail, hydrogen, carbon dioxide, and methane showed a constant concentration of around 10%, 35%, and 55%, respectively (Micolucci et al. 2014). The raw food waste that is diluted to the OLR is fed to the CSTR. Subsequently, the effluent coming from CSTR is fed to AFBR for converting to methane. The recirculation rate of AFBR effluent to CSTR, which is calculated as the ratio of AFBR effluent returned volume to the volume of CSTR influent. Hence, it was in the range of 0.6–1, 0.5–0.8, and 0.24–0.48 in Periods III, II, and I, respectively (Yeshanew et al. 2016). It should be taken into account that because of the basic difference of organic matter of the raw food waste and the fed substrate fed to CSTR, which is described in VS concentration (Cavinato et al. 2012), the COD value is used for quantification of organic matter content of AFBR influent, which is CSTR effluent (Ghimire et al. 2015). A semi-continuous process of CSTR was initiated at a hydraulic retention time of 6 days and OLR of about 2.0 kg VS/m^3.day, while the AFBR was started at a hydraulic retention time of 20 days and 0.1 kg COD/m^3.day OLR (Yeshanew et al. 2016). The study shows total organic acid, COD removal efficiency, pH, and total alkalinity of the AFBR effluent. The pH was determined in the range of 7.2–7.7, and the total alkalinity was in the range of 1,000–2,000 mg/L throughout all periods that were operative and showed that these are optimal surroundings for methanogenic activity (Yeshanew et al. 2016). Qin et al. (2019) has studied the effects of paper waste and food waste co-digestion upon the energy yield. They varied TS of paper waste and food waste from about 0% to 50%, and recorded a methane and hydrogen yield of 657 and 157 NL-H_2/kg-VS_{FW}, respectively. The maximum yielded energy of 28.1 MJ/kg VS_{FW} is recorded for the co-digestion ratio 1:1. Meena and co-workers (Meena et al. 2020) have similarly studied of energy yield in waste-activated sludge and food waste co-digestion, and even achieved the maximum methane (353 mL CH_4/g-VS) and the maximum hydrogen (106 mL H_2/g-VS) yield for 85% food to 15% waste-activated sludge. Liu et al. (2019) has obtained the energy yield at changing OLRs in two-stage

AD. At 2 g VS/L of OLR, they obtain the maximum energy yield of 15.1 MJ/kg VS. They have studied on the recovery of energy from the mixed organic waste. It was certain from this study that the yield of energy from the methane and hydrogen reactor was found out to be 12.3 ± 0.69 and 0.4 ± 0.05 kJ/g VS, respectively (Meena et al. 2020). The overall recovery of energy was noted to be 12.7 ± 0.72 kJ/g VS from the anaerobic digester. Chu et al. (2008) have carried out the digestion of 1,000 kg of wet food waste with 33.8% of TS in the two-stage anaerobic digester. The energy recovery from methanogenic and hydrogenic reactors were obtained as 1,722,925 and 139,333 kcal/t-wet, respectively (Meena et al. 2020).

14.4.2 Municipal Solid Wastes

The treatment organic fraction of municipal solid wastes by the two-phase anaerobic digestion process produces 43 mL of H_2/g of VS, which are fed in acidogenic reactor, and 500 mL CH_4/g of volatile solids was recorded in the methanogenic reactor (Cavinato et al. 2010). Subsequently, Zuo et al. (2013) used organic fraction of municipal solid wastes for the production of acidogenic hydrogen and then production of CH_4 with effluents of reactor that is acidogenic as substrate for enhancing the energy recovery efficiency for the methanogenic reactor. The source OFSMW, substrate, is sorted in two CSTRs with volume (working) of 0.2 and 0.76 m^3 are operated for a duration of 90 days. The Biohythane with a percentage composition of 38% CO_2, 53% CH_4, and 9% of H_2 was produced (Cavinato et al. 2012). Bolzonella et al. (2018) examined the anaerobic co-digestion of organic fraction of municipal solid wastes and waste-activated sludge in a single-and two-stage processes. The two-stage AD produced 24 L/kg of fed VS, while the single-stage AD reported 490 L/kg VS of H_2 production in the first stage also production of 570 L/kg VS of CH_4 in the second stage. Numerous studies exhibited that the single-stage process produced biohythane from model substrates only, for example, glucose (Kumar et al. 2019) (Table 14.1).

14.5 BIOHYTHANE FROM WASTEWATER

Industrial effluents and wastewaters are becoming more complex with involvement of intricate processes for various industrial operations. These wastewaters pose a health hazard to the ecosystem – water, air, and soil. Anaerobic technologies for treatment of wastewaters have been gaining increased attention owing to environment-friendly and cost-efficient features as well as the ability to recover clean bioenergy. The production of biohythane from these industrial wastewaters could fulfill the purpose of generation of clean energy as well as management of toxicity by reduction in chemical oxygen demand (COD) and biological oxygen demand (BOD). Several industrial effluents are strong organic wastewaters enriched with carbohydrates in the form of sugars, cellulose, and starch. Literature suggests that carbohydrate-rich substrates are paramount for hydrogen production, therefore, making wastewaters promising for biohydrogen production. The effluents from hydrogen production are rich in volatile fatty acid (VFA), suitable for methane production. In the following section, this sequential production of biohythane is discussed for different types of wastewaters based on their availability, feasibility, and potential.

TABLE 14.1
Substrates with Fuel Production Efficiencies

Sr. no	Substrate/Agriculture Waste Used	COD Reduction	H_2 Yield (%)	CH_4 Yield (%)	Technique Used	Remarks	References
1	Cassava Plant	82%	27.7		Two-stage anaerobic digester	-	Jiang et al. (2018)
2	Organic fraction of municipal solid waste (OFMSW)	64%	5	25	Single-stage anaerobic digester	34% CO_2 was produced	Kumar et al. (2019)
3	Organic fraction of municipal solid waste (OFMSW)	77%	6	15	Two-stage anaerobic digester	56% CO_2 was emitted	Kumar et al. (2019)
4	Mix Food Waste [Cooked Rice - 45%–50% Boiled vegetable-15%–20% Vegetable Peels -10%–15% Waste veg & fruits -5%–10% and Egg and Meat waste – 5%–10%]	71%	43	6	(Dark-fermentation)	Untreated waste was used	Sarkar et al. (2021)
5	Mix Food Waste [Cooked Rice - 45%–50% Boiled vegetable-15%–20% Vegetable Peels -10%–15% Waste veg & fruits -5%–10% and Egg and Meat waste – 5%–10%]	84%	9	57	Dark-fermentation	Pre-treated (heat-shock) Waste was used	Sarkar et al. (2021)
6	L. digitata and micro-algae (Macro and Micro algae)	70.9%	-	-	Two-stage batch co-fermentation	This two-stage fermentation increases ECE by 2%	Ding et al. (2018)

(*Continued*)

TABLE 14.1 (*Continued*)
Substrates with Fuel Production Efficiencies

Sr. no	Substrate/Agriculture Waste Used	COD Reduction	H_2 Yield (%)	CH_4 Yield (%)	Technique Used	Remarks	References
7	**Mix Food waste** 79% vegetables and fruits; 5% cooked pasta and rice; 6.0% bread and bakery; 8.0% meat and fish; and 2.0% dairy products	84%	4.9	79.5	The reactor arrangements were a CSTR and anaerobic fixed bed reactor (AFBR) for the first and second stage,	The maximum production of biohythane was attained at the shortest chosen hydraulic retention time HRTs	Yeshanew et al. (2016)
8	**Food Waste**	95.6% (first phase) & 15.5 (second phase)	6	58	Dark fermentation coupled with anaerobic digestion process	The process produced 36% CO_2 which is relatively low	Cavinato et al. (2012)
9	**Household food waste**	-	10%–30%	70%–90%	Two-stage AD	-	Bolzonella et al. (2018)

14.5.1 Textile Wastewater (TWW)

The textile industry is one of the major industries contributing to about 7% of global export (Lellis et al. 2019). The broad spectrum of dyes used in fabric coloring and large consumption of water in the manufacturing process consequently contributes to extensive amounts of harmful colored wastewater. Textile wastewaters (TWW) contain considerable concentrations of inorganic and organic compounds in the form of dyes, acids, alkalis, soap, metal, and surfactants (Bhatia et al. 2021). Different processes encompass the manufacturing processes like desizing, dyeing, bleaching, printing, and finishing. Major concerns with TTW originate due to presence of unutilized dyes that include some toxic agents of carcinogenic or mutagenic nature, which may enter through the food chain to affect living organisms in the environment (Bhatia et al. 2021; Djafer et al. 2017). The dyes hinder aquatic life by causing low light penetration and consumption of oxygen. Another issue revolves around desizing operations that use sizing agents such as starch and carboxymethyl cellulose (CMC), and release effluents with BOD in the range of 300–450 ppm (Gadow and Li 2020b; Bhatia et al. 2021). However, these substances under anaerobic conditions act as electron donors that affects the process of dye decomposition (Gadow and Li 2020a). Textile desizing water (TDW) contains COD of up to 20–40 g/L, proving to have good potential for bioenergy recovery via dark fermentation (Lin et al. 2017). Hydrogen fermentation effluents, which contain high soluble microbial products (SMP) concentrations of mainly ethanol, acetate, and butyrate, are usually methane fermentation ready. This creates an avenue for two-stage fermentation for bioenergy recovery as biohythane. The fermentation of TDW pretreated with coagulant (dosage 1 g/L) results in hydrogen production rate (HPR) of 3.9 L/L/day and effluent with SMP of 10–20 g COD/L (Lin et al. 2017). Similarly, fermentation of starch-based textile wastewater with inoculums like sewage sludge, soil, and cow dung result in HPR of 1.14 L/L/day, and effluents with SMP composition of acetate (28%–64%) and butyrate (17%–60%) (Lay et al. 2012). These results are indicative of potential biohythane production from addition of methanogenic fermentation to the process. The work of Gadow and Li (2020a) on development of continuous two-stage mesophilic anaerobic system to treat TWW-containing C.I Acid Red 88 dye—a toxic azo dye elucidates on bioenergy recovery in the form of biohythane. A two-stage, an acidogenic CSTR, and a methanogenic UASB anaerobic system was used for the study. The first stage, CSTR, showed the maximum of 30.7% COD conversion to biohydrogen at a 24 hours HRT. And the second stage, UASB, removed 94.8% COD and showed the maximum methane production of 1.01 l/l/day at 12 hours HRT. The system exhibited an overall efficiency of 94.8% at OLR of 6.49 g COD/l/day for conversion of initial COD to biohythane and produced a maximum hythane energy of 121.5 MJ/m^3 at 12 hours HRT.

14.5.2 Palm Oil Mill Effluent (POME)

Palm oil is a leading agricultural crop globally, as it makes up for 34% of the global market (Aziz et al. 2020). However, the production of palm oil mill effluent (POME) has negative impacts on the environment. Therefore, several researchers have gained

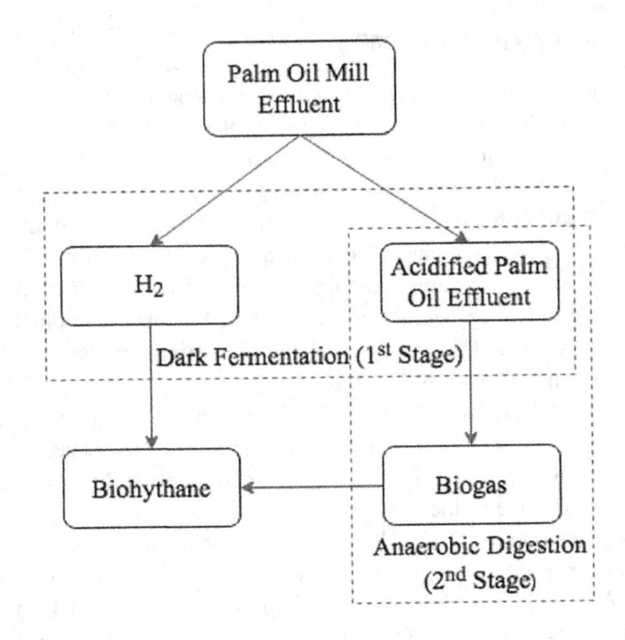

FIGURE 14.3 Schematic diagram of two-stage biohythane production from POME (Aziz et al. 2020).

interest to utilize palm oil mill effluent (POME) as feedstock in order to control production of agricultural waste from palm oil industries. POME has high organic content that makes waste to value-added conversions possible. Additionally, the applicability of thermophilic process on POME is feasible, as it is discharged at high temperatures (80°C–90°C) (Abd Nasir et al. 2019). This also eliminates the need for a cooling system for discharged POME. Under thermophilic conditions, the microbial fermentation for hydrogen production can be achieved with less variation in fermentation end products and low inhibition of hydrogen partial pressure but with 7.5%–15% conversion of organic waste energy to hydrogen (Mamimin et al. 2015). The low substrate conversion efficiency observed makes the conversion of POME to hydrogen economically unfeasible, since most of the energy remains in volatile fatty acids (VFA). Therefore, production of biohythane has been shown to be more energy efficient by conversion of the VFA into methane via methanogenesis after conventional fermentation. The POME with VFA (mostly butyrate and acetate) after first-stage dark fermentation is often referred to as acidified POME (Figure 14.3).

The influence of VFA present in acidified POME on biohythane synthesis is noteworthy. A low VFA loading (0.9–1.8 g/L) reportedly shows 15%–20% greater methane yield than high VFA loading (3.6–4.7 g/L). The presence of high concentration (8 g/L) butyric acids negatively impacts the methane production process individually; furthermore, an interaction of lactic, acetic and butyric acid, at concentrations higher than 0.5 g/L, with propionic acid develops an inhibition condition (Mamimin et al. 2017). Correspondingly, the feasibility of biomethane production by thermophilic process from acidified POME is boosted by high rates, i.e., low HRT and higher

OLR than optimum biomethane processes (Abd Nasir et al. 2019). Similarly, Lay et al. (2020) reported that an optimum pH range is 5–6 and HRT of 1–3 days for the first stage and HRT of 10–15 days, while a pH of 7–8 favors methanogenic bacteria in the second stage. Mamimin et al. (2015) have developed a two-stage thermophilic fermentation in an ASBR and mesophilic methanogenic process in a UASB, where a combined bioenergy recovery and COD removal in POME occurs. The study reported 34% increase in energy yield than a single-stage methane synthesis and 90% increase in energy yield than a single-stage hydrogen synthesis. The yields of methane and hydrogen from POME are 315 l CH_4/kgCOD and 210 l H_2/kgCOD, respectively.

In a subsequent study by Mamimin et al. (2019) and Seengnyoung et al. (2019), a pilot scale analysis of a similar process focused on biohythane production was performed. The first stage was operated at HRT of 2 days and OLR of 27.5 gCOD/L-d, and the second stage at HRT of 10 days and OLR of 5.5 gCOD/L-d. The biohythane was produced at a rate of 1.93L-gas/L-d with composition of 11% H_2, 37% CO_2, and 52% CH_4. The pH in the first stage was controlled at 5–6.5 by recirculation of methane effluent (from second stage) with POME at a ratio of 1:1. From a microbial community analysis, it was evident that the hydrogen stage process involves a dominant *Thermoanaerobacterium sp.*, and the methane stage process involves *Methanosarcina sp.* Biohythane from the process can be a potential cleaner biofuel with a flexible as well as controllable H_2/CH_4 ratio in the range of 0.13–0.18.

A similar two-stage process with a CSTR (first stage) and UASB (second stage) with 30% recirculation of methanogenic effluent into hydrogen reactor has been examined by O-Thong et al. (2016). The recirculation reportedly increases hydrogen production by two times, since the methanogenic effluent compensates for required alkalinity by first-stage hydrogen reactor. The hydrogen and methane yields from the process were 135 and 414 mL CH_4/g VS, respectively. The biohythane produced had a composition of 13.3% H_2, 32.2% CO_2, and 54.4% CH_4. Krishnan et al. (2016) investigated a different configuration with same reactor types, where a UASB was used as the hydrogen reactor and a CSTR in the subsequent methane production stage. A recirculation was introduced to maintain pH levels in UASB without buffers. The recirculation also maintained hydrogen production at high concentrations in UASB. The CSTR was supplied with a propylene stream that enhanced microbial growth of methanogens and reduced HRT of methane production. A total energy recovery from the process is 15.43 MJ/kgCOD, while a 94% COD removal is achieved. Hydrogen and methane yields obtained are 215 l H_2/kgCOD and 320 l CH_4 l/kgCOD, respectively.

The digestion of wastes, wastewaters, or effluents in certain cases is not feasible owing to absence of nutrients in the basic anaerobic medium—recalcitrance that may cause long fermentation times and low gas production. Addition of supplement nutrients may ensure optimal cultivation environment; however, the process then will not be cost effective for industrial applications. Therefore, a co-digestion of such wastes potentially provides a positive synergistic effect, improved nutrients balance for microbial growth, and dilutes toxic inhibition effects for economical and effective biohythane production. Palm oil mill effluent (POME) is a carbohydrate-rich substrate consisting of large amount of nutrients (Kongjan et al. 2018). The digestion of POME individually without addition of nutrients, as stated earlier, is a promising route for biohythane production. Therefore, the use of POME as a co-digesting substrate proves to be simple and suitable approach to enhance biohythane yields.

14.5.2.1 Skin Latex Serum and POME

Concentrated latex processes generate large quantities of sulfate-rich high-strength wastewaters. The mixed wastewaters majorly comprise of wash water and skin latex serum (SLS). Skin latex serum (SLS) is an acidic yellow wastewater with high amounts of organic matter, sulfate, and ammonia nitrogen. Under regular anaerobic digestion conditions, high amounts of sulfide generated by sulfate-reducing bacteria (SRB) could inhibit activity of methanogens even at concentrations (0.002–0.003 M) (Batstone et al. 2002). The two-stage process of acidogenesis and methanogenesis scheme reduces sulphate concentration in the first stage by conversion to sulphide, which can be removed before second-stage methanogenesis. Jariyaboon et al. (2015) evaluated biohydrogen and biomethane production from SLS, initial concentration of 37.5%–75% v/v, via two-stage thermophilic process. SLS comprises of major nitrogen and phosphorus macronutrients, while lacking carbohydrate sources for hydrogen production. Therefore, a load shock, pretreated anaerobic digested sludge was added as inoculum to favor production of hydrogen, and a basic anaerobic medium was maintained to ensure optimal cultivation growth. The microbial analysis for the process revealed that the hydrogen-stage dominant bacteria are *Clostrdium* sp. and *Thermoanaerobacterium* sp., while the methane stage dominant bacteria are *Methanothermobacter defluvii* and *Methanosarcina mazei*. The consequent biomethane, biohydrogen and biohythane potentials for SLS were 12.2, 1.57 and 13.77 l/l SLS, i.e. 321.5, 41.3 and 362.37 mL/g VS, respectively. Furthermore, the use of POME as a carbon-rich substrate (9 g/L carbohydrate concentration in POME while 0.6 g/L in SLS) (Kongjan et al. 2018) along with SLS having a nitrogen-rich substrate is a suitable way to enhance biohythane production, economically. The concern of the primary factor of C/N ratio is also addressed in the co-digestion regime, as SLS has nine times lower C/N ratio than POME. Kongjan et al. (2018) studied co-digestion of SLS with POME to eliminate addition of nutrients through a medium to provide optimal conditions. A VS mixing ratio of 55 SLS:45 POME for initial concentration of 7 g VS/l could produce around 85 mL H_2/g-VS added and around 315 mL CH_4/g-VS added. The produced biohythane accounts for an equivalent 85% COD removal. The total energy of 1.76×10^6 GJ/year could be produced from $2.03 \times 10^6 m^3$ of POME and $1.85 \times 10^6 m^3$ of SLS via two-stage anaerobic digestion without nutrient addition.

14.5.2.2 Solid Residues and POME

The palm oil industry generates large amounts of solid wastes like empty fruit bunches (EFB), palm press fibre (PPF), decanter cake (DC), oil palm trunk (OPT), and oil palm frond (OPF) in replantation operations along with palm oil mill effluent (POME) as liquid waste from mills (Mamimin et al. 2019; Suksong et al. 2015). The composition of the palm oil solid wastes includes 40%–50% cellulose, 20%–30% hemicellulose, and 16%–29% lignin (Mamimin et al. 2019). The present cellulose and hemicellulose are pentose- and hexose-rich that can be potentially used for production of biohythane (Chew and Bhatia 2008). Solid wastes possess high content of volatile solids that results in higher biogas production in comparison to liquids (O-Thong et al. 2012). The biological degradation of solid wastes takes a long time due to lignin content present in the plant cell wall. Therefore, anaerobically co-digesting solid wastes with easily degradable waste materials prove to be an energetically efficient method of pretreatment for solid waste (Park and Li 2012). Digestion of solid wastes with liquid waste anaerobically could create a positive synergistic effect as well as enhance microbial growth with improved balance of nutrients in culture (Mata-Alvarez et al. 2011). Suksong et al. (2015) investigated co-digestion of empty fruit branches (EFB) and decanter cake (DC) with palm oil effluent (POME) via a two-stage anaerobic digestion. The first stage co-digestion of POME with 10% EFB and 10% DC resulted in hydrogen yield of 16 mL H_2/g VS for both the cases. The effluent from hydrogen stage in the second stage produced methane yields of 391 mL CH_4/g VS for POME with 10% DC and 240 mL CH_4/g VS for 10% POME with EFB. The process recovered thermal energy content of hydrogen to be 0.1 MJ/kg VS and 8.82–13.9 MJ/kg VS of methane. The total removal efficiency of the two-stage digestion for cellulose, hemicellulose, and lignin was 57%–59%, 35%–40%, and 16%–27%, respectively. Furthermore, Mamimin et al. (2019) examined the co-digestion of solid wastes—decanter cake (DC), palm press fibre (PPF), oil palm frond (OPF), oil palm trunk (OPT), and empty fruit branches (EBF), with POME. The range of biohythane production via the two-stage thermophilic co-digestion of solid wastes with POME is 26.5–34 m^3/ton waste, where the highest was observed for OPT and EFB. The co-digestion increased biohythane production by 67%, 80.6%, 104.2%, 72.4% and 113.9% for DC, OPF, EFB, PPF and OPT, respectively, compared to sole digestion of POME. An improvement in hydrolysis constant (k_h), gas production rate, and lag phase were also observed owing to the synergistic effect. The hydrogen production stage involved predominantly *Clostridium* sp., while the methane-production stage majorly involved *Methanosphaera* sp. archaea.

14.5.3 Distillery Effluent

The distilleries use sugarcane molasses and other agro-products to produce alcoholic beverages. Distillery wastes are produced in the form of spent wash or stillage, commonly regarded as complex high-strength organic effluents (Thakur 2006). Several strict norms on discharge of distillery effluents have been established by regulatory authorities like CPCB and WHO. Distillery industry is a major one of the 17 most polluting industries in India with discharge of 30–40 billion liters annually (Fito et al. 2017). It constitutes a high COD (60–120 g/L) as well as BOD (45–60 g/L), and has a peculiar dark brown color. The brown color is due to melanoidins ($C_{17-18}H_{26-27}O_{10}N$)—a color-causing pigment (Thakur 2006; Sridevi et al. 2014). The presence of high-density nutrients along with the unpleasant smell indicate that distillery effluents cannot be directly discharged, as it may contribute to eutrophication of water bodies (Thakur 2006). Distillery wastewater comprises of large quantity of carbohydrates and essential elements like calcium, potassium, chlorides, and sulfates but is deficient in micronutrients and trace elements like nickel and iron that are paramount for coenzymes to aid biohydrogen production by bacteria (Sridevi et al. 2014; Saranga et al. 2020; Naaz and Kumar 2021). However, the addition of macronutrient and micronutrient supplements in the form of pure nitrogen and mineral salts to distillery effluents increases the cost of biohydrogen production process (Da Silva et al. 2014; Saharan et al. 2011). Saranga et al. (2020) studied a two-stage fermentation of distillery-spent wash effluent diluted with sewage wastewater to reduce high organic content and use of leachate was as a nutrient source for effective production of biohythane. The maximum biohythane production for optimized leachate concentration (30% at 60 g/L substrate concentration) under mesophilic conditions (35°C) was 67 mmol H_2/L biohydrogen production and 42 mmol CH_4/L biomethane production. Similarly, a study of two-stage anaerobic fermentation of distillery effluent with groundnut deoiled cake (GDOC), distillers' dried grain with solubles (DDGS), mustard deoiled cake (MDOC), and algal biomass (AB) was done by Mishra et al. (2017). Distillers' grain with solubles (DDGS) is made from whole stillage, the water and solids remaining after distillation of ethanol, with fiber, protein, and fat as main constituents (Li et al. 2002). GDOC and MDOC, deoiled cakes, are enriched sources of nitrogen with other minerals to be used as co-substrates and also abundant due to high production of oil seeds (Barnwal and Sharma 2005). Algal biomass comprises of high protein and mineral content, making it suitable as co-substrate for biohythane production. Additionally, algal biomass has low lignin content (< 2%) (Mishra et al. 2017). Excess growth of algae in natural water bodies can be fatal for aquatic life due to eutrophication. Therefore, utilization of wastewater-grown algal biomass for bioenergy recovery could be efficient and save costs (Chiu et al. 2015). The maximum gaseous energy recovered from each two-stage biohythane production regime was 23.44, 16.63, 16.21 and 23.93 kcal/L for DDGS, MDOC, AB and GDOC as co-substrates, respectively. The maximum cumulative production was achieved for GDOC, 150 mmol/L of hydrogen, and 64 mmol/L of methane. The co-supplementation technique also reduced substrate input costs (SIC) by 98% for GDOC and even 100% for algal biomass in comparison to tryptone being used as supplement. Furthermore, a semi-pilot scale, single-stage biohythane production process was evaluated using spent wash distillery effluent by Pasupuleti and Venkata Mohan (2015). A single-stage production process facilitates bioenergy generation without an additional reactor cost. A maximum biohythane production of 147.5 L was observed along with production rate of 4.7 L/h after five operation cycles. The fraction of H_2 in biohythane varies from 0.6 to 0.23 during 48 hours of operation cycle, where the first 12 hours exhibit a higher H_2 fraction, while the rest 36 hours produced CH_4 majorly. This is attributed to availability of carbon and buffer capacity of the system. The major driving factor for the single-stage system is the synergistic effect between VFA production and consumption that plays a critical role for simultaneous production of H_2 and CH_4. An effective 60% COD removal (29 kg COD/m^3-day) was observed for the process. The single-stage biosystem can also be realized by upgrading an existing biomethane process as few modifications will be required. The process also features fewer control parameters than individual acidogenic stage and methanogenic stage in the process.

14.5.4 Cellulosic Paper Wastewater (CPW)

The paper industry contributes to large quantities of organically contaminated cellulosic wastewater. The cellulosic paper wastewater (CPW) is often of a dark color, owing to the presence of small components of tree wood and crusts, cellulosic fibers and lignin. Anaerobic treatment of CPW is a sustainable treatment option due to pollution reduction, odor release prevention, and recovery of energy. However, pretreatment of cellulosic wastewater is a major challenge while catering to the socio-economic constraints (Liu et al. 2013). Several cellulosic degrading bacteria are mentioned in literature like *Enterobacter cloacae* and *Clostridium* sp. (Jiang et al. 2015). Similarly, noncellulolytic microflorae also aid bioenergy recovery from cellulose by consumption of metabolite and pH maintenance (Wongwilaiwalin et al. 2010). Therefore, the selection of microflora source is crucial, since the principal limiting factor for energy recovery optimization is hydrolysis of cellulose.

Gadow and Li (2020b) carried out an optimization study for energy recovery as biohythane from cellulosic wastewater using a single-stage bioreactor under mesophilic conditions. A CSTR was utilized at HRT of 10 days with mixed microflora for the study without pretreatment. More than 45% COD conversion to biohythane was reported. Stable hydrogen and methane yields of 5.6 and 18.2 L/kg VS were observed for a period of 240 days. The total energy recovered from the cellulosic wastewater was 4.54 MJ/m^3. This study realized a feasible single-stage mesophilic process with improved degradation efficiency (Table 14.2).

14.6 CONCLUSION

The production of biohythane in essence of its mechanisms and potential feedstocks have been summarized in the book chapter. Bioenergy has been gaining amplified interest in the current scenario of depleting fossil fuels; this has given rise to extensive research on the concept of waste to energy cultivation. Biohythane as a source of biofuel is beneficial in terms of reduced carbon content due to presence of hydrogen, which makes lesser contribution to greenhouse emissions. Feedstocks for production of biohythane in the form of agricultural wastes, food wastes, and industrial effluents have been studied to provide a comparative outlook. The treatment of waste feedstocks using dark fermentation has shown promising results in a twofold manner—treatment and processing of waste as well as recovery of bioenergy. The production process of biohythane has several meticulous parameter requirements—pH, temperature, HRT, OLR, microflora, feedstock characteristics, and reactor configurations that have been studied by several researchers to maximize yields. Emphasis in several works has been thrown on critical selection of microbial communities in the form of sludges as well as adequate nutrient supplements. However, the use of co-digestion regimes and dilution with a secondary substrate have proven to be synergistically advantageous to maintain optimal conditions for microbial growth by improved nutrient balance and toxicity mitigation. Several potential co-digestion regimes to tackle recalcitrance can be explored along with mixing ratio variations. Various pre-treatment strategies have been utilized to serve the same purpose as well. However, formation of byproducts or inhibitory products along with cost effectiveness must be considered to evaluate feasibility.

Specifically, the production of biohythane has been described in a two-stage setup with a coupled mesophilic and thermophilic scheme. The major challenge in

TABLE 14.2
Substrates of Different Biosystems with Energy Recoveries

Substrate	Biosystem	Comments	COD Conversion	Hydrogen Yield	Methane Yield	Energy Recovery	Reference
Synthetic starch wastewaters	Two-stage thermophilic	Cassava starch showed highest production	Cassava-84.9% Corn – 84.9% Rice – 88.3%	For Cassava-81.5 L H_2/kgCOD	For Cassava-310.5 L CH_4/kgCOD	For Cassava-13,363 kJ/kgCOD	Khongkliang et al. (2015)
Glycerol waste with canned sardine wastewater	1-Stage mesophilic, 1-stage thermophilic and 2-stage Mesophilic	2-Stage mesophilic process showed considerable synergism.	> 90%	43 mL H_2/gCOD	306.69 mL CH_4/gCOD	8.07 MJ/kgCOD	Panpong et al. (2017)
Post-hydrothermal liquefaction wastewater	Two-stage fermentation (TF) and Catalytic hydrothermal gasification (CHG)	TF with conventional reactors had a higher net energy return but higher cost than CHG	-	TF – 29 mL H_2/gCOD CHG – 116 mL H_2/gCOD	TF – 254 mL CH_4/gCOD CHG – 65 mL CH_4/gCOD	-	Si et al. (2019)
Rubber sheet wastewater	Two-stage mesophilic	Simultaneous sulphate reduction and biohythane production	96%	101 mL H_2/g VS	629 mL CH_4/g VS	-	Promnuan et al. (2020)
Petrochemical wastewater	Two-stage: stepping baffled anaerobic (SAB) reactor	SAB reactor exhibited good organic loading distribution	95%	359 mL H_2/gCOD	159 mL CH_4/gCOD	-	Elreedy et al. (2015)
Polyester resin wastewater	Up-flow anaerobic self-separation gases (UASG) reactor	UASG reactor provides microbial consortium separation and ability to produce different gas compositions	-	0.019 L H_2/L/day	0.152 L CH_4/L/day	-	Mahmoud et al. (2017)

(Continued)

TABLE 14.2 (*Continued*)
Substrates of Different Biosystems with Energy Recoveries

Substrate	Biosystem	Comments	COD Conversion	Hydrogen Yield	Methane Yield	Energy Recovery	Reference
Cheese whey	Mesophilic: expanded granular sludge bed (EGSB) reactor	HRT was a major factor of control for hythane production	-	3.2 mL H_2/g COD (applied) @ 0.25 hours HRT	9.8 mL CH_4/g COD (applied) @ 4 hours HRT	-	Ramos et al. (2020)
Olive mill wastewater (co-digested – cattle manure, waste activated sludge and fruits & vegetable waste)	Two- stage mesophilic [Continuous stirred tank reactor (CSTR) + anaerobic sequencing batch reactor (ASBR)]	The recirculation of CO_2 and H_2 from acidogenic stage to methaniser improves energy recovery by 56%–65%	-	79.4 mL H_2/g VS inlet	530 mL CH_4/g VS inlet	21.1 kJ/g VS inlet	Farhat et al. (2018)

biohythane synthesis lies in efficient hydrogen production, since it influences methanogensis in subsequent stage. The re-circulation of methane effluent strategies in two-stage systems to substitute use of buffers has been utilized for pH maintenance to increase hydrogen yields; however, long-term operations may hamper the methanogenesis process. More studies on regime development with microbial community evaluation along with enhancements in fermentative pathways could be key. Few pilot-scale studies suggest that development of biohythane production on industrial scales are possible. Furthermore, recent works have explored the utilization of a single-stage biosystem. In spite facing hurdles of low and unstable yields, the single-stage system, however, holds an economic advantage and reduced set of parameters to control, making it lucrative for further development. Robust technology in terms of reactors and efficient microbes are paramount to make the process economically viable.

REFERENCES

Aziz, M.M.A., K.A. Kassim, M. ElSergany, S. Anuar, M.E. Jorat, H. Yaacob, A. Ahsan, and M.A. Imteaz. 2020. "Recent advances on palm oil mill effluent (POME) pretreatment and anaerobic reactor for sustainable biogas production." *Renewable and Sustainable Energy Reviews* 119 (February): 109603. doi: 10.1016/j.rser.2019.109603.

Abd Nasir, M.A., J.M. Jahim, P.M. Abdul, H. Silvamany, R.M. Maaroff, and M.F.M. Yunus. 2019. "The use of acidified palm oil mill effluent for thermophilic biomethane production by changing the hydraulic retention time in anaerobic sequencing batch reactor." *International Journal of Hydrogen Energy*, 3373–81. doi: 10.1016/j.ijhydene.2018.06.149.

Abreu, A.A., Tavares, F., Alves, M.M., Pereira M.A. 2016. 'Boosting dark fermentation with co-cultures of extreme thermophiles for biohythane production from garden waste.' *Bioresource Technology* 219: 132–138. doi: https://doi.org/10.1016/j.biortech.2016.07.096.

Barnwal, B.K., and M.P. Sharma. 2005. "Prospects of biodiesel production from vegetable oils in India." *Renewable and Sustainable Energy Reviews* 9 (4): 363–78. doi: 10.1016/j.rser.2004.05.007.

Batstone, D.J., J. Keller, I. Angelidaki, S.V. Kalyuzhnyi, S.G. Pavlostathis, A. Rozzi, W.T. Sanders, H. Siegrist, and V.A. Vavilin. 2002. "The IWA anaerobic digestion model no 1 (ADM1)." *Water Science and Technology : A Journal of the International Association on Water Pollution Research* 45 (10): 65–73. doi: 10.2166/wst.2002.0292.

Bhatia, S.K., S. Mehariya, R.K. Bhatia, M. Kumar, A. Pugazhendhi, M.K. Awasthi, A. E. Atabani, et al. 2021. "Wastewater based microalgal biorefinery for bioenergy production: Progress and challenges." *Science of the Total Environment* 751: 141599. doi: 10.1016/j.scitotenv.2020.141599.

Bolzonella, D., F. Battista, C. Cavinato, M. Gottardo, F. Micolucci, G. Lyberatos, and P. Pavan. 2018. "Recent developments in biohythane production from household food wastes: A review." *Bioresource Technology* 257: 311–19. doi: 10.1016/j.biortech.2018.02.092.

Cao, J., E. Sanganyado, W. Liu, W. Zhang, and Y. Liu. 2019. "Decolorization and detoxification of direct blue 2B by indigenous bacterial consortium." *Journal of Environmental Management* 242 (July): 229–37. doi: 10.1016/j.jenvman.2019.04.067.

Cavinato, C., F. Fatone, D. Bolzonella, and P. Pavan. 2010. "Thermophilic anaerobic co-digestion of cattle manure with agro-wastes and energy crops: Comparison of pilot and full scale experiences." *Bioresource Technology* 101 (2): 545–50. doi: 10.1016/j.biortech.2009.08.043.

Cavinato, C., A. Giuliano, D. Bolzonella, P. Pavan, and F. Cecchi. 2012. "Bio-hythane production from food waste by dark fermentation coupled with anaerobic digestion process: A long-term pilot scale experience." *International Journal of Hydrogen Energy* 37 (15): 11549–55. doi: 10.1016/j.ijhydene.2012.03.065.

Chen, H., R. Huang, J. Wu, W. Zhang, Y. Han, B. Xiao, D. Wang, Y. Zhou, B. Liu, and G. Yu. 2021. "Biohythane production and microbial characteristics of two alternating mesophilic and thermophilic two-stage anaerobic co-digesters fed with rice straw and pig manure." *Bioresource Technology* 320: 124303. doi: 10.1016/j.biortech.2020.124303.

Chew, T.L., and S. Bhatia. 2008. "Catalytic processes towards the production of biofuels in a palm oil and oil palm biomass-based biorefinery." *Bioresource Technology* 99 (17): 7911–22. doi: 10.1016/j.biortech.2008.03.009.

Chiu, S.Y., C.Y. Kao, T.Y. Chen, Y.B. Chang, C.M. Kuo, and C.S. Lin. 2015. "Cultivation of microalgal chlorella for biomass and lipid production using wastewater as nutrient resource." *Bioresource Technology* 184: 179–89. doi: 10.1016/j.biortech.2014.11.080.

Chu, C.-F., Y.-Y. Li, K.-Q. Xu, Y. Ebie, Y. Inamori, and H.-N. Kong. 2008. "A PH- and temperature-phased two-stage process for hydrogen and methane production from food waste." *International Journal of Hydrogen Energy* 33 (18): 4739–46. doi: 10.1016/j.ijhydene.2008.06.060.

Ding, L., E.C. Gutierrez, J. Cheng, A. Xia, R. O'Shea, A.J. Guneratnam, and J.D. Murphy. 2018. "Assessment of continuous fermentative hydrogen and methane co-production using macro- and micro-algae with increasing organic loading rate." *Energy* 151: 760–70. doi: 10.1016/j.energy.2018.03.103.

Djafer, A., L. Djafer, B. Maimoun, A. Iddou, S. Kouadri Mostefai, and A. Ayral. 2017. "Reuse of waste activated sludge for textile dyeing wastewater treatment by biosorption: Performance optimization and comparison." *Water and Environment Journal* 31 (1): 105–12. doi: 10.1111/wej.12218.

Ellabban, O., H. Abu-Rub, and F. Blaabjerg. 2014. "Renewable energy resources: Current status, future prospects and their enabling technology." *Renewable and Sustainable Energy Reviews* 39: 748–64. doi: 10.1016/j.rser.2014.07.113.

Elreedy, A., A. Tawfik, K. Kubota, Y. Shimada, and H. Harada. 2015. "Hythane (H_2+CH_4) production from petrochemical wastewater containing mono-ethylene glycol via stepped anaerobic baffled reactor." *International Biodeterioration and Biodegradation* 105: 252–61. doi: 10.1016/j.ibiod.2015.09.015.

Farghaly, A., and A. Tawfik. 2017. "Simultaneous hydrogen and methane production through multi-phase anaerobic digestion of paperboard mill wastewater under different operating conditions." *Applied Biochemistry and Biotechnology* 181 (1): 142–56. doi: 10.1007/s12010-016-2204-7.

Farhat, A., B. Miladi, M. Hamdi, and H. Bouallagui. 2018. "Fermentative hydrogen and methane co-production from anaerobic co-digestion of organic wastes at high loading rate coupling continuously and sequencing batch digesters." *Environmental Science and Pollution Research* 25 (28): 27945–58. doi: 10.1007/s11356-018-2796-2.

Fito, J., N. Tefera, and S.W. Van Hulle. 2017. "Adsorption of distillery spent wash on activated bagasse fly ash: Kinetics and thermodynamics." *Journal of Environmental Chemical Engineering* 5 (6): 5381–88. doi: 10.1016/j.jece.2017.10.009.

Gadow, S.I., and Y.Y. Li. 2020a. "Efficient treatment of recalcitrant textile wastewater using two-phase mesophilic anaerobic process: Bio-hythane production and decolorization improvements." *Journal of Material Cycles and Waste Management* 22 (2): 515–23. doi: 10.1007/s10163-019-00944-z.

Gadow, S.I., and Y.Y. Li. 2020b. "Optimization of energy recovery from cellulosic wastewater using mesophilic single-stage bioreactor." *Waste and Biomass Valorization* 11 (11): 6017–23. doi: 10.1007/s12649-019-00837-0.

Galanakis, C.M. 2018. "12- food waste recovery: Prospects and opportunities." In *Sustainable Food Systems from Agriculture to Industry*, edited by C.M. Galanakis, 401–19. Academic Press. doi: 10.1016/B978-0-12-811935-8.00012-3.

Ghimire, A., L. Frunzo, F. Pirozzi, E. Trably, R. Escudie, P.N.L. Lens, and G. Esposito. 2015. "A review on dark fermentative biohydrogen production from organic biomass: Process parameters and use of by-products." *Applied Energy* 144: 73–95. doi: 10.1016/j.apenergy.2015.01.045.

Hans, M., and S. Kumar. 2019. "Biohythane production in two-stage anaerobic digestion system." *International Journal of Hydrogen Energy* 17363–80. doi: 10.1016/j.ijhydene.2018.10.022.

Holkar, C.R., A.J. Jadhav, D.V. Pinjari, N.M. Mahamuni, and A.B. Pandit. 2016. "A critical review on textile wastewater treatments: Possible approaches." *Journal of Environmental Management* 182: 351–66. doi: 10.1016/j.jenvman.2016.07.090.

Ito, T., Y. Nakashimada, T. Kakizono, and N. Nishio. 2004. "High-yield production of hydrogen by enterobacter aerogenes mutants with decreased alpha-acetolactate synthase activity." *Journal of Bioscience and Bioengineering* 97 (4): 227–32. doi: 10.1016/S1389-1723(04)70196-6.

Jariyaboon, R., S. O-Thong, and P. Kongjan. 2015. "Bio-hydrogen and bio-methane potentials of skim latex serum in batch thermophilic two-stage anaerobic digestion." *Bioresource Technology* 198: 198–206. doi: 10.1016/j.biortech.2015.09.006.

Jehlee, A., P. Khongkliang, W. Suksong, S. Rodjaroen, J. Waewsak, A. Reungsang, and S. O-Thong. 2017. "Biohythane production from *Chlorella* Sp. biomass by two-stage thermophilic solid-state anaerobic digestion." *International Journal of Hydrogen Energy* 42 (45): 27792–800. doi: 10.1016/j.ijhydene.2017.07.181.

Jehlee, A., S. Rodjaroen, J. Waewsak, A. Reungsang, and S. O-Thong. 2019. "Improvement of biohythane production from *Chlorella* Sp. TISTR 8411 biomass by co-digestion with organic wastes in a two-stage fermentation." *International Journal of Hydrogen Energy* 44 (32): 17238–47. doi: 10.1016/j.ijhydene.2019.03.026.

Ji, C., C.-X. Kong, Z.-L. Mei, and J. Li. 2017. "A review of the anaerobic digestion of fruit and vegetable waste." *Applied Biochemistry and Biotechnology* 183 (3): 906–22. doi: 10.1007/s12010-017-2472-x.

Jiang, H., S.I. Gadow, Y. Tanaka, J. Cheng, and Y.Y. Li. 2015. "Improved cellulose conversion to bio-hydrogen with thermophilic bacteria and characterization of microbial community in continuous bioreactor." *Biomass and Bioenergy* 75: 57–64. doi: 10.1016/j.biombioe.2015.02.010.

Jiang, H., Y. Qin, S.I. Gadow, A. Ohnishi, N. Fujimoto, and Y.Y. Li. 2018. "Bio-hythane production from cassava residue by two-stage fermentative process with recirculation." *Bioresource Technology* 247: 769–75. doi: 10.1016/j.biortech.2017.09.102.

Khongkliang, P., P. Kongjan, and S. O-Thong. 2015. Hydrogen and methane production from starch processing wastewater by thermophilic two-stage anaerobic digestion. *Energy Procedia* 79. doi: 10.1016/j.egypro.2015.11.573.

Kim, S., J. Bae, O. Choi, D. Ju, J. Lee, H. Sung, S. Park, B.-I. Sang, and Y. Um. 2013. "A pilot scale two-stage anaerobic digester treating food waste leachate (FWL): Performance and microbial structure analysis using pyrosequencing." *Process Biochemistry* 49. doi: 10.1016/j.procbio.2013.10.022.

Kongjan, P., K. Sama, K. Sani, R. Jariyaboon, and A. Reungsang. 2018. "Feasibility of biohythane production by co-digesting skim latex serum (SLS) with palm oil mill effluent (POME) through two-phase anaerobic process." *International Journal of Hydrogen Energy* 43 (20): 9577–90. doi: 10.1016/j.ijhydene.2018.04.052.

Krishnan, S., L. Singh, M. Sakinah, S. Thakur, Z.A. Wahid, and M. Alkasrawi. 2016. "Process enhancement of hydrogen and methane production from palm oil mill effluent using two-stage thermophilic and mesophilic fermentation." *International Journal of Hydrogen Energy* 41 (30): 12888–98. doi: 10.1016/j.ijhydene.2016.05.037.

Kumar, C.P., A. Meenakshi, A.S. Khapre, S. Kumar, A. Anshul, L. Singh, S.H. Kim, B.D. Lee, and R. Kumar. 2019. "Bio-hythane production from organic fraction of municipal solid waste in single and two stage anaerobic digestion processes." *Bioresource Technology* 294: 122220. doi: 10.1016/j.biortech.2019.122220.

Kvesitadze, G., T. Sadunishvili, T. Dudauri, N. Zakariashvili, G. Partskhaladze, V. Ugrekhelidze, G. Tsiklauri, B. Metreveli, and M. Jobava. 2012. "Two-stage anaerobic process for bio-hydrogen and bio-methane combined production from biodegradable solid wastes." *Energy* 37 (1): 94–102. doi: 10.1016/j.energy.2011.08.039.

Lay, C.H., G. Kumar, A. Mudhoo, C.Y. Lin, H.J. Leu, S. Shobana, and M.L.T. Nguyen. 2020. "Recent trends and prospects in biohythane research: An overview." *International Journal of Hydrogen Energy* 45 (10): 5864–73. doi: 10.1016/j.ijhydene.2019.07.209.

Lay, C.H., S.Y. Kuo, B. Sen, C.C. Chen, J.S. Chang, and C.Y. Lin. 2012. "Fermentative biohydrogen production from starch-containing textile wastewater." *International Journal of Hydrogen Energy* 37 (2): 2050–57. doi: 10.1016/j.ijhydene.2011.08.003.

Lellis, B., C.Z. Fávaro-Polonio, J.A. Pamphile, and J.C. Polonio. 2019. "Effects of textile dyes on health and the environment and bioremediation potential of living organisms." *Biotechnology Research and Innovation* 3 (2): 275–90. doi: 10.1016/j.biori.2019.09.001.

Li, Y., D. Li, J. Xing, S. Li, Y. Han, and I.K. Han. 2002. "Effects of supplementing different combinations of nitrogen supplements on digestibility and performance in sheep fed wheat straw diets." *Asian-Australasian Journal of Animal Sciences* 15 (10): 1428–32. doi: 10.5713/ajas.2002.1428.

Lin, C.Y., C.C. Chiang, M.L.T. Nguyen, and C.H. Lay. 2017. "Enhancement of fermentative biohydrogen production from textile desizing wastewater via coagulation-pretreatment." *International Journal of Hydrogen Energy* 42 (17): 12153–58. doi: 10.1016/j.ijhydene.2017.03.184.

Liu, Z., B. Si, J. Li, J. He, C. Zhang, Y. Lu, Y. Zhang, and X.H. Xing. 2018. "Bioprocess engineering for biohythane production from low-grade waste biomass: Technical challenges towards scale up." *Current Opinion in Biotechnology* 50: 25–31. doi: 10.1016/j.copbio.2017.08.014.

Liu, Z., C. Zhang, Y. Lu, X. Wu, L. Wang, L. Wang, B. Han, and X.H. Xing. 2013. "States and challenges for high-value biohythane production from waste biomass by dark fermentation technology." *Bioresource Technology* 135: 292–303. doi: 10.1016/j.biortech.2012.10.027.

Luongo, M., A. Cristina, M. Bernardi, D. Fino, and B. Ruggeri. 2015. "Multistep anaerobic digestion (MAD) as a tool to increase energy production via H_2+CH_4." *International Journal of Hydrogen Energy* 40 (15): 5050–61. doi: 10.1016/j.ijhydene.2015.02.068.

Maaroff, R.M., J.M. Jahim, A.M. Azahar, P.M. Abdul, M.S. Masdar, D. Nordin, and M.A.A. Nasir. 2019. "Biohydrogen production from palm oil mill effluent (POME) by two stage anaerobic sequencing batch reactor (ASBR) system for better utilization of carbon sources in POME." *International Journal of Hydrogen Energy* 3395–406. doi: 10.1016/j.ijhydene.2018.06.013.

Mahmoud, M., A. Elreedy, P. Pascal, L.R. Sophie, and A. Tawfik. 2017. "Hythane (H2 and CH4) production from unsaturated polyester resin wastewater contaminated by 1,4-dioxane and heavy metals via up-flow anaerobic self-separation gases reactor." *Energy Conversion and Management* 152 (August): 342–53. doi: 10.1016/j.enconman.2017.09.060.

Mamimin, C., P. Kongjan, S. O-Thong, and P. Prasertsan. 2019. "Enhancement of biohythane production from solid waste by co-digestion with palm oil mill effluent in two-stage thermophilic fermentation." *International Journal of Hydrogen Energy* 44 (32): 17224–37. doi: 10.1016/j.ijhydene.2019.03.275.

Mamimin, C., P. Prasertsan, P. Kongjan, and S. O-Thong. 2017. "Effects of volatile fatty acids in biohydrogen effluent on biohythane production from palm oil mill effluent under thermophilic condition." *Electronic Journal of Biotechnology* 29: 78–85. doi: 10.1016/j.ejbt.2017.07.006.

Mamimin, C., A. Singkhala, P. Kongjan, B. Suraraksa, P. Prasertsan, T. Imai, and S. O-Thong. 2015. "Two-stage thermophilic fermentation and mesophilic methanogen process for biohythane production from palm oil mill effluent." *International Journal of Hydrogen Energy* 40 (19): 6319–28. doi: 10.1016/j.ijhydene.2015.03.068.

Martinez-Jimenez, F.D., M.P.M. Pinto, A. Mudhoo, T.D.A. Neves, M.A. Rostagno, and T. Forster-Carneiro. 2017. "Influence of ultrasound irradiation pre-treatment in biohythane generation from the thermophilic anaerobic co-digestion of sugar production residues." *Journal of Environmental Chemical Engineering* 5 (4): 3749–58. doi: 10.1016/j.jece.2017.07.030.

Massanet-Nicolau, J., R. Dinsdale, A. Guwy, and G. Shipley. 2013. "Use of real time gas production data for more accurate comparison of continuous single-stage and two-stage fermentation." *Bioresource Technology* 129 (February): 561–67. doi: 10.1016/j.biortech.2012.11.102.

Mata-Alvarez, J., J. Dosta, S. Macé, and S. Astals. 2011. "Codigestion of solid wastes: A review of its uses and perspectives including modeling." *Critical Reviews in Biotechnology* 31 (2): 99–111. doi: 10.3109/07388551.2010.525496.

Meena, R.A.A., J.R. Banu, R.Y. Kannah, K.N. Yogalakshmi, and G. Kumar. 2020. "Biohythane production from food processing wastes – Challenges and perspectives." *Bioresource Technology* 298: 122449. doi: 10.1016/j.biortech.2019.122449.

Micolucci, F., M. Gottardo, D. Bolzonella, and P. Pavan. 2014. "Automatic process control for stable bio-hythane production in two-phase thermophilic anaerobic digestion of food waste." *International Journal of Hydrogen Energy* 39 (31): 17563–72. doi: 10.1016/j.ijhydene.2014.08.136.

Mishra, P., G. Balachandar, and D. Das. 2017. "Improvement in biohythane production using organic solid waste and distillery effluent." *Waste Management* 66: 70–78. doi: 10.1016/j.wasman.2017.04.040.

Naaz, S., and S. Kumar. 2021. "ScienceDirect enhancement effect of zero-valent iron nanoparticle and iron oxide nanoparticles on dark fermentative hydrogen production from molasses- based distillery wastewater." *International Journal of Hydrogen Energy* doi: 10.1016/j.ijhydene.2021.06.125.

National Horticulture Board. 2018. *Educational Statistics at a Glance 2018*, 1–127. https://www.mhrd.gov.in/sites/upload_files/mhrd/files/statistics-new/ESAG-2018.pdf.

O-Thong, S., K. Boe, and I. Angelidaki. 2012. "Thermophilic anaerobic co-digestion of oil palm empty fruit bunches with palm oil mill effluent for efficient biogas production." *Applied Energy* 93: 648–54. doi: 10.1016/j.apenergy.2011.12.092.

O-Thong, S., W. Suksong, K. Promnuan, M. Thipmunee, C. Mamimin, and P. Prasertsan. 2016. "Two-stage thermophilic fermentation and mesophilic methanogenic process for biohythane production from palm oil mill effluent with methanogenic effluent recirculation for PH control." *International Journal of Hydrogen Energy* 41 (46): 21702–12. doi: 10.1016/j.ijhydene.2016.07.095.

Panpong, K., T. Srimachai, K. Nuithitikul, P. Kongjan, S. O-Thong, T. Imai, and N. Kaewthong. 2017. "Anaerobic co-digestion between canned sardine wastewater and glycerol waste for biogas production: Effect of different operating processes." *Energy Procedia* 138: 260–66. doi: 10.1016/j.egypro.2017.10.050.

Park, S., and Y. Li. 2012. "Evaluation of methane production and macronutrient degradation in the anaerobic co-digestion of algae biomass residue and lipid waste." *Bioresource Technology* 111: 42–48. doi: 10.1016/j.biortech.2012.01.160.

Pasupuleti, S.B., and S.V. Mohan. 2015. "Single-stage fermentation process for high-value biohythane production with the treatment of distillery spent-wash." *Bioresource Technology* 189: 177–85. doi: 10.1016/j.biortech.2015.03.128.

Promnuan, K., T. Higuchi, T. Imai, P. Kongjan, A. Reungsang, and S. O-Thong. 2020. "Simultaneous biohythane production and sulfate removal from rubber sheet wastewater by two-stage anaerobic digestion." *International Journal of Hydrogen Energy* 45 (1): 263–74. doi: 10.1016/j.ijhydene.2019.10.237.

Qin, Y., J. Wu, B. Xiao, M. Cong, T. Hojo, J. Cheng, and Y.Y. Li. 2019. "Strategy of adjusting recirculation ratio for biohythane production via recirculated temperature-phased anaerobic digestion of food waste." *Energy* 179: 1235–45. doi: 10.1016/j.energy.2019.04.182.

Rai, P.K, S.P. Singh, R.K. Asthana, and S. Singh. 2014. "Biohydrogen production from sugarcane bagasse by integrating dark- and photo-fermentation." *Bioresource Technology* 152: 140–46. doi: 10.1016/j.biortech.2013.10.117.

Ramos, L.R., C.A. de Menezes, L.A. Soares, I.K. Sakamoto, M.B.A. Varesche, and E.L. Silva. 2020. "Controlling methane and hydrogen production from cheese whey in an EGSB reactor by changing the HRT." *Bioprocess and Biosystems Engineering* 43 (4): 673–84. doi: 10.1007/s00449-019-02265-9.

Ren, N., A. Wang, G. Cao, J. Xu, and L. Gao. 2009. "Bioconversion of lignocellulosic biomass to hydrogen: Potential and challenges." *Biotechnology Advances* 27 (6): 1051–60. doi: 10.1016/j.biotechadv.2009.05.007.

Rena, K.M.B. Zacharia, S. Yadav, N.P. Machhirake, S.H. Kim, B.D. Lee, H. Jeong, L. Singh, S. Kumar, and R. Kumar. 2020. "Bio-hydrogen and bio-methane potential analysis for production of bio-hythane using various agricultural residues." *Bioresource Technology* 309 (March): 123297. doi: 10.1016/j.biortech.2020.123297.

Rivière, D., V. Desvignes, E. Pelletier, S. Chaussonnerie, S. Guermazi, J. Weissenbach, T. Li, P. Camacho, and A. Sghir. 2009. "Towards the definition of a core of microorganisms involved in anaerobic digestion of sludge." *The ISME Journal* 3 (6): 700–714. doi: 10.1038/ismej.2009.2.

Roy, S., and D. Das. 2016. "Biohythane production from organic wastes: Present state of art." *Environmental Science and Pollution Research International* 23 (10): 9391–9410. doi: 10.1007/s11356-015-5469-4.

Saharan, B.S., R.K. Sahu, and D. Sharma. 2011. "A review on biosurfactants : Fermentation, applications, current" *Journal of Genetic Engineering and Biotechnology* 2011: 1–42.

Saranga, V.K., P.K. Kumar, K. Verma, D. Bhagawan, V. Himabindu, and M.L. Narasu. 2020. "Effect of biohythane production from distillery spent wash with addition of landfill leachate and sewage wastewater." *Applied Biochemistry and Biotechnology* 190 (1): 30–43. doi: 10.1007/s12010-019-03087-x.

Sarkar, O., J. Santhosh, A. Dhar, and S.V. Mohan. 2021. "Green hythane production from food waste: Integration of dark-fermentation and methanogenic process towards biogas up-gradation." *International Journal of Hydrogen Energy* 46 (36): 18832–43. doi: 10.1016/j.ijhydene.2021.03.053.

Schievano, A., A. Tenca, S. Lonati, E. Manzini, and F. Adani. 2014. "Can two-stage instead of one-stage anaerobic digestion really increase energy recovery from biomass?" *Applied Energy* 124: 335–42. doi: 10.1016/j.apenergy.2014.03.024.

Seengenyoung, J., C. Mamimin, P. Prasertsan, and S. O-Thong. 2019. "Pilot-scale of biohythane production from palm oil mill effluent by two-stage thermophilic anaerobic fermentation." *International Journal of Hydrogen Energy* 44 (6): 3347–55. doi: 10.1016/j.ijhydene.2018.08.021.

Si, B., J. Watson, A. Aierzhati, L. Yang, Z. Liu, and Y. Zhang. 2019. "Biohythane production of post-hydrothermal liquefaction wastewater: A comparison of two-stage fermentation and catalytic hydrothermal gasification." *Bioresource Technology* 274 doi: 10.1016/j.biortech.2018.11.095.

Silva, T.L.D., L. Gouveia, and A. Reis. 2014. "Integrated microbial processes for biofuels and high value-added products: The way to improve the cost effectiveness of biofuel production." *Applied Microbiology and Biotechnology* 98 (3): 1043–53. doi: 10.1007/s00253-013-5389-5.

Singh, A., A. Kuila, S. Adak, M. Bishai, and R. Banerjee. 2012. "Utilization of vegetable wastes for bioenergy generation." *Agricultural Research* 1 (3): 213–22. doi: 10.1007/s40003-012-0030-x.

Sridevi, K., E. Sivaraman, and P. Mullai. 2014. "Back propagation neural network modelling of biodegradation and fermentative biohydrogen production using distillery wastewater in a hybrid upflow anaerobic sludge blanket reactor." *Bioresource Technology* 165 (C): 233–40. doi: 10.1016/j.biortech.2014.03.074.

Suksong, W., P. Kongjan, and S. O-Thong. 2015. "Biohythane production from co-digestion of palm oil mill effluent with solid residues by two-stage solid state anaerobic digestion process." *Energy Procedia* 79 (Dc): 943–49. doi: 10.1016/j.egypro.2015.11.591.

Ta, D.T., C.Y. Lin, T.M.N. Ta, and C.Y. Chu. 2020. "Biohythane production via single-stage anaerobic fermentation using entrapped hydrogenic and methanogenic bacteria." *Bioresource Technology* 300 (December 2019): 122702. doi: 10.1016/j.biortech.2019.122702.

Thakur, I.S. 2006. "Distillery industry 2.1." *Industrial Biotechnology: Problems and Remedies*, 44–68.

Tiwari, A., and A. Pandey. 2012. "Cyanobacterial hydrogen production – A step towards clean environment." *International Journal of Hydrogen Energy* 37: 139–50.

Vo, T.P., C.H. Lay, and C.Y. Lin. 2019. "Effects of hydraulic retention time on biohythane production via single-stage anaerobic fermentation in a two-compartment bioreactor." *Bioresource Technology* 292 (July): 121869. doi: 10.1016/j.biortech.2019.121869.

Wongwilaiwalin, S., U. Rattanachomsri, T. Laothanachareon, L. Eurwilaichitr, Y. Igarashi, and V. Champreda. 2010. "Analysis of a thermophilic lignocellulose degrading microbial consortium and multi-species lignocellulolytic enzyme system." *Enzyme and Microbial Technology* 47 (6): 283–90. doi: 10.1016/j.enzmictec.2010.07.013.

Xu, N., S. Liu, F. Xin, J. Zhou, H. Jia, J. Xu, M. Jiang, and W. Dong. 2019. "Biomethane production from lignocellulose: Biomass recalcitrance and its impacts on anaerobic digestion." *Frontiers in Bioengineering and Biotechnology* 7 (Aug): 1–12. doi: 10.3389/fbioe.2019.00191.

Yang, G., G. Zhang, and H. Wang. 2015. "Current state of sludge production, management, treatment and disposal in China." *Water Research* 78: 60–73. doi: 10.1016/j.watres.2015.04.002.

Yaseen, D.A., and M. Scholz. 2019. "Textile dye wastewater characteristics and constituents of synthetic effluents: A critical review." *International Journal of Environmental Science and Technology* 16. doi: 10.1007/s13762-018-2130-z.

Yeshanew, M.M., L. Frunzo, F. Pirozzi, P.N.L. Lens, and G. Esposito. 2016. "Production of biohythane from food waste via an integrated system of continuously stirred tank and anaerobic fixed bed reactors." *Bioresource Technology* 220: 312–22. doi: 10.1016/j.biortech.2016.08.078.

Zhang, Z., C. Xu, Y. Zhang, S. Lu, L. Guo, Y. Zhang, Y. Li, B. Hu, C. He, and Q. Zhang. 2020. "Cohesive strategy and energy conversion efficiency analysis of bio-hythane production from corncob powder by two-stage anaerobic digestion process." *Bioresource Technology* 300 (December 2019): 122746. doi: 10.1016/j.biortech.2020.122746.

Zuo, Z., S. Wu, W. Zhang, and R. Dong. 2013. "Effects of organic loading rate and effluent recirculation on the performance of two-stage anaerobic digestion of vegetable waste." *Bioresource Technology* 146: 556–61. doi: 10.1016/j.biortech.2013.07.128.

15 Waste-Derived Biohydrogen Enriched CNG/Biohythane

Research Trend and Utilities

Omprakash Sarkar
Luleå University of Technology

J. Santosh and S. Venkata Mohan
CSIR—Indian Institute of Chemical Technology (CSIR—IICT)

Young-Cheol Chang
Muroran Institute of Technology

CONTENTS

DOI: 10.1201/9781003197737-21

15.1 INTRODUCTION

Fossil fuels such as coal, oil, and natural gas are the source of over 80% of the world's energy, burning these resulting with CO_2 emissions, which is 89% of human derived. A critical issue facing by humankind is increasing CO_2 emissions (Sarkar et al. 2022a; Birchall & Bonnett 2021). The Intergovernmental Panel on Climate Change (IPCC) reflected the severity of climate change and highlighted the issues related with rising global temperature to avoid the 1.5°C of temperature rise (Do et al. 2022; Birchall & Bonnett 2021). To prevent global warming, government bodies and energy agencies are taking actions to minimize carbon footprints (Do et al. 2022). In the recent COP26 (Glasgow, Scotland), political leaders from different countries were committed and planned the strategies that assist the world with a clear and viable pathway to keep temperature rise to 1.5°C. However, to do so requires concrete policies like deploying clean energy or transitioning to vehicles with less/zero emissions. Reduction of air emissions can play a crucial role supporting sustainable development, as the greenhouse gases such as carbon dioxide (CO_2), nitrous oxide (NO_2), hydrofluorocarbon, etc., are responsible for climate change. Green House gases (GHG) and carbon dioxide equivalent describes the gaseous contribution to climate change (Graham et al. 2008). About 76% of the global GHG emission is contributed by CO_2. Concerns about sustainability and the environment have grown because of the transportation sector's reliance on fossil fuels. Growing petroleum shortages and the effects of greenhouse gas emissions on the climate have accelerated the development of renewable and low-carbon energy sources.

The transportation sector is the second-largest source of CO_2 emissions, after electricity and heat generation, accounting for about 25% of global emissions. Therefore, emissions reduction strategies, as well as alternate feedstock, are required to meet climate change targets. Emission-free hydrogen (H_2) can be adopted towards low-carbon economy over fossil-based petrol and diesel utilization by powering fuel-cell electric vehicles (FCEVs) like cars, trucks, and trains or as a feedstock for synthetic fuels for ships and airplanes (Dahiya et al. 2021). As an alternative, compressed natural gas (CNG), which has traces of CO_2 content of only 0.15%, is a perfect blend for hybrid fuels for diesel or gasoline engines. Additionally, this blend lowers 25% in greenhouse gas production (GHG). CNG is economical, less flammable than LPG, and has a higher output delivery, all of which have contributed to the US and Indian markets for CNG-based vehicles gaining attention.

Currently, H_2 is blended with compressed natural gas (CNG) because it is 99.9% free of greenhouse emissions and because the high-octane of H_2 improves engine performance. Because H_2 is highly combustible in atmospheric air, the system is insecure and prone to accidental explosion. Upon blending with CNG, the risk of explosion is reduced, the flame rate of this blend (HCNG) is raised, and a higher combustion rate takes place compared to biogas and conventional gasoline. Hydrogen (H_2)-enriched compressed natural gas (CNG)/hythane is emerging as an energy carrier for combustion engines. Hythane/HCNG as a fuel is beneficial because of (1) improved heat efficiency, (2) good ignition ability in engines with the least input energy, (3) cleaner emissions compared to methane (Wang et al. 2009; Zhang et al. 2009; Hans & Kumar 2019; Mamimin et al. 2015). Hydrogen (H_2)-enriched compressed natural gas (CNG)/biohythane-powered vehicles offer a key advantage such as (1) widens the range of flammability limits, (2) enhances the flame propagation speed, and (3) increases the rates of pressure rise and deflagration index (Zheng et al. 2017; Wang et al. 2019; Sandalcı et al. 2019; Prashanth Kumar et al. 2019). Methane (CH_4) has been an attractive source of energy for years (Li et al. 2021; Sravan et al. 2020b). With economic and environmental benefits, biohythane is expected to take a long journey, since the blend (H_2+CH_4) is in direct use with the existing vehicle engines and has been commercialized as vehicle fuel. Hydrogen-enriched methane/CNG/biohythane has been used as a fuel in buses (Ridell 2004; Patil & Brown 2008), fuel trapped vortex combustor (Ghenai 2019), spark ignition engine (Huang et al. 2006), and methanol production (Patel et al. 2017) (Figure 15.1).

Moreover, the hydrogen being used in the blend is mostly derived from fossils through steam reforming, which is an environment-impacting process resulting in the emission of >10 kg CO_2/kg H_2 produced (Cetinkaya et al. 2012; Sarkar et al. 2021c). Therefore, an alternate and green process should be considered for the production of this low-carbon hydrogen in order to hold global warming at 1.5°C. Acidogenic dark fermentation is a versatile platform that potentially coverts high strength organic substrates (waste/wastewater) to hydrogen and significant volatile fatty acids (VFA) (Ghimire et al. 2022; Sarkar et al. 2016, 2017). However, to gain a significant amount of hydrogen, the process needs to be regulated, making acidogenesis the final step instead of methanogenesis, when mixed culture is inoculum source (Santhosh et al.

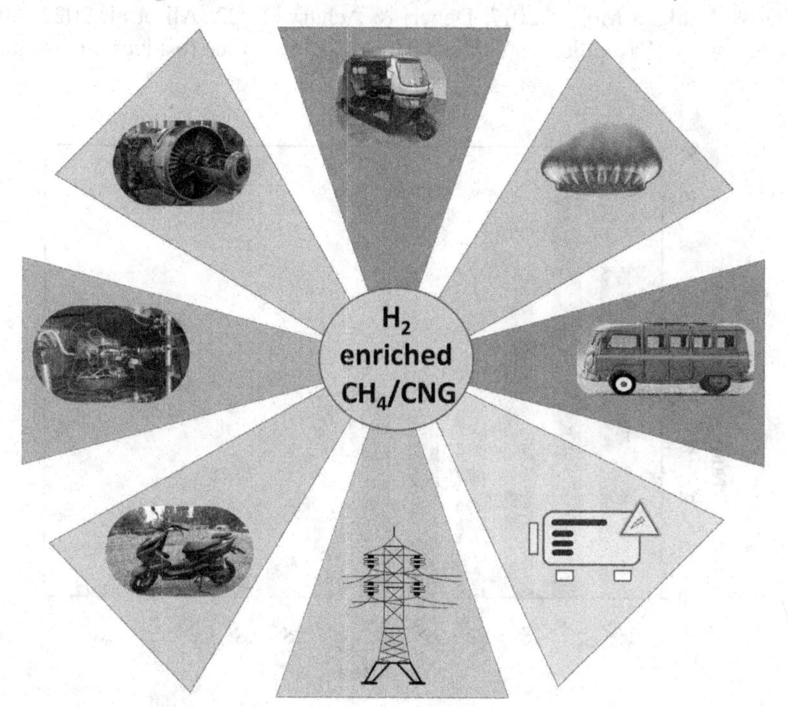

FIGURE 15.1 Application of Hydrogen (H_2) enriched CNG/compressed natural gas (CNG) as hythane as fuel in different vehicle and heat/electricity generating engines.

2021; Sarkar et al. 2020, 2021a). In case, if not regulated, the methanogens feed on the accumulated carboxylic acids and hydrogen, converting them to methane by methanogens (Sarkar et al. 2022a). Strategies like pretreatment of the mixed culture have been successfully adopted over years to regulate the process aiming with enrichment of acidogens, as this strategy enables spore-forming bacteria, which has the potential to generate molecular hydrogen. Further, the hydrogen production is combined with methane from anaerobic digestion either in single or dual stage for the formation of biohythane (Lay et al. 2020). This chapter reviewed many components of waste valorization for the production of biohythane-/biohydrogen-enriched CNG (bioH_2-CNG) in the context of its application as a fuel for vehicles. In addition, this chapter also highlighted the developments of bioH_2-CNG production and its utilization in the biorefinery framework. Finally, different bioprocesses that can be employed to produce biohythane were discussed advocating that an environment-friendly path to a cleaner future would be the novel method of creating biofuel sources using the biogenic waste/wastewater as potential feedstock.

15.2 RESEARCH TRENDS OF BIOHYTHANE PRODUCTION

In the past 10 years, research on biohythane production from various biogenic waste/wastewater has significantly increased (Figure 15.2). Many studies explored biohythane production from various feedstock such as food waste (Sarkar et al. 2021a; Sarkar & Venkata Mohan 2017; Deheri & Acharya 2022; Ali et al. 2022), distillery spent wash (Pasupuleti & Venkata Mohan 2015), coffee residues, and sugarcane

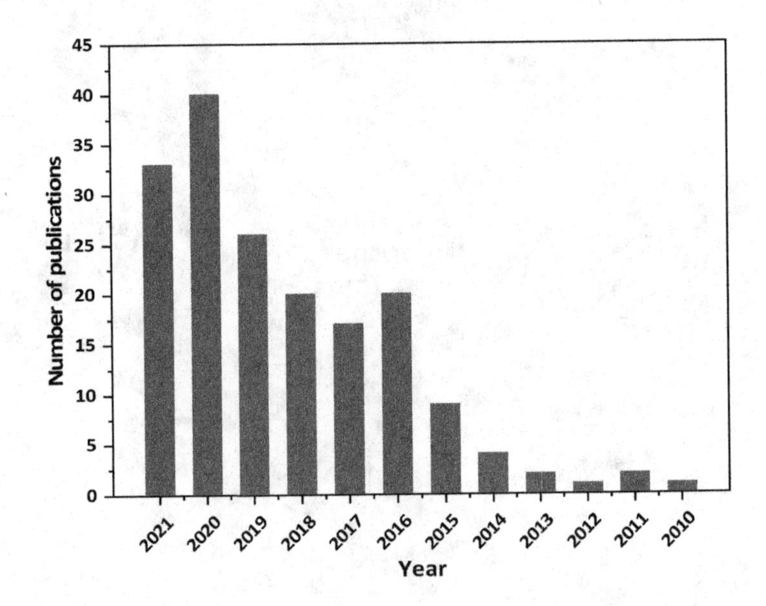

FIGURE 15.2 Statistics of research articles published each year on biohythane from different waste/wastewater as substrate that were published in the past 12 years based on the "Web of Science" database.

vinasse (Pinto et al. 2018), co-digestion of rice straw and pig manure (Liu et al. 2022), swine manure (Nguyen et al. 2022), brewery spent grains (Sarkar et al. 2021b), corn stalk (Li et al. 2020), rubber sheet wastewater (Promnuan et al. 2020), birch sawdust (Sarkar et al. 2022b), and algal biomass (Chen et al. 2020; Jehlee et al. 2017). With 50% biodegradable organic content, food waste followed by municipal waste, algal biomass has been mostly studied towards biohythane production, which provides an environment-friendly approach, avoiding dump it to landfill with simultaneously production of high-value products. From a geographical point of view, China followed by India, USA, Thailand, Malaysia, Italy, Vietnam, South Korea, Taiwan, Japan, Brazil, Netherlands, France, Indonesia, Sweden, Wales, Australia, Egypt, Mauritius, and Mexico are the countries that have published articles on biohythane production, which might be due to realizing the strength and scope of this source of energy for future. The research in this area has emerged with huge economy body.

15.3 PARAMETERS INFLUENCING BIOHYTHANE PRODUCTION

Many studies have identified the factors that enhances biohythane from various organic substrates. Factors that influence the production of biohythane include reactor type, pH, and type of biocatalyst, substrate organic load, temperature, and hydraulic/solid retention time. pH and nature of biocatalyst was mostly studied, which directly influences the bioprocess. Mixed culture fermentation has been extensively studied for biohythane production due to its practical advantage over pure culture (Dahiya et al. 2021). Mixed culture was found to be appropriate even for large-scale application overcoming the associated hurdles of medium sterilization especially for pure culture fermentation, which affects the overall economics of the process (Pasupuleti et al. 2014). A wide range of substrates can be converted to biohythane employing mixed microbial culture. However, the process needs to predetermine for either hydrogen or methane. Specifically, for biohydrogen production, different pretreatment methods can be adopted targeting enrichment of acidogens, simultaneously inhibiting hydrogen-consuming microorganisms such as hydrogenotropic methanogens, homoacetogens, lactic acid bacteria, propionate-producing bacteria, and sulfate reducers in the mixed microbial culture (Rafieenia et al. 2018; Sarkar et al. 2016; Dahiya et al. 2015). Different pretreatment methods such as acid shock, heat shock, fermentation monitoring, pH control, and addition of suitable chemicals (BESA and iodopropane), inhibits methanogens (Sarkar & Venkata Mohan 2016, 2017; Sarkar et al. 2022a). Acidic pH ranging from 3 to 4 or alkaline from 9 to 12 can significantly inhibit methanogenic activity. Heating the mixed culture at 80°C–100°C suppresses methanogens by selectively enriching the spore-forming acidogenic bacteria (*Clostridium spp.*). Chemical substance such as BESA (Sodium 2 bromoethanesulfonate) and iodopropane also inhibits the growth of methanogens. However, it is to be noted that the intensive pretreatment may lead to the partial suppression of bacteria responsible for the hydrolyzing the substrate, the presence of this particular microbes is important to enhance the overall yields when fermenting complex substrates (Rafieenia et al. 2018; Venkata Mohan et al. 2008; Shanthi Sravan et al. 2018). Culture pretreatment strategies for biohydrogen production is mainly associated with microbes, which are spore forming and sporulate when they are subjected to harsh environmental conditions of pH, temperature, irradiation, chemicals, as these microbes are resistant to aggressive physical and chemical conditions (Kannaiah Goud et al. 2014). Whereas on the other side, the non-spore-forming hydrogen-consuming microbes in such conditions is destroyed. Moreover, the spore-forming bacteria are able to germinate and grow when the environmental conditions become suitable.

15.4 BIOHYTHANE PRODUCTION FROM BIOGENIC WASTE/WASTEWATER

15.4.1 Single-Stage Process

In single-stage process, both bio-H_2 and bio-CH_4 are being simultaneously produced in a single reactor. Factors that are need to be critically considered in single-stage process is reactor design, operation strategy, and the most crucial is the mechanisms for a balanced bio-H_2 and bio-CH_4, which need to be retrieved without phase separation (Ta et al. 2020; Pasupuleti & Venkata Mohan 2015; Sarkar & Venkata Mohan 2016). All the four steps (hydrolysis, acidogenesis, acetogenesis, and methanogenesis) occurs in a single reactor/digester, resulting with production of more methane (60–65%), CO_2, and traces of H_2. The major associated issue with single-stage fermentation is acidification of reactor due to accumulation of carboxylic acids produced during acidogenic/acetogenic phase (Dahiya et al. 2018). Carboxylic acids, which generally dominates the reactors, are acetic, propionic, and butyric acids. Accumulation of these metabolites in the reactor steeply decreases the pH that interrupts the process, as the bacteria responsible for H_2 production cannot tolerate excessive accumulation of carboxylic acids. Though single stage biohythane production is challenging, many studies have successfully demonstrated its production from organic waste degradation. Pasupuleti and Venkata Mohan demonstrated single-stage fermentation process for biohythane production from distillery spent-wash (2015). Using an encapsulated hydrogenic and methanogenic bacteria supported by a combination of κ-carrageenan and gelatin (2%/2% w/w), Ta et al. (2020) demonstrated a single-stage dark fermentation process generating H_2-enriched CH_4 (Ta et al. 2020). The strategy was to adjust biomass ratio of the two encapsulated bacteria, which assisted to achieve a H_2 and CH_4 production of 64.6 and 395 mL/L/day, respectively. Vo et al. reported single-stage anaerobic fermentation in two-compartment bioreactor, separating dark fermentation for hydrolysis followed by methanogenesis for biomethane for an enhanced biohythane production (Vo et al. 2019).

15.4.2 Two-Stage Biohythane Production

Production of biohythane in two-stage process is becoming an attracting attention due to its advantage of both waste treatment and biogas production (Hans & Kumar 2019). Two-stage process for biohydrogen and methane production can significantly prepare a gas composition favorable as hythane (H_2: 10%–15%, CH_4: 50%–55%). Over conventional anaerobic digesters, two fermentation systems at two different stages are important, because the dark fermentation that occurs in the first stage (chamber/reactor) produces a significant quantity of biogas rich in hydrogen. Here in this stage, the *Clostridium spp., Thermoanaerobacterium spp., Enterobacter,* and *Bacillus* plays a major role toward accumulation of biohydrogen in the reactor. On the other side, the second stage i.e., acetogenesis followed by methanogenesis occurs producing biogas rich in methane mostly by *Methanosarcina barkeri* and *Methanococcus* along with traces of hydrogen (Battista et al. 2016). The acidogenic phase for biohydrogen production via the acetate and butyrate pathways is optimum in a pH ranging from 5 to 6 with a hydraulic retention time (HRT) range of 1–3 days. The accumulated acetate and butyrate in the fermenting media is then converted to CH_4 and CO_2 by acetoclastic methanogens pH range of 7–8 and optimal HRT of 10–15 days, whereas H_2 is consumed by hydrogenotrophic methanogens (Equations 15.1–15.3).

$$4H_2 + 2CO_2 \rightarrow CH_3COOH + 2H_2O \tag{15.1}$$

$$CH_3COOH \rightarrow CH_4 + CO_2CH_3COOH \rightarrow CH_4 + CO_2 \tag{15.2}$$

$$2CH_3CH_2OH + CO_2 \rightarrow 2CH_3COOH + CH_4 \tag{15.3}$$

The two-stage process can be beneficial to overcome the negative impacts associated with single stage, such as inhibitive compounds in the feedstock, high organic loading rates (OLR), and reduced fermentation time for the overall processes. Towards higher energy recovery and cleaner transport biofuel, organic waste conversion to biohythane via two-stage anaerobic fermentation process could be a promising technology. A traditional anaerobic digester is not very efficient at maximizing the recovery of energy, because biogas contains abundant CO_2 (30%–50%). This way the two-stage process that supports biogas upgradation converting CO_2 via hydrogenotrophic methanogenesis, in particular, has attracted considerable attention (Annie Modestra et al. 2020; Deppenmeier 2002).

15.4.3 Reactors for Biohythane

Several different bioreactor configurations have been used targeting an improved biohythane production. The often-used bioreactors in laboratory are batch reactors; however, these reactors are not practical at industrial scale, as it demands larger and long-term production cycles. On the other side, continuously stirred tank reactor (CSTR) allows a steady-state operation at a long run. Microbial electrolysis cells (MECs) are an emerging technology for the production of hydrogen from degradation of organic wastes at an applied small voltage ranging between 0.4 and 0.8 V (Luo et al. 2017). On the other side, Microbial fuel cell (MFC) is a simple and robust technology that offers direct conversion of organic substrate to electrical energy with simultaneous generation of bioelectricity (Sarkar et al. 2016; Krishna et al. 2014; Sevda et al. 2013). An up-flow air–cathode chamber microbial fuel cell biosensor was used for *in-situ* evaluation of biohydrogen and biomethane towards biohythane production (Huang et al. 2021). Alongside, MFC-based biosensors have been also used to evaluate various parameters related to wastewater, such as dissolved oxygen, biochemical oxygen demand, chemical oxygen demand, volatile fatty acids (VFAs), heavy metals, organic contaminants, and microbial activity (Luo et al. 2021; Zhou et al. 2017; Sravan & Venkata Mohan 2021; Sravan et al. 2020a). In a newly designed tubular MEC reactor with circles stainless steel mesh towards energy recovery, the observed biogas was composed of hydrogen and methane representing biohythane (hydrogen accounted for 12.3 in mixture of hydrogen and methane) (Ma et al. 2021). Microbial electrolysis cell (MEC) with an alkali-pretreated sludge showed biohythane production rate of 0.148 L/L/day, which was about 40% and 80% higher than the MECs operated with untreated raw sludge and anaerobic digestion (Liu et al. 2016). Evaluating two different system configurations for biohythane production for a sustainable energy recovery from wastewater treatment, dual-chamber MEC was found with a stable biohythane composition guaranteeing energy recovery in a long-term operation. Whereas in single-chamber MEC, biocathode favored the growth of hydrogenotrophic methanogen with higher methane and lower hydrogen production, rate which was not appropriate for biohythane composition (Luo et al. 2021). To control the composition of biohythane, a dual cathode bioelectrochemical system was developed through adjusting its external resistance (Li et al. 2020). This BES system was capable of producing both methane in cathode chamber loaded with methanogens and hydrogen in abiotic cathode, which was facilitated with electrons from anode. This kind of BES-based biohythane system can significantly address the energy loss along with CO_2 capture and (Li et al. 2020).

15.5 BIOGAS UPGRADATION

Raw biogas is produced during decomposition of organics is generally composed a of CH_4 (40%–65% v/v), CO_2 (35%–55% v/v) and traces of H_2S (0.1%–3% v/v) and water vapor (Mishra et al. 2021). Utilization of upgraded biogas in fuel cells provides an opportunity to meet the market demand for producing electricity. Two metabolic pathways are involved in biogas upgradation, (1) where hydrogenotrophic methanogens converts CO_2 to methane; (2) homoacetogens converts CO_2 to acetate by Wood-Ljungdahl pathway, which further transformed to methane by acetoclasetic methanoges (Sarkar et al. 2021a) (Equations 15.4–15.7).

$$CO_2 + 4H_2 \rightarrow CH_4 + 2H_2O \tag{15.4}$$

$$CH_3COO^- + H^+ \rightarrow CH_4 + CO_2 \tag{15.5}$$

$$\begin{gathered} 2CO_2 + 4H_2 \rightarrow CH_3COO^- + H^+ + 2H_2O \\ \text{leading to } CH_3COO^- + H^+ \rightarrow CH_4 + CO_2 \end{gathered} \tag{15.6}$$

$$\begin{gathered} CH_3CH_2COO^- + 3H_2O \rightarrow CH_3COO^- + HCO_3^- + H^+ + 3H_2 \\ \text{leading to } CH_3COO^- + H^+ \rightarrow CH_4 \end{gathered} \tag{15.7}$$

This approach not only utilize CO_2 for CH_4 but also enhances substrate conversion and adds to the process returns. For example, the volatile acids produced in high quantity along with CO_2 in during acidogenic fermentation can be upgraded to CH_4 (Sarkar et al. 2021a; Sravan et al. 2021). Moreover, the conversion of CO_2 to fuels and high value chemicals through biological and chemical process is a current trend of research. However, the associated issues with these strategies are supply of hydrogen. Here, in biological systems, an *in-situ* biohydrogen can be used for the process. Moreover, the high CO_2 content in the biogas lowers the heat value of the biogas, and thus it requires an additional treatment to be used as fuel as hythane. To enrichment of the energy of the fuel gas, removal of unimportant gases is crucial, especially when the gas is needed to be transported. While upgrading biogas, Huang et al. (2017) found that electrochemical oxidation lowered the CO_2 concentration from 40% to 15%, while increasing the hydrogen content, increasing the gas's heating value by 25%. Recently technologies such as absorption, membrane filtration, adsorption, cryogenic separation, etc., has been developed for biogas upgradation (Mishra et al. 2021). On the other side, supplementation of conductive materials such as biochar and carbon-based cloth was recently studied (Tang et al. 2020; Zhang et al. 2018). Towards upgrading the biogas into vehicle fuel, Cheng et al. (2016) evaluated the adsorption performance of molecular sieve impregnated with phosphonium-based IL and found that the IL-loaded molecular sieve exhibited greater CO_2 adsorption compared to the pure analogous IL in both pure CO_2 and biohythane atmospheres (Cheng et al. 2016).

15.6 BIOHYTHANE AS FUEL

15.6.1 Engine Fuel

Towards adopting a clean fuel, utilization of natural gas (CNG) has been proposed over conventional diesel, as it mainly consists of methane and a small fraction of light hydrocarbons that does not form harmful emissions. In actual, the source of particulate emissions from vehicles fueled by natural gas vehicles is through combustion of lubricant. In the last decade, a large R&D has been carried exploring hydrogen addition to methane (hydrogen-enriched methane) as well as the design of engine architectures, retaining the capability of the engine performance similar when operated with traditional fuels. In order to reduce vehicular particulate emission, diesel vehicles were started getting replaced with CNG. For instance, many Indian cities such as Mumbai, Delhi started investing in CNG-based buses (Sarkar et al. 2021a). Later, the CNG was replaced with hydrogen enriched CNG (HCNG), in order to enhance the efficiency and reduce the pollution levels. Currently many nations are experimenting to adopt hydrogen enriched CNG (hythane) over natural gas engines. Over the last few years, there has been growing interest in the possibilities offered by biological hydrogen-blending natural gases. The utilization of biohythane in transportation sector is increasing as an alternate to fossil based HCNG. The hydrogen in the hythane significantly increases the burning speed, reduces the carbon emission, improves flame range, and enhances combustion assisting combustion of methane in the engines (Luo et al. 2017; Sarkar et al. 2021a). For the first time, biohythane as transportation fuel was used in 1995 under a project named as Montreal Hythane Bus Project, which showed promising results in NO_X reduction compared to conventional CNG (Hans & Kumar 2019; Bolzonella et al. 2018). Similarly, Sweden and China demonstrated utilization of biohythane in Beijing Hythane Bus Project (Bolzonella et al. 2018). Biohythane has the potential to be used as vehicle fuel and alternative to replace the hythane. The increase in utilization of hythane as fuel is due to (1) the blend mostly doesn't require any changes in the vehicle layout or likely a small modification in the engine, (2) high combustion efficiency, (3) emission of CO_2, and (4) zero emission of NO_x and SO_x, contributing towards a clean environment. Apart from this, biohythane has been used as a fuel in different systems. A combustor modeled by Meziane and Bentebbiche (2019) could effectively decrease NO and CO to 14% and 60%, respectively, from a blend of natural gas and hydrogen in a micro gas turbine (Meziane & Bentebbiche 2019). Panagi et al. (2019), investigated biohythane as a fuel in solid oxide fuel cells (SOFCs) for the first time for generation of electricity and synthesis gas. The input fuel was derived from a two-stage anaerobic digestion process, and the gases were composed of $CH_4/CO_2/H_2$ in vol% of 60/30/10 accounting for an improved performance with 77% increased electrical energy yields due to the presence of H_2 in biohythane (Panagi et al. 2019). Dimopoulos et al. (2008) demonstrated that utilization of hythane improved the efficiency of natural gas internal combustion engines and lowered the well-to-wheel (WTW) emissions (Dimopoulos et al. 2008). The emissions of CO and CH_4 can potentially reduce with H_2 addition. Assessing the effects of different HCNG concentrations on SI engine performance, Hao et al. (2020) observed that CNG with 20% H_2 provided the excellent efficiency. Adding 20% H_2 to a single-cylinder diesel engine improved the combustion of biogas and diesel fuel with 30% less CO release (Bouguessa et al. 2020). Tangoz et al. (2015) tested an Isuzu 3.9 L diesel engine with CNG and HCNG at 9.6, 12.5, and 15 compression rates at full load and 1,500 rpm. Although it is well known that H_2 produces NO_x, interestingly, Tangoz et al. (2015) obtained a better performance at 12.5 compression rates due to reduction in NO_x upon utilization of HCNG in comparison to other CRs (Tangöz et al. 2015). The presence of hydrogen in H_2CNG/hythane plays an important role as electron source for enzymatic reactions in biological methanol production pathway. Hydrogen lowers the inhibiting effects of ammonia, CO_2, and H_2S (Patel et al. 2017). A Hyundai sonata SI commercial engine with a wide range of engine speeds (1,500–6,000 rpm) was tested with HCNG as fuel, comparing its performance with pure CNG and Gasohol-91. Here Wongwuttanasatian et al. (2022) noticed that although torque and brake power are greatly reduced, HCNGs' specific fuel consumption levels are lower than those of pure CNG. Additionally, HCNG fuel greatly reduced the CO, CO_2, and NO_x compared to conventional fuels (Wongwuttanasatian et al. 2022) indicating the advantage of H_2-blended CNG, as fuel utilization of these green fuels can significantly aid the effort to reduce greenhouse gases. When biohythane was used as feed for methanol production by *Methylocella tundrae* immobilized on a solid support, the production was 1.9-fold higher compared to pure CH_4 as a feed (Patel et al. 2017). Targeting conversion of agricultural/municipal wastes to biogas/bio-CNG, Government of India initiated a program to recover biogas/enriched biogas/bioCNG as energy source (Sharma et al. 2021). Indian Government bodies such as Ministry of Petroleum & Natural Gases encouraging blending of H_2 with CNG towards decarbonization, whereas Environment Pollution Prevention & Control Authority (EPCA) is considered transforming the CNG-based buses to H-CNG with an addition of H_2 of 18% to reduced volumetric calorific value (CV) and increased gravimetric CV of the CNG fuel (Sharma et al. 2021). Patel et al. (2021) suggested that biohythane improves gas flame temperature by 0.6%–4.05% when the hydrogen content increased from 25% to 75%, a decrease in CO_2 emission was observed compared to CH_4 or CNG (Patel et al. 2021). The literature indicates that because hydrogen has a higher-octane number than pure CNG, it improves the SI engine's efficiency higher than that of pure CNG. Furthermore, utilizing hydrogen as a blended fuel smooths engine operation, lowering noise and emission levels, and making hydrogen a dependable fuel for storage and transportation.

15.6.2 Household Appliances

Not only in motor vehicles, the blend (hydrogen and methane) has also been used in residential purposes such as cooking and heating connecting with stove and boilers contributing towards decarbonization (Özçelep et al. 2021; Sarkar et al. 2021a; Ozturk et al. 2021; Santhosh et al. 2021). With development of hydrogen production and utilization technologies, its injection into the existing natural gas pipelines is becoming attractive step replacing carbon (Deymi-Dashtebayaz et al. 2019). Currently, combustion devices consume the majority of pipeline gas for electricity generation in power plants, or to provide heat in industrial/commercial and residential applications. When hydrogen was injected into the natural gas pipeline integrated with cooking appliances, the emissions were reduced with increasing hydrogen flow. Here, Zhao et al (2019) found that the cooktop burners were safe with a hydrogen up to 20% by volume (Zhao et al. 2019). Enabling household appliances to use the blend natural gas and hydrogen effectively, a novel blending was developed, which showed a reduced CO_2 emission upon using renewable hydrogen up to 20% (Ozturk & Dincer 2020). Moreover, the varying combustion properties between hydrogen and natural gas, such as wide flammability and high flame speed of hydrogen, the practical possibility of replacing the natural gas with hydrogen directly is restricted. Jones et al. used 34.7% hydrogen into the cooktop burner and suggested that 30 mol% of the natural gas supply in the UK may be replaced, without any modification of the appliance infrastructure (2018). Many studies have examined the complications associated with the blending of hydrogen with natural gas and also suggested the way to utilize the blend (Ogden et al. 2018; Levinsky 2021; Yan et al. 2018). The research suggests that the hydrogen between 5% and 15% require minor modifications to adapt for natural gas. However, from a safety point of view, the fire and explosion risk is similar to natural gas for mixtures less than 20%, and it concerns when the H_2 exceeds above 50%. The recent studies indicated the impacts of this blend (H_2+ natural gas) on the performance and safety of residential appliances. Considering safety, De Vries and coworkers (2017) developed a computational interchangeability analysis that potentially recommends the limits of hydrogen addition, further suggesting that hydrogen addition also depends on the composition of natural gas (de Vries et al. 2017).

15.7 BIOREFINERY FOR BIOHYTHANE: INTEGRATION OF PROCESSES

A biorefinery can be described as a facility that integrates different biomass conversion processes and equipment for the production of fuels and value-added chemicals (Venkata Mohan et al. 2016; Sekoai et al. 2018; Sarkar et al. 2021c; Dahiya et al. 2018; Naresh Kumar et al. 2022). A biorefinery setup is as similar as to the petro refinery dedicated to produce multiple fuels and products from petroleum-based feedstock from different processes (Ghimire et al. 2017). The circular bioeconomy is gaining momentum globally, and biorefineries will be a crucial part of the sustainability framework. Biorefineries will play a significant role as essential components in the circular bioeconomy in the next years. However, it is crucial to design and develop waste-based self-sustaining biorefinery that is sustainable, economically viable, and renewable into account. The biorefinery can potentially convert various feedstock (waste/wastewater) by taking advantage of various processes, thereby maximizing the value of the biomass feedstock (Dahiya et al. 2021; Hemalatha et al. 2019). Integration of process is crucial to enhance volumetric biohydrogen production from conversion of biomass (Figure 15.3). For example, the dark fermentative hydrogen yield seldom exceeds 2 mol/mol glucose, representing that only 7%–15% of the energy from organic waste is converted to hydrogen. The limited hydrogen production is due to formation of carboxylic acids (mainly acetate, propionate, and butyrate) in the fermenting media (Mamimin et al. 2017; Zhang et al. 2011; Luo et al. 2017; Sarkar et al. 2021c). Upon integrating acidogenic fermentation with anaerobic digestion, the biohythane production was 8–14 L of biohythane-documenting energy content of 8–9.08 kJ/g VS from organosolv pretreated birch sawdust (Sarkar et al.

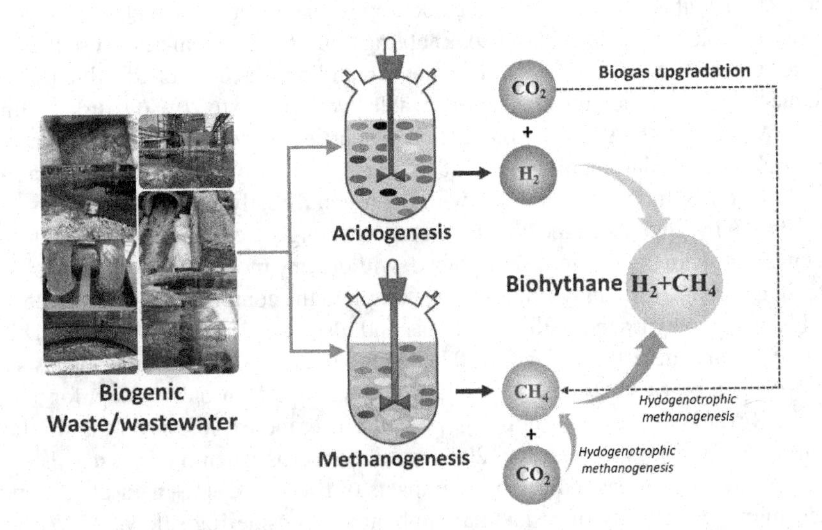

FIGURE 15.3 Biogenic waste/wastewater conversion to biohydrogen (acidogenesis/dark fermentation) and biomethane (methanogenic process) and its integration to biohythane production.

2022b). Here, initially the biohydrogen production was 121.44 mL H_2/g VS_{load}, the untreated carbon (VS) along with a fraction of carboxylic acids was integrated with anaerobic digestion producing 246.01 mLCH_4/g VS_{load}. Here integration offered conversion of acidogenic-fermented solids and carboxylic acids to bioCH_4 by methanogens with an additional energy gain (Sarkar et al. 2022b). This study also signified one of the most prominent ways to utilize the acidogenic carboxylic acids to biomethane contributing to biohythane. Refining the residual biomass from anaerobic solid-state fermentation of *Chlorella sp,* the hydrogen and methane yield was 12.5 and 81 mL/g VS, with biohythane containing 13.4% of hydrogen with energy recovery of total energy of 3.03 kJ/g VS (Lunprom et al. 2019). Coupling of dark fermentation with anaerobic digestion, Ghimire et al. (2017) demonstrated biohythane production alternate to thermochemical process (Ghimire et al. 2017). Proposing a novel strategy, Si et al. (2016) combined biohydrogen and methane production for biohythane production from cornstalk integrating hydrothermal liquefaction (HTL) and two-stage fermentation. Here the cornstalk was initially liquid phase from hydrothermal liquefaction (HTL) was fed into an upflow anaerobic sludge blanket (UASB) and packed bed reactor for biohythane production with an energy recovery was 79% (Si et al. 2016). Biorefinery has the potential to utilize organic wastes—a wide range of organic wastes resolving the issues of waste management and greenhouse gases GHGs emissions. For a large-scale application, the waste/wastewater can be conjugated with the existing anaerobic digestion/sewage treatment plants modifying the input parameters (such as pH, nature of biocatalyst, and substrate load) in a biorefinery approach to have the greatest impact to produce biohydrogen, biomethane, and biohythane, along with platform chemicals.

15.8 TECHNO-ECONOMIC AND LIFE CYCLE ASSESSMENT OF BIOH$_2$-CNG/BIOHYTHANE PRODUCTION

Life cycle assessment (LCA) is a tool for evaluating and identifying the most relevant environmental impacts and hotspots during the life cycle of any product or process (Yadav et al. 2021; Katakojwala & Mohan 2021; Prasad et al. 2020). LCA examines the individual life cycle stage (from raw material to end-of-life product) of the process/system or product concerning specific environmental impacts, based on which the corresponding hotspots are identified (Sarkar et al. 2021c; Singh et al. 2020). As per the International Organization for Standardization (ISO), sustainability analysis offers a consistent tool for assessing any significant or distinctive impact on the outcome. This information can subsequently influence decision-making and affect the viability of the process or product. LCA helps evaluate the cost and environmental benefits of gas fermentation and to addresses its sustainability (Yadav et al. 2021; Sarkar et al. 2021c). Converting the post-hydrothermal liquefaction wastewater generated from human feces to biohythane through two-stage fermentation and catalytic hydrothermal gasification process, and its techno economic analysis depicted that the two-stage fermentation had a higher net energy return but higher cost than catalytic hydrothermal gasification (Si et al. 2019). Life-cycle analysis showed that fossil-based hydrogen production pathways are associated with relatively high carbon footprints compared to green hydrogen production technologies (Valente et al. 2021; Sarkar et al. 2021c). In the future, sustainability assessment results will help identify and optimize process parameters, minimizing costs, and enhancing production efficiency.

15.9 CONCLUSIONS AND PERSPECTIVES

Right from vehicles to industries, hydrogen enriched natural gas/methane/CNG as hythane is demonstrating excellent performance in their combustion devices as a promising low-carbon fuel. The added advantage with utilization with this blend is easy adaption in the existing natural gas-fueled engine infrastructure with minor modification, making it economical compared to hydrogen infrastructure. Thus, utilization of biohydrogen coupled with biomethane in the form of bioH_2-enriched CNG/hythane can be a step toward achieving hydrogen economy. Moreover, the hydrogen currently used in the H-CNG/hythane is derived from fossil-based resources, which is unsustainable. Acidogenic fermentation offers sustainable solutions with renewable hydrogen converting biogenic waste and wastewater with an added advantage of cogeneration of carboxylic acids as platform chemicals. Hydrogen production through biological routes has advanced in the last decade, which need to demonstrate at semi or pilot scale with life cycle and techno-economic analysis for its utilization as fuel. The benefit of producing bioH_2-CNG/biohythane from biogenic waste/wastewater is a sustainable, affordable, and environmentally beneficial process. BioH_2-CNG/biohythane as vehicular fuel results in a greener economy due to its efficient combustion, greater heat efficiency, and clean design. Currently fuels are blended in small proportions with hydrogen to promote efficient engine operation and to overcome the associated drawback, such as lower heating value, lower octane number, and lower specific heat capacity compared to hydrogen. An appropriate biohydrogen blend in bioH_2-CNG assists lowering greenhouse gas emissions. Based on advantages of bioH_2-CNG over fossil-derived fuels in terms of fuel efficiency and pollution reduction, bioH_2-CNG is expected to serve as a next-generation clean biofuel. Despite its several advantage over other fossil fuels such as gasoline and diesel, the current gas distribution infrastructure is the greatest hurdle towards utilization of bioH_2-CNG vehicle fuel.

REFERENCES

Ali, M.M. et al. 2022. Combination of ultrasonic and acidic pretreatments for enhancing biohythane production from tofu processing residue via one-stage anaerobic digestion. *Bioresource Technology* 344: 126244. doi: 10.1016/j.biortech.2021.126244. https://www.sciencedirect.com/science/article/pii/S0960852421015868.

Annie Modestra, J., Katakojwala, R., & Venkata Mohan, S. 2020. CO_2 fermentation to short chain fatty acids using selectively enriched chemolithoautotrophic acetogenic bacteria. *Chemical Engineering Journal* 394: 124759. doi: 10.1016/j.cej.2020.124759. https://www.sciencedirect.com/science/article/pii/S1385894720307506.

Battista, F. et al. 2016. Mixing in digesters used to treat high viscosity substrates: The case of olive oil production wastes. *Journal of Environmental Chemical Engineering* 4 (1): 915–923. doi: 10.1016/j.jece.2015.12.032. https://www.sciencedirect.com/science/article/pii/S2213343715301214.

Birchall, S.J. & Bonnett, N. 2021. Climate change adaptation policy and practice: The role of agents, institutions and systems. *Cities* 108: 103001. doi: 10.1016/j.cities.2020.103001. https://www.sciencedirect.com/science/article/pii/S0264275120313494.

Bolzonella, D. et al. 2018. Recent developments in biohythane production from household food wastes: A review. *Bioresource Technology* 257: 311–319. doi: 10.1016/j.biortech.2018.02.092. https://www.sciencedirect.com/science/article/pii/S096085241830289X.

Bouguessa, R. et al. 2020. Experimental investigation on biogas enrichment with hydrogen for improving the combustion in diesel engine operating under dual fuel mode. *International Journal of Hydrogen Energy* 45(15): 9052–9063.

Cetinkaya, E; Dincer, I & Naterer, G F. 2012. Life cycle assessment of various hydrogen production methods. *International Journal of Hydrogen Energy* 37 (3): 2071–2080. doi: 10.1016/j.ijhydene.2011.10.064. https://www.sciencedirect.com/science/article/pii/S036031991102430X.

Chen, C. et al. 2020. Sustainable biohythane production from algal bloom biomass through two-stage fermentation: Impacts of the physicochemical characteristics and fermentation performance. *International Journal of Hydrogen Energy* 45 (59): 34461–34472. doi: 10.1016/j.ijhydene.2020.02.027. https://www.sciencedirect.com/science/article/pii/S0360319920305085.

Cheng, J. et al. 2016. CO_2 adsorption performance of ionic liquid [P66614][2-Op] loaded onto molecular sieve MCM-41 compared to pure ionic liquid in biohythane/pure CO_2 atmospheres. *Energy & Fuels* 30 (4) (apirilak 21): 3251–3256. doi: 10.1021/acs.energyfuels.5b02857.

Dahiya, S. et al. 2015. Acidogenic fermentation of food waste for volatile fatty acid production with co-generation of biohydrogen. *Bioresource Technology* 182: 103–113. doi: 10.1016/j.biortech.2015.01.007. https://www.sciencedirect.com/science/article/pii/S0960852415000176.

Dahiya, S. et al. 2018. Food waste biorefinery: Sustainable strategy for circular bioeconomy. *Bioresource Technology* 248 (July 2017): 2–12. doi: 10.1016/j.biortech.2017.07.176.

Dahiya, S. et al. 2021. Renewable hydrogen production by dark-fermentation: Current status, challenges and perspectives. *Bioresource Technology* 321: 124354. doi: 10.1016/j.biortech.2020.124354. https://www.sciencedirect.com/science/article/pii/S096085242031628X.

Deheri, C. & Acharya, S.K. 2022. Purified biohythane (biohydrogen+biomethane) production from food waste using $CaO_2 + CaCO_3$ and NaOH as additives. *International Journal of Hydrogen Energy* 47 (5): 2862–2873. doi: 10.1016/j.ijhydene.2021.10.232. https://www.sciencedirect.com/science/article/pii/S0360319921043147.

Deppenmeier, U. 2002. The unique biochemistry of methanogenesis. *Progress in Nucleic Acid Research and Molecular Biology* 71: 223–283. doi: 10.1016/S0079-6603(02)71045-3. https://www.sciencedirect.com/science/article/pii/S0079660302710453.

Deymi-Dashtebayaz, M. et al. 2019. Investigating the effect of hydrogen injection on natural gas thermo-physical properties with various compositions. *Energy* 167: 235–245. doi: 10.1016/j.energy.2018.10.186. https://www.sciencedirect.com/science/article/pii/S0360544218321832.

Dimopoulos, P. et al. 2008. Hydrogen–natural gas blends fuelling passenger car engines: Combustion, emissions and well-to-wheels assessment. *International Journal of Hydrogen Energy* 33 (23): 7224–7236. doi: 10.1016/j.ijhydene.2008.07.012. https://www.sciencedirect.com/science/article/pii/S0360319908008185.

Do, T.N., You, C., & Kim, J. 2022. A CO_2 utilization framework for liquid fuels and chemical production: Techno-economic and environmental analysis. *Energy Environment Science* doi: 10.1039/D1EE01444G.

Ghenai, C. 2019. Combustion of sustainable and renewable biohythane fuel in trapped vortex combustor. *Case Studies in Thermal Engineering* 14: 100498. doi: 10.1016/j.csite.2019.100498. https://www.sciencedirect.com/science/article/pii/S2214157X19301108.

Ghimire, A. et al. 2017. Bio-hythane production from microalgae biomass: Key challenges and potential opportunities for algal bio-refineries. *Bioresource Technology* 241: 525–536. doi: 10.1016/j.biortech.2017.05.156. https://www.sciencedirect.com/science/article/pii/S0960852417308246.

Ghimire, A. et al. 2022. Biohythane production from food waste in a two-stage process: Assessing the energy recovery potential. *Environmental Technology* 43(14): 2190–2196.

Graham, L.A. et al. 2008. Greenhouse gas emissions from heavy-duty vehicles. *Atmospheric Environment* 42 (19): 4665–4681. doi: 10.1016/j.atmosenv.2008.01.049. https://www.sciencedirect.com/science/article/pii/S1352231008001003.

Hans, M., & Kumar, S. 2019. Biohythane production in two-stage anaerobic digestion system. *International Journal of Hydrogen Energy* 44 (32): 17363–17380. doi: 10.1016/j.ijhydene.2018.10.022. https://www.sciencedirect.com/science/article/pii/S0360319918331975.

Hao, D. et al. 2020. Experimental study of hydrogen enriched compressed natural gas (HCNG) engine and application of support vector machine (SVM) on prediction of engine performance at specific condition. *International Journal of Hydrogen Energy* 45(8): 5309–5325.

Hemalatha, M., Sarkar, O., & Venkata Mohan, S. 2019. Self-sustainable azolla-biorefinery platform for valorization of biobased products with circular-cascading design. *Chemical Engineering Journal* 373: 1042–1053. doi: 10.1016/j.cej.2019.04.013. https://www.sciencedirect.com/science/article/pii/S1385894719307910.

Huang, S. et al. 2021. Long-term in situ bioelectrochemical monitoring of biohythane process: Metabolic interactions and microbial evolution. *Bioresource Technology* 332: 125119. doi: 10.1016/j.biortech.2021.125119. https://www.sciencedirect.com/science/article/pii/S0960852421004582.

Huang, Z. et al. 2017. Electrochemical hythane production for renewable energy storage and biogas upgrading. *Applied Energy* 187: 595–600. doi: 10.1016/j.apenergy.2016.11.099. https://www.sciencedirect.com/science/article/pii/S0306261916317305.

Huang, Z. et al. 2006. Experimental study on engine performance and emissions for an engine fueled with natural gas–hydrogen mixtures. *Energy & Fuels* 20 (5) (irailak 1): 2131–2136. doi: 10.1021/ef0600309.

Jehlee, A. et al. 2017. Biohythane production from *Chlorella* sp. biomass by two-stage thermophilic solid-state anaerobic digestion. *International Journal of Hydrogen Energy* 42 (45): 27792–27800. doi: 10.1016/j.ijhydene.2017.07.181. https://www.sciencedirect.com/science/article/pii/S0360319917330173.

Jones, D.R., Al-Masry, W.A., & Dunnill, C.W. 2018. Hydrogen-enriched natural gas as a domestic fuel: An analysis based on flash-back and blow-off limits for domestic natural gas appliances within the UK. *Sustainable Energy & Fuels* 2 (4): 710–723. doi:10.1039/C7SE00598A.

Kannaiah Goud, R.., Sarkar, O., & Venkata Mohan, S. 2014. Regulation of biohydrogen production by heat-shock pretreatment facilitates selective enrichment of *Clostridium* sp. *International Journal of Hydrogen Energy* 39 (14). doi: 10.1016/j.ijhydene.2013.10.046.

Katakojwala, R. & Mohan, S.V. 2021. A critical view on the environmental sustainability of biorefinery systems. *Current Opinion in Green and Sustainable Chemistry* 27: 100392. doi: 10.1016/j.cogsc.2020.100392. https://www.sciencedirect.com/science/article/pii/S2452223620300882.

Krishna, K.V., Sarkar, O. & Venkata Mohan, S. 2014. Bioelectrochemical treatment of paper and pulp wastewater in comparison with anaerobic process: Integrating chemical coagulation with simultaneous power production. *Bioresource Technology* 174: 142–151. doi: 10.1016/j.biortech.2014.09.141. https://www.sciencedirect.com/science/article/pii/S096085241401400X.

Lay, C.-H. et al. 2020. Recent trends and prospects in biohythane research: An overview. *International Journal of Hydrogen Energy* 45 (10): 5864–5873. doi: 10.1016/j.ijhydene.2019.07.209. https://www.sciencedirect.com/science/article/pii/S0360319919328459.

Levinsky, H. 2021. Why can't we just burn hydrogen? Challenges when changing fuels in an existing infrastructure. *Progress in Energy and Combustion Science* 84: 100907. doi: 10.1016/j.pecs.2021.100907. https://www.sciencedirect.com/science/article/pii/S0360128521000058.

Li, J. et al. 2020. A pilot study of biohythane production from cornstalk via two-stage anaerobic fermentation. *International Journal of Hydrogen Energy* 45 (56): 31719–31731. doi: 10.1016/j.ijhydene.2020.08.253. https://www.sciencedirect.com/science/article/pii/S0360319920333097.

Li, X., Liu, G., & He, Z. 2020. Flexible control of biohythane composition and production by dual cathodes in a bioelectrochemical system. *Bioresource Technology* 295: 122270. doi: 10.1016/j.biortech.2019.122270. https://www.sciencedirect.com/science/article/pii/S0960852419315007.

Li, Z. et al. 2021. Methane production from wheat straw pretreated with CaO_2/cellulase. *RSC Advances* 11 (33): 20541–20549. doi:10.1039/D1RA02437J.

Liu, Q. et al. 2016. Multiple syntrophic interactions drive biohythane production from waste sludge in microbial electrolysis cells. *Biotechnology for Biofuels* 9 (1): 162. doi:10.1186/s13068-016-0579-x.

Liu, R. et al. 2022. Effect of mixing ratio and total solids content on temperature-phased anaerobic codigestion of rice straw and pig manure: Biohythane production and microbial structure. *Bioresource Technology* 344: 126173. doi: 10.1016/j.biortech.2021.126173. https://www.sciencedirect.com/science/article/pii/S0960852421015157.

Lunprom, S. et al. 2019. Bio-hythane production from residual biomass of *Chlorella* sp. biomass through a two-stage anaerobic digestion. *International Journal of Hydrogen Energy* 44 (6): 3339–3346. doi: 10.1016/j.ijhydene.2018.09.064. https://www.sciencedirect.com/science/article/pii/S0360319918329100.

Luo, S. et al. 2017. Effective control of biohythane composition through operational strategies in an innovative microbial electrolysis cell. *Applied Energy* 206: 879–886. doi: 10.1016/j.apenergy.2017.08.241. https://www.sciencedirect.com/science/article/pii/S0306261917312266.

Luo, S. et al. 2021. Onset investigation on dynamic change of biohythane generation and microbial structure in dual-chamber versus single-chamber microbial electrolysis cells. *Water Research* 201: 117326. doi: 10.1016/j.watres.2021.117326. https://www.sciencedirect.com/science/article/pii/S0043135421005248.

Ma, X. et al. 2021. Energy recovery from tubular microbial electrolysis cell with stainless steel mesh as cathode. *Royal Society Open Science* 4 (12) (abenduak 22): 170967. doi:10.1098/rsos.170967.

Mamimin, C. et al. 2015. Two-stage thermophilic fermentation and mesophilic methanogen process for biohythane production from palm oil mill effluent. *International Journal of Hydrogen Energy* 40 (19): 6319–6328. doi: 10.1016/j.ijhydene.2015.03.068. https://www.sciencedirect.com/science/article/pii/S0360319915006679.

Mamimin, C. et al. 2017. Effects of volatile fatty acids in biohydrogen effluent on biohythane production from palm oil mill effluent under thermophilic condition. *Electronic Journal of Biotechnology* 29: 78–85. doi: 10.1016/j.ejbt.2017.07.006. https://www.sciencedirect.com/science/article/pii/S0717345817300465.

Meziane, S., & Bentebbiche, A. 2019. Numerical study of blended fuel natural gas-hydrogen combustion in rich/quench/lean combustor of a micro gas turbine. *International Journal of Hydrogen Energy* 44 (29): 15610–15621. doi: 10.1016/j.ijhydene.2019.04.128. https://www.sciencedirect.com/science/article/pii/S0360319919315587.

Mishra, A. et al. 2021. Multidimensional approaches of biogas production and up-gradation: Opportunities and challenges. *Bioresource Technology* 338: 125514. doi: 10.1016/j.biortech.2021.125514. https://www.sciencedirect.com/science/article/pii/S0960852421008543.

Naresh Kumar, A. et al. 2022. Upgrading the value of anaerobic fermentation via renewable chemicals production: A sustainable integration for circular bioeconomy. *Science of The Total Environment* 806: 150312. doi: 10.1016/j.scitotenv.2021.150312. https://www.sciencedirect.com/science/article/pii/S0048969721053894.

Nguyen, T.-T. et al. 2022. Biohythane production from swine manure and pineapple waste in a single-stage two-chamber digester using gel-entrapped anaerobic microorganisms. *International Journal of Hydrogen Energy* doi: 10.1016/j.ijhydene.2022.05.259. https://www.sciencedirect.com/science/article/pii/S0360319922024612.

Ogden, J. et al. 2018. Natural gas as a bridge to hydrogen transportation fuel: Insights from the literature. *Energy Policy* 115: 317–329. doi: 10.1016/j.enpol.2017.12.049. https://www.sciencedirect.com/science/article/pii/S0301421517308741.

Özçelep, Y., Bekdaş, G., & Apak, S. 2021. Investigation of photovoltaic-hydrogen power system for a real house in Turkey: Hydrogen blending to natural gas effects on system design. *International Journal of Hydrogen Energy* 46 (74): 36678–36686. doi: 10.1016/j.ijhydene.2021.08.186. https://www.sciencedirect.com/science/article/pii/S0360319921034315.

Ozturk, M., & Dincer, I. 2020. Development of renewable energy system integrated with hydrogen and natural gas subsystems for cleaner combustion. *Journal of Natural Gas Science and Engineering* 83: 103583. doi: 10.1016/j.jngse.2020.103583. https://www.sciencedirect.com/science/article/pii/S1875510020304376.

Ozturk, M., Midilli, A., & Dincer, I. 2021. Effective use of hydrogen sulfide and natural gas resources available in the Black Sea for hydrogen economy. *International Journal of Hydrogen Energy* 46 (18): 10697–10707. doi: 10.1016/j.ijhydene.2020.12.186. https://www.sciencedirect.com/science/article/pii/S0360319920348217.

Panagi, K. et al. 2019. Highly efficient coproduction of electrical power and synthesis gas from biohythane using solid oxide fuel cell technology. *Applied Energy* 255: 113854. doi: 10.1016/j.apenergy.2019.113854. https://www.sciencedirect.com/science/article/pii/S0306261919315417.

Pasupuleti, S.B., Sarkar, O., & Venkata Mohan, S. 2014. Upscaling of biohydrogen production process in semi-pilot scale biofilm reactor: Evaluation with food waste at variable organic loads. *International Journal of Hydrogen Energy* 39 (14). doi: 10.1016/j.ijhydene.2014.02.034.

Pasupuleti, S.B., & Venkata Mohan, S. 2015. Single-stage fermentation process for high-value biohythane production with the treatment of distillery spent-wash. *Bioresource Technology* 189: 177–185. doi: 10.1016/j.biortech.2015.03.128.

Patel, S.K.S. et al. 2017. Biological methanol production by immobilized Methylocella tundrae using simulated biohythane as a feed. *Bioresource Technology* 241: 922–927. doi: 10.1016/j.biortech.2017.05.160. https://www.sciencedirect.com/science/article/pii/S0960852417308283.

Patel, S.K.S. et al. 2021. Integrating strategies for sustainable conversion of waste biomass into dark-fermentative hydrogen and value-added products. *Renewable and Sustainable Energy Reviews* 150: 111491. doi: 10.1016/j.rser.2021.111491. https://www.sciencedirect.com/science/article/pii/S1364032121007723.

Patil, A.C., & Brown, K. 2008. Trends in emission standards and the implications for bus fleet management: Technology assessment for Brisbane Transport. in. *2008 First International Conference on Infrastructure Systems and Services: Building Networks for a Brighter Future (INFRA)*, 1–6. doi:10.1109/INFRA.2008.5439637.

Pinto, M.P.M. et al. 2018. Co–digestion of coffee residues and sugarcane vinasse for biohythane generation. *Journal of Environmental Chemical Engineering* 6 (1): 146–155. doi: 10.1016/j.jece.2017.11.064. https://www.sciencedirect.com/science/article/pii/S221334371730622X.

Prasad, S. et al. 2020. Sustainable utilization of crop residues for energy generation: A life cycle assessment (LCA) perspective. *Bioresource Technology* 303: 122964. doi: 10.1016/j.biortech.2020.122964. https://www.sciencedirect.com/science/article/pii/S0960852420302339.

Prashanth Kumar, C. et al. 2019. Bio-Hythane production from organic fraction of municipal solid waste in single and two stage anaerobic digestion processes. *Bioresource Technology* 294: 122220. doi: 10.1016/j.biortech.2019.122220. https://www.sciencedirect.com/science/article/pii/S0960852419314506.

Promnuan, K. et al. 2020. Simultaneous biohythane production and sulfate removal from rubber sheet wastewater by two-stage anaerobic digestion. *International Journal of Hydrogen Energy* 45 (1): 263–274. doi: 10.1016/j.ijhydene.2019.10.237. https://www.sciencedirect.com/science/article/pii/S0360319919341345.

Rafieenia, R., Lavagnolo, M.C., & Pivato, A. 2018. Pre-treatment technologies for dark fermentative hydrogen production: Current advances and future directions. *Waste Management* 71: 734–748. doi: 10.1016/j.wasman.2017.05.024. https://www.sciencedirect.com/science/article/pii/S0956053X1730346X.

Ridell, B. 2004. Malmo hydrogen and CNG/hydrogen filling station and Hythane bus project. https://www.osti.gov/etdeweb/biblio/20577224.

Sandalcı, T. et al. 2019. Effect of hythane enrichment on performance, emission and combustion characteristics of an ci engine. *International Journal of Hydrogen Energy* 44 (5): 3208–3220. doi: 10.1016/j.ijhydene.2018.12.069. https://www.sciencedirect.com/science/article/pii/S0360319918340011.

Santhosh, J., Sarkar, O., & Venkata Mohan, S. 2021. Green Hydrogen-Compressed Natural Gas (bio-H-CNG) production from food waste: Organic Load Influence on Hydrogen and Methane Fusion. *Bioresource Technology*: 125643. doi: 10.1016/j.biortech.2021.125643. https://www.sciencedirect.com/science/article/pii/S0960852421009846.

Sarkar, O. et al. 2016. Regulation of acidogenic metabolism towards enhanced short chain fatty acid biosynthesis from waste: Metagenomic profiling. *RSC Advances* 6 (22): 18641–18653. doi:10.1039/c5ra24254a.

Sarkar, O. et al. 2020. Salinity induced acidogenic fermentation of food waste regulates biohydrogen production and volatile fatty acids profile. *Fuel* 276: 117794. doi: 10.1016/j.fuel.2020.117794. https://www.sciencedirect.com/science/article/pii/S0016236120307900.

Sarkar, O. et al. 2021a. Green hythane production from food waste: Integration of dark-fermentation and methanogenic process towards biogas up-gradation. *International Journal of Hydrogen Energy* 46 (36): 18832–18843. doi: 10.1016/j.ijhydene.2021.03.053. https://www.sciencedirect.com/science/article/pii/S0360319921008995.

Sarkar, O. et al. 2021b. Influence of initial uncontrolled pH on acidogenic fermentation of brewery spent grains to biohydrogen and volatile fatty acids production: Optimization and scale-up. *Bioresource Technology* 319: 124233. doi: 10.1016/j.biortech.2020.124233. https://www.sciencedirect.com/science/article/pii/S0960852420315078.

Sarkar, O. et al. 2022a. Green hydrogen and platform chemicals production from acidogenic conversion of brewery spent grains co-fermented with cheese whey wastewater: Adding value to acidogenic CO_2. *Sustainable Energy & Fuels.* doi:10.1039/D1SE01691A.

Sarkar, O. et al. 2022b. Organosolv pretreated birch sawdust for the production of green hydrogen and renewable chemicals in an integrated biorefinery approach. *Bioresource Technology* 344: 126164. doi: 10.1016/j.biortech.2021.126164. https://www.sciencedirect.com/science/article/pii/S0960852421015066.

Sarkar, O., Butti, S.K., & Venkata Mohan, S. 2017. Acidogenesis driven by hydrogen partial pressure towards bioethanol production through fatty acids reduction. *Energy* 118: 425–434. doi: 10.1016/j.energy.2016.12.017. https://www.sciencedirect.com/science/article/pii/S0360544216318217.

Sarkar, O., Katakojwala, R., & Venkata Mohan, S. 2021. Low carbon hydrogen production from a waste-based biorefinery system and environmental sustainability assessment. *Green Chemistry* 23 (1): 561–574. doi: 10.1039/D0GC03063E.

Sarkar, O. & Venkata Mohan, S. 2016. Deciphering acidogenic process towards biohydrogen, biohythane, and short chain fatty acids production: Multi-output optimization strategy. *Biofuel Research Journal* 3 (3): 458–469. doi:10.18331/BRJ2016.3.3.5. https://www.biofueljournal.com/article_31936.html.

Sarkar, O. & Venkata Mohan, S. 2017. Pre-aeration of food waste to augment acidogenic process at higher organic load: Valorizing biohydrogen, volatile fatty acids and biohythane. *Bioresource Technology* 242: 68–76. doi: 10.1016/j.biortech.2017.05.053. https://www.sciencedirect.com/science/article/pii/S0960852417307071.

Sekoai, P.T. et al. 2018. Integrated system approach to dark fermentative biohydrogen production for enhanced yield, energy efficiency and substrate recovery. *Reviews in Environmental Science and Bio/Technology* 17 (3): 501–529. doi:10.1007/s11157-018-9474-1.

Sevda, S. et al. 2013. High strength wastewater treatment accompanied by power generation using air cathode microbial fuel cell. *Applied Energy* 105: 194–206. doi: 10.1016/j.apenergy.2012.12.037. https://www.sciencedirect.com/science/article/pii/S0306261912009282.

Shanthi Sravan, J et al. 2018. Electrofermentation of food waste – Regulating acidogenesis towards enhanced volatile fatty acids production. *Chemical Engineering Journal* 334: 1709–1718. doi: 10.1016/j.cej.2017.11.005. https://www.sciencedirect.com/science/article/pii/S1385894717319174.

Sharma, H.; Dhir, A., & Mahla, S.K. 2021. Application of clean gaseous fuels in compression ignition engine under dual fuel mode: A technical review and Indian perspective. *Journal of Cleaner Production* 314: 128052. doi: 10.1016/j.jclepro.2021.128052. https://www.sciencedirect.com/science/article/pii/S0959652621022708.

Si, B.-C. et al. 2016. Continuous production of biohythane from hydrothermal liquefied cornstalk biomass via two-stage high-rate anaerobic reactors. *Biotechnology for Biofuels* 9 (1): 254. doi:10.1186/s13068-016-0666-z.

Si, B. et al. 2019. Biohythane production of post-hydrothermal liquefaction wastewater: A comparison of two-stage fermentation and catalytic hydrothermal gasification. *Bioresource Technology* 274: 335–342. doi: 10.1016/j.biortech.2018.11.095. https://www.sciencedirect.com/science/article/pii/S0960852418316328.

Singh, A.D. et al. 2020. Life-cycle assessment of sewage sludge-based large-scale biogas plant. *Bioresource Technology* 309: 123373. doi: 10.1016/j.biortech.2020.123373. https://www.sciencedirect.com/science/article/pii/S0960852420306453.

Sravan, J.S., et al. 2020a. Cathodic selenium recovery in bioelectrochemical system: Regulatory influence on anodic electrogenic activity. *ACS ES&T Water* 399 (abenduak 6): 122843. doi:10.1021/acsestwater.1c00224.

Sravan, J.S., Sarkar, O., & Venkata Mohan, S. 2020b. Electron-regulated flux towards biogas upgradation – triggering catabolism for an augmented methanogenic activity. *Sustainable Energy & Fuels* 4 (2): 700–712. doi:10.1039/C9SE00604D.

Sravan, J.S., Tharak, A., & Venkata Mohan, S. 2021. Chapter 1- Status of biogas production and biogas upgrading: A global scenario. In Aryal, N., et al. (arg.), 3–26. Academic Press. doi: 10.1016/B978-0-12-822808-1.00002-7. https://www.sciencedirect.com/science/article/pii/B9780128228081000027.

Sravan, J.S., & Venkata Mohan, S. 2021. Bioelectrocatalytic reduction of tellurium oxyanions toward their cathodic recovery: Concentration dependence and anodic electrogenic activity. *ACS ES&T Water* (abenduak 6). doi: 10.1021/acsestwater.1c00224.

Ta, D.T. et al. 2020. Biohythane production via single-stage anaerobic fermentation using entrapped hydrogenic and methanogenic bacteria. *Bioresource Technology* 300: 122702. doi: 10.1016/j.biortech.2019.122702. https://www.sciencedirect.com/science/article/pii/S0960852419319315.

Tang, S. et al. 2020. The role of biochar to enhance anaerobic digestion: A review. *Journal of Renewable Materials*. doi: 10.32604/jrm.2020.011887.

Tangöz, S. et al. 2015. Effects of compression ratio on performance and emissions of a modified diesel engine fueled by HCNG. *international Journal of Hydrogen Energy* 40 (44): 15374–15380.

Valente, A., Iribarren, D., & Dufour, J. 2021. Harmonised carbon and energy footprints of fossil hydrogen. *International Journal of Hydrogen Energy* 46 (33): 17587–17594. doi: 10.1016/j.ijhydene.2020.03.074. https://www.sciencedirect.com/science/article/pii/S0360319920310107.

Venkata Mohan, S. et al. 2016. Waste biorefinery models towards sustainable circular bioeconomy: Critical review and future perspectives. *Bioresource Technology* 215. doi: 10.1016/j.biortech.2016.03.130.

Venkata Mohan, S., Lalit Babu, V., & Sarma, P.N. 2008. Effect of various pretreatment methods on anaerobic mixed microflora to enhance biohydrogen production utilizing dairy wastewater as substrate. *Bioresource Technology* 99 (1): 59–67. doi: 10.1016/j.biortech.2006.12.004.

Vo, T.-P., Lay, C.-H., & Lin, C.-Y. 2019. Effects of hydraulic retention time on biohythane production via single-stage anaerobic fermentation in a two-compartment bioreactor. *Bioresource Technology* 292: 121869. doi: 10.1016/j.biortech.2019.121869. https://www.sciencedirect.com/science/article/pii/S0960852419310995.

de Vries, H., Mokhov, A.V., & Levinsky, H.B. 2017. The impact of natural gas/hydrogen mixtures on the performance of end-use equipment: Interchangeability analysis for domestic appliances. *Applied Energy* 208: 1007–1019. doi: 10.1016/j.apenergy.2017.09.049. https://www.sciencedirect.com/science/article/pii/S0306261917313302.

Wang, J. et al. 2009. Numerical study of the effect of hydrogen addition on methane–air mixtures combustion. *International Journal of Hydrogen Energy* 34 (2): 1084–1096. doi: 10.1016/j.ijhydene.2008.11.010. https://www.sciencedirect.com/science/article/pii/S0360319908014997.

Wang, L.-Q. et al. 2019. The influence of an orifice plate on the explosion characteristics of hydrogen-methane-air mixtures in a closed vessel. *Fuel* 256: 115908. doi: 10.1016/j.fuel.2019.115908. https://www.sciencedirect.com/science/article/pii/S0016236119312608.

Wongwuttanasatian, T., Jankoom, S., & Velmurugan, K. 2022. Experimental performance investigation of an electronic fuel injection-SI engine fuelled with HCNG (H_2+CNG) for cleaner transportation. *Sustainable Energy Technologies and Assessments* 49: 101733.

Yadav, P. et al. 2021. Environmental impact and cost assessment of a novel lignin production method. *Journal of Cleaner Production* 279: 123515. doi: 10.1016/j.jclepro.2020.123515. https://www.sciencedirect.com/science/article/pii/S0959652620335605.

Yan, F., Xu, L., & Wang, Y. 2018. Application of hydrogen enriched natural gas in spark ignition IC engines: From fundamental fuel properties to engine performances and emissions. *Renewable and Sustainable Energy Reviews* 82: 1457–1488. doi: 10.1016/j.rser.2017.05.227. https://www.sciencedirect.com/science/article/pii/S136403211730864X.

Zhang, C., Lv, F.-X., & Xing, X.-H. 2011. Bioengineering of the Enterobacter aerogenes strain for biohydrogen production. *Bioresource Technology* 102 (18): 8344–8349. doi: 10.1016/j.biortech.2011.06.018. https://www.sciencedirect.com/science/article/pii/S0960852411008212.

Zhang, L., Zhang, J., & Loh, K.-C. 2018. Activated carbon enhanced anaerobic digestion of food waste – Laboratory-scale and Pilot-scale operation. *Waste Management* 75: 270–279. doi: 10.1016/j.wasman.2018.02.020. https://www.sciencedirect.com/science/article/pii/S0956053X18300862.

Zhang, Y., Wu, J., & Ishizuka, S. 2009. Hydrogen addition effect on laminar burning velocity, flame temperature and flame stability of a planar and a curved CH_4–H_2–air premixed flame. *International Journal of Hydrogen Energy* 34 (1): 519–527. doi: 10.1016/j.ijhydene.2008.10.065. https://www.sciencedirect.com/science/article/pii/S0360319908014018.

Zhao, Y., McDonell, V., & Samuelsen, S. 2019. Influence of hydrogen addition to pipeline natural gas on the combustion performance of a cooktop burner. *International Journal of Hydrogen Energy* 44 (23): 12239–12253. doi: 10.1016/j.ijhydene.2019.03.100. https://www.sciencedirect.com/science/article/pii/S0360319919311061.

Zheng, K. et al. 2017. Experimental study on premixed flame propagation of hydrogen/methane/air deflagration in closed ducts. *International Journal of Hydrogen Energy* 42 (8): 5426–5438. doi: 10.1016/j.ijhydene.2016.10.106. https://www.sciencedirect.com/science/article/pii/S0360319916331780.

Zhou, T. et al. 2017. Microbial fuels cell-based biosensor for toxicity detection: A review. *Sensors*. doi: 10.3390/s17102230.

16 Effect of Co-Digestion and Pretreatment on the Bio-Hythane Production

Palas Samanta and Sukanta Mahavidyalaya
University of North Bengal

Sukhendu Dey
The University of Burdwan

Debajyoti Kundu
CSIR—National Environmental Engineering Research Institute (CSIR—NEERI)

Apurba Ratan Ghosh
The University of Burdwan

CONTENTS

DOI: 10.1201/9781003197737-22

16.1 INTRODUCTION

Fossil fuels are the backbone of a country for meeting the energy demands. Currently, fossil fuels contribute over 90% of world's energy consumption. This dependence caused environmental deterioration and climate change because of greenhouse gas (GHGs) emissions, and simultaneously exhausts natural sources/fuel reserves (Guo et al. 2015). According to an estimate, if current fossil energy exploration practice persists, among fossil fuels only coal will remain after 2042 (Shanmugam et al. 2021). Accordingly, it is expected that carbon dioxide level in atmosphere would increase up to 445 ppm by 2050 from 415 ppm in 2020 and may even reach up to 700 ppm by 2100 from manufacturing industries and electricity production units only, which will contribute additional 3.7°C temperature elevation on global climate (Islam et al. 2020). This has triggered to quest for alternative sustainable energy source to meet the ever-increasing global energy demands.

Bio-hythane, in this regard, has emerged as a promising alternative to fossil-based fuels due to their cleanliness, high calorific value, carbon neutrality, and reduced environmental impacts, in particularly global climate change and to ensure energy security (Romagnoli et al. 2011). Bio-hythane is a hydrogen–methane blend, with typical mixture of 10%–15% H_2, 30%–40% CO_2, and 50%–55% CH_4. The hydrogen concentration in hydrogen–methane blend sometimes can reach up to 30% by v/v. The term *hythane* (methane and hydrogen mixture) was first coined by Hydrogen Component Inc. (HCI) in early 90s and was trademarked as HCNG or methagen by Eden Innovations (2010), while the term "bio-hythane" was used to denote hythane production especially from organic substrates like agricultural residues and food wastes (Liu et al. 2018). Hythane is considered as a backbone for the development of

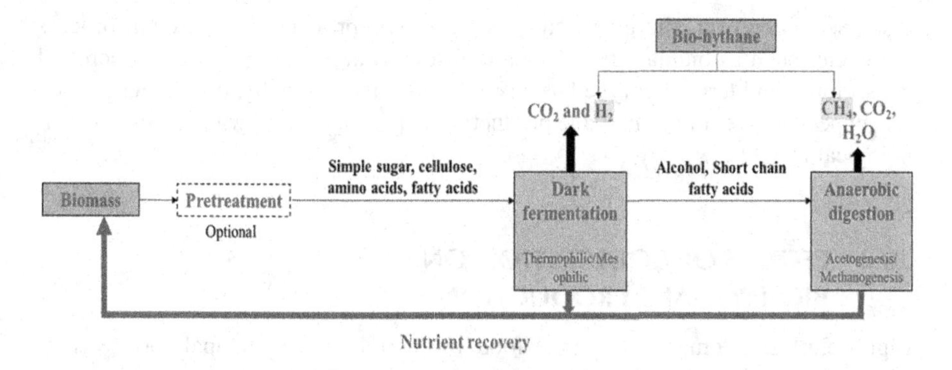

FIGURE 16.1 Schematic presentation of bio-hythane production process.

hydrogen-based society (Liu et al. 2013). Bio-hythane can be produced by combined dark fermentation and anaerobic digestion techniques, and this method is considered as the most promising method of bio-hythane production (Shanmugam et al. 2021). Generally, bio-hythane production is carried out following two-stage system. Figure 16.1 schematically represents bio-hythane production process. In the first stage, H_2 fermentation is achieved by dark fermentation using acidogenic bacteria through acidogenesis pathway, while the next stage is employed for methanogenic digestion of resultant volatile fatty acids (VFAs) mixture in presence of using methanogenic bacteria and archaea consortium (Krishnan et al. 2018). The prime objective of using two-stage system was to separate acidogenesis and methanogenesis in order to maximize the yield. Recently, single-stage reactor system is also used for bio-hythane production, but the process has several drawbacks like lower yield, need additional energy input, and instability of the process (Suksong et al. 2015; Krishnan et al. 2018).

Additionally, a wide range of substrates ranging from first-generation substrates to fourth-generation substrates can be used by two-stage system to produce bio-hythane (Liu et al. 2013). In particularly, lignocellulosic biomass and algal biomass has recently gained scientific attention and considered as potential alternatives for sustainable energy recovery due to their several advantageous characteristics (Monlau et al. 2012; Ghimire et al. 2017). Bio-hythane production still is dependent on substrate type, substrate's biochemical composition, anaerobic biodegradability, and process scale up (Meena et al. 2019). Besides, the bio-hythane yield is regulated by factors like organic loading rate, pH and temperature changes, microbial consortium, fermenter type, *etc.* Accordingly, researchers and scientists have suggested different treatment options, which has not yet received attention as it deserves such as co-digestion and biomass pretreatment. Co-digestion improves nutrient requirement for sustainable bio-hythane production, while biomass pretreatment improves the biodegradability of substrate (Sen et al. 2016). Accordingly, this book chapter firstly summarizes the role of codigestion for improving the efficiency of bio-hythane yield. Secondly, this book chapter addressed the role of different pretreatment methods to improve the bio-hythane yield either independently or combinedly. Additionally, their suitability with regard to energy efficiency, yield, and environmental sustainability was discussed. Moreover, the major advancement of improving yield through process parameter such as biomass type, organic loading rate, pH, temperature, microbial consortium, and fermenter type has also been discussed. Finally, the challenges and recent perspectives of bio-hythane production will be addressed with respect to technological insight.

16.2 EFFECTS OF CODIGESTION ON BIO-HYTHANE PRODUCTION

Lignocellulosic biomass, household food wastes (HFWs), municipal solid wastes, sewage sludge, and algae has emerged as promising substrates for bio-hythane production due to their continuous stable supply and higher (> 60% dry matter) organic content (Yang et al. 2015). However, the commercialization of bio-hythane production from diverse substrates and chemical composition remains questionable, as substrates mono-digestion is not enough to optimize two-stage anaerobic digestion (AD) process (Zhang et al. 2013, 2017). The mono-digestion of substrates in many AD plants produced relatively lower yield efficiency due to unbalanced nutrient profiles, VFAs accumulation, and AD product inhibition. It is very important to supply nutrients continuously for efficient metabolism of microorganism in order to improve the bio-hythane yield. To solve these problems codigestion emerged as promising tool. Codigestion is a very common traditional strategy to optimize the production process, especially two-stage AD system. Codigestion is a process of simultaneous anaerobic decomposition of two or more organic substrates mixture (Gottardo et al. 2015). Codigestion continuously supply nutrients to the microorganisms for their efficient metabolism (Bolzonella et al. 2018). Additionally, even in presence of seasonality substrates the codigestion is an excellent technique to provide continuous feed to the reactors. Besides, the prime advantages of anaerobic codigestion are reduction of toxicity of substrates, improves nutrient balance, enhances organic loading rate, establishes positive synergism among microorganisms, and finally improves bio-hythane yield (Gottardo et al. 2015).

Sewage sludge (activated) is emerged as an excellent substrate to improve the bio-hythane production due to their stable source and high organic content. Sewage generally originates from three community sources: (1) industrial wastewater, (2) domestic wastewater, originated from toilets, bathrooms, washing, cooking, *etc.*, and (3) rain water (Seghezzo 1998). Hydrogen production from sewage sludge fermentation is not commercially acceptable, as hydrogen yield (10 and 90 mL/gVS) is comparatively lower than HFW (100250 mL/gVS), macroalga (29.5–158 mL/gVS), and crude glycerol (29.2–219.1 mL/gVS) (Yang et al. 2015). The prime constraints are lower C/N value (4–9 *vs* standard optimal, 12–17), lower carbohydrate level (<10% dry weight), and high ammonia formation (Yang et al. 2015; Gottardo et al. 2015). Likely, Goberna et al. (2018) demonstrated lower AD inhibition by municipal sewage sludge compared with industrial sludge due to lower heavy metals presence, but the microbial composition is not influenced by sludge type, rather it depends on HRT and ammonia concentration. Contrarily, anaerobically codigestion of organic waste with activated sewage sludge is a standard technique for generating bio-hythane in

Europe currently. For example, codigestion of activated sludge anaerobically with OFMSW (organic fraction of municipal solid waste) biowaste is proved to be advantageous, as nitrogen (N) content in activated sludge caused nutrient deficit condition in other substrates, *i.e.*, OFMSW, which lead to higher degradation of OFMSW for bio-hythane production (Mata-Alvarez et al. 2011). On the other hand, several authors used two-stage AD system feeding with food waste and activated sludge to improve the bio-hythane yield and to remove organics (Zhang et al. 2017; Liu et al. 2019). Náthia-Neves et al. (2018) developed mesophilic codigestion vinasse—a sugarcane waste and restaurant food waste. The results demonstrated that codigestion is very effective for lowering total solids (52%) and total volatile solids (64%) as well as effective substrate for bio-hythane production (72.7%). Recently, codigestion with household food waste has emerged as the finest method to improve the bio-hythane yield due to high carbohydrates content (Cheng et al. 2016). For instance, Maragkaki et al. (2018) demonstrated that codigestion of HFWs (addition @ 5%) with activated sewage improved biogas yield approx. 150 v/v with elevated methane level (60%–70% v/v) and simultaneously improved removal efficiency of organic matter from 45% to 55%. Cheng et al. (2016) demonstrated that anaerobically codigestion of HFW with activated sludge under mesophilic condition improved the sewage sludge AD yield about 174 L/kg VS hydrogen and 264 L/kg VS methane.

More recently, palm oil mill effluent (POME) has emerged as a promising substrate. The POME codigestion with activated sewage sludge anaerobically improved the bio-hythane production, approximately 67%–114% compared with POME digestion alone (Mamimin et al. 2019). On the other hand, Suksong et al. (2015) co-digested POME with oil palm industry-derived products like empty fruit bunch (EFB) and decanter cake (DC) for higher economical production of bio-hythane.

16.3 EFFECTS OF PRETREATMENT ON BIO-HYTHANE PRODUCTION

In the previous section we have discussed that lignocellulosic biomass, household food wastes (HFWs), municipal solid wastes, sewage sludge, and algae serve as promising substrates for bio-hythane production. Typically, the use of lignocellulosic biomass (hemicelluloses, celluloses, hexoses, pentoses are the main constituents) for bio-hythane production is increasing due to presence of 70%–80% polysaccharides sugars. However, to extract polysaccharide special treatment is very essential. Recently, algae have emerged as a potential feedstock material for bioenergy production due to high lipid content and its homogeneous composition. Similar to lignocellulosic biomass, algae have some problem like salinity, sand accumulation, and more interestingly they grow very slowly. Accordingly, pretreatment is obligatory to maximize the yield through breakdown of substrates' recalcitrance cell wall matrix. Additionally, pretreatment method improves the porosity of biomass and subsequent surface area, resulting higher efficiency. The common pretreatment methods for bio-hythane production are as follows: physical (thermal), mechanical (microwave, ultrasonication), chemical (acid or alkaline), chemico-physical (thermochemical, hot compressed water), and biological (microbes or their enzymes). Figure 16.2

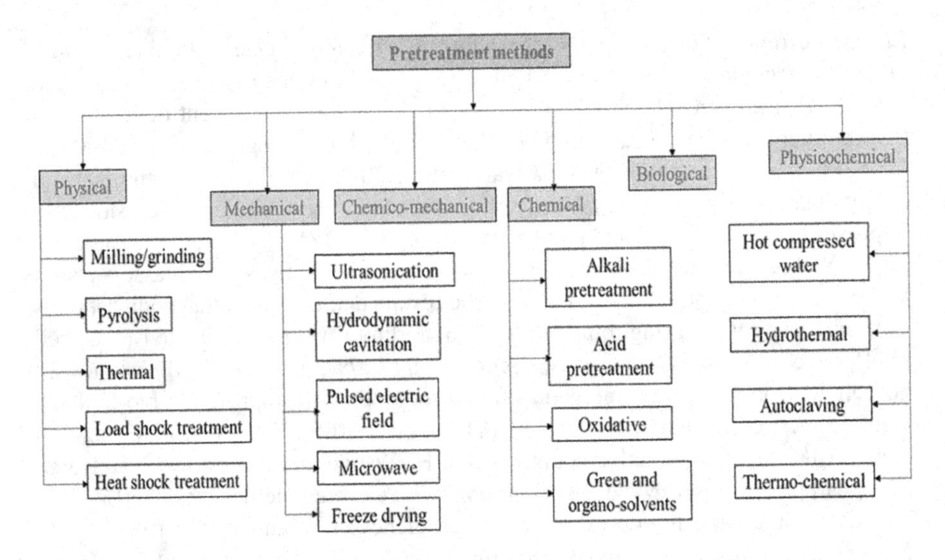

FIGURE 16.2 Different pretreatment methods for bio-hythane production.

represents different pretreatment methods for bio-hythane production. The role of different pretreatment methods for bio-hythane production are discussed below:

16.3.1 Physical Pretreatment Methods

16.3.1.1 Milling/Grinding

Physical treatment is generally used for size reduction. Milling, also called grinding, is the most widely used physical method, and it can be performed by either ball-, hammer-, roll-, colloid-, two-roll-, disk- milling, or vibratory milling. A high-speed shear forces is generally applied in the presence of steel or quartz beads or ceramic or glass to disrupt the biomass (Onumaegbu et al. 2018). Milling method yielded products usually of 0.2–2 mm size ranges. Among different milling methods, vibratory ball milling proved to be excellent method to break down cellulose crystallinity and biomass digestibility (Bhatia et al. 2012). Additionally, milling method produced less amount of inhibitory compounds like furfural, hydroxy methyl furfural, *etc.* The efficiency of this process can be improved by selecting small-sized beads, 0.3–0.4 mm. Because small-sized beads yield more VFA, which inhibits bio-hythane production (Shanmugam et al. 2021). Additionally, the power requirement of milling methods depends on biomass type and particle size (Bhatia et al. 2012). Generally, high energy is required for *milling/grinding* methods. Accordingly, it is very essential to choose suitable milling methods based on biomass type for efficient bio-hythane production.

16.3.1.2 Thermal Pretreatment

Thermal pretreatment is another common technique to hydrolyze a wide range of biomass. Thermal treatment induces biomass liquification process in order to improve anaerobic digestion process (Onumaegbu et al. 2018). The prime advantages of thermal pretreatment are higher disintegration capacity of cell membranes, pathogen removal capacity, reduced viscosity of substrate, improved dewatering performance, and enhanced organic compounds solubilization (Ariunbaatar et al. 2014; Onumaegbu et al. 2018). Generally, two temperature ranges are used to disintegration cell membranes, namely, low-temperature thermal treatment (<110°C) and high-temperature thermal treatment (>110°C) (Ariunbaatar et al. 2014). Low-temperature thermal treatment generally induces deflocculation of macromolecules instead of degradation of biomass. Likely, surplus sludge pretreatment at 90°C is not efficient to break down filaments, but affected by thermal pretreatment and has profound effect on pathogen removal (Barjenbruch and Kopplow 2003). Appels et al. (2010) documented 20 times higher bio-hythane for sludge pretreatment at 90°C compared with sludge pretreatment at 70°C. On the other hand, high temperature thermal treatment induces both biomass degradation and biomass solubilization. The combined thermal pretreatment and dark fermentation improves hydrogen production and solubility of nonhydrogen-producing microbes, but simultaneously lowered inoculum supplementary dosages. Additionally, high-temperature method enhances the growth of spore-forming bacteria but reduces rapid accumulation of organic acids (Karthikeyan et al. 2018). Ma et al. (2011) documented enhanced biomethane production (24%) by food waste pretreatment at 120°C compared to without treatment. Pagliaccia et al. (2018) documented that thermal pretreatment (at 134°C for 20 min) enhanced hydrogen production up to 30%. This treatment also enhances the soluble fraction of chemical oxygen demand (68.4–92 g/L), carbohydrates (12–19 gCOD/L), and protein (8–12 gCOD/L) concentrations. Besides these advantageous applications, thermal pretreatment has some drawbacks such as it reduces volatile organics and bio-hythane production potential in case of biodegradable substrates. Therefore, it is clear that the effects of thermal pretreatment on bio-hythane production depends on substrate type and temperature range.

16.3.1.3 Pyrolysis

Pyrolysis is another widely used physical pretreatment method based on thermal treatment. Biomass in this process is treated around 300°C (or above 300°C) to hydrolyze them for bio-hythane production. The end-products of this process are again acid treated to hydrolyze carbohydrates fraction into glucose, which is fermented to produce bio-hythane. For example, sludge pyrolysis enhanced hydrogen yield during fermentation (Stanislaus et al. 2018). The pyrolytic treatment (at 108°C for 30 minutes) of blue-green microalgae *Arthrospira platensis* and *Chlorella* sp. In the presence of 0.5 atm pressure enhanced fermentation speed and subsequently elevated hydrogen yield approximately 60% (Efremenko et al. 2012; Giang et al. 2019). During pyrolytic treatment of cellulose-rich feedstock 1,6-anhydro-b-D-glucopyranose is produced and used as fermentable feed. Accordingly, combined pyrolysis and fermentation technique used to make bio-hythane production process economical. Although the method is widely applicable for degradation of large number of substrates, the major drawback of this method is high energy and consumption and operating cost. Accordingly, to scale up the process economically further studies are needed to minimize the operating cost and to consider this method as effective treatment option for enhanced bio-hythane production.

16.3.1.4 Heat Shock Treatment (HST)

Heat shock treatment is another physical method. HST refers to short-term exposure to high temperature and subsequent cooling to ambient temperature. Formation of more spore-forming bacteria instead of non-spore-forming bacteria is the basis of HST (Ueno et al. 1996). Generally, H_2-forming microbes such as *Clostridium* sp. and *Bacillus* sp. are used for HST (Lay et al. 2004). The vegetative parts of non-spore-forming microbes are eliminated in HST; otherwise they might encompass methanogens, non-H_2 producers, H_2 consumers, and even can kill H_2-forming microbes (Watanabe et al. 1997). In addition to this, the HST-assisted bioenergy production efficiency depends on temperatures ranges (80°C–104°C) and exposure time (15–120 minutes).

16.3.1.5 Load Shock Treatment (LST)

Load shock treatment method is employed to enrich inoculum (Fang et al. 2002). Seed culture is kept in ambient environment to modify the volumetric organic load, *i.e.,* microbial diversity. This method is more effective than HST, as there is no requirement of chemical/physical treatment. Additionally, due to high microbial diversity induce accumulation of organic acids, resulting more acidic condition (pH drops to 4.6 from 5.5), which leads to loss of methanogens (Fang et al. 2002). Further, LST technique is efficient to enrich H_2-producing seeds compared to other pretreatment techniques like acid/base treatment and HST methods (O-Thong et al. 2009).

16.3.2 Mechanical Pretreatment Methods

16.3.2.1 Ultrasonication

Ultrasonication is the most widely used mechanical method and has gained increasing attention in bio-hythane production due to their ability to solubilize a wide number of substrates. It uses ultrasonic energy waves (10 kHz–20 MHz) to disrupt the cells through cavitational bubbles generation. During bubble breaking by cavitational force, it develops temporary bursts of high pressure (103 mpa) in the solution, and ultimately this burst helps to disrupt cell membrane (Banu et al. 2020; Shanmugam et al. 2021). The prime advantages are that it greatly improves the biomass solubility and microbial activity. Additionally, heat generation during wave actions creates an additional force to disintegrate cellular matrix. Further, ultrasonic pretreatment ensures maximum substrate release, which favors enhanced bio-hythane yield. For example, Gadhe et al. (2014) documented that ultrasonic pretreatment of distillery wastewater (DWW) showed 1.4-fold enhanced biohydrogen production compared with untreated DWW. Kumar et al. (2016) also documented that ultrasonic pretreatment of microalgae *Scendesmus* sp. and *Chlorella* sp. showed enhanced hydrogen production. Although ultrasonic waves range 10 kHz–20 MHz was recommended as effective range by several authors, some controversy still remain exists. For instance, researchers noticed that ultrasonic wave frequency of 3.3 MHz for *Chaetoceros calcitrans*, 4.3 MHz for *Nannochloropsis* sp., 3.3 MHz for *Scenedesmus* sp., and 2.2 MHz for *C. gracilis* is more effective to disrupt cell (Wang et al. 2014; Kurokawa et al. 2016). Recently, this technique is integrated with dark fermentation to improve the bio-hythane yield. For instance, Jeon et al. (2013) used integrated ultrasonic treatment (40 kHz frequency) with dark fermentation to enhance bio-hythane production from microalgae. However, the major drawback of this method is oxidative damage beyond optimal wave energy and subsequent effects on fermentation process, leading to lower bio-hythane yield.

16.3.2.2 Microwave Irradiation

Microwave irradiation is another important technique to break down polysaccharides (lignin layer) and partly eliminates inhibitory compounds. It uses electromagnetic waves (0.3–300 GHz frequency) to degrade the biomass structure. Upon interaction microwaves absorb intracellular components and start vibrating, which ultimately result in enormous amount of heat and pressure. This heat energy breaks hydrogen bonds of biomass cell matrix, resulting in easy accessibility of substrates to microorganisms (Shanmugam et al. 2021). For example, Jackowiak et al. (2011) documented wheat straw treated with microwave showed enhanced methane production (28% higher) than untreated wheat straw. The major advantages of microwave irradiation are uniform heating of substrates and improved substrate solubility. On the other hand, intense microwave heating above 160°C produced inhibitory compounds, which reduce the bio-hythane yield. The prime inhibitory compounds are phenolic and furfural for lignocellulosic wastes, and ethanol and propionic acid for other substrates (Banu et al. 2020).

16.3.2.3 Hydrodynamic Cavitation

This is very common mechanical method and its functional mechanism is similar to ultrasonication method. Considering the orifice model in hydrodynamic cavitation, shear forces were used instead of ultrasonic energy waves to produce microbubbles. When the pressure drops microbubbles break to produce hydrodynamic forces, *i.e.*, temporary shock waves. The hydrodynamic force disrupts cell matrix to release intracellular compounds under normal temperature and pressure (Shanmugam et al. 2021). For instance, hydrodynamic cavitation treated *Chlorella* biomass under pressure condition of 5 bar for duration of 180 minutes showed enhanced cellular disruption compared with untreated *Chlorella* biomass (Waghmare et al. 2019). The mechanism of cell disruption by hydrodynamic cavitational forces is still under infancy stage with regard to energy consumption, which demands further study for efficient use of this technique in future bio-hythane production. This method is still more efficient than ultrasonication pretreatment method in terms of energy consumption (Shanmugam et al. 2021).

16.3.2.4 Pulsed Electric Field

Pulsed electric field method is recently used to treat biomass. It uses controlled high-voltage electric field to disrupt electrical potential of cell membrane to generate suitable pore development in cell wall. Generally, it is used to extract different proteins, pigments, antioxidants, *etc.* In case of dry biomass, no precautionary steps are needed, but internal compositions should be concerned. Besides, this method does not produce cell debris, which is extra beneficial compared with other pretreatment methods (Carullo et al. 2018). Goettel et al. (2013) documented that pulsed electric field enhanced bioenergy production from microalgae. On the other hand, Lam et al. (2017) demonstrated that pulsed electric field-treated *C. vulgaris* and *N. oleoabundans* showed poor performance for release of proteins compared with carbohydrates in microalgal biomass, *i.e.*, bioenergy production. The findings also showed that pulsed electric field showed poor performance compared to bead milling in terms of protein release. The prime advantage of pulsed electric field method is that the method does not need chemical treatment, which indicated that for commercialized bio-hythane production proper pulsed electric field method could play crucial role in near future. Accordingly, the method should be standardized based on biomass shape and size, voltage, time, electric power, and pulsed cycles (Shanmugam et al. 2021). In addition to this, spark and arc formation might occur during pulsed electric field treatment, which might affect cell disruption. Therefore, deionizing the solution medium is very essential to control arc formation for efficient bio-hythane production (Shanmugam et al. 2021).

16.3.2.5 Freeze Drying

Freeze drying technique is recently adapted in biofuel industries. During freezing water molecules expanded to generate ice crystals on biomass, resulting in cellular disruption physically, while intracellular ingredients come out through disrupted cells during thawing (Shanmugam et al. 2021). The combined use of freezing and thawing is more beneficial for excess cell disruption, and it is performed in lyophilizer. It increases digestibility for wide ranges of rigid biomass. Additionally, the method does not require any chemicals, have higher efficiency, and are relatively safe. However, the method requires huge energy for operational purposes. Chang et al. (2011) used freezing technique to treat rice straw for improving the yield of fermentable sugars, and the findings showed that enzyme digestibility of rice straw enhanced to 84% from 48%. Smichi et al. (2016) used combined freezing and thawing technique to produce enhanced bioenergy from halophyte *Juncus maritimus*. Further, it is a promising method to disrupt microalgal structures and lignocellulosic biomass. Accordingly, it should be adopted as efficient pretreatment method for satisfactory bio-hythane production after proper optimization.

16.3.3 Chemical Pretreatment Methods

Chemical pretreatment method has gained much attention among various treatment methods due to their effective treatment of complex biomass. Chemical treatments methods generally use acids like H_2SO_4, HCl, CH_3COOH, oxidants like H_2O_2, alkylating agents like KOH, NaOH, $Ca(OH)_2$, NH_3H_2O, and different solvents.

16.3.3.1 Acid Pretreatment

Acid pretreatment method is widely used for easy operation. In acidic pretreatment, the milled/dried biomass is submerged in acidic solution for specific time period at defined temperature. During acidic submersion, acids degrade bonds like van der Waals forces, hydrogen/covalent bonds to hydrolyze them into cellulose/hemicellulose, and unhydrolyzed parts are washed to extract sugars. Finally, before saccharification for production of bio-hythane neutralization is performed (Amin et al. 2017).

Acid treatment enhanced surface area and reduced polymerization degree. It has been documented that dilute acid treatment (0.4%–4% at 160°C–220°C) is the most preferable for maximum carbohydrate hydrolysis. More specifically, dilute acid treatment attacks hemicellulose and improves hemicellulose degradation into monomeric sugars (85%–95%) (Shanmugam et al. 2021). Acids like sulphuric acid/hydrochloric acid/phosphoric acid are commonly used either in lower strength at high temperature or higher strength at low temperature (Rajendran et al. 2017). Park et al. (2013) reported that H_2SO_4 (0.5%) treated red algal biomass at elevated temperature (164°C) produced enhanced biohydrogen @ 0.51 L H_2/L/h. Likely, Song et al. (2014) documented enhanced methane yield about 175.6 (74.6%) and 163.4 (62.4%) mL/gVS for H_2SO_4 (2%) and HCl (2%) treated straw than untreated straw (74.6% and 62.4%, respectively). Vavouraki et al. (2013) recorded enhanced (120%) soluble sugars levels for 1.12% HCl treatment for 94 min compared to control. Roy and Das (2015) reported that pH 6.8–7.2 is more efficient for production of bioenergy. Besides, the critical issue encountered during strong acidic pretreatment is inhibitors formation like furfural, acetic acid, 5-hydroxymethyl furfural, carboxylic acids and phenolic, which inhibits bio-hythane production (Shanmugam et al. 2021). Additionally, other drawbacks of acidic pretreatment are loss of fermentable sugar due to enhanced biomass degradation, need high-volume acids, higher cost for acid purchase, and additional cost for acid neutralization (Ariunbaatar et al. 2014). Accordingly, optimization is mandatory to prevent inhibitor formations in order to improve bio-hythane production.

16.3.3.2 Alkali Pretreatment

Alkali pretreatment is very simplistic method to treat biomass. Solvation followed by saphonication is the prime reaction for alkali treatment. Base addition induces biomass surface to swell, which increases surface area and makes the substrates easily accessible to anaerobic microbes, *i.e.*, it decreased substrates' polymerization as well as crystallinity (Ariunbaatar et al. 2014). Usually, saponification of esters of uronic acids and acetyl esters helps to increase pore numbers in cell wall to access intracellular components by hydrolytic enzymes as well as accelerate saccharification process for enhanced bio-hythane production. The most commonly used alkylating agents are aqueous ammonia ($NH_3.H_2O$), KOH, NaOH, $Ca(OH)_2$, and $Mg(OH)_2$. NaOH among them are widely used for biomass treatment followed by KOH. Alkali pretreatment is generally carried out at either low temperature (55°C–65°C) and pressure for longer duration (weeks), or higher temperature (100°C–160°C) and pressure (13 bar) for shorter duration (hours) (Sierra et al. 2009). Cheng et al. (2010) reported that 3% NaOH treated hyacinth leaves showed enhanced methane (143.4 mL–CH_4/gTVS) and hydrogen (51.7 mL-H_2/g-TVS) yield compared to untreated hyacinth leaves. Likely, Zhao et al. (2011) documented that alkali pretreatment (pH 13) of kitchen food waste showed enhanced solubilization rates for proteins, lipids, carbohydrates, and soluble COD by about 203%, 259%, 283%, and 108%, respectively, compared with control, and approximately 2.66 times higher hydrogen production (105.38 mL/gVS). Kumari et al. (2015) noted about 44.8% improved bio-hythane production (93.4 and 221.8 mL/g-VS for biohydrogen and biomethane, respectively) due to alkali pretreatment (0.25 N NaOH) of sugarcane bagasse (remove 60% w/w lignin). Jang et al. (2015)

documented that optimum bio-hythane yield can be achieved at pH 11–12. Besides several advantages, alkali pretreatments recently become obsolete due to high-quantity chemical requirements for neutralization (Banu et al. 2020).

16.3.3.3 Oxidative Treatment

Oxidative treatment is another commonly used chemical pretreatment method. It uses oxidizing agents namely oxygen, ozone, and H_2O_2 to treat biomass. Oxidative treatment is generally categorized as wet oxidation, oxidative agents, and ozonolysis. In wet oxidation, oxygen is used as an oxidizer to disintegrate lignocellulosic material of substrates (Banerjee et al. 2009). H_2O_2 is the prime oxidizing agent for delignification of cellulose, while ozone is used in ozonolysis technique for biomass disintegration. In ozonation, ozone first dissociates into radicals, and after that reacts directly or indirectly with substrates (Ariunbaatar et al. 2014). Direct application is dependent on reactant structure, while hydroxyl radicals performed indirect reactions. The prime advantages of oxidative treatments are pathogen disinfection capability, continuous formation of salt, and no chemical remaining compared to other chemical methods (Ariunbaatar et al. 2014). On the other hand, integration of oxidative treatment with other chemical pretreatment methods showed enhanced hydrolysis of biomass. For example, oxidizing agents such as H_2O_2 or $C_2H_4O_3$ in combination with alkaline pretreatment (NaOH) under mild temperature condition improves biomass digestibility compared with NaOH-treated crop residue (Nwosu-Obieogu et al. 2016).

16.3.3.4 Green and Organo-Solvents

Green and organo-solvents have recently gained scientific attention for treating lignocellulosic biomass due to tunable solvent chemistry. Dadi et al. (2007) documented that those ionic liquids (IL) treatment significantly enhanced cellulose hydrolysis. *N*-methyl morpholine *N*-oxide (NMMO) solvent is successfully used to dissolve a variety of substrates to enhance bioenergy yield, and additionally over >99% solvent can be recovered (Perepelkin 2007). Acidic organic solvents (0.5% HCl, 19.5% water, 80% ethylene glycol mixing @ 178°C for 90 minutes) is used for treating lignocellulosic materials (Yamashita et al. 2010). Recently, acetylene, iodopropane, and 2-bromoethanesulfonic acid (BESA) are successfully used to treat different biomass (Roy and Das 2015).

16.3.4 Physicochemical Pretreatment Methods

16.3.4.1 Hot Compressed Water (HCW) Method

Hot compressed water (HCW) technique is widely used physicochemical pretreatment method for treating biomass structure. In HCW technique, liquid water heated above 180°C temperature under high pressure is used for biomass hydrolysis, especially hemicelluloses, cellulose, and lignin (Harmsen et al. 2010). At certain point, liquid and gaseous phases become indistinguishable, which makes them identical for use (Xiao et al. 2011). However, the reaction time varied from few minutes to hour depending on biomass types and quantity of sugar formed (Yu et al. 2010).

HCW is the best option for starch hydrolysis than other treatment methods because of semicrystalline amorphous configuration and simultaneously enhanced the yield of fermentable sugars (Banu et al. 2020). In addition to this, HCW is also efficient for hydrolyzing polysaccharides, but simultaneously produced degradation products that have potential to affect microbial activity. The technique is one of the prime methods for a number of reasons, like no acid catalysts are required, cost-effective operating system, degradation products do not require neutralization, and no corrosion problems. The only one disadvantage is generation of inhibitory products like phenolics, aliphatic acids, and furans (Banu et al. 2020).

16.3.4.2 Hydrothermal Method

Hydrothermal pretreatment is very economical method for treating biomass, as the process requires only water to hydrolyze biomass. The water is directly fed into digestion chamber for fuel production through liquefaction process (Shen et al. 2016). However, the process requires high temperature to produce fuel. Acetic and formic acids generated in hydrothermal technique transform the biomass into simple sugars and finally produce biohydrogen. The process is directly dependent upon reaction rate and activation. Ding et al. (2017) documented that hydrothermal pretreatment (at 500 rpm for 20 minutes and temperature 100°C–200°C) of food waste significantly enhanced biohydrogen (205.7%), biomethane, and soluble protein production than untreated food waste. Li et al. (2013) observed that hydrothermal treatment of food waste and associated packaging materials at elevated temperature (100°C–180°C) showed enhancement of soluble COD up to 70.38% from 54.32.

16.3.4.3 Autoclaving Method

Autoclave technique is recently used to treat biomass under controlled heat and pressure conditions. Generally, the technique is used to sterilize bacterial growth medium either automatically or manually (at 121°C and 1 bar pressure). Currently, autoclave technique is used to treat municipal solid wastes for generating energy. Similar condition is suitable for efficient pretreatment of biomass (Shanmugam et al. 2021). For instance, autoclave pretreated microalgae *Scenedesmus* sp. integrated with dark fermentation in presence of *Clostridium butyricum* showed higher biohydrogen yield, *i.e.*, higher fermentation efficiency (Batista et al. 2014). Likely, Ferreira et al. (2013) demonstrated that autoclave pretreated microalgal biomass released the maximum fermentable sugars from substrates to be used in dark fermentation for bio-hythane production with higher yield. Autoclave technique also have the potentiality to hydrolyze hemicellulose and cellulose, thereby reducing cellulose crystallinity but enhancing biomass porosity (Abubackar et al. 2019). It also improves carbon content as well as C/N ratio for efficient bio-hythane production (Ma et al. 2018). Accordingly, autoclaving technique is proved to be an ideal method than others, as no chemicals are needed and preserves nutrient level within biomass (Banu et al. 2020).

16.3.4.4 Thermo-Chemical Technique

Thermo-chemical technique is generally conducted in minimal moisture level, usually less than 30% total weight. Combined thermo-chemical method generally produced higher energy (7–9 times) than single pretreatment, as well as it removes degradable and nondegradable ingredients efficiently (Banu et al. 2020). Philosophy and Prabhudessai (2013) reported that combined thermal–alkaline pretreatment (NaOH - Temperature at 120°C for 20 minutes) significantly enhanced biofuel yield (441 mL/g) than chemical (293 mL/g) and thermal (357 mL/g VS) treatment. Likely, Rafique et al. (2010) reported that combined alkaline–thermal (70°C) pretreatment resulted in higher biogas production (78%) with 60% higher methane content compared with combined alkaline–thermal (110°C) (28% biogas with 50% methane) treatment. Junoh et al. (2015) documented that NaOH-coupled thermal treatment enhanced solubilization of organic food waste, resulting in higher bioenergy production.

16.3.5 Chemico-Mechanical Pretreatment Methods

Thermo-mechanical pretreatment has recently been used to enhance degradability of biomass. Microwave-assisted acid treatment enhanced algal biomass *Nannochlorpsis oceanica* efficiency as well as total hydrogen yield (Xia et al. 2013). Schieder et al. (2000) documented that combined microwave and thermal treatment (160°C–200°C at 40 bars for 60 minutes) of organic wastes improves the hydrolysis as well as biogas yield approximately 70% higher compared with untreated organic wastes. Similarly, Yin et al. (2018) reported enhanced biohydrogen yield from *Laminaria japonica* by microwave-assisted acid treatment compared with untreated seaweeds. Shahriari et al. (2012) documented that combined microwave heating (115°C–145°C) with chemical treatment (hydrogen peroxide) of municipal organic waste showed enhanced biogas production, but heating above 145°C hindered biogas yield because of refractory fraction formation. Cheng et al. (2014) documented that microwave-coupled acid treatment enhanced biohydrogen production in integrated dark and photo fermentation process. Wett et al. (2010) reported that combined ultrasonic and alkaline pretreatment (19–21 bar and 160°C–180°C for 1 hour) of sludge enhanced biogas production by about 75% as well as improves dewatering characteristics, which reduced disposal cost by 25%. Montingelli et al. (2016) noted that bead milling coupled with microwave treatment decreased biohydrogen production. Therefore, it is clear that combined use of different pretreatment methods is more efficient in treating biomass to produce enhanced bio-hythane.

16.3.6 Biological Pretreatment Methods

Biological pretreatment method is the most preferable method for enhancing the bioenergy yield, as it doesn't need any chemicals and doesn't produce inhibitory substances. It uses microorganisms namely bacteria and fungi and their enzymes to degrade feedstock materials (Nwosu-Obieogu et al. 2016). *Aspergillus fumigatus, Sphingomonas paucimobilis, Baillus circulans* are commonly used bacteria to treat large variety of biomass (Bhatia et al. 2012). Among fungi, white-, brown-, and soft-rot are predominantly employed to break down lignocellulosic substances. In particularly, white-rot fungi more effective for biological degradation and it includes *Phlebia radiata, Penicillium chrysosporium, P. floridensis* and *Daedalea flavida* (Bhatia et al. 2012). The only disadvantage is that the method required highly restricted control, and rate of hydrolysis is very low. Generally, the pretreated cellulose and hemicelluloses are hydrolysed using cellulases, ligninases, and hemicellulases.

Kumari and Das (2016) demonstrated that biological treatment of food waste yield maximum 147.5 ± 2.4 L- bio-hythane within 36 hours residence time. Zhao et al. (2012) reported that *Pleurotus pulmonarium* (MTCC 1805)-treated sugarcane top at 28°C produced maximum 180.86 mL/g methane and 0.3 L/kg hydrogen. Shanmugam et al. (2018) noted that *Trichoderma asperellum* (BPLMBT1)-derived laccase have the potentiality to remove maximum 76.93% lignin as well as hydrogen production (402.01 mL) within anaerobic incubation of 84 hours. An et al. (2020) observed that *Clostridium thermocellum* assisted-anaerobic fermentation generates 63 L H_2/kg paper sludge and 122 L CH_4, while An et al. (2018) noted 64.32 mM H_2 yield from 7.4% paper sludge using anerobic fermentation. Recently, Islam et al. (2017) used combined *C. thermocellum* and *Clostridium thermosaccharolyticum* bacterial strain to check the co-culture efficiency. The results showed that method enhanced yield significantly likely 5.1 mmol/g H_2, 1.05 g/L butyric acid, and 1.27 g/L acetic acid. Abreu et al. (2016) recorded that bio-hythane production from xylose, cellobiose, and garden waste was significantly enhanced in co-culture method than monoculture of *Caldicellulosiruptor saccharolyticus* (2.7 mol/mol H_2) and *Thermotoga maritima* (4.8 mol/mol H_2). Combined method produced 68.2 mL/g hydrogen, which is 94.2% higher than monoculture incubation. Similarly, combined used of cellulolytic bacterium (*C. thermocellum,* DSM 1237 and noncellulolytic bacterium (*C. thermopalmarium)* significantly enhanced biohydrogen production (1387 mL/L) (Geng et al. 2010; Li et al. 2012). Similarly, Seppälä et al. (2011) recorded enhanced hydrogen production (maximum 5.85 mmol) due to codigestion of *Clostridium butyricum* and *Escherichia coli*. Additionally, carbon source variation significantly regulates biohydrogen production. For example, glycerol byproduct from biodiesel production serves as an excellent enriched carbon source for bio-hythane production (Hu and Wood 2010). Recently, Maru et al. (2016) investigated impact of microflora and *E. coli* CECT434, *E. coli* CECT432, and *E. cloacae* MCM2/1 mixture during hydrogen generation from glycerol. The findings showed that addition of extra carbon source significantly enhanced co-culture-induced bio-hythane production to maximum 1.53 mol (1307 mL/L) H_2.

16.4 FACTORS AFFECTING BIO-HYTHANE PRODUCTION

16.4.1 Temperature

Temperature is the most influencing factor (Figure 16.3) for bio-hythane production, as it regulates acidogenesis and methanogenesis (Abreu et al. 2019). Actually, the cellular mechanism of microorganisms during AD process is enzyme dependent, whose functions are controlled by temperature. Any variation from optimum ranges caused either enzymes denaturation or inactivation. Generally, enzymatic activity increases two folds if temperature increases by factor 10°C. Microorganisms responsible for bio-hythane production are generally found in varied temperature ranges. Accordingly, it is expected that temperature might influence metabolite formation, nutritional requirement, and microbial characteristics (Sanjay Kumar et al. 2021).

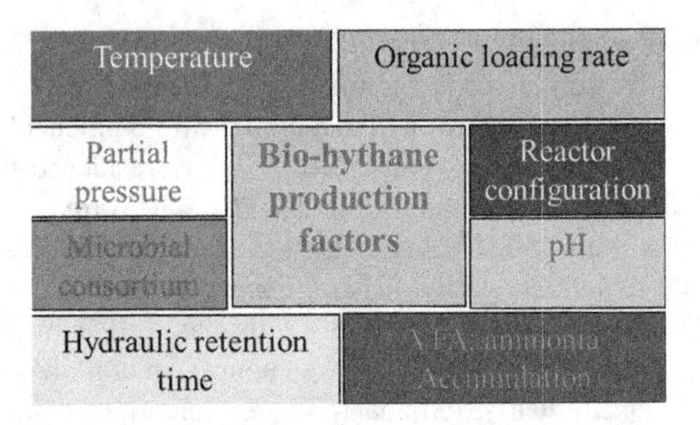

FIGURE 16.3 Factors affecting bio-hythane production.

However, the operational temperature of bioreactor depends on choice of microorganism. For example, fermentative microbes are able to produce hydrogen at different temperature, namely, thermophilic (42°C–75°C), mesophilic (20°C–42°C), and psycrophilic (0°C–20°C). Vindis (2009) reported optimal temperature of about 37°C–45°C for mesophilic bacterial strains like *Clostridium* and *Enterobacter*. Achmon et al. (2019) recorded enhanced yield of H_2 (73.17±32.76 $mL^{-1}g$) and CH_4 (201 $mL^{-1}g$) due to combined mesophilic (30°C) and thermophilic (55°C) condition during AD process. In particularly, the role of temperature is very vital for methanogens. Even a minute temperature variation (2°C–3°C) inhibits the activity of methanogens and subsequently facilitates acidogens to accumulate VFAs within the system and even decrease pH level below 7 (Roy and Das 2015). The performance of all anaerobic microbes is adversely affected by even minute temperature changes, which ultimately disrupt methane production. Therefore, it is clear that temperature directly regulates bio-hythane production.

16.4.2 pH

The pH is most important chemical factor that influences not only efficiency of enzymatic machinery of microbes but also oxidation–reduction potentiality of the cells, as enzymes have an optimal pH range for maximum activity. Generally, enzymes like formate lyase, Fe–Fe hydrogenase, and others are prime regulator involved H_2 production pathways (Roy and Das 2015). VFAs formed during dark fermentation is primarily responsible for pH variation. This variation adversely affects enzymatic machinery of H_2 production pathways due to disturbances in cell membrane's integrity, and at very low pH value (pH 3.8–4.2) the generation of H_2 stopped (Sanjay Kumar et al. 2021). Additionally, a metabolic shift from acidogenesis to solventogenesis may happen at low pH value. Calli (2008) recorded enhanced H_2 production (1.72 mol/mol xylose) at pH 6.5 compared to pH 5.5. Park et al. (2010) noted the maximum H_2 (2.8 LH_2/L/d) and CH_4 (1.48 LCH_4/L/d) yield at pH 5.5 and 7.0, respectively, during molasses treatment. Likely, pH values 5.5 and 7.2 is considered as optimal H_2 (14.8 mL/g VS) and CH_4 (274 mL/g VS) production, respectively, from vinasse (Fu et al. 2017). These findings concluded that neutral to alkaline pH for CH_4 (methanogenesis) and acidic pH for H_2 (acidogenesis) enhanced bio-hythane production (Dhar et al. 2016; Li et al. 2018).

16.4.3 Partial Pressure

Hydrogen partial pressure is another important factor that controls bio-hythane production as gaseous products (hydrogen and methane) dissolved in fermentation broth. Hydrogen partial pressure is continuously increased in course of fermentation inside reactor chamber as gaseous products get accumulated in head space, resulting in metabolic shift (Sanjay Kumar et al. 2021). Simultaneously, the H_2 accumulation inhibits product formation like ethanol, butanol, propionate, acetone, and lactate (Hawkes et al. 2007). Accordingly, it is very necessary to remove H_2 from fermentation system instantly after its production. Mizuno et al. (2000) proposed intermittent N_2-sparging technique to reduce the partial pressure for improving the hydrogen yield by about 68%.

16.4.4 Hydraulic Retention Time (HRT)

Hydraulic retention time (HRT) is another most important factor that greatly controls bio-hythane production. It refers to time retained by cells and soluble nutrients inside reactor, and is governed by reactor volume and feed flow rate. Low HRT is very essential for optimized bio-hythane production. Generally, at low HRT all active cells wash out from the reactor, resulting optimized HRT. In case of mixed consortia, *i.e.*, containing acidogenic and methanogenic H_2 producers, the HRT optimization caused shift in microbial profile. Lower HRT coupled with acidic pH (6–6.5) completely wash out methanogens from mixed consortia, but increases acidogenic H_2 producers inside reactor (Roy and Das 2015). Additionally, HRT also plays crucial role in end metabolites formation, which linked with variation in microbial profile coupled with HRT. Buitrón et al. (2014) noted optimal H_2 and CH_4 production at 6 and 24 hours HRT, respectively, from vinasse treating produced from tequila processing. Likely, Corona and Razo-Flores (2018) recorded the maximum H_2 (105 mL H_2/g) and CH_4 (225 mL CH_4/g) production at 6 and 24 hours HRT, respectively, for agave bagasse derived from tequilana processing. These findings concluded that short and longer HRT for H_2 and CH_4, respectively, enhanced bio-hythane production efficiency (Anzola-Rojas et al. 2016). On the other hand, Zhang et al. (2006) reported that lowering HRT from 10 to 6 h significantly enhanced H_2 yield in *Clostridium* sp.-mediated AD process. Moreover, the propionate level reduced HRT reduction (Zhang et al. 2006). In addition to this, HRT regulates microbial consortia as well as expelling slow growing microorganisms (methanogens and other H_2-consuming microbes) from reactor. Therefore, shorter HRTs are very crucial for selective H_2-producing microbes. Moreover, the reactor configuration and organic loading rate (OLR) regulates the HRT, which necessitates HRT optimization. Further, the HRT must be twofold higher than generation time of the slowest growing microorganism during methanogenesis of substrates (Roy and Das 2015).

16.4.5 Microbial Consortium

Preparation of enriched inoculum is very crucial for achieving desirable bioenergy production. The hydrolysis of substrates is generally governed by microbes.

Sometimes, single microbe is not enough to degrade substrates due to lack of all hydrolytic enzymes, which forced scientists to move towards use of mixed microbial consortium. This mixed microbial consortium significantly induced desirable bio-hythane production. H_2 producible microorganisms occurred in both artificial and natural habitats, which includes in sludge, sewage sludge, acclimated sludge, animal manure, compost, hot springs, soil, oceanic sediments, *etc.* (Lin and Lay 2005; Roy and Das 2015). The prime drawback of mixed consortia is that consortia might contain methanogens, non-H_2 producible microorganisms, homoacetogens, H_2-consumable microbes, and lactic acid-producible bacteria. Achmon et al. (2019) demonstrated that strains of *Clostridium, Syntrophomonas, Steroidobacter, Caldicoprobacter,* and *Thermacetogenium* enhanced degradation of fruit processing wastes to produce higher H_2, while *Methanosarcina, Methanoculleus, Methanobrevibacter,* and *Methanomassiliicoccus* strains are responsible for higher CH_4 yield. Besides, abovementioned species several authors decoded that *Clostridium butyricum, Methanothrix soehngenii, Clostridium thermocellum, Thermococcus kodakaraensi* and *Methanobacterium beijingense, Anaeroturncus, Coprothermobacter, Pectinatus,* and *Lactobacillus* also enhanced bio-hythane production (Hung et al. 2011; Júnior et al. 2014; Jariyaboon et al. 2015; Algapani et al. 2018). Alexandropoulou et al. (2018) reported that *Ruminococcaceae/Clostridiaceae* and *Enterobacteriaceae* strains from food waste significantly improved H_2 production (101.75 ± 3.71 L/kg). Veeramalini et al. (2019) reported that combined use of *Enterobacter aerogenes* and *Rhodobacter* M 19 potentially improved biohydrogen yield (1.96 mol H_2/sugar) using brewery industry effluent. Finally, the selection of proper inoculum is very important for getting the maximum efficiency.

16.4.6 Reactor Configuration

Reactor configuration is another important factor that influence bio-hythane production greatly. Commonly used bio-hythane reactors are continuously stirred tank reactor (CSTR), fluidized bed reactor, packed-bed/trickling filter-based, membrane bioreactor, up-flow anaerobic sludge blanket (UASB) reactor, and immobilized/granulated-type reactor (Roy and Das 2015). Different reactors have different effects on bio-hythane production. For example, for maintaining higher biomass granulated/immobilized-type reactors are better. HRT is directly related with reactor configuration. These reactors work through either batch mode or continuous mode. Continuous mode is more advantageous than batch mode due to its steady and continuous production, and engineering design (Pinto et al. 2018). These reactors use two units either similar one or different head for bio-hythane generation (Dareioti and Kornaros 2015). For example, Nualsri et al. (2016) used two different reactors, namely CSTR (1 L volume) and UASB (24 L volume) to produce bio-hythane separately; CSTR for H_2 production and UASB for CH_4 production. On the other hand, Banks et al. (2010) reported that use of similar reactors, *i.e.,* CSTR showed profound impacts on bio-hythane during treating dairy waste permeate; first one for acidogenic bacterial growth, while the second one is designed for methanogenic process (Banks et al., 2010). Similarly, Chavadej et al. (2019) reported that use of similar-type reactors (UASB) enhanced bio-hythane production during treating cassava wastewater.

16.4.7 Organic Loading Rate (OLR)

Organic loading rate play a crucial role in bio-hythane production. Additionally, OLR improves microbial consortia in the reactor, regulates HRT, and have capability to predict system stability. Organic loading rate about 16 g COD/L is able to yield 57.4 ± 4.0 mL/L H_2, while 1636 mg COD/L OLR yield 257.9 ± 13.8 mL/g CH_4 from vinasse (Buitrón et al. 2014). Likely, Wang et al. (2013) used 6 and 2.1 g COD/L/d OLR during treating sugary wastewater to produce maximum H_2 and CH_4 yield, respectively. It has been postulated by several authors that to maximize bio-hythane yield longer and shorter OLRs for H_2 and CH_4 is very essential (Koutrouli et al. 2009; Corona and Razo-Flores 2018).

16.4.8 Accumulation of VFA and Ammonia Concentration

VFA accumulation in reactor greatly influenced bio-hythane production. It is the intermediate metabolic products of acidogenesis and acetogenesis phases. There are several causes of VFA formation and subsequent accumulation withing the reactors. The prime causes are switching to short reaction time after acidogenesis, prolonged HRT, mesophilic temperature condition, pH range, and OLR (Jiang et al. 2013; Algapani et al. 2018). This excess VFA accumulation hindered the bio-hythane yield. Corona and Razo-Flores (2018) reported butyric and acetic acids accumulation during bio-hythane production from agave bagasse. Accumulation of propionic, palmitic, formic, and lactic acids have been reported by different scientists (Park et al. 2010; Estevam et al. 2018). Apart from VFA, the AD process produced different amounts of ammonia (*e.g.*, NH^{4+} and NH_3) and finally accumulated within reactor. Ammonia accumulation caused inhibitory effect, *i.e.*, reduced bio-hythane yield as well as makes the system unstable (Giuliano et al. 2013; Polizzi et al. 2018). Additionally, high ammonia concentration significantly changes transcriptional profile of microorganisms, caused toxicity, and kills microbial population resulting reduced yield. High OLR and neutral pH is proved to be prime reason behind this phenomenon (Jiang et al. 2013).

16.5 FEASIBILITY FOR FULL-SCALE APPLICATION

The potential feasibility of different treatment methods for bio-hythane production at full scale is discussed in this section. The prime regulators that hindered full-scale applications in bio-hythane production is high capital cost and energy consumption, excessive chemical requirements, maintenance of treatment facility, and operating conditions. The energy requirement generally depends on pretreatment methods and most proportion of it is used to maintain process temperature. For example, if temperature crossed 100°C in thermal method, most of the energy is utilized for evaporation, which makes the method less feasible (Perez-Elvira et al. 2009). Microwave pretreatments are considered as more efficient in comparison with thermal pretreatments and traditional heating methods, as no external heating required and heat loss happened (Perez-Elvira et al. 2009). However, neither microwave nor ultrasound energy is sensitive to treat biomass, as enhanced methane production do not require energy compensation (Mottet et al. 2009). Yang et al. (2010) documented that thermal treatment enhanced methane production significantly, and integration of heat exchanger reduces the operating cost. Two-stage anaerobic treatment is proved to be more energy efficient over on-stage reactor during bio-hythane production (Escamilla-Alvarado et al. 2012).

Economic feasibility is another important part for full-scale applications of different pretreatment methods during bio-hythane production. It generally depends on biomass availability, its cost, and type of pretreatment method used. Rittmann et al. (2008) reported that full-scale AD process (3,300 m^3 capacity) could be economically feasible, if they are able to treat 380 m^3 sludge per day, which can generate benefit about 540,000 USD/y. A full-scale ultrasound pretreatment method could be economically feasible, if the reactor is able to treat 6 kWh/m^3 sludge (Perez-Elvira et al. 2009). Han et al. (2016) reported that combined bioprocess and dark fermentation could generate benefit about 115,717 USD/y if the facility has ability to treat 3 tons food waste per day (annual productivity 42,858 m^3). Bordeleau and Droste (2011) conducted cost estimation of different pretreatment techniques and depicted approximate cost about 0.0162 USD/m^3 for microwave treatment, while for thermal pretreatment it was 0.0187 USD/m^3, 0.0264 USD/m^3 for ultrasound pretreatment, and 0.0358 US Dollar/m^3 for chemical pretreatment. However, to boost the bio-hythane production economically as well as cheaper way, the biological pretreatment methods are the most preferable (Park et al. 2005).

Similar to energetic and economic feasibility, considering environmental aspects and pretreatment methods, sustainability is very essential for full-scale applications of different pretreatment methods during bio-hythane production. Environmental aspects such as sustainable use of chemicals and biomass, pathogen removal capacity, human health risk, and environmental deterioration should be considered during selection of pretreatment method (Ariunbaatar et al. 2014; Banu et al. 2020). In addition to this, soil type (as anaerobic residues is used as soil fertilizers) and potential gaseous emissions should be considered (Thorin et al. 2012). Pretreatment methods that have higher efficiencies and are environment friendly should be considered first to treat biomass. Thermal, chemical, and thermochemical pretreatment methods have higher efficiency to produce bio-hythane under high-temperature condition compared to untreated substrates, but capital cost and environmental aspects make it less desirable (Ariunbaatar et al. 2014). Likely, Carballa et al. (2011) investigated environmental aspects of methods like chemical, freeze–thaw, ozonation, pressurize–depressurize, and thermal with regard to Global Warming Potential (GWP), Abiotic Resources Depletion Potential (ARDP), Eutrophication Potential (EP), Human and Terrestrial Toxicity Potential (HTTP) through life cycle assessment (LCA) study. They concluded that chemical pretreatment and pressurize–depressurize are more efficient than freeze–thaw, ozonation, and thermal methods.

16.6 FUTURE PERSPECTIVES FOR FULL-SCALE SCALE-UP

High capital investment, excessive energy consumption and chemical requirements, maintenance of treatment facility, and operating conditions are the major constraints of biomass pretreatment methods for bio-hythane production. Accordingly, the use of environment friendly and economically feasible pretreatment methods is mandatory for sustainable production of bio-hythane. Biomass composition is another constraint for efficient production of bio-hythane. Besides, microbial consortium's rational control and consistent design is considered as prime issue for effective bio-hythane production. Because indigenous consortia are less effective than pure H_2 producers' culture. Accordingly, designing and controlling of microbial consortia are very essential to make the process commercially feasible. Additionally, another key issue of bio-hythane product chain is dark fermentation-based technology. Process integration such as integration of two-stage system is the best way to produce bio-hythane efficiently, which controls H_2:CH_4 ratio. Accordingly, the following suggestions should be taken very carefully for sustainable production of bio-hythane: (1) selection of proper pretreatment method based on chemical composition of biomass, (2) selection of environment-friendly method to ensure environmental sustainability, (3) should perform pilot scale optimization study of pretreatment method before commercial production, (4) should focus on microbial consortium's rational control and consistent design so that unwanted microbes can be minimized to enhance bio-hythane yield, (5) should adopt integrated process with recent technological advancement, and finally (6) repeated use of digestate to reduce the environmental burden.

16.7 CONCLUSIONS

This book chapter addresses the role of codigestion and different pretreatment methods for improving the bio-hythane yield. Different pretreatment methods, namely, physical, mechanical, chemical, physicochemical, physicomechanical, and biological are discussed elaborately. Additionally, advantages and disadvantages of each method is discussed very carefully with regard to environmental sustainability. Both operational and environmental factors are reviewed for full-scale technological scale-up to boost bio-hythane production. In conclusion, this book chapter also suggested solution in terms of methods optimization to improve their performances for producing higher bio-hythane yield.

REFERENCES

Abreu, A.A., F. Tavares, M.M. Alves, A.J. Cavaleiro, and M.A. Pereira. "Garden and food waste co-fermentation for biohydrogen and biomethane production in a two-step hyperthermophilic-mesophilic process." *Bioresource Technology* 278 (2019): 180–86. doi:10.1016/j.biortech.2019.01.085.

Abreu, A.A., F. Tavares, M.M. Alves, and M.A. Pereira. "Boosting dark fermentation with co-cultures of extreme thermophiles for biohythane production from garden waste." *Bioresource Technology* 219 (2016): 132–38. doi:10.1016/j.biortech.2016.07.096.

Abubackar, H.N., T. Keskin, O. Yazgin, B. Gunay, K. Arslan, and N. Azbar. "Biohydrogen production from autoclaved fruit and vegetable wastes by dry fermentation under thermophilic condition." *International Journal of Hydrogen Energy* 44, no. 34 (2019): 18776–8784. doi:10.1016/j.ijhydene.2018.12.068.

Achmon, Y., J.T. Claypool, S. Pace, B.A. Simmons, S.W. Singer, and C.W. Simmons. "Assessment of biogas production and microbial ecology in a high solid anaerobic digestion of major California food processing residues." *Bioresource Technology Reports* 5 (2019): 1–11. doi:10.1016/j.biteb.2018.11.007.

Alexandropoulou, M., G. Antonopoulou, E. Trably, H. Carrere, and G. Lyberatos. "Continuous Biohydrogen production from a food industry waste: Influence of operational parameters and microbial community analysis." *Journal of Cleaner Production* 174(2018): 1054–63.

Algapani, D.E., W. Qiao, F.D. Pumpo, D. Bianchi, S.M. Wandera, F. Adani, and R. Dong. "Long-term bio-H_2 and bio-CH_4 production from food waste in a continuous two-stage system: Energy efficiency and conversion pathways." *Bioresource Technology* 248 (2018): 204–13. doi:10.1016/j.biortech.2017.05.164.

Amin, F.R., H. Khalid, H. Zhang, S.U. Rahman, R. Zhang, G. Liu, and C. Chen. "Pretreatment methods of lignocellulosic biomass for anaerobic digestion." *AMB Express* 7, no. 1 (2017). doi:10.1186/s13568-017-0375-4.

An, Q., J.-L. Wang, Y.-T. Wang, Z.-L. Lin, and M.-J. Zhu. "Investigation on hydrogen production from paper sludge without inoculation and its enhancement by *Clostridium thermocellum*." *Bioresource Technology* 263 (2018): 120–27. doi:10.1016/j.biortech.2018.04.105.

An, Q., J.-R. Cheng, Y.-T. Wang, and M.-J. Zhu. "Performance and energy recovery of single and two stage biogas production from paper sludge: *Clostridium Thermocellum* augmentation and microbial community analysis." *Renewable Energy* 148(2020): 214–22.

Anzola-Rojas, M.D.P., M. Zaiat, and H. De Wever. "Improvement of hydrogen production via ethanol-type fermentation in an anaerobic down-flow structured bed reactor." *Bioresource Technology* 202 (2016): 42–49. doi:10.1016/j.biortech.2015.11.084.

Appels, L., J. Degrève, B.V. Der Bruggen, J. Van Impe, and R. Dewil. "Influence of low temperature thermal pre-treatment on sludge solubilisation, heavy metal release and anaerobic digestion." *Bioresource Technology* 101, no. 15 (2010): 5743–748. doi:10.1016/j.biortech.2010.02.068.

Ariunbaatar, J., A. Panico, G. Esposito, F. Pirozzi, and P.N.l. Lens. "Pretreatment methods to enhance anaerobic digestion of organic solid waste." *Applied Energy* 123 (2014): 143–56. doi:10.1016/j.apenergy.2014.02.035.

Banerjee, S., R. Sen, R. Pandey, T. Chakrabarti, D. Satpute, B.S. Giri, and S. Mudliar. "Evaluation of wet air oxidation as a pretreatment strategy for bioethanol production from rice husk and process optimization." *Biomass and Bioenergy* 33, no. 12 (2009): 1680–686. doi:10.1016/j.biombioe.2009.09.001.

Banks, C.J., E.A. Zotova, and S. Heaven. "Biphasic production of hydrogen and methane from waste lactose in cyclic-batch reactors." *Journal of Cleaner Production* 18 (2010). doi:10.1016/j.jclepro.2010.04.018.

Banu, J.R., J. Merrylin, T.M. Mohamed Usman, R. Yukesh Kannah, M. Gunasekaran, S.-H. Kim, and G. Kumar. "Impact of pretreatment on food waste for biohydrogen production: A review." *International Journal of Hydrogen Energy* 45, no. 36 (2020): 18211–8225. doi:10.1016/j.ijhydene.2019.09.176.

Barjenbruch, M., and O. Kopplow. "Enzymatic, mechanical and thermal pre-treatment of surplus sludge." *Advances in Environmental Research* 7, no. 3 (2003): 715–20. doi:10.1016/s1093–0191(02)00032-1.

Batista, A.P., P. Moura, P.A.S.S. Marques, J. Ortigueira, L. Alves, and L. Gouveia. "Scenedesmus obliquus as feedstock for biohydrogen production by *Enterobacter aerogenes* and *Clostridium butyricum*." *Fuel* 117 (2014): 537–43. doi:10.1016/j.fuel.2013.09.077.

Bhatia, L., S. Johri, and R. Ahmad. "An economic and ecological perspective of ethanol production from renewable agro waste: A review." *AMB Express* 2, no. 1 (2012): 65. doi:10.1186/2191-0855-2-65.

Bolzonella, D., F. Battista, C. Cavinato, M. Gottardo, F. Micolucci, G. Lyberatos, and P. Pavan. "Recent developments in biohythane production from household food wastes: A review." *Bioresource Technology* 257 (2018): 311–19. doi:10.1016/j.biortech.2018.02.092.

Bordeleau, É.L., and R.L. Droste. "Comprehensive review and compilation of pretreatments for mesophilic and thermophilic anaerobic digestion." *Water Science and Technology* 63, no. 2 (2011): 291–96. doi:10.2166/wst.2011.052.

Buitrón, G., G. Kumar, A. Martinez-Arce, and G. Moreno. "Hydrogen and methane production via a two-stage processes (H_2-SBR+CH_4-UASB) using *Tequila vinasses*." *International Journal of Hydrogen Energy* 39, no. 33 (2014): 19249–9255. doi:10.1016/j.ijhydene.2014.04.139.

Calli, B. "Dark fermentative H_2H_2 production from xylose and lactose—Effects of on-line PH control." *International Journal of Hydrogen Energy* 33, no. 2 (2008): 522–30. doi:10.1016/j.ijhydene.2007.10.012.

Carballa, M., C. Duran, and A. Hospido. "Should we pretreat solid waste prior to anaerobic digestion? An assessment of its environmental cost." *Environmental Science & Technology* 45, no. 24 (2011): 10306–0314. doi:10.1021/es201866u.

Carullo, D., B.D. Abera, A.A. Casazza, F. Donsì, P. Perego, G. Ferrari, and G. Pataro. "Effect of pulsed electric fields and high pressure homogenization on the aqueous extraction of intracellular compounds from the microalgae *Chlorella vulgaris*." *Algal Research* 31 (2018): 60–69. doi:10.1016/j.algal.2018.01.017.

Chang, K.-L., J. Thitikorn-Amorn, J.-F. Hsieh, B.-M. Ou, S.-H. Chen, K. Ratanakhanokchai, P.-J. Huang, and S.-T. Chen. "Enhanced enzymatic conversion with freeze pretreatment of rice straw." *Biomass and Bioenergy* 35, no. 1 (2011): 90–95. doi:10.1016/j.biombioe.2010.08.027.

Chavadej, S., T. Wangmor, K. Maitriwong, P. Chaichirawiwat, P. Rangsunvigit, and P. Intanoo. "Separate production of hydrogen and methane from cassava wastewater with added cassava residue under a thermophilic temperature in relation to digestibility." *Journal of Biotechnology* 291 (2019): 61–71. doi:10.1016/j.jbiotec.2018.11.015.

Cheng, J., B. Xie, J. Zhou, W. Song, and K. Cen. "Cogeneration of H_2 and CH_4 from water hyacinth by two-step anaerobic fermentation." *International Journal of Hydrogen Energy* 35, no. 7 (2010): 3029–35.

Cheng, J., L. Ding, R. Lin, L. Yue, J. Liu, J. Zhou, and K. Cen. "Fermentative biohydrogen and biomethane co-production from mixture of food waste and sewage sludge: Effects of physiochemical properties and mix ratios on fermentation performance." *Applied Energy* 184 (2016): 1–8. doi:10.1016/j.apenergy.2016.10.003.

Cheng, J., Y. Liu, R. Lin, A. Xia, J. Zhou, and K. Cen. "Cogeneration of hydrogen and methane from the pretreated biomass of algae bloom in Taihu lake." *International Journal of Hydrogen Energy* 39, no. 33 (2014): 18793–8802. doi:10.1016/j.ijhydene.2014.09.056.

Corona, V.M., and E. Razo-Flores. "Continuous hydrogen and methane production from *Agave tequilana* bagasse hydrolysate by sequential process to maximize energy recovery efficiency." *Bioresource Technology* 249 (2018): 334–41. doi:10.1016/j.biortech.2017.10.032.

Dadi, A.P., C.A. Schall, and S. Varanasi. "Mitigation of cellulose recalcitrance to enzymatic hydrolysis by ionic liquid pretreatment." *Applied Biochemistry and Biotecnology*, 2007, 407–21. doi:10.1007/978-1-60327-181-3_35.

Dareioti, M.A., and M. Kornaros. "Anaerobic mesophilic co-digestion of ensiled sorghum, cheese whey and liquid cow manure in a two-stage CSTR system: Effect of hydraulic retention time." *Bioresource Technology* 175 (2015): 553–62. doi:10.1016/j.biortech.2014.10.102.

Dhar, H., P. Kumar, S. Kumar, S. Mukherjee, and A.N. Vaidya. "Effect of organic loading rate during anaerobic digestion of municipal solid waste." *Bioresource Technology* 217 (2016): 56–61. doi:10.1016/j.biortech.2015.12.004.

Ding, L., J. Cheng, D. Qiao, L. Yue, Y.-Y. Li, J. Zhou, and K. Cen. "Investigating hydrothermal pretreatment of food waste for two-stage fermentative hydrogen and methane co-production." *Bioresource Technology* 241 (2017): 491–99. doi:10.1016/j.biortech.2017.05.114.

Eden Innovations. "Eden Annual Report." Accessed August 21, 2021. http://www.edenenergy.com.au.

Efremenko, E.N., A.B. Nikolskaya, I.V. Lyagin, O.V. Senko, T.A. Makhlis, N.A. Stepanov, O.V. Maslova, F. Mamedova, and S.D. Varfolomeev. "Production of biofuels from pretreated microalgae biomass by anaerobic fermentation with immobilized *Clostridium acetobutylicum* cells." *Bioresource Technology* 114 (2012): 342–48. doi:10.1016/j.biortech.2012.03.049.

Escamilla-Alvarado, C., E. Ríos-Leal, M.T. Ponce-Noyola, and H.M. Poggi-Varaldo. "Gas biofuels from solid substrate hydrogenogenic–methanogenic fermentation of the organic fraction of municipal solid waste." *Process Biochemistry* 47, no. 11 (2012): 1572–587. doi:10.1016/j.procbio.2011.12.006.

Estevam, A., M.K. Arantes, C. Andrigheto, A. Fiorini, E.A. Da Silva, and H.J. Alves. "Production of biohydrogen from brewery wastewater using *Klebsiella pneumoniae* isolated from the environment." *International Journal of Hydrogen Energy* 43, no. 9 (2018): 4276–283. doi:10.1016/j.ijhydene.2018.01.052.

Fang, H.H. P., Hong Liu, and T. Zhang. "Characterization of a hydrogen-producing granular sludge." *Biotechnology and Bioengineering* 78, no. 1 (2002): 44–52. doi:10.1002/bit.10174.

Ferreira, A.F., J. Ortigueira, L. Alves, L. Gouveia, P. Moura, and C.M. Silva. "Energy requirement and CO_2 emissions of $BioH_2$ production from microalgal biomass." *Biomass and Bioenergy* 49 (2013): 249–59. doi:10.1016/j.biombioe.2012.12.033.

Fu, S.-F., X.-H. Xu, M. Dai, X.-Z. Yuan, and R.-B. Guo. "Hydrogen and methane production from vinasse using two-stage anaerobic digestion." *Process Safety and Environmental Protection: Transactions of the Institution of Chemical Engineers, Part B* 107(2017): 81–86.

Gadhe, A., S.S. Sonawane, and M.N. Varma. "Ultrasonic pretreatment for an enhancement of biohydrogen production from complex food waste." *International Journal of Hydrogen Energy* 39, no. 15 (2014): 7721–729. doi:10.1016/j.ijhydene.2014.03.105.

Geng, A., Y. He, C. Qian, X. Yan, and Z. Zhou. "Effect of key factors on hydrogen production from cellulose in a co-culture of *Clostridium thermocellum* and *Clostridium thermopalmarium*." *Bioresource Technology* 101, no. 11 (2010): 4029–033. doi:10.1016/j.biortech.2010.01.042.

Ghimire, A., G. Kumar, P. Sivagurunathan, S. Shobana, G.D. Saratale, H.W. Kim, V. Luongo, G. Esposito, and R. Munoz. "Bio-hythane production from microalgae biomass: Key challenges and potential opportunities for algal bio-refineries." *Bioresource Technology* 241 (2017): 525–36. doi:10.1016/j.biortech.2017.05.156.

Giang, T.T., S. Lunprom, Q. Liao, A. Reungsang, and A. Salakkam. "Improvement of hydrogen production from *Chlorella* Sp. biomass by acid-thermal pretreatment." *Peer Journal* 7, no. e6637 (2019): e6637.

Giuliano, A., D. Bolzonella, P. Pavan, C. Cavinato, and F. Cecchi. "Co-digestion of livestock effluents, energy crops and agro-waste: Feeding and process optimization in mesophilic and thermophilic conditions." *Bioresource Technology* 128 (2013): 612–18. doi:10.1016/j.biortech.2012.11.002.

Goberna, M., P. Simón, M.T. Hernández, and C. García. "prokaryotic communities and potential pathogens in sewage sludge: Response to wastewater origin, loading rate and treatment technology." *Science of The Total Environment* 615 (2018): 360–68. doi:10.1016/j.scitotenv.2017.09.240.

Goettel, M., C. Eing, C. Gusbeth, R. Straessner, and W. Frey. "Pulsed electric field assisted extraction of intracellular valuables from microalgae." *Algal Research* 2, no. 4 (2013): 401–08. doi:10.1016/j.algal.2013.07.004.

Gottardo, M., F. Micoluccib, A. Mattiolib, S. Faggiana, C. Cavinatoa, and P. Pavan. "Hydrogen and methane production from biowaste and sewage sludge by two phases anaerobic codigestion." *Chemical Engineering Transactions* 43(2015): 379–384.

Guo, M., W. Song, and J. Buhain. "Bioenergy and biofuels: History, status, and perspective." *Renewable and Sustainable Energy Reviews* 42 (2015): 712–725.

Han, W., J. Fang, Z. Liu, and J. Tang. "Techno-economic evaluation of a combined bioprocess for fermentative hydrogen production from food waste." *Bioresource Technology* 202 (2016): 107–12. doi:10.1016/j.biortech.2015.11.072.

Hawkes, F., I. Hussy, G. Kyazze, R. Dinsdale, and D. Hawkes. "Continuous dark fermentative hydrogen production by mesophilic microflora: principles and progress." *International Journal of Hydrogen Energy* 32, no. 2 (2007): 172–84. doi:10.1016/j.ijhydene.2006.08.014.

Hu, H., and T.K. Wood. "An evolved *Escherichia coli* strain for producing hydrogen and ethanol from glycerol." *Biochemical and Biophysical Research Communications* 391, no. 1 (2010): 1033–038. doi:10.1016/j.bbrc.2009.12.013.

Hung, C.-H., Y.-T. Chang, and Y.-J. Chang. "Roles of microorganisms other than *Clostridium* and *Enterobacter* in anaerobic fermentative biohydrogen production systems – A review." *Bioresource Technology* 102, no. 18 (2011): 8437–444. doi:10.1016/j.biortech.2011.02.084.

Islam, M.S., C. Zhang, K.-Y. Sui, C. Guo, and C.-Z. Liu. "Coproduction of hydrogen and volatile fatty acid via thermophilic fermentation of sweet sorghum stalk from co-culture of *Clostridium thermocellum* and *Clostridium thermosaccharolyticum*." *International Journal of Hydrogen Energy* 42, no. 2 (2017): 830–37. doi:10.1016/j.ijhydene.2016.09.117.

Islam, M.S., S. Begum, M.L. Malcolm, M.S.J. Hashmi, and M.S. Islam. "The role of engineering in mitigating global climate change effects: Review of the aspects of carbon emissions from fossil fuel-based power plants and manufacturing industries." *Encyclopedia of Renewable and Sustainable Materials* (2020): 750–62. doi:10.1016/b978-0-12–803581-8.11274-3.

Jackowiak, D., D. Bassard, A. Pauss, and T. Ribeiro. "Optimisation of a microwave pretreatment of wheat straw for methane production." *Bioresource Technology* 102, no. 12 (2011): 6750–756. doi:10.1016/j.biortech.2011.03.107.

Jang, S., D.-H. Kim, Y.-M. Yun, M.-K. Lee, C. Moon, W.-S. Kang, S.-S. Kwak, and M.-S. Kim. "Hydrogen fermentation of food waste by alkali-shock pretreatment: Microbial community analysis and limitation of continuous operation." *Bioresource Technology* 186 (2015): 215–22. doi:10.1016/j.biortech.2015.03.031.

Jariyaboon, R., S. O-Thong, and P. Kongjan. "Bio-hydrogen and bio-methane potentials of skim latex serum in batch thermophilic two-stage anaerobic digestion." *Bioresource Technology* 198 (2015): 198–206. doi:10.1016/j.biortech.2015.09.006.

Jeon, B.-H., J.-A. Choi, H.-C. Kim, J.-H. Hwang, R.A. Abou-Shanab, B.A. Dempsey, J.M. Regan, and J.R. Kim. "Ultrasonic disintegration of microalgal biomass and consequent improvement of bioaccessibility/bioavailability in microbial fermentation." *Biotechnology for Biofuels* 6, no. 1 (2013): 37. doi:10.1186/1754-6834-6-37.

Jiang, J., Y. Zhang, K. Li, Q. Wang, C. Gong, and M. Li. "Volatile fatty acids production from food waste: Effects of PH, temperature, and organic loading rate." *Bioresource Technology* 143 (2013): 525–30. doi:10.1016/j.biortech.2013.06.025.

Júnior, A.D.N.F., M. Zaiat, M. Gupta, E. Elbeshbishy, H. Hafez, and G. Nakhla. "Impact of organic loading rate on biohydrogen production in an up-flow anaerobic packed bed reactor (UAnPBR)." *Bioresource Technology* 164 (2014): 371–79. doi:10.1016/j.biortech.2014.05.011.

Junoh, H., K. Palanisamy, C.H. Yip, and F.L. Pua. "Optimization of NaOH thermo-chemical pretreatment to enhance solubilisation of organic food waste by response surface methodology." *International Journal of Chemical and Molecular Engineering* 2015. doi:10.5281/zenodo.1110279.

Karthikeyan, O.P., E. Trably, S. Mehariya, N. Bernet, J. W.C. Wong, and H. Carrere. "Pretreatment of food waste for methane and hydrogen recovery: A review." *Bioresource Technology* 249 (2018): 1025–039. doi:10.1016/j.biortech.2017.09.105.

Koutrouli, E.C., H. Kalfas, H.N. Gavala, I.V. Skiadas, K. Stamatelatou, and G. Lyberatos. "Hydrogen and methane production through two-stage mesophilic anaerobic digestion of olive pulp." *Bioresource Technology* 100, no. 15 (2009): 3718–723. doi:10.1016/j.biortech.2009.01.037.

Krishnan, S., M.F.M. Din, S.M. Taib, Y.E. Ling, E. Aminuddin, S. Chelliapan, P. Mishra, S. Rana, M. Nasrullah, M. Sakinah, Z.A. Wahid, and L. Singh. "Utilization of micro–algal biomass residues (MABRS) for bio–hythane production– A perspective." *Journal of Applied Biotechnology & Bioengineering* 5, no. 3 (2018). doi:10.15406/jabb.2018.05.00133.

Kumar, G., P. Sivagurunathan, N.B.D. Thi, G. Zhen, T. Kobayashi, S.-H. Kim, and K. Xu. "Evaluation of different pretreatments on organic matter solubilization and hydrogen fermentation of mixed microalgae consortia." *International Journal of Hydrogen Energy* 41, no. 46 (2016): 21628–1640. doi:10.1016/j.ijhydene.2016.05.195.

Kumari, S., and D. Das. "Biohythane production from sugarcane bagasse and water hyacinth: A way towards promising green energy production." *Journal of Cleaner Production* 207 (2019): 689–701. doi:10.1016/j.jclepro.2018.10.050.

Kumari, S., and D. Das. "Biologically pretreated sugarcane top as a potential raw material for the enhancement of gaseous energy recovery by two stage biohythane process." *Bioresource Technology* 218 (2016): 1090–097. doi:10.1016/j.biortech.2016.07.070.

Kumari, S., and D. Das. "Improvement of gaseous energy recovery from sugarcane bagasse by dark fermentation followed by biomethanation process." *Bioresource Technology* 194(2015): 354–63.

Kurokawa, M., P.M. King, X. Wu, E.M. Joyce, T.J. Mason, and K. Yamamoto. "Effect of sonication frequency on the disruption of algae." *Ultrasonics Sonochemistry* 31 (2016): 157–62. doi:10.1016/j.ultsonch.2015.12.011.

Lam, G.P., P.R. Postma, D.A. Fernandes, R.A.H. Timmermans, M.H. Vermuë, M.J. Barbosa, M.H.M. Eppink, R.H. Wijffels, and G. Olivieri. "Pulsed electric field for protein release of the microalgae *Chlorella vulgaris* and *Neochloris oleoabundans*." *Algal Research* 24 (2017): 181–87. doi:10.1016/j.algal.2017.03.024.

Li, L., R. Diederick, J.R.V. Flora, and N.D. Berge. "Hydrothermal carbonization of food waste and associated packaging materials for energy source generation." *Waste Management* 33, no. 11 (2013): 2478–492. doi:10.1016/j.wasman.2013.05.025.

Li, Q., and C.-Z. Liu. "Co-culture of *Clostridium thermocellum* and *Clostridium thermosaccharolyticum* for enhancing hydrogen production via thermophilic fermentation of cornstalk waste." *International Journal of Hydrogen Energy* 37, no. 14 (2012): 10648–0654. doi:10.1016/j.ijhydene.2012.04.115.

Li, Y., F. Xu, Y. Li, J. Lu, S. Li, A. Shah, X. Zhang, H. Zhang, X. Gong, and G. Li. "Reactor performance and energy analysis of solid state anaerobic co-digestion of dairy manure with corn stover and tomato residues." *Waste Management* 73 (2018): 130–39. doi:10.1016/j.wasman.2017.11.041.

Lin, C., and C. Lay. "A nutrient formulation for fermentative hydrogen production using anaerobic sewage sludge microflora." *International Journal of Hydrogen Energy* 30, no. 3 (2005): 285–92. doi:10.1016/j.ijhydene.2004.03.002.

Liu, X., R. Li, and M. Ji. "Effects of two-stage operation on stability and efficiency in co-digestion of food waste and waste activated sludge." *Energies* 12, no. 14 (2019): 2748. doi:10.3390/en12142748.

Liu, Z., B. Si, J. Li, J. He, C. Zhang, Y. Lu, Y. Zhang, and X.-H. Xing. "Bioprocess engineering for biohythane production from low-grade waste biomass: Technical challenges towards scale up." *Current Opinion in Biotechnology* 50 (2018): 25–31. doi:10.1016/j.copbio.2017.08.014.

Liu, Z., C. Zhang, Y. Lu, X. Wu, L. Wang, L. Wang, B. Han, and X.-H. Xing. "States and challenges for high-value biohythane production from waste biomass by dark fermentation technology." *Bioresource Technology* 135 (2013): 292–303. doi:10.1016/j.biortech.2012.10.027.

Ma, C., J. Liu, M. Ye, L. Zou, G. Qian, and Y.-Y. Li. "Towards utmost bioenergy conversion efficiency of food waste: Pretreatment, co-digestion, and reactor type." *Renewable and Sustainable Energy Reviews* 90 (2018): 700–09. doi:10.1016/j.rser.2018.03.110.

Ma, J., T.H. Duong, M. Smits, W. Verstraete, and M. Carballa. "Enhanced biomethanation of kitchen waste by different pre-treatments." *Bioresource Technology* 102, no. 2 (2011): 592–99. doi:10.1016/j.biortech.2010.07.122.

Mamimin, C., P. Kongjan, S. O-Thong, and P. Prasertsan. "Enhancement of biohythane production from solid waste by co-digestion with palm oil mill effluent in two-stage thermophilic fermentation." *International Journal of Hydrogen Energy* 44, no. 32 (2019): 17224–7237. doi:10.1016/j.ijhydene.2019.03.275.

Maragkaki, A.E., I. Vasileiadis, M. Fountoulakis, A. Kyriakou, K. Lasaridi, and T. Manios. "Improving biogas production from anaerobic co-digestion of sewage sludge with a thermal dried mixture of food waste, cheese whey and olive mill wastewater." *Waste Management* 71 (2018): 644–51. doi:10.1016/j.wasman.2017.08.016.

Maru, B.T., F. López, S.W.M. Kengen, M. Constantí, and F. Medina. "Dark fermentative hydrogen and ethanol production from biodiesel waste glycerol using a co-culture of *Escherichia coli* and *Enterobacter* Sp." *Fuel* 186 (2016): 375–84. doi:10.1016/j.fuel.2016.08.043.

Mata-Alvarez, J., J. Dosta, S. Macé, and S. Astals. "Codigestion of solid wastes: A review of its uses and perspectives including modeling." *Critical Reviews in Biotechnology* 31, no. 2 (2011): 99–111. doi:10.3109/07388551.2010.525496.

Meena, R.A.A., J.R. Banu, R.Y. Kannah, K.N. Yogalakshmi, and G. Kumar. "Biohythane production from food processing wastes – Challenges and perspectives." *Bioresource Technology* 298 (2020): 122449. doi:10.1016/j.biortech.2019.122449.

Mizuno, O., R. Dinsdale, F.R. Hawkes, D.L. Hawkes, and T. Noike. "Enhancement of hydrogen production from glucose by nitrogen gas sparging." *Bioresource Technology* 73, no. 1 (2000): 59–65. doi:10.1016/s0960-8524(99)00130-3.

Monlau, F., C. Sambusiti, A. Barakat, X.M. Guo, E. Latrille, E. Trably, J.-P. Steyer, and H. Carrere. "Predictive models of biohydrogen and biomethane production based on the compositional and structural features of lignocellulosic materials." *Environmental Science & Technology* 46, no. 21 (2012): 12217–2225. doi:10.1021/es303132t.

Montingelli, M.E., K.Y. Benyounis, J. Stokes, and A.G. Olabi. "Pretreatment of macroalgal biomass for biogas production." *Energy Conversion and Management* 108 (2016): 202–09. doi:10.1016/j.enconman.2015.11.008.

Mottet, A., J.P. Steyer, S. Déléris, F. Vedrenne, J. Chauzy, and H. Carrère. "Kinetics of Thermophilic batch anaerobic digestion of thermal hydrolysed waste activated sludge." *Biochemical Engineering Journal* 46, no. 2 (2009): 169–75. doi:10.1016/j.bej.2009.05.003.

Náthia-Nevesorcid, G., T.D.A. Neves, M. Berni, G. Dragone, S.I. Mussatto, and T. Forster-Carneiro. "Start-up phase of a two-stage anaerobic co-digestion process: Hydrogen and methane production from food waste and vinasse from ethanol industry." *Biofuel Research Journal* 5, no. 2 (2018): 813–20. doi:10.18331/brj2018.5.2.5.

Nualsri, C., P. Kongjan, and A. Reungsang. "Direct integration of CSTR-UASB reactors for two-stage hydrogen and methane production from sugarcane syrup." *International Journal of Hydrogen Energy* 41, no. 40 (2016): 17884–7895. doi:10.1016/j.ijhydene.2016.07.135.

Nwosu-Obieogu, K., L.I. Chiemenem, and K.F. Adekunle. "Utilization of agricultural waste for bioethanol production- A review." *International Journal of Current Research and Review* 8, no. 19 (2016): 1–5.

Onumaegbu, C., J. Mooney, A. Alaswad, and A.G. Olabi. "Pre-treatment methods for production of biofuel from microalgae biomass." *Renewable and Sustainable Energy Reviews* 93 (2018): 16–26. doi:10.1016/j.rser.2018.04.015.

O-thong, S., C. Mamimin, and P. Prasertsan. "Effect of temperature and initial PH on biohydrogen production from palm oil mill effluent: Long-term evaluation and microbial community analysis." *Electronic Journal of Biotechnology: EJB* 14, no. 5 (2011). doi:10.2225/vol14-issue5-fulltext-9.

Pagliaccia, P., A. Gallipoli, A. Gianico, D. Montecchio, and C.M. Braguglia. "Single stage anaerobic bioconversion of food waste in mono and co-digestion with olive husks: Impact of thermal pretreatment on hydrogen and methane production." *International Journal of Hydrogen Energy* 41, no. 2 (2016): 905–15. doi:10.1016/j.ijhydene.2015.10.061.

Park, C., C. Lee, S. Kim, Y. Chen, and H.A. Chase. "Upgrading of anaerobic digestion by incorporating two different hydrolysis processes." *Journal of Bioscience and Bioengineering* 100, no. 2 (2005): 164–67. doi:10.1263/jbb.100.164.

Park, J.-H., H.-C. Cheon, J.-J. Yoon, H.-D. Park, and S.-H. Kim. "Optimization of batch dilute-acid hydrolysis for biohydrogen production from red algal biomass." *International Journal of Hydrogen Energy* 38, no. 14 (2013): 6130–136. doi:10.1016/j.ijhydene.2013.01.050.

Park, M.J., J.H. Jo, D. Park, D.S. Lee, and J.M. Park. "Comprehensive study on a two-stage anaerobic digestion process for the sequential production of hydrogen and methane from cost-effective molasses." *International Journal of Hydrogen Energy* 35, no. 12 (2010): 6194–202. doi:10.1016/j.ijhydene.2010.03.135.

Perepelkin, K.E. "Lyocell fibres based on direct dissolution of cellulose in N-methylmorpholine N-oxide: Development and prospects." *Fibre Chemistry* 39, no. 2 (2007): 163–72. doi:10.1007/s10692-007-0032-9.

Pérez-Elvira, S., M. Fdz-Polanco, F. I. Plaza, G. Garralón, and F. Fdz-Polanco. "Ultrasound pre-treatment for anaerobic digestion improvement." *Water Science and Technology: A Journal of the International Association on Water Pollution Research* 60, no. 6 (2009): 1525–1532.

Pinto, M.P.M., A. Mudhoo, T.D.A. Neves, M.D. Berni, and T. Forster-Carneiro. "Co–digestion of coffee residues and sugarcane vinasse for biohythane generation." *Journal of Environmental Chemical Engineering* 6, no. 1 (2018): 146–55. doi:10.1016/j.jece.2017.11.064.

Polizzi, C., F. Alatriste-Mondragón, and G. Munz. "The role of organic load and ammonia inhibition in anaerobic digestion of tannery fleshing." *Water Resources and Industry* 19 (2018): 25–34. doi:10.1016/j.wri.2017.12.001.

Prabhudessai, V. *Anaerobic Digestion of Food Waste in A Horizontal Plug Flow Reactor.* PhD Diss. Birla Institute of Technology and Science Pilani (Rajasthan) India, 2013.

Rafique, R., T.G. Poulsen, A.-S. Nizami, Z.-U.-Z. Asam, J.D. Murphy, and G. Kiely. "Effect of thermal, chemical and thermo-chemical pre-treatments to enhance methane production." *Energy* 35, no. 12 (2010): 4556–561. doi:10.1016/j.energy.2010.07.011.

Rajendran, K., E. Drielak, V.S. Varma, S. Muthusamy, and G. Kumar. "Updates on the pretreatment of lignocellulosic feedstocks for bioenergy production–A review." *Biomass Conversion and Biorefinery* 8, no. 2 (2017): 471–83. doi:10.1007/s13399-017-0269-3.

Rittmann, B.E., H.-S. Lee, H. Zhang, J. Alder, J.E. Banaszak, and R. Lopez. "Full-scale application of focused-pulsed pre-treatment for improving biosolids digestion and conversion to methane." *Water Science and Technology* 58, no. 10 (2008): 1895–901. doi:10.2166/wst.2008.547.

Romagnoli, F., D. Blumberga, and I. Pilicka. "Life cycle assessment of biohydrogen production in photosynthetic processes." *International Journal of Hydrogen Energy* 36, no. 13 (2011): 7866–871. doi:10.1016/j.ijhydene.2011.02.004.

Roy, S., and D. Das. "Biohythane production from organic wastes: Present state of art." *Environmental Science and Pollution Research* 23, no. 10 (2015): 9391–410. doi:10.1007/s11356-015-5469–4.

Sanjay Kumar, K., R. Reshma, S. Manokaran and A.H.M. Reddy. "Biohythane: An emerging future fuel." *Journal of Pharmaceutical Sciences and Research* 13, no. 5 (2021): 238–246.

Schieder, D., R. Schneider, and F. Bischof. "Thermal hydrolysis (TDH) as a pretreatment method for the digestion of organic waste." *Water Science and Technology* 41, no. 3 (2000): 181–87. doi:10.2166/wst.2000.0070.

Seghezzo, L., G. Zeeman, J.B. Van Lier, H.V.M. Hamelers, and G. Lettinga. "A review: The anaerobic treatment of sewage in UASB and EGSB reactors." *Bioresource Technology* 65, no. 3 (1998): 175–90. doi:10.1016/s0960-8524(98)00046-7.

Sen, B., J. Aravind, P. Kanmani, and C.-H. Lay. "State of the art and future concept of food waste fermentation to bioenergy." *Renewable and Sustainable Energy Reviews* 53 (2016): 547–57. doi:10.1016/j.rser.2015.08.065.

Seppälä, J.J., J.A. Puhakka, O. Yli-Harja, M.T. Karp, and V. Santala. "Fermentative hydrogen production by *Clostridium butyricum* and *Escherichia coli* in pure and cocultures." *International Journal of Hydrogen Energy* 36, no. 17 (2011): 10701–0708. doi:10.1016/j.ijhydene.2011.05.189.

Shahriari, H., M. Warith, M. Hamoda, and K.J. Kennedy. "Anaerobic digestion of organic fraction of municipal solid waste combining two pretreatment modalities, high temperature microwave and hydrogen peroxide." *Waste Management* 32, no. 1 (2012): 41–52. doi:10.1016/j.wasman.2011.08.012.

Shanmugam, S., A. Hari, P. Ulaganathan, F. Yang, S. Krishnaswamy, and Y.-R. Wu. "Potential of biohydrogen generation using the delignified lignocellulosic biomass by a newly identified thermostable laccase from *Trichoderma asperellum* strain BPLMBT1." *International Journal of Hydrogen Energy* 43, no. 7 (2018): 3618–628. doi:10.1016/j.ijhydene.2018.01.016.

Shanmugam, S., M. Sekar, R. Sivaramakrishnan, T. Raj, E.S. Ong, A.H. Rabbani, E.R. Rene, T. Mathimani, K. Brindhadevi, and A. Pugazhendhi. "Pretreatment of second and third generation feedstock for enhanced biohythane production: Challenges, recent trends and perspectives." *International Journal of Hydrogen Energy* 46, no. 20 (2021): 11252–1268. doi:10.1016/j.ijhydene.2020.12.083.

Shen, D., K. Wang, J. Yin, T. Chen, and X. Yu. "Effect of phosphoric acid as a catalyst on the hydrothermal pretreatment and acidogenic fermentation of food waste." *Waste Management* 51 (2016): 65–71. doi:10.1016/j.wasman.2016.02.027.

Sierra, R., C.B. Granda, and M.T. Holtzapple. "Lime pretreatment." *Methods in Molecular Biology (Clifton, N.J.)* 581(2009): 115–124.

Smichi, N., Y. Messaoudi, N. Moujahed, and M. Gargouri. "Ethanol production from halophyte juncus maritimus using freezing and thawing biomass pretreatment." *Renewable Energy* 85 (2016): 1357–361. doi:10.1016/j.renene.2015.07.010.

Song, Z., X. Liu, Z. Yan, Y. Yuan, and Y. Liao. "Comparison of seven chemical pretreatments of corn straw for improving methane yield by anaerobic digestion." *PLoS One* 9, no. 4 (2014). doi:10.1371/journal.pone.0093801.

Stanislaus, M.S., N. Zhang, Y. Yuan, H. Zheng, C. Zhao, X. Hu, Q. Zhu, and Y. Yang. "Improvement of biohydrogen production by optimization of pretreatment method and substrate to inoculum ratio from microalgal biomass and digested sludge." *Renewable Energy* 127 (2018): 670–77. doi:10.1016/j.renene.2018.05.022.

Suksong, W., P. Kongjan, and S. O-Thong. "Biohythane production from co-digestion of palm oil mill effluent with solid residues by two-stage solid state anaerobic digestion process." *Energy Procedia* 79 (2015): 943–49. doi:10.1016/j.egypro.2015.11.591.

Thorin, E., J. Lindmark, E. Nordlander, M. Odlare, E. Dahlquist, J. Kastensson, N. Leksell, and C.-M. Pettersson. "Performance optimization of the Växtkraft biogas production plant." *Applied Energy* 97 (2012): 503–08. doi:10.1016/j.apenergy.2012.03.007.

Ueno, Y., S. Otsuka, and M. Morimoto. "Hydrogen production from industrial wastewater by anaerobic microflora in chemostat culture." *Journal of Fermentation and Bioengineering* 82, no. 2 (1996): 194–97. doi:10.1016/0922-338x(96)85050-1.

Vavouraki, A.I., E.M. Angelis, and M. Kornaros. "Optimization of thermo-chemical hydrolysis of kitchen wastes." *Waste Management* 33, no. 3 (2013): 740–45. doi:10.1016/j.wasman.2012.07.012.

Veeramalini, J.B., I.A.E. Selvakumari, S. Park, J. Jayamuthunagai, and B. Bharathiraja. "Continuous production of biohydrogen from brewery effluent using co-culture of mutated *Rhodobacter* M 19 and *Enterobacter aerogenes*." *Bioresource Technology* 286 (2019): 121402. doi:10.1016/j.biortech.2019.121402.

Vindis, P., B. Mursec, J. Marjan and F. Cus. "The impact of mesophilic and thermophilic anaerobic digestion on biogas production." *Journal of Achievements of Materials and Manufacturing Engineering* 36 (2009):192–198.

Waghmare, A., K. Nagula, A. Pandit, and S. Arya. "Hydrodynamic cavitation for energy efficient and scalable process of microalgae cell disruption." *Algal Research* 40 (2019): 101496. doi:10.1016/j.algal.2019.101496.

Wang, B., Y. Li, D. Wang, R. Liu, Z. Wei, and N. Ren. "Simultaneous coproduction of hydrogen and methane from sugary wastewater by an "ACSTRH–UASBMet" System." *International Journal of Hydrogen Energy* 38, no. 19 (2013): 7774–779. doi:10.1016/j.ijhydene.2013.04.065.

Wang, M., W. Yuan, X. Jiang, Y. Jing, and Z. Wang. "Disruption of microalgal cells using high-frequency focused ultrasound." *Bioresource Technology* 153 (2014): 315–21. doi:10.1016/j.biortech.2013.11.054.

Watanabe, H., T. Kitamura, S. Ochi, and M. Ozaki. "Inactivation of pathogenic bacteria under mesophilic and thermophilic conditions." *Water Science and Technology* 36, no. 6–7 (1997): 25–32. doi:10.2166/wst.1997.0571.

Wett, B., P. Phothilangka, and A. Eladawy. "Systematic comparison of mechanical and thermal sludge disintegration technologies." *Waste Management* 30, no. 6 (2010): 1057–062. doi:10.1016/j.wasman.2009.12.011.

Xia, A., J. Cheng, R. Lin, H. Lu, J. Zhou, and K. Cen. "Comparison in dark hydrogen fermentation followed by photo hydrogen fermentation and methanogenesis between protein and carbohydrate compositions in *Nannochloropsis oceanica* Biomass." *Bioresource Technology* 138 (2013): 204–13. doi:10.1016/j.biortech.2013.03.171.

Xiao, L.-P., Z.-J. Sun, Z.-J. Shi, F. Xu, and R.-C. Sun. "Impact of hot compressed water pretreatment on the structural changes of woody biomass for bioethanol production." *BioResources* 6(2011): 1576–1598.

Yamashita, Y., C. Sasaki, and Y. Nakamura. "Effective enzyme saccharification and ethanol production from Japanese cedar using various pretreatment methods." *Journal of Bioscience and Bioengineering* 110, no. 1 (2010): 79–86. doi:10.1016/j.jbiosc.2009.12.009.

Yang, G., G. Zhang, and H. Wang. "Current state of sludge production, management, treatment and disposal in China." *Water Research* 78 (2015): 60–73. doi:10.1016/j.watres.2015.04.002.

Yang, X., X. Wang, and L. Wang. "Transferring of components and energy output in industrial sewage sludge disposal by thermal pretreatment and two-phase anaerobic process." *Bioresource Technology* 101, no. 8 (2010): 2580–584. doi:10.1016/j.biortech.2009.10.055.

Yin, Y., and J. Wang. "Pretreatment of macroalgal *Laminaria japonica* by combined microwave-acid method for biohydrogen production." *Bioresource Technology* 268 (2018): 52–59. doi:10.1016/j.biortech.2018.07.126.

Zhang, C., G. Xiao, L. Peng, H. Su, and T. Tan. "The anaerobic co-digestion of food waste and cattle manure." *Bioresource Technology* 129 (2013): 170–76. doi:10.1016/j.biortech.2012.10.138.

Zhang, J., W. Li, J. Lee, K.-C. Loh, Y. Dai, and Y.W. Tong. "Enhancement of biogas production in anaerobic co-digestion of food waste and waste activated sludge by biological co-pretreatment." *Energy* 137 (2017): 479–86. doi:10.1016/j.energy.2017.02.163.

Zhang, Z.-P., K.-Y. Show, J.-H. Tay, D.T. Liang, D.-J. Lee, and W.-J. Jiang. "Effect of hydraulic retention time on biohydrogen production and anaerobic microbial community." *Process Biochemistry* 41, no. 10 (2006): 2118–123. doi:10.1016/j.procbio.2006.05.021.
Zhao, L., G.-L. Cao, A.-J. Wang, H.-Y. Ren, D. Dong, Z.-N. Liu, X.-Y. Guan, C.-J. Xu, and N.-Q. Ren. "Fungal pretreatment of cornstalk with *Phanerochaete chrysosporium* for enhancing enzymatic saccharification and hydrogen production." *Bioresource Technology* 114 (2012): 365–69. doi:10.1016/j.biortech.2012.03.076.
Zhao, M.-X., Q. Yan, W.-Q. Ruan, H.-F. Miao, H.-Y. Ren, and Y. Xu. "Enhancement of substrate solubilization and hydrogen production from kitchen wastes by pH pretreatment." *Environmental Technology* 32, no. 2 (2011): 119–25. doi:10.1080/09593330.2010.482596.

Index

Note: **Bold** page numbers refer to tables and *italic* page numbers refer to figures.

Printed in the United States
by Baker & Taylor Publisher Services

Printed in the United States
by Baker & Taylor Publisher Services